Point-of-Care Testing

Peter B. Luppa
Ralf Junker
(*Eds.*)

Point-of-Care Testing

Principles and Clinical Applications

With 100 figures in color

Springer

Editors
Peter B. Luppa
Klinikum rechts der Isar
München, Germany

Ralf Junker
Universitätsklinikum Schleswig-Holstein
Kiel, Germany

ISBN 978-3-662-54496-9 ISBN 978-3-662-54497-6 (eBook)
https://doi.org/10.1007/978-3-662-54497-6

Library of Congress Control Number: 2018945534

Springer

Cover illustration: © Kzenon/fotolia.com
Cover Design: deblik Berlin

This Springer imprint is published by Springer, part of Springer Nature
The registered company is Springer-Verlag GmbH, DE
The registered company address is: Heidelberger Platz 3, 14197 Berlin, Germany

Preface

Point-of-care Testing (POCT) enables health care professionals and caregivers perform clinical laboratory testing in close proximity to or directly at the patient's side. POCT also allows suitable patients to perform self-monitoring in indications like diabetes, coagulation disorders and pregnancy. The underlying technology and range of test parameters available are evolving at a very rapid pace. Because delays are no longer caused by clinical sample preparation, transport or central laboratory analysis, innumerable clinical applications are possible that shorten the clinical decision-making time with implications for additional testing or therapy. Tests in a POC environment are currently performed by many medical disciplines including endocrinology, diabetology, cardiology, nephrology, critical care medicine, gynecology, hematology, hemostaseology, infectiology to name a few. Indications for a more patient-centered health screening include tests for cholesterol, lipids, drugs of abuse, coagulation monitoring, drugs of abuse etc.

The notion of conventional and POCT laboratory services residing within the same hospital may seem contradictory, but these two are, in fact, complementary. Together, POCT and the central laboratory are pivotal to the optimal functioning of diagnostic processes. They complement each other, provided that dedicated POCT coordination integrates the quality assurance of POCT into the overall quality management system of the central laboratory.

The motivation for this first English edition of the POCT book by Luppa and Junker, adapted from the third German edition, was to explore and describe – from the European perspective – clinically relevant analytical techniques, organizational concepts for applications and future development of POCT. The European approach to the clinical role of POCT is clearly emphasized herein, and the reader is encouraged to compare and contrast these perspectives with the current practices they are familiar with.

The book presents an overview of the opportunities that POCT can provide while identifying the limitations to be considered when choosing whether to implement POCT or not. Current technologies, clinical applications, networking issues and quality issues and regulatory aspects are all addressed. In addition, the authors also present a survey of future technologies which hold great promise.

The editors have spent considerable efforts to highlight the latest developments, e.g., novel POCT applications of nucleic acid testing for the rapid identification of infectious agents. In addition, we, the editors, have included a comparison of European quality assurance recommendations drafted by a team of international experts.

This book aims to be a field guide and resource for anyone looking to adopt or manage POCT. It should also prove useful to assay developers and clinical experts as it highlights relevant aspects and describes the complexities to be considered when designing new POCT products and applications.

The editors hope our readers will gain unique insights into the intricacies of this multifaceted bioanalytical technology. We hereby encourage them to reflect on the role POCT could play in their own settings and on how to best harness this powerful clinical tool.

We would like to thank all the IVD companies listed below for their valuable support in facilitating this book project.

Peter B. Luppa and Ralf Junker
Munich and Kiel, February 2018

Kindly supported by

ABAXIS Global Diagnostics
Alere
Axonlab connecting ideas
Cepheid. A better way.
concile innovation meets diagnostics
INSTAND
nova biomedical World Leader In Biosensor Technology
OSM Gruppe
RADIOMETER
Roche
SARSTEDT
SIEMENS Healthineers
sysmex
ThermoFisher SCIENTIFIC
Werfen Diagnostic Solutions for Life

About the editors

Prof. Dr. Peter B. Luppa

Peter B. Luppa is a clinical pathologist in the core laboratory of the university hospital Klinikum rechts der Isar in Munich, Germany. He received appointment as Associate Professor by the Technische Universität München in 1997. Since 2006 he is also heading the working group for POCT of the German Society for Clinical Chemistry and Laboratory Medicine (DGKL). His particular interests concern organizational and analytical challenges of new POCT technologies and their influence on the hospital and public healthcare sectors.

Prof. Dr. Ralf Junker

Ralf Junker is a medical specialist in Laboratory Medicine and Master of Business Administration. After completing his education, he held executive positions in hospitals and private laboratories. Since 2009 he is appointed Director of the Diagnostic Centre and head of the Institute of Clinical Chemistry at the University Hospital Schleswig-Holstein in Germany. Since 2010 he is also CMO of the Dialog Diagnostiklabor GmbH, a company that provides laboratory services to hospitals. Furthermore he is the head for the degree program "hospital management" in Kiel. Various research projects focus on POCT.

Table of Contents

I Introduction

II Methodology and analytical techniques

III Clinical applications

V Areas of application

VII Development trends

List of authors

Editors

Prof. Dr. Peter B. Luppa
Institut für Klinische Chemie und Pathobiochemie
Klinikum rechts der TU München
Ismaninger Str. 22
81675 München

Prof. Dr. Ralf Junker
Institut für Klinische Chemie
Universitätsklinikum Schleswig-Holstein
Arnold-Heller-Str. 3
24105 Kiel

Authors

Dr. Silvia Achtzehn
Institut für Trainingswissenschaft und Sportinformatik
Deutsche Sporthochschule Köln
Am Sportpark 6
50933 Köln

Dr. Frauke Adams
in.vent Diagnostica GmbH
Neuendorfstr. 17
16761 Hennigsdorf

Franziska Amiet
Inselspital - Universitätsspital Bern
Zentrum für Labormedizin
Universitätsinstitut für Klinische Chemie
INO F 610, 3010 Bern
Schweiz

Benedict Beneteau-Burnat
Service de Biochimie
Hôpital Saint Antoine
184, rue du Faubourg Saint-Antoine
75012 Paris
France

Dr. Andreas Bietenbeck
Institut für Klinische Chemie und Pathobiochemie
Klinikum rechts der TU München
Ismaninger Str. 22
81675 München

Christoph Braun
synlab Services GmbH
Informationstechnologie
Gubener Straße 39
86156 Augsburg

Miguel Cantero
Area de Laboratorios
Hospital Costa del Sol. Autovía Nacional II, Km 187
29601 Marbella. Málaga
España

Prof. Paul Collinson
St George's Hospital
Blood Sciences
Jenner Wing
London, SW17 0QT
United Kingdom

Dr. Philipp Deetjen
Klinik für Anästhesiologie und Operative Intensivmedizin
Stenglinstraße 2
86156 Augsburg

Bert Dikkeschei
Department of Clinical Chemistry
Isala klinieken
Dr. Van Heesweg 2
8025 AB Zwolle
The Netherlands

Gerhard Eiselen
Social Care Services Europe – SoCaSE GmbH
Kompetenzzentrum Demenz München
Landsberger Str. 369
80687 München

Prof. Dr. Martin Fiedler
Inselspital – Universitätsspital Bern
Zentrum für Labormedizin
Universitätsinstitut für Klinische Chemie
INO F 610, 3010 Bern
Schweiz

Univ. Doz. Prim. Dr. Peter Fraunberger
Medizinisches Zentrallabor (MZL) und
Blutspendedienst des ÖRK Vorarlberg
Carinagasse 41
6800 Feldkirch/Vorarlberg
Österreich

Prof. Dr. Norbert Gässler
Zentrum für Labordiagnostik,
St. Bernward-Krankenhaus
Treibestr. 9
31134 Hildesheim

Prof. Dr. Ulrich M. Gassner
Forschungsstelle für Medizinprodukterecht (FMPR)
Universitätsstr. 24
86159 Augsburg

Prof. Dr. med. Evangelos Giannitsis
Thoraxklinik am Universitätsklinikum Heidelberg
Amalienstraße 5
69126 Heidelberg

Sylvia Gruber
Medizinisches Zentrallabor (MZL) und
Blutspendedienst des ÖRK Vorarlberg
Carinagasse 41
6800 Feldkirch/Vorarlberg
Österreich

Dr. Eva Gruner
BZH GmbH
Deutsches Beratungszentrum für Hygiene
Schnewlinstraße 10
79098 Freiburg/Breisgau

Dr. Jörg-M. Hollidt
in.vent Diagnostica GmbH
Neuendorfstr. 17
16761 Hennigsdorf

Maria Luisa Hortas
Area de Laboratorios
Hospital Costa del Sol. Autovía Nacional II, Km 187
29601 Marbella, Málaga
España

Prof. Dr. Frank Hufert
Institut für Mikrobiologie und Virologie
Medizinische Hochschule Brandenburg
Theodor Fontane
Standort Senftenberg, B-TU Campus 15
Großenhainer-Str. 57
01968 Senftenberg

Prof. Dr. Michael Imhoff
Buchenweg 18c–d
23568 Lübeck

Dr. med. Siegfried Jedamzik
Oberer Grasweg 50a
85055 Ingolstadt

Priv.-Doz. Dr. Jürgen Koszielny
Institut für Transfusionsmedizin
Charité-Universitätsmedizin Berlin
Charitéplatz 1
10117 Berlin

Prof. Dr. Axel Kramer
Institut für Hygiene und Umweltmedizin
Universitätsmedizin Greifswald
Walther-Rathenau-Straße 49A
17489 Greifswald

Dr. Dorthe Kixmüller
Institut für Klinische Chemie
Universitätsklinikum Schleswig-Holstein/
Campus Kiel
Arnold-Heller-Straße 3, Haus 17
24105 Kiel

Prof. Dr. Theodor Koschinsky
Vertreter der Deutschen Diabetes Gesellschaft
Heilmannstr. 25
81479 München

Dr. Claus Langer
Zentrum für Labormedizin und Mikrobiologie
Herwarthstraße 100
45138 Essen

Dr. Elizabeth Lee-Lewandrowski
Department of Pathology
Massachusetts General Hospital
Gray 5 Chemistry
55 Fruit St
Boston MA 02114-2696
USA

Dr. med. Tobias Lindner
Rettungsstellen CVK/CCM
Charité-Universitätsmedizin Berlin
Augustenburger Platz 1
13363 Berlin

Francisco Javier Lirón-Hernández
Clinical Laboratory.
Hospital Universitario de la Ribera
Crta. Corbera, Km. 1
46600 Alzira (Valencia)
España

Prof. Dr. John H. T. Luong
Department of Chemistry and Analytical
Biological Chemistry
Research Facility (ABCRF)
University College Cork, Cork
Ireland

Dr. Jan Martin
Klinik für Anästhesiologie
Klinikum rechts der TU München
Ismaninger Str. 22
81675 München

Prof. Dr. Joachim Mester
Institut für Trainingswissenschaft und
Sportinformatik
Deutsche Sporthochschule Köln
Am Sportpark 6
50933 Köln

Prof. Dr. Martin Möckel
Rettungsstellen CVK / CCM
Charité-Universitätsmedizin Berlin
Augustenburger Platz 1
13363 Berlin

Dr. rer. nat. Christian Müller
Rettungsstellen CVK/CCM
Charité-Universitätsmedizin Berlin
Augustenburger Platz 1
13363 Berlin

James H. Nichols, Ph.D., DABCC, FACB
Vanderbilt University School of Medicine
4918D TVC (The Vanderbilt Clinic)
1301 Medical Center Drive
Nashville, TN 37232-5310
USA

Thomas Norgall
Fraunhofer Institut für Integrierte
Schaltungen IIS/BMT
Am Wolfsmantel 33
91058 Erlangen

Priv.-Doz. Dr. Dirk Peetz
Institut für Laboratoriumsmedizin
HELIOS Klinikum Berlin-Buch
Schwanebecker Chaussee 50
13125 Berlin

Pascal Pernet
Service de Biochimie
Hôpital Saint Antoine
184, rue du Faubourg Saint-Antoine
75012 Paris
France

Paloma Oliver
Laboratorio de Urgencias.
Servicio de Laboratorio. Hospital Universitario
Planta Baja Hospital Maternal,
Paseo de la Castellana 261
28046 Madrid
España

Priv.-Doz. Dr. Astrid Petersmann
Institut für Klinische Chemie
und Laboratoriumsmedizin
Universitätsmedizin Greifswald
Ferdinand-Sauerbruchstr.
17475 Greifswald

Prof. Mario Plebani
Department of Laboratory Medicine
University Hospital of Padova
35128 Padova
Italia

Alexander Pröbstl
Pflegedirektion
Universitätsklinikum Bonn
Sigmund-Freud-Straße 25
53127 Bonn

Dr. Günther Proll
Eberhard Karls University Tübingen
Institut für Physikalische und Theoretische Chemie
Auf der Morgenstelle 18
72076 Tübingen

Prof. Dr. Udo Reischl
Institut für Medizinische Mikrobiologie
und Hygiene
Universitätsklinikum Regensburg
Franz-Josef-Strauss-Allee 11
93053 Regensburg

Christina Rode-Schubert
TCI GmbH
Wildbader Straße 9
68239 Mannheim

Daniel Romann
Deepenstegen 29
22952 Lütjensee

Dr. Hannelore Rott
Gerinnungszentrum Rhein-Ruhr
Königstraße 13
47051 Duisburg

Prof. Dr. Walter Schaffartzik
Klinik für Anästhesiologie, Intensivmedizin und Schmerztherapie
Unfallkrankenhaus Berlin
Warener Str. 7
12683 Berlin

Birgit Schäfer
VDGH - Verband der Diagnostica-Industrie e. V.
Neustädtische Kirchstr. 8
10117 Berlin

Prof. Dr. Ingolf Schimke
Medizinische Klinik (Kardiologie), Charité-Universitätsmedizin Berlin
Schumannstr. 20/21
10098 Berlin

Prof. Dr. Harald Schlebusch
Nadistr. 14
80809 München

Aline Schröder
Med-Lab Medizinische Dienstleistungs GmbH
Holzmühlenstraße 86
22041 Hamburg

Dr. Lutz Schwettmann
Helse Møre og Romsdal HF
Klinikk for diagnostikk
Postboks 1600
6026 Ålesund
Norge

Dr. med. Julia Searle
Rettungsstellen CVK/CCM
Charité-Universitätsmedizin Berlin
Augustenburger Platz 1
13363 Berlin

Priv.-Doz. Dr. Vanadin Seifert-Klauss
Frauenklinik und Poliklinik
Klinikum rechts der TU München
Ismaninger Str. 22
81675 München

Prof. Dr. Naoto Shimetani
Laboratory Technology department
IUHW Atami Hospital
13-1, Higashi-Kaigan-cho
Atami-city, Shizuoka
413-0012
Japan

Dr. Robbert Slingerland
Department of Clinical Chemistry
Isala klinieken
Dr. Van Heesweg 2
8025 AB Zwolle
The Netherlands

Oswald Sonntag
Schulstr. 32
82223 Eichenau

Prof. Dr. Michael Spannagl
Klinikum der Universität München
Ziemssenstr. 1
80336 München

Dr. Folker Spitzenberger
Zentralstelle der Länder für Gesundheitsschutz
bei Arzneimitteln und Medizinprodukten (ZLG)
Heinrich-Böll-Ring 10
53119 Bonn

Priv.-Doz. Dr. Enno Stürenburg
Dr. Staber & Kollegen
Holzmühlenstraße 84–86
22041 Hamburg

Dr. Laura Tooth
St George's Hospital
Blood Sciences
Jenner Wing
London, SW17 0QT
United Kingdom

Dr. Sandeep K. Vashist
IDS Immunodiagnostic Systems
Rahmhofstr. 2–4
60313 Frankfurt am Main

Dr. Michel Vaubourdolle
Service de Biochimie
Hôpital Saint Antoine
184, rue du Faubourg Saint-Antoine
75012 Paris
France

Priv.-Doz. Dr. Dr. Hans Günther Wahl, MBA
Medizinisches Labor Wahl
Paulmannshöher Str. 14
58515 Lüdenscheid

Lars Wilhelm
LADR GmbH, Medizinisches Versorgungszentrum
Lauenburger Straße 67
21502 Geesthacht

Dr. Dr. Christof Winter
Institut für Klinische Chemie und Pathobiochemie
Klinikum rechts der TU München
Ismaninger Str. 22
81675 München

Abbreviations

The list below is exclusive of commonly known abbreviations for parameters in clinical chemistry, hematology, immunology and other laboratory disciplines

AACC	American Association for Clinical Chemistry
AAL	Ambient assisted living
ABDA	Bundesvereinigung Deutscher Apothekerverbände (Federal Union of German Associations of Pharmacists)
ACC	American College of Cardiology
ACS	Acute coronary syndrome
ACT	Activated clotting time
ADP	Adenosine diphosphate
AG	Anion gap
AHA	American Heart Association
ApBetrO	Apothekenbetriebsordnung (German Ordinance on the Operation of Pharmacies)
ASSURED	WHO criteria for the application of POCT devices
AST	Alternate site testing
ASTM	American Society for Testing and Materials
ATP	Adenosine triphosphate
ATR	Attenuated total reflection
AWMF	Arbeitsgemeinschaft der Wissenschaftlichen Medizinischen Fachgesellschaften (Association of the Scientific Medical Societies in Germany)
BCSH	British Committee for Standards in Haematology
BE	Base excess
BEE	Base excess extracellular fluid
BfArM	Bundesinstitut für Arzneimittel und Medizinprodukte (Federal Institute for Drugs and Medical Devices)
BFR	Blood flow restriction
BGA	Blood gas analysis
BGB	Bürgerliches Gesetzbuch (German Civil Code)
BGSC	Blood glucose self-control
BL	Blood volume
BNP	Brain natriuretic peptide
BUN	Blood urea nitrogen
CAD	Coronary artery disease
CAP	College of American Pathologists
CARBA-R	Carbapenem-resistant Enterobacteriaceae
CDC	Centers of Disease Control and Prevention
CDI	Clostridium difficile
CE	Conformité Européenne
CEN	European Committee for Standardization
CENELEC	European Committee for Electrotechnical Standardization
CFT	Clotting formation time
CGM	Continuous glucose monitoring
CIC	Connectivity Industry Consortium
CK	Creatine kinase
CLIA	Clinical Laboratory Improvement Amendments
CLSI	Clinical and Laboratory Standards Institute
CMS	Center for Medicare Services
CO	Carbon monoxide
CO	Cardiac output
COFRAC	Comité français d'accréditation (French National Accreditation Body)
COLA	Commission on Office Laboratory Accreditation
CREA	Creatinine
CRP	C-reactive protein
CT	Chlamydia trachomatis
CT	Clotting time
CTG	Cardiotocography
cTnI	Cardiac troponin I
cTnT	Cardiac troponin T
CV	Coefficient of variation
CRC	Colorectal cancer
CTG	Cardiotocography
CVD	Chemical vapor deposition
DAkkS	Deutsche Akkreditierungsstelle (Accreditation Body for the Federal Republic of Germany)
D-BIL	Conjugated bilirubin
DCCT	Diabetes Control and Complication Trial
DCT	Direct-to-consumer testing
DDAVP	1-Desamino-8D-arginine vasopressin
DDG	Deutsche Diabetes Gesellschaft (German Diabetes Association)
DGKH	Deutsche Gesellschaft für Krankenhaushygiene (German Society of Hospital Hygiene)

DGKL	Deutsche Gesellschaft für Klinische Chemie und Laboratoriumsmedizin (German Society for Clinical Chemistry and Laboratory Medicine)
DI	Device interface
DIN	Deutsches Institut für Normung (German Institute for Standardization)
DKG	Deutsche Krankenhausgesellschaft (German Hospital Federation)
DMS	Data management system
DRG	Diagnosis-related group(s)
DVT	Deep vein thrombosis
EA	European Cooperation of Accreditation
EBM	Einheitlicher Bewertungsmaßstab (German Uniform Assessment Standard)
EBV	Epstein-Barr virus
ECT	Ecarin clotting time
ED	Emergency department
EDMA	European Diagnostic Manufacturer Association
EHEC	Enterohemorrhagic E. coli
ELBW	Extreme low birth weight
ELISA	Enzyme-linked immunosorbent assay
EMS	Electromyostimulation
EQA(P)	External quality assurance (program)
EQC	External quality control
ESC	European Society of Cardiology
EU	European Union
EWDTS	European Workplace Drug Testing Society
FACS	Fluorescence-activated cell sorting
FAD	Flavin adenine dinucleotide
FDA	U.S. Food and Drug Administration
FER	Ferritin
FET	Field effect transistor
FMCG	Fast-moving consumer goods
FOBT	Fecal occult blood test
FRED	Förster resonance energy transfer
GAS	Group A streptococcus
G-BA	Gemeinsamer Bundesausschuss (Federal Joint Committee)
GBS	Group B streptococcus
GDH	Glucose dehydrogenase
GDM	Gestational diabetes mellitus
GHTF	Global Harmonization Task Force
GKV	Gesetzliche Krankenversicherung (National Association of Statutory Health Insurance Funds)
GLORIA	Gold-labelled optical-read rapid immunoassay
GMR	Giant magnetoresistance
GOÄ	Gebührenordnung für Ärzte (German physicians fee schedule)
GOD	Glucose oxidase
Hb	Hemoglobin
HBV	Hepatitis B virus
Hct	Hematocrit
HCV	Hepatitis C virus
HDA	Helicase-dependent amplification
HIS	Hospital information system
HIT	High intensive training
HITECH	Health Information Technology for Economic and Clinical Health
HIV	Human immunodeficiency virus
HL7	Health Level 7
HPLC	High-performance liquid chromatography
HPV	Human papilloma virus
HRP	Horseradish peroxidase
hsCRP	High-sensitive C-reactive protein
HTLV	Human T-cell leukemia virus
HVT	High-volume training
ICG	Indocyanine green
ICSH	International Council for Standardization in Haematology
ICT	Immunochromatographic test
ICU	Intensive care unit
IEC	International Electrotechnical Commission
IEEE	Institute of Electrical and Electronics Engineers
IFCC	International Federation of Clinical Chemistry
iFOBT	Quantitative fecal immunochemical test
IGeL	Individuelle Gesundheitsleistung (Individual healthcare service under according to the German physicians' fee schedule)
INEK	Institut für das Entgeltsystem im Krankenhaus (Institute for the Hospital Remuneration System)
INR	International normalized ratio
INSTAND	Gesellschaft zur Förderung der Qualitätssicherung in medizinischen Laboratorien e.V. (Society for Promoting External Quality Assurance in Medical Laboratories)
IoT	Internet of things
IQC	Internal quality control
IQCP	Individualized quality control plan
IR	Infrared (spectroscopy)
ISE	Ion-selective electrode
ISFET	Ion-sensitive field effect transistor
ISO	International Organization for Standardization
IVD	In-vitro diagnostics

JC	Joint Commission
KBV	Kassenärztliche Bundesvereinigung (National Association of Statutory Health Insurance Physician)
KRINKO	Kommission für Krankenhaushygiene und Infektionsprävention (Commission for Hospital Hygiene and Infection Prevention)
LAMP	Loop-mediated isothermal amplification
LBWI	Low birth weight infant
LDH	Lactate dehydrogenase
LED	Light-emitting diode
LFA	Lateral flow assay
LFD	Lateral flow device
LGS	Low glucose suspend
LIS	Laboratory information system
LOD	Limit of detection
LOINC	Logical observation identifiers names and codes
MA	Maximal amplitude
Mb	Myoglobin
MCF	Maximal clot firmness
MDD	Medical Device Directive
MDMC	Medical devices and management committee
MEMS	Micromechanical systems
MG	Mycoplasma genitalium
MHRH	Medicines and Healthcare Products Regulatory Agency
MI	Myocardial infarction
MIR	Mid-infrared (spectroscopy)
ML	Maximal lysis
MPBetreibV	Medizinprodukte-Betreiberverordnung (German Medical Devices Operator Ordinance)
MPG	Medizinproduktegesetz (German Medical Devices Act)
MRSA	Methicillin-resistant staphylococcus aureus
µTAS	Micro total analysis system
MTB/RIF	Mycobacterium tuberculosis/ rifampicin resistance
MVZ	Medizinisches Versorgungszentrum (Medical Service Center)
MW	Mean value
NACB	National Academy of Clinical Biochemistry
NAD	Nicotinamide adenine dinucleotide
NAT	Nucleic acid testing
NCCLS	National Committee on Clinical Laboratory Standards
NG	Neisseria gonorrhoeae
NGM	Noninvasive glucose monitoring
NGSP	National glycohemoglobin standardization program
NHS	National Health Service (UK)
NICE	National Institute for Clinical and Health Care Excellence (UK)
NIR	Near-infrared (spectroscopy)
NKK	Norwegian clinical chemistry EQA program
NOAC	Non-vitamin K antagonist oral anticoagulants
NOKLUS	Norwegian Quality Improvement of Primary Health Care Laboratories
NPV	Negative predictive value
NQAAP	National Quality Assessment Advisory Panels
NSTEMI	Non-ST segment elevation myocardial infarction
NT-proBNP	N-terminal proBNP
NVKC	Dutch Society for Clinical Chemistry
ÖQUASTA	Österreichische Gesellschaft für Qualitätssicherung und Standardisierung medizinisch-diagnostischer Untersuchungen (Austrian Association for Quality Assurance and Standardization of Medical and Diagnostic Tests)
oGTT	Oral glucose tolerance test
OI	Oxygenation index
OR	Operating room
ORI	Observation reporting interface
PCHA	Personal Connected Health Alliance
PCR	Polymerase chain reaction
PDMS	Patient data management system
PEI	Paul-Ehrlich-Institut (Paul Ehrlich Institute)
PEP	Post-exposure prophylaxis
PFA	Platelet function analyzer
PLGM	Predictive low glucose management
POC	Point-of-Care
POCT	Point-of-Care Testing
PPM	Provider-performed microscopy
PPP	Public private partnership
PPV	Positive predictive value
PQQ	Pyrroloquinoline quinone
ProdHaftG	Produkthaftungsgesetz (German Product Liability Act)
PV	Plasma volume
QC	Quality control
QCM	Quartz crystal microbalance
QI	Quality indicator
QM	Quality management
QMS	Quality management system
QUALAB	Schweizerische Kommission für Qualitätssicherung im medizinischen Labor (Swiss Commission for Quality Assessment in Medical Laboratories)

RfB	Referenzinstitut für Bioanalytik der DGKL (German Reference Institute for Bioanalytics)
RFID	Radio frequency identification
RI	Reflection index
RI	Respiratory index
RIfS	Reflectometric interference spectroscopy
RiliBÄK	Richtlinie der Bundesärztekammer zur Qualitätssicherung laboratoriumsmedizinischer Untersuchungen (Guideline of the German Medical Association on Quality Assurance in Medical Laboratory Examinations)
ROC	Receiver operating characteristic
ROTEM	Rotation thromboelastometry
RPA	Recombinase polymerase amplification
RPFA	Rapid platelet function analysis
RQ	Respiratory quotient
RSV	Respiratory Syncytial Virus
sAA	Salivary alpha amylase
SAM	Self-assembling monolayer
SAMSHA	Substance Abuse and Mental Health Services Administration
SAW	Surface acoustic wave
SBE	Standard base excess
sC	Salivary cortisol
SD	Standard deviation
SELEX	Systematic evolution of ligands by exponential enrichment
SEQC	Spanish Society of Clinical Biochemistry and Molecular Pathology
SERS	Surface-enhanced Raman spectroscopy
SID	Strong ion difference
SIG	Strong ion gap
sIgA	Salivary immunoglobulin A
SKUP	Scandinavian Evaluation of Laboratory Equipment for Primary Health Care
SLA	Service level agreement
SLE	Systemic lupus erythematosus
SOP	Standard operating procedure
SPR	Surface plasmon resonance
STEMI	Segment elevation myocardial infarction
SaP	Sensor-assisted pump therapy
SaT	Sensor-assisted therapy decision
TAT	Turn-around time
T-BIL	Total bilirubin
TEG	Thrombelastography
tHb	Total hemoglobin mass
THC	Tetrahydrocannabinol
TIRF	Total internal reflection fluorescence
TIRS	Total internal reflection spectroscopy
TMB	Tetramethylbenzidine
TRBA	Technische Regeln für Biologische Arbeitsstoffe (Technical Rules for Biological Agents)
TTR	Time in therapeutic range
TV	Trichomonas vaginalis
UDI	Unique device identification
ULN	Upper limit of normal (99th percentile of healthy controls)
UREA	Urea
UTI	Urinary tract infection
VDGH	Verband der Diagnostica-Industrie (German Diagnostic Manufacturers Association)
VKA	Vitamin K antagonists
VLBW	Very low birth weight
WBC	White blood count
WPA2	Wi-Fi protected access 2
ZLG	Zentralstelle der Länder für Gesundheitsschutz bei Arzneimitteln und Medizinprodukten (Central Authority of the Länder for Health Protection with regard to Medicinal Products and Medical Devices)
ZLS	Zentralstelle der Länder für Sicherheitstechnik (Central Authority of the German Federal States for Safety)

Introduction

Content

Definitions and areas of application

Peter B. Luppa, Ralf Junker, Claus Langer

P. B. Luppa, R. Junker (Eds.), *Point-of-Care Testing*
https://doi.org/10.1007/978-3-662-54497-6_1

1.1 Introduction

Throughout the history of laboratory medicine, there has always been concern about the reliability of results. The trend towards centralized laboratories where high-volume complex testing is reliable and cost-effective has ultimately been driven by the realization that the implementation of a quality management system is a necessary part of reliable and accurate laboratory diagnostics. However, one disadvantage of centralized diagnostics is that the well-functioning logistics required to transport samples and ensure quick processing is not always available. In contrast, **POCT – laboratory diagnostics performed near the patient** – can usually ensure short-term analysis. In most cases, POCT is associated with fewer pre-analytical problems (e.g. specimen transport, specimen stability) [3]. Whether a test is carried out in a central laboratory or at a patient's bedside has always been and is a complex organizational decision, guided by the principle that the best outcome for the patient is the decisive criteria. That said, there is still a paucity of data on patient outcomes which are both dependent on the time-to-result and the quality of the analysis [10].

1.2 Terminology and definitions

The term "near-patient diagnostics" or "point-of-care testing" (POCT) is used to describe clinical laboratory testing that is carried out at the patient's bedside or in the direct proximity of the patient. A uniform, generally accepted definition has still not been agreed. Rather, a variety of terms is used that can sometimes – but not always – refer to different definitions.

In addition to the phrases "near-patient laboratory testing" and "remote rapid testing", the international nomenclature "point-of-care testing" (POCT) has become most widely accepted. Furthermore, descriptive synonyms such as "bedside testing", "ancillary testing" or "decentralized testing" are also used in the vernacular. "Patient self-management" is also a POCT concept. In the literature, numerous definitions or descriptions of POCT are given, together with evaluations of technical solutions or process sequences [4, 9].

The most important characteristics attributed to POCT are summarized here. In exceptional cases, one might refer to a test as "point-of-care" even if it does not meet all the criteria listed in the overview below (e.g. no sample preparation or pipetting steps), provided it still meets the typical description of a test of that nature (particularly with regard to 1., 8. and 9.) [2].

Typical characteristics of POCT

1. Laboratory testing is performed in the direct proximity of the patient
2. Laboratory tests are performed outside of a central or satellite laboratory
3. No sample preparation, i.e. mostly whole blood is used as test material
4. No pipetting steps
5. "Ready-to-use" reagents, e.g. as cassettes or unit-use devices
6. Special measuring devices intended or used exclusively for single sample measurement
7. No pertinent medical technical qualifications needed for operating the measuring device
8. Results available quickly
9. Results lead to a rapid diagnosis or consequences for treatment

According to Kost [4], a key factor for using POCT is to bring the test directly to the patient. This is more convenient and increases the probability of a fast result.

However, there are also many other POCT definitions, some of which should be mentioned here:

Definition from the Guideline of the German Medical Association (RiliBÄK) According to the Guideline of the German Medical Association (RiliBÄK) [1], the use of POCT in hospitals is only feasible when the POC test is used as a single determination that produces immediate consequences impacting treatment. For example, a series of regular, possibly automated tests, carried out close to the patient should not be classified as POCT. Within its national purview, RiliBÄK has more narrowly defined POC methods in their actual sense by the term "unit-use reagents". With the aim of assuring quality control for optimal patient safety, this regulatory document excludes from the POCT concept for rapid near-patient diagnostics those more complex devices that do not operate with unit-use reagents. This simply means that it is the operator's own responsibility to check conventional laboratory test equipment such as clinical chemical or hematology analyzers in full compliance with the regulations of RiliBÄK.

Definition by the National Academy of Clinical Biochemistry (NACB) The Laboratory Medicine Practice Guidelines (LMPG) introduced by the NACB [6] define POCT as "clinical laboratory testing conducted close to the site of patient care, typically by clinical personnel whose primary training is not in the clinical laboratory sciences or by patients (self-testing). POCT refers to any testing performed outside of the traditional, core or central laboratory."

Definition by the College of American Pathologists (CAP) This group pragmatically applies various definitions of POCT, depending on its **geography** (hospital, outpatient services etc.), **function** (in hospital intensive care units, outpatient departments etc.), its **technology** (simple hand-held devices, complex multi-parameter analyzers etc.) or **operational** context (nurse, patient etc. as user).

Pragmatic definition by the Irish working group of O'Kelly et al. [7] These authors define POCT pragmatically, describing the pivotal scenario as: "Point of care testing is defined as a quality-assured pathology service using analytical devices (including test kits and analyzers such as blood gas and critical care analyzers and meters for glucose, urinalysis and other metabolites) provided near to the patient rather than in the traditional environment of a clinical laboratory."

POCT was originally almost exclusively used to check blood glucose and measure vital parameters with blood gas analyzers in operating theaters and intensive care units, but its range and potential uses have continually expanded in recent years. There are multiple reasons for this. It can be anticipated that existing analytical principles and devices will continue to evolve and become more simplified, while new processes are also certain to emerge (catchphrase: "lab-on-a-chip"). The panoply of measurable parameters will expand and POCT applications will not be limited to medicine alone (catchphrase: "direct-to-consumer testing").

1.3 Areas of application

▣ Tab. 1.1 gives an overview of the meanwhile very extensive POCT spectrum; it focuses on

▣ **Tab. 1.1** Areas of POCT application

Within the hospital setting		Outside the hospital setting
Area of application	Intensive care unit Operating room/recovery room Delivery room/neonatal ward Lung function tests Invasive radiology Emergency room Specialized outpatients Diabetic care ward	Emergency physician (also for disaster control or in the military setting) Private practitioners (practice, house calls) Medical services Sports medicine Outpatient care Home care Pharmacy For forensic drug screening For patient self-monitoring (blood glucose, clotting)
Criteria for use	Outside the central laboratory's regular service hours Hospitals without their own laboratory	

Tab. 1.2 Key clinical parameters that can be measured using POCT

Parameters and clinical application	Parameters
Acid-base status, blood gases	pH, pCO_2, pO_2 (often combined with electrolytes, metabolites and CO-Oximetry)
Electrolytes	Na^+, K^+, Cl^-, ionized Ca^{2+}, ionized Mg^{2+}
Metabolites	Cholesterol, HDL-cholesterol, triglyceride, creatinine, urea, uric acid, bilirubin, lactate, ammonia
Enzymes	Amylase, alkaline phosphatase, creatine kinase, aspartate aminotransferase, alanine aminotransferase, γ-glutamyl transferase
Hemostaseology	Activated clotting time, partial thromboplastin time, thromboplastin time (Quick test, INR), D-dimer, platelet function, bleeding time
Hematology	Hemoglobin, hematocrit, erythrocytes, leukocytes, platelets, differential blood count
Hemoglobin fractions	CO-oximetry
Cardiac markers	Troponin T, troponin I, myoglobin, creatine kinase (CK-M muscle type, CK-B, brain type), brain natriuretic peptide (BNP), N-terminal pro-BNP, interleukin 1 receptor-like 1 (ST2)
Diabetes mellitus	Glucose, HbA_{1c}, minimal-invasive continuous glucose monitoring, β-hydroxybutyrate (in urine and capillary blood)
Acute phase protein	C-reactive protein
Allergy diagnostics	Allergen specific IgE
Medication levels and drug screening	Medication, alcohol, amphetamine, barbiturate, benzodiazepines, cannabinoids, cocaine, methadone, opiates
Infectious diseases	HIV, infectious mononucleosis, Chlamydia trachomatis, Trichomonas vaginalis, Plasmodium falciparum, Plasmodium vivax, influenza viruses type A and B, group A and B streptococcus
Fertility	Human chorionic gonadotropin, luteinizing hormone, follicle-stimulating hormone, sperm count in ejaculate, pregnanediol glucuronide, estriol 3-glucuronide (E3G)
Urine diagnostics	Test strips (pH, protein, glucose, ketones, pregnanediol glucuronide, estriol glucuronide, bilirubin, urobilinogen, nitrite, leukocytes, blood, specific gravity), microalbumin, bacteria
Stool diagnostics	Occult blood in stool
Non-invasive measurements	Transcutaneous pCO_2 and pO_2 values, neonatal bilirubin
Patient self-monitoring	Blood glucose, urine glucose, thromboplastin time (INR)

medical uses, although it is difficult to prevent crossover to applications outside of medicine.

In addition, there are manifold applications outside of clinical medicine, e.g. in veterinary medicine, but also homeopathy, for fitness studios and industry. Currently, there are more than 100 POCT analytes available and further tests are being developed. Tab. 1.2 focuses on the clinically relevant analytes or parameters already for POCT on the market. These include products with several generally available technical solutions, while excluding non-standard ones.

The quality of user information about individual tests varies significantly. Some tests still under development or still in the prototype stage have already been published as established tests. However, even for established tests the manufacturer's information is often incomplete, at times missing valid data for analytical quality (precision, accuracy, sensitivity, specificity) to be able to compare results with standard laboratory tests and judge their practicability. This is even more evident regarding information on diagnostic sensitivity and specificity. Before introducing a test, it is therefore essential to conduct a literature search, to do your own evaluation and/or to discuss it with specialists and experts in the field. Evidence-based recommendations for a specific analyte may also be useful [5]. In general, it is difficult to decide if a test is useful or not. This is only possible if all medical, technical, economic, human resources-, equipment- and facilities-related aspects and constraints relevant to the user's situation are considered. A POCT committee should be set up at a healthcare facility to make decisions about the selection, verification and evaluation of relevant tests (▶ Chapter 30 and ▶ Chapter 36).

Point-of-care laboratory testing (POCT) is being increasingly used in developing countries and emerging markets. Future developments of this technology need to be considered globally [8]. More detailed information is given in ▶ Chapter 35.

References

1. Bundesärztekammer (German Medical Association) (2014) Guideline of the German Medical Association for the Quality Assurance of Clinical Laboratory Testing (Rili-BÄK 2014). Dtsch Ärztebl 111:A1583–1618
2. Ehrmeyer SS, Laessig RH (2007) Point-of-care testing, medical error, and patient safety: A 2007 assessment. Clin Chem Lab Med 45: 766–773
3. Junker R, Schlebusch H, Luppa PB (2010) Point-of-care testing in hospitals and primary care. Dtsch Ärztebl Int 107:561–7
4. Kost GJ (2002) Goals, guidelines, and principles for point-of-care testing. In: Kost GJ (ed) Principles and practice of point-of-care testing. Lippincott Williams & Wilkins, Philadelphia, pp 3–12
5. National Academy of Clinical Biochemistry (NACB) (2007) Practice guidelines: Evidence-based practice for point-of-care testing. Point of care 6: 213–217
6. National Academy of Clinical Biochemistry (NACB) (2014) https://www.aacc.org/community/national-academy-of-clinical-biochemistry/practice-guidelines
7. O'Kelly RA, Brady JJ, Byrne E, Hooley K, Mulligan C, Mulready K, et al. (2011) A survey of point of care testing in Irish hospitals. Ir J Med Sci 180:237–240
8. Plebani M, Lippi G (2014) Point of care testing: evolving scenarios and innovative perspectives. Clin Chem Lab Med 52: 309–311
9. Price CP, St John A, Hicks JM (2004) Point-of-Care Testing: What, Why, When, and Where? In: Price CP, St John A, Hicks JM (eds) Point-of-Care Testing, 2nd edition. AACC Press, Washington, pp 3–12
10. Wyer LA et al. (2009) Quality management: Approaches to reducing errors at the Point of Care; Proposed Guideline. CLSI document POCT07-P

The relevance of POCT in healthcare

Ralf Junker, Astrid Petersmann, Peter B. Luppa

P. B. Luppa, R. Junker (Eds.), *Point-of-Care Testing*
https://doi.org/10.1007/978-3-662-54497-6_2

2.1 Introduction

Laboratory diagnostics performed directly at the hospital bedside are an efficient form of laboratory medicine whose development has benefited greatly from the miniaturization of measuring equipment and procedures and promoted by their integration into information technology. The polarization that can occasionally be seen between the organizational structure of the central laboratory and POCT confirms this impression: Bedside or near-patient diagnostics is an important option in the on-going interplay between diagnostics and treatment [18].

In retrospect, the principles of medical action have not changed for hundreds of years: An experienced physician still acts based on detailed observation and targeted questioning supported by manual and technical examination of patients and their organs. This analytical approach enabled clinicians to answer their patients' questions and to confirm their own suspicions whereupon therapeutic interventions could then be initiated, checked and optimized. The need to render a medical diagnosis promptly and, where possible, at the point of care is as old as medicine itself.

Hence, every era had its tests that were simple and robust enough to be applied directly on or around the patient. There were, nevertheless, equally valid reasons not to carry out certain tests at the point of care. The term "laboratory test" stems from a time when the technical complexity of diagnostic methods was linked to special instruments as well as spatial requirements and to the increasing need for laboratory medical expertise. With the exception of those laboratory tests primarily required in life-threatening situations, results made available within 30 minutes to 2 hours for a typical small to middle-sized hospital laboratory or 24 to 48 hours for a private practitioner seemed generally satisfactory for medical decision-making. This situation has now at least partially changed. Nowadays, when clinical practice is confronted with life-threatening emergencies, immediately available laboratory results directly impact treatment decisions. There are also advantages in outpatient settings, e.g. for the medical care of patients with chronic conditions such as diabetes mellitus and for patient self-management in homecare settings. The different areas of application for POCT are depicted in ◘ Fig. 2.1.

The analytic process of POCT in healthcare is both innovative, while offering novel options for prevention, diagnosis and patient monitoring. According to Price and St John [17], real innovation is characterized by the fact that the invention offers new benefits. In the healthcare system this means that innovation must have direct benefits for patients receiving treatment. In the face of the many different challenges in developing countries, it is easy to recognize that POCT can offer answers more easily than conventional laboratory medicine. One notable example is that the increasing number of malaria tests carried out have already helped to significantly reduce the administration of anti-malaria drugs to those without the disease [7].

By contrast, in the hospital setting, the advantages of POCT process management can only be realized in close collaboration with the central laboratory and its core competencies [1]. As long as POCT specialists and central laboratory experts complement rather than contradict each other, they can reshape clinical guidelines and bring about significant, positive results in patient care (▶ Chapter 30).

Nowadays blood glucose and blood gas analysis are usually carried out at the POC. A growing number of biochemical, hematological and hemostaseological tests are also used. On the one hand this development is in contrast to growing centralization of laboratory medicine in recent years. On the other hand, thanks to the ability to carry out emergency tests at the point of care using simple, quick processes, it is laying the foundations for stronger centralized laboratory analytics. For example, emergency and rapid diagnostics can be carried out in smaller clinics using POCT, whereas, routine and specialized analyses are performed in larger regional hospital laboratories or in laboratory physicians' practices.

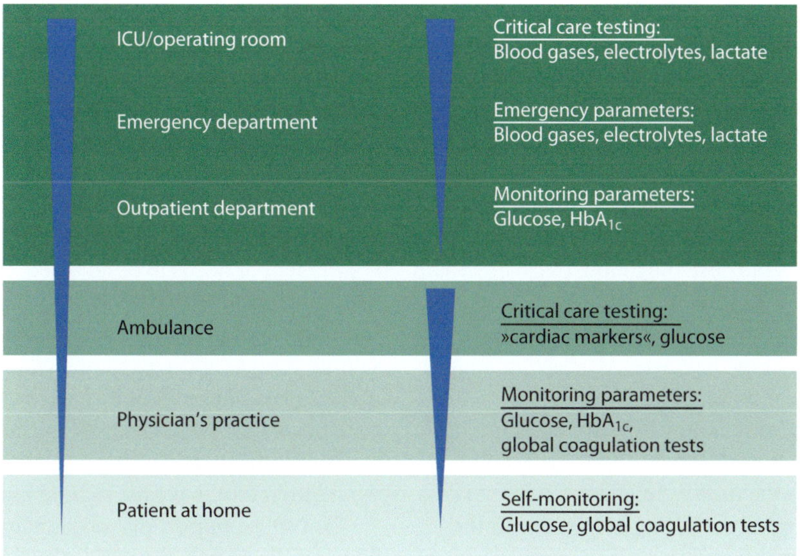

Fig. 2.1 Hierarchy of POCT applications

Several American studies have shown that the increasing use of laboratory tests that are easy to perform can be advantageous in areas away from hospitals and medical practices. It was possible to prove that the offer by pharmacy staff to check cholesterol levels of their customers using simple tests and to offer lipid lowering drug treatment, if appropriate, was positively received and led to a significant reduction in cholesterol levels within this customer group [14, 16, 21]. These tests are also often classified under the term POCT. It is conceivable that the knowledge of an abnormal finding leads to a higher health consciousness in the affected person with all the resulting consequences, including a necessary medical consultation. Implications for the whole health care economy are possible.

In principle, therefore, immediate point-of-care diagnostics and high-throughput analysis at central or large laboratories only appear to be a contradiction in terms [4]. Immediate point-of-care diagnostics or the more broadly categorized POCT currently seems to be a helpful adjunct in patient care which under no circumstances means that we can do without conventional laboratory analysis. Although conventional laboratory analytics is still important at present, it is clear that technological development and the mass production of POCT devices and processes will lead to its widespread use in everyday medical care. Whether multifunctional devices that are easy to use are more widespread in hospitals and medical practices in the next few years does not depend on the availability of the technology. The extent and speed of these developments depends much more on how far health insurance companies are prepared to cover them.

2.2 The medical and financial aspects of POCT diagnostics

Medical and financial considerations are closely linked when talking about POCT. It is beyond doubt that POCT plays a significant medical role regarding the provision of emergency analyses in life threatening situations. In intensive care, medical as well as organizational aspects are important whereas for outpatients and admission units the organizational aspects are the most important. Also for general practice, POCT is at present primarily used for organizational reasons. It is not possible to generalize about the medical or financial benefits from any of its possible applications alone. As each case is different, the evaluation criteria for its cost benefit ratio are also different [10]. However, the following applies; using a new process at the same cost the medical performance should improve, using equivalent medical care costs are reduced or, in an ideal situation, medical performance should be improved at lower cost.

For this purpose the following goals should be strived for [11]:

- To improve or change clinical strategies, to render early diagnoses, to speed up diagnostic or therapeutic processes.
- To shorten the stay in hospital, in intensive care and in the operating theatre, and to reduce the time a clinician or member of nursing/care staff has to spend with the patient.
- To reduce treatment costs using appropriate treatment structures in optimal patient monitoring and prevention of complications.
- To improve general satisfaction of clinicians, nursing/care staff and patients.

An overview of previous studies about the medical and/or financial benefits of POCT is given in the reference citations [8] and [9]. The need for information and recommendations about the appropriate use of point-of-care rapid diagnostics has spawned evidence-based guidelines for POCT like those developed and published by the National Academy of Clinical Biochemistry (NACB) [15]. There are now detailed recommendations for commonly used POCT processes with a clear rationale for their application. These guidelines can be useful for staff involved in point-of-care rapid diagnostics as well as for decision makers.

2.2.1 Medical aspects

Analysis processing time

The analytical and diagnostic advantages of using POCT are mainly attributed to the shorter turn-around time (TAT); meaning the time from ordering the test to receiving the result. This time is considerably shorter thanks to the elimination of sample transport and sample preparation as the tests are performed near the patient, usually on whole blood. Measurement of the activated clotting time (ACT) in cardiac catheter interventions, for example, is required by the guidelines. The ACT shows whether the heparin administered during the intervention is sufficient or needs to be adjusted. On the one hand, the result is needed very quickly and on the other hand it is not possible to send the sample away as adding an anticoagulant would make the sample unusable. It is, however, impossible to transport the sample without adding an anticoagulant, as clotting occurs quickly and would therefore make the sample unusable.

The most important advantage of POCT diagnostics, namely the rapid availability of test results, is not necessarily a medical or economic advantage in the hospital setting [19, 22]. This is frequently attributable to the fact that a TAT shortened by POCT still has an unchanged, long therapeutic TAT (TTAT), i.e. the time from blood draw to the treatment decision made based on the test results [20]. There are still only a limited number of concrete studies asking whether the many advantages attributed to POCT really bring about improvements in patient care. In one comprehensive English study, van Heyningen et al. [22] explored the question of clinical outcomes as impacted by POCT diagnostics versus conventional laboratory analytics. Included were 1,728 surgical and medical patients in the emergency department of a teaching hospital. Patients were randomly selected for the type of biochemical analytics (POCT or central laboratory). A range of indicators were recorded such as TTAT, the length of stay in the emergency department, the rate of patients admitted as well as mortality rate and length of hospital stay. Although there was a considerably shorter TAT for the POCT group, no significant differences in outcome indicators could be demonstrated. The authors concluded that the diagnostic TAT in the clinical management of their emergency department patients was not the decisive factor impacting the medical benefit.

Note

It is therefore necessary for each hospital to work with the central laboratory to develop an integrated plan for the sites where POCT is used and, furthermore, to safeguard the quality of results by implementing quality assurance measures.

This concept must be adapted in each case to the space and resources available in each hospital. For example, in hospitals with modern pipe delivery or cassette systems, a very short analytical TAT is usually ensured by the central laboratory, without any comprehensive POCT. An ideal combination is the "order-entry system" where the required laboratory tests are requested directly on the PC and transferred on-line to the laboratory. The results can then be returned on-line immediately after being released.

Quality assurance

For the medical feasibility of POCT, it is important that quality standards are adhered to: in Germany, such standards have been effective in conventional medical laboratory diagnostics for decades and are set forth in the quality assurance guideline for tests in laboratory medicine testing issued by the German Medical Association (RiliBÄK) [2] (▶ Chapter 36).

RiliBÄK requires the same quality standards for analysis by central laboratories as for POCT. From a patient's perspective, this is an important aspect, as time-critical analyses with POCT are carried out and applied to the treatment of patients. If the number of analyte testing devices in a hospital expands, the risk of systematic differences between the measuring systems increases. This applies to both central laboratory analytics and POCT diagnostics. Systematic differences between POCT and central laboratory test results are often noticed by ward personnel, particularly when POCT analysis is not supervised by the central laboratory. In general, systematic differences in devices are an ongoing topic in laboratory analytics and it is the responsibility of the specialized staff to ensure that variations in device readings do not lead to misinterpreted results. The employees have to check for these variations before test procedures are introduced to either find alternative tests or to develop relevant corrective measures. The aim is to ensure the best comparability of results generated and reported for patient care.

In the clinical setting, there are two basic options for the appropriate, guideline-compliant regulation of responsibilities for quality assurance in decentralized laboratory diagnostics [12]:

Option 1 Laboratory management is responsible for the organization and monitoring of quality assurance on all wards and outpatient departments that carry out POCT analysis in the hospital. Specially trained laboratory staff and personnel are frequently employed for this purpose. Even if there is no generally recognized name for these qualified staff, they are usually medical technical laboratory assistants and have become known as "POCT coordinators" or as the "persons responsible for POCT". The equipment and tests are chosen in consultation with the hospital management or a decision-making committee, e.g. a POCT committee that trains operators and monitors and records internal and external quality control measurements. In the case of quality assurance issues, laboratory management initiates the necessary corrective steps.

Option 2 The individual hospital departments act independently of the central laboratory and are responsible for carrying out all internal and external quality assurance measures prescribed by RiliBÄK. It is neither possible nor permitted for the hospital management to pass the responsibility for organizing and monitoring quality assurance to the POCT coordinator. This option is mandatory when laboratory tests are carried out at the point of care, but do not fully meet POCT criteria as laid out in the RiliBÄK guideline due to technical and organizational processes.

Private practitioners are also bound by and obliged to follow the RiliBÄK guideline. However, unlike hospitals, doctors in private practice themselves are responsible for performing analyses correctly.

2.2.2 Economic aspects

The demand for implementing or expanding POCT diagnostics needs to be scrutinized in each case, particularly given the high direct costs [5, 20]. POCT reagents can be several times the cost of analysis by the central laboratory. Using POCT devices adds a further burden for nursing and medical staff. In individual cases, the introduction of POCT methods and the related increased workload requires an increased number of nursing jobs without being able to cut posts in the central laboratory. Similarly, quality assurance also adds to costs. It has been repeatedly described that when the introduction of POCT procedures is poorly coordinated across the whole organization of a hospital, it can lead to considerably higher costs, not least when POCT results still need to be checked retrospectively in the central laboratory ("parallel diagnostics") [3].

In order to make a clear statement about the economic impacts of POCT, the following questions need to be clarified. Should, for example, costs only be calculated per test or should general cost efficiency be considered? How can the medical benefit be measured? Is the length of stay a satisfactory criterion? Very different results can be produced, depending on the level at which costs are calculated (e.g. analysis, patient, hospital, healthcare system). It holds true for hospitals, but also for other medical facilities, that a shorter inpatient stay, e.g. improved organizational processes or improved medical care can lead to a reduction in surplus beds. Also, a higher number of treated patients per time unit can economically benefit the hospital. This explains why in the short term the costs for the wards can increase while the hospital or healthcare system can reap long-term savings through improved patient care.

Very little concrete data have been published to date about economic issues. In some specific studies, the benefit of POCT was demonstrated from a medical-economical perspective. Examples include the use of HIV rapid tests for screening pregnant women just before birth or CRP checks in the doctor's practice to prevent unnecessary prescribing of antibiotics in patients with lower respiratory conditions [6]. In other settings, like in doctor's practices or pharmacies, the potential economic impacts of POCT are much more difficult to evaluate as they are mainly linked to monitoring the clinical course of chronic diseases where impacts are not measurable until years later when they can often no longer be linked to a triggering event. Important for the economic use of POCT in hospitals is the comprehensive recording of all POC tests to be able to map performance in internal performance data. However, this can only be achieved if all devices are connected to a central server and patient identification data are entered mandatorily.

2.3 The POCT market

POCT systems make up an important segment of the in-vitro diagnostics (IVD) market that had worldwide sales of some $14.5 billion in 2014 (approx. 26°% of the total IVD market). According to TriMark Publications (http://www.trimarkpublications.com) or Boston Biomedical Consultants (http://www.bostonbiomed.com) approximately $5 billion was from Europe, of which $1.2 billion was from Germany. Over the last 10 years, the POCT share of the IVD market has increased rapidly in leading European countries such as Germany, France, Italy and Great Britain and has, in the meantime, reached more than a third in Germany. The largest part of the POCT market (approx. $ 700 million) is taken by blood glucose strips and test meters for diabetes self-testing.

The global POCT systems and reagents market is anticipated to show **rapid growth** over the next years. The reasons for this are:

- Advances in analytical, microfluidic and readout methods
- A demand for continuous metabolite measurements
- A rapidly increasing demand from developing countries
- A rapidly increasing demand in the outpatient and nursing care sector ("assisted ambient living")
- Increasing patient and consumer interest (increasing health awareness)

2.3.1 Problems with market valuation

While the sales figures relating to established parameters of rapid point-of-care testing, such as blood glucose or blood gases, are relatively well documented, a considerable uncertainty is to be expected when making a valuation of the total market. The product range is very diverse, not exactly defined and not fully documented. Moreover, a significant number of new procedures are coming onto the market. A further uncertainty lies in the correct classification of tests such as blood gas analysis (including electrolyte and metabolite measurements) carried out in part by hospitals' central laboratories, but increasingly also at the POC. For example, the Association of the German Diagnostics Industry (VDGH) has only previously allocated these tests to the laboratory sector.

Given a wider interpretation of the POCT concept, the market volume could in fact even be larger than the official figures show. Despite the collection and assignment problems, these facts stress the importance of POCT for today's diagnostic market. At least every third IVD euro is generated at the patient's bedside or in the pharmacy (▶ Chapter 1).

2.3.2 POCT categories

To date there is no mandatory classification of POCT systems; a classification can basically be made, using different criteria such as technology, area of application or place of use [13]. It makes sense to start with a differentiation between the two large areas of **self-testing by the patient** (home testing) and **POCT diagnostics performed at the hospital**, primarily carried out by the medical staff. The test menu for the first category is mainly limited to glucose and global coagulation assays for monitoring diabetics or patients on anticoagulants. However, in the second category the menu seems to be almost endless: There are hardly any laboratory tests that could not, in theory, also be made available at the point of care. In intensive care units, POCT devices are partially integrated into the larger monitoring systems so that laboratory values like blood gases and electrolytes as well as vital parameters like ECG and central venous pressure are continuously available. Most POCT systems are now used as benchtop analyzers and are connected to patient data management systems.

Devices are categorized systematically in ▶ Chapter 3.

The areas of application for POCT in hospitals can be further divided into typical places of use such as operating rooms and recovery rooms, intensive care units and emergency departments, specialist outpatient departments etc. Acute laboratories represent a mixed version of this for hospitals without a central laboratory. They are not necessarily areas of application for POCT in a traditional sense, but still come under the wider definition of "point-of-care rapid testing performed outside of a central laboratory" [6]. There are many different devices and IT solutions available for each area.

Blood glucose measurements are leading in numbers, followed by blood gas analysis and electrolytes, blood coagulation, hematology, measurement of metabolites and enzymes, medications and drugs, as well as multiple specialized tests e.g. for allergies or pathogen diagnostics. From a technical perspective, a classification by parameter spectrum is useful. As well as being used in hospitals and for patient self-testing, the former are also used in doctor's practices and pharmacies. Urine test strips, test cards for occult blood in stool or pregnancy tests, for example, are also classed as "single test systems without a readout meter". Larger devices offer test profiles for clinical or organizational problems, e.g. cardiovascular markers – troponin, CK-MB, (NT-pro)BNP, myoglobin, D-dimer – or for intensive care monitoring – hemoglobin, hematocrit, blood gases and electrolytes. The designs range from 2 channel devices for global coagulation tests to miniaturized multichannel devices with up to 25 parameters. This means that they represent a considerable cross-section of the range of tests offered in a conventional central laboratory (► Chapter 3).

2.3.3 Future market trends

In the future, new markers and the on-going centralization of the conventional laboratory market should also add to interest in the POCT systems and further significant growth can be expected in some areas. According to the market research company Frost & Sullivan (www.frost.com), rapid tests, in particular for cardiovascular and infectious diseases as well as ovulation and pregnancy testing hold considerable future potential. A new area for POCT growth, called **companion diagnostics**, may also come to the fore. Companion diagnostics refers to (stratified) tests that help to decide if a planned treatment is suitable for patients in a particular disease stage or have a known predisposition to it. Predictions that POCT would overtake conventional laboratory diagnostics or even replace it should not be taken seriously nevertheless. POCT and the central laboratory can be seen as complementary pillars of laboratory diagnostics with a different focus in patient care: POCT will continue to garner market share for flexible rapid testing in critical care settings and at decentralized institutions. In the meantime, however, the central laboratory will continue to dominate the processing of large-scale analysis series and complex specialized tests.

The enormous further developments in **chip** and **microfluidic technologies** are heralds that multiplexed analysis will also be used for POCT. However, their clinical diagnostic value has still to be proven in studies. Long-term development will also be influenced by new diagnostic technologies, such as the continuous "inline" measurement of metabolites. For example, continuous glucose testing in subcutaneous tissue has developed from clinical studies into a useful additional diagnostic tool for type 1 diabetics (► Chapter 13). These analysis systems are also easily classified as POCT and could have great significance in the future, particularly in intensive care.

References

1. Bietenbeck A, Junker R, Luppa PB (2015) Central laboratory service and Point-of-Care Testing in Germany – From conflicting notions to complementary understandings. Point Care J Near-Patient Test Technol 14: 1–11
2. Bundesärztekammer (German Medical Association) (2014) Guideline of the German Medical Association for the Quality Assurance of Clinical Laboratory Testing (Rili-BÄK 2014). Dtsch Ärztebl 111:A1583–1618
3. Doyle NM, Levison JE, Gardner MO (2005) Rapid HIV versus enzyme-linked immunosorbent assay screening in a low-risk Mexican American population presenting in labor: a cost-effectiveness analysis. Am J Obstet Gynecol 193: 1280–1285
4. Drenck N (2001) Point of care testing in critical care medicine: The clinician's view. Clin Chim Acta 307: 3–7
5. Gruszecki AC, Hortin G, Lam J et al. (2003) Utilization, reliability, and clinical impact of point-of-care testing during critical care transport: Six years of experience. Clin Chem 49: 1017–1019
6. Hänecke P, Haeckel R, Koschinsky T, Luppa P, Schlebusch H, Wahl HG (2004) Qualitätssicherung in der patientennahen Sofortdiagnostik (Point-of-Care Testing) im Krankenhaus: Muster für eine hausinterne Richtlinie. J Lab Med 28: 256–263
7. Jani IV, Peter TF (2013) How Point-of-Care Testing could drive innovation in global health. N Engl J Med 368: 2319–2324
8. Kost GJ, Tran NK. Effects of point-of-care testing on time, process, decision making, treatment and outcome. In: Kost GJ (ed) (2002) Principles and practice of point-of-care testing. Lippincott Williams & Wilkins, Philadelphia, pp 605–611
9. Kost GJ, Tran NK (2002) The economics of point-of-care testing. In: Kost GJ (ed) Principles and practice of point-of-care testing. Lippincott Williams & Wilkins, Philadelphia, pp 613–622

10. Kost GJ (2002) Controlling economics, preventing errors, and optimizing outcome in point-of-care testing. In: Kost GJ (ed) Principles and practice of point-of-care testing. Lippincott Williams & Wilkins, Philadelphia, pp 577–600
11. Kost GJ (2002) Goals, guidelines, and principles for point-of-care testing. In: Kost GJ (ed) Principles and practice of point-of-care testing. Lippincott Williams & Wilkins, Philadelphia, pp 3–12
12. Luppa P, Junker R (2010) Verantwortung für die Durchführung der Qualitätssicherung einer dezentral organisierten Laboratoriumsdiagnostik im Krankenhaus. Klin. Chemie Mitteilungen 41:205–8
13. Luppa PB, Müller C, Schlichtiger A, Schlebusch H (2011) Point-of-care testing (POCT): Current techniques and future perspectives. Tr Anal Chem 30: 887–898
14. Mahtabjafari M, Masih M, Emerson AE (2001) The value of pharmacist involvement in a point-of-care service, walk-in lipid screening program. Pharmacotherapy 21: 1403–1406
15. National Academy of Clinical Biochemistry (NACB) (2007) Practice guidelines: Evidence-based practice for point-of-care testing. Point of care 6: 213–217
16. Peterson GM, Fitzmaurice KD, Naunton M, Vial JH, Stewart K, Krum H (2004) Impact of pharmacist-conducted home visits on the outcomes of lipid-lowering drug therapy. J Clin Pharm Ther 29: 23–30
17. Price CP, St. John A (2014) Innovation in healthcare. The challenge for laboratory medicine. Clin Chim Acta 427: 71–78
18. Robert LS, Edward HL (2007) Point-of-care testing: Guidelines and challenges. NC Med J 68: 132–135
19. Schlüter B, Junker R (2003) Labordiagnostik – Schneller ist nicht immer besser. Dtsch Ärztebl 100: A87–A89
20. Scott MG (2000) Faster is better – It's rarely that simple! Clin Chem 46: 441–442
21. Till LT, Voris JC, Horst JB (2003) Assessment of clinical pharmacist management of lipid-lowering therapy in a primary care setting. J Manag Care Pharm 9: 269–273
22. van Heyningen C, Watson ID, Morrice AE (1999) Point-of-care testing outcomes in an emergency department. Clin Chem 45: 437–438

Device classes

Peter B. Luppa

P. B. Luppa, R. Junker (Eds.), *Point-of-Care Testing*
https://doi.org/10.1007/978-3-662-54497-6_3

3

3.1 Introduction

Myriad devices are offered on the in vitro diagnostics (IVD) market – ranging from compact hand-held to benchtop devices that embody complicated analytical systems for use in cuvette tests or carrier-bound test methods. The spectrum of methods spans a range across simple test strips to complex immunochemical assays. In general, these POCT devices tend to be almost fully automated and require only a few simple maneuvers on the part of the user to go from sample preparation to test result.

In the past few years, the following **instrument designs** have prevailed:

- Hand-held devices, e.g. glucose measuring devices (corresponding to a "unit-use" system; see below)
- Other "unit-use" or "multi-use" cassette systems (e.g. i-STAT for blood gas analyses or Hemochron Jr. Signature+ for coagulation analyses)
- Stationary benchtop devices, e.g. blood gas, electrolyte and substrate analyzers
- Lab-on-a-chip systems

Lab-on-a-chip systems will not be discussed here because the development of such chips is currently undergoing very rapid advances, whereas applications in medicine have not been sufficiently evaluated. This chapter aims to provide a categorization of the analyzers [10] that is oriented along the following system characteristics: Sensor characteristics, system complexity, measurement principles, sample matrix and practical benefits.

POCT device classes

- Type 1
  - 1a: Qualitative POCT methods
  - 1b: Unit-use POCT systems
- Type 2: Benchtop POCT instruments
- Type 3: Viscoelastic coagulation analyzers
- Type 4: Continuous POCT measurement methods
- Type 5: Molecular biological POCT analyzers
- Type 6: Direct-to-consumer testing (DCT)

Examples of the respective systems are described extensively in the individual chapters of this book.

3.2 Type 1a – Qualitative POCT methods

These qualitative tests discriminate between plus/minus results and are usually encountered as test strips. The measuring signal can either be read off directly as a visual display or recorded on a simple read-out device. The detection principles range from chemical indicator reactions to immunological antigen-antibody interactions (such as "lateral flow assays", LFA, ▶ Chapter 9).

The strips are made of a solid carrier material and a porous matrix to which lyophilized reagents have been added. The patient's sample (urine, blood, stool, cerebrospinal fluid, smear etc.) is applied to the strip. The analytical reaction is triggered when the strip layer is penetrated and wetted. Typical applications include pregnancy tests, tests for blood in stool, urine dipsticks and a variety of quick tests for infectious pathogens in material from smears.

3.3 Type 1b – "Unit-use" POCT systems

"Unit-use" POCT devices represent the simplest type of quantitative POCT devices. The actual analytical reaction takes place on the test strip; the read-out device only generates the readable measured value. In general, the unit-use devices have the following characteristics:

- Individually packaged reagents/strips are used for each measurement. "Unit-use" reagents are only suitable for single use.

- Unit-use reagents rely on unprocessed whole blood for testing, where the spectrum of available tests varies greatly and depends on the sensors employed.
- Typically, the sensors are incorporated into the test strip and not into the device itself. Commonly, dry chemical methods are implemented, e.g. glucose-converting enzymes that are immobilized on reagent strips. Any calibration in these devices is usually replaced by electronic or physical standards.
- By contrast, more complex cassette devices are equipped with automated calibration programs that run at set time intervals.
- There are devices which are designed for a singular parameter only (e.g. Bayer's Ascensia Contour for glucose, Roche Coaguchek for INR). Frequently, however, device-based systems can measure multiple parameters using different strips (e.g. Nova's StatStrip for glucose and creatinine, Alere's Triage for multiple parameter groups).

Although the blood glucose test continues to be the most common application of unit-use devices, the past years in particular have seen a panoply of systems developed, e.g. for coagulation analyses, myocardial infarction markers, HbA_{1c} and blood gases. This trend will certainly intensify over the course of the coming years.

The applied technologies for unit-use devices range from electrochemical methods (blood glucose meters) through immunoassay analyzers (DCA Vantage by Siemens) up to thin-film sensors (i-STAT of Abbott). The latter device offers a good example for the use of chip-based microfabrication technologies to manufacture biosensor chips. The aim is to incorporate the actual analytical sensors directly into a microchip and connect them through multiple channels to the transducer by electrical means as well as to a flow cell construction by means of microfluidics. Using thin-film technology, SiO_2 chips first undergo milling; next a 200-nm film, e.g. with Si_3N_4, is applied to the underlying structure (e.g. 1 µm); then, electrode structures are sputtered on to this layer. The intermediate layers and/or microchannels are then isolated, e.g. with thin titanium layers (15 nm). The coating to isolate the chips against the outside is accomplished by means of substances like polyethylene and Si_3N_4 using a chemical-vapor deposition (CVD) process [7]. In the case of i-STAT (Abbott), a series of biosensors are applied to a chip as thin-film electrodes. This is incorporated into a single-use test cartridge that provides calibrator solutions, a sample handling system and contact pads for the analyzer. ◻ Fig. 3.1 shows the explosion diagram of such a cartridge [4].

Recently, thin-film biosensing has been employed experimentally as well as in holographic optic sensors for online glucose monitoring [8].

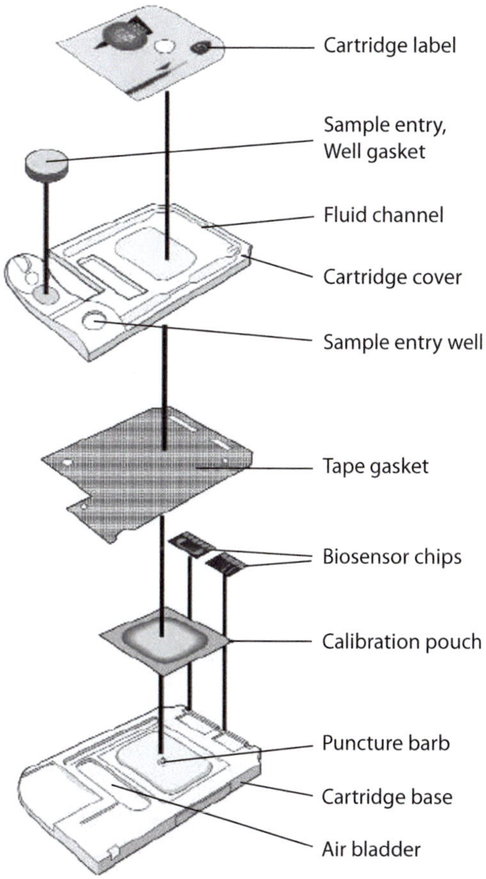

◻ **Fig. 3.1** Setup of the i-STAT single-use test cartridge. (By kind permission of Abbott GmbH & Co. KG, Wiesbaden, Germany)

3

3.4 Type 2 – Benchtop POCT instruments

In general, benchtop devices are complex analytical systems not located directly at the bedside, but in functional areas of intensive care units, in outpatient centers or doctors' practices. Benchtop devices are differentiated into:

- Clinical chemistry, spectrophotometric multichannel devices (e.g. Abaxis Piccolo Xpress),
- Clinical chemistry test strips/cartridge devices (e.g. Roche Reflotron),
- Hematologic multichannel devices (e.g. Sysmex pocH-100i),
- Blood gas analyzers with/without oximetry and electrolyte/substrate measurements (e.g. from Werfen/IL, Keller Medical, Nova, Radiometer, Roche and Siemens),
- Immunological multichannel devices (e.g. Radiometer AQT90).

The technology used in the different systems mostly derives from the mechanized analytical systems used in central laboratories, but optimized with respect to miniaturization, speed of analysis and user-friendliness.

By contrast, the blood gas analyzers were designed primarily for POCT analysis many decades ago and since then evolved into more consummate devices with constantly expanding user menus.

However, new technologies such as that used in thick-film sensors have been specifically developed for POCT applications alongside established POCT techniques like reflectance photometry for dry chemistry reagent carriers. Thin-film technology involves a special method for manufacturing electronic circuits. To integrate sensor elements, the necessary conductor lines, condensers, resistors and inductivities are plated on an insulator (substrate) by screen printing. For this purpose, 10–50-µm thick film layers of organic paste systems containing conductive/dielectric substances are applied and permanently fixated. The QBC Star from Drucker Diagnostics (Port Matilda, PA, USA) follows an alternative technological approach to conventional hematological multichannel devices in that it employs dry hematology analysis [3]. At the heart of this device is a special 65-µl blood collection tube which is pre-coated with staining reagents on the inside. During centrifugation of the test tube, the cell types separate into color-differentiated packed cell layers. The separation into platelets, lymphocytes, monocytes and granulocytes is supported by precisely suspended particles in the test tube which facilitates optical detection.

By offering improved computing performance, many benchtop devices relieve the user of control tasks like calibration and quality assurance. One of the most important accomplishments that these complex devices achieve is the mastery of whole blood sensor technology that permit continuous flow operations and very short analysis time with the most minimum sample volumes. This is accomplished by employing highly developed **microfluidics** which enable and process the handling, transport and mixing of fluids within the smallest of space. During this process, the flow can be controlled by capillary forces or electroosmotic effects [14]. Moreover, microfluidic technologies consist of microvalves and/or of micropumps. Here, the valves involved are passive valves controlled by hydrodynamic pressure and should be distinguished from active micro-valves that run on principles like piezoceramic technology. Nowadays, micro-valves are fabricated from silicon wafers. Among the micropumps, there are likewise a large number of different types. Oscillating membranes are often used, but peristaltic systems as well. In the case of non-mechanical pumps that allow further miniaturization, high electrical and/or magnetic field strengths are applied and electrokinetic or electrohydrodynamic effects exploited for transporting fluids. However, centrifugal forces are also employed to transport the minutest amounts of fluid, e.g. on centrifugal analyzers.

One important component of a benchtop device is its control subsystem, which controls all analytical process steps including reagent loading, temperature setting, injection, incubation, time sequence etc. That said, data storage options (including calibration and quality control data) are also important prerequisites for POCT benchtop devices. The following overview summarizes the key device characteristics specific to blood gas analysis systems that minimize the risk of erroneous measurements and establish robust procedures [15].

Key characteristics of blood gas analyzers

- Maintenance-free sensors
- Touchscreen as user interface with built-in training videos
- Software that assists the user and aids in the identification of patient sample to be tested with built-in barcode scanners
- Sample aspiration instead of injection (important!)
- Clot-detection in the flow cell
- Volume detection in order to allocate low-volume samples to analyses that are still possible
- Hemolysis detection
- Fluid calibration system instead of gas bottles
- Automatic (re-)calibration and quality control
- Interconnectivity with information systems

3.5 Type 3 – Viscoelastic coagulation analyzers

Viscoelastic coagulation analysis refers to the integral testing of the interplay between plasmatic coagulation, platelet function and fibrinolysis [5]. Viscoelastic coagulation analyzers exhibit a high degree of complexity and are therefore only conditionally feasible in POCT settings. As a rule, they are operated on-site by specially trained personnel (e.g. in the operating room). Alternatively, the analysis is conducted in the central laboratory and graphic results are transferred directly to the clinical arena in real-time. Examples include rotation thromboelastometry using ROTEM (TEM International, Munich) and Sonoclot by Sienco Inc. (Arvada, CO, USA).

Platelet function can also be analyzed by measuring the bleeding time in-vitro or by optical aggregometry. Product examples here include the PFA 100 from Siemens Healthcare Diagnostics (Eschborn, Germany) or VerifyNow from Accriva Diagnostics (San Diego, CA, USA). The Multiplate from Roche Diagnostics features platelet impedance aggregometry.

3.6 Type 4 – Continuous POCT methods

Continuous measurement methods were particularly developed for glucose monitoring, but meanwhile have become available commercially as well [1]. The FreeStyle Libre from Abbott Diabetes Care (Alameda, CA, USA) is one example that has been very successful since 2014. Continuous measurement over several days can be accomplished using a minimally invasive microdialysis catheter in the subcutaneous tissue (▶ Chapter 13). In recent years, other non-invasive methods such as microporation or optical techniques for the direct transcutaneous measurement of metabolic parameters have not been able to establish themselves. This is mainly because the human skin is not uniform in terms of thickness, pigmentation, hair coverage, but also with regard to physiological phenomena like moisture or salinity. The currently available systems like Enlite by Medtronic/Minimed (Minneapolis, MN, USA), or FreeStyle Libre (Abbott Laboratories, Wiesbaden, Germany) are presented in ▶ Chapter 5 and 13.

3.7 Type 5 – Molecular biological POCT analyzers

Molecular biology amplification techniques (usually based on the polymerase chain reaction (PCR)) are currently undergoing intensive development for the POCT sector. Such systems are sophisticated in terms of preanalytical technology (DNA or RNA extraction), but also the PCR-related analysis time will be above those of other POCT methods in future. Besides PCR, isothermal amplification methods can also be applied, the process guidance of which is technically easier to master. As examples of such methods, helicase-dependent amplification (HDA) [6] and recombinase polymerase amplification (RPA) [12] can be cited.

Given the complexity involved with performing these tests and interpreting their results, fast nucleic-acid testing (NAT) methods are unlikely to be found in many clinical settings in the future. Similar to type 3, such devices are often operated in the central laboratory [13]. Nevertheless, the fast and quantifying detection of the DNA/RNA of bacterial and viral pathogens supplies clinicians with valuable diagnostic information. The GeneXpert system by Cepheid (Sunnyvale, CA, USA) was an innovation in terms of full automation and the integration of all real-time PCR-based NAT steps (sample preparation, DNA/RNA amplification and DNA/RNA detection) (▶ Chapter 10). Another device to mention is the Biofire FilmArray system (bioMérieux, Marcy l'Etoile, France). A multiplex PCR enables the simultaneous testing of various pathogens. The following two systems are suitable for pure POCT applications: One is the Liat (lab in a tube) device by Roche. It operates on the principle of PCR amplification and features extremely fast analysis. Another is the Alere i analyzer from Abbott for detecting influenza A and B viruses based on the fast isothermal RPA reaction.

Whether the POCT use of such systems in industrialized countries makes sense still remains controversial. To date, there continues to be a lack of large-scale observational studies that have scrupulously compared the NAT systems described with conventional microbiological diagnostics (including cultivation and identification). Moreover, Dark et al. [2] recommend conducting clinical efficacy studies aimed at documenting the therapeutic efficacy and cost effectiveness. The background to this proposal is that NAT tests cannot additionally test for antibiotic resistances when microbes are detected.

3.8 Type 6 – Direct-to-consumer testing (DTC)

Besides the physician-ordered clinical laboratory diagnostic tests, a new healthcare market has emerged in Europe over the past years: that of direct-to-consumer testing (DTC). By eliminating the physician, the patient can choose from a selection of generally available DTC tests. The "quantified-self" movement allows knowledgeable and empowered patients and consumers to answer questions about their own health and exercise-related habits. In this respect, POCT technology offers exactly that option of delivering minimally invasive performance data about one's own body (as per the motto: "self-knowledge through numbers"). The potential opportunities and disadvantages of these novel applications must be viewed carefully [9], considering that the patients themselves are responsible for carrying out and interpreting the tests. DCT tests are to be regarded with equal importance whether they are over-the-counter pregnancy tests or specimen collection kits for genetic analyses, purchased on the internet (◘ Fig. 3.2) [11].

Fig. 3.2 Overlapping definitions of POCT and DTC terms with examples

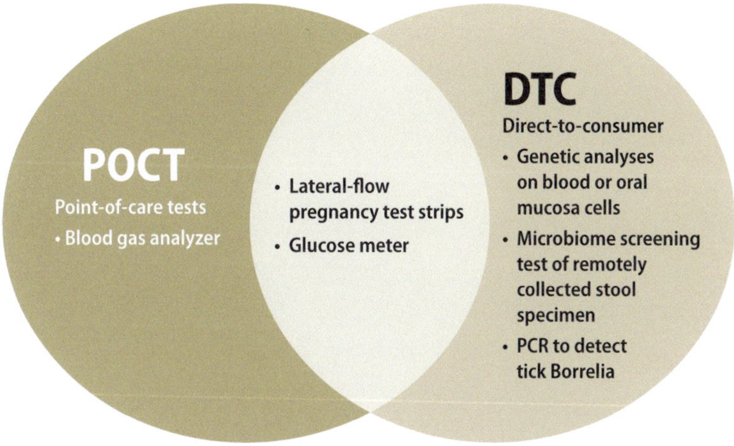

References

1. Corstjens AM, Ligtenberg JJ, van der Horst IC, Spanjersberg R, Lind JS, Tulleken JE, et al. (2006) Accuracy and feasibility of point-of-care and continuous blood glucose analysis in critically ill ICU patients. Crit Care 10:R135
2. Dark P, Dean P, Warhurst G (2009) Bench-to-bedside review: the promise of rapid infection diagnosis during sepsis using polymerase chain reaction-based pathogen detection. Crit Care 13: 1–6
3. Erhabor O, Richardson G, Mohammed I et al. (2013) Evaluation of the QBC Star centrifugal three-part differential haematology system. Br J Biomed Sci 70: 67–74
4. Erickson KA, Wilding P (1993) Evaluation of a novel point-of-care system, the i-STAT portable clinical analyzer. Clin Chem 39: 283–287
5. Ganter MT, Hofer CK (2008) Coagulation monitoring: current techniques and clinical use of viscoelastic point-of-care coagulation devices. Anesth Analg 106:1366–1375
6. Goldmeyer J, Kong H, Tang W (2007) Development of a novel one-tube isothermal reverse transcription thermophilic helicase-dependent amplification platform for rapid RNA detection. J Mol Diagn 9: 639–644
7. Herrmann S, Vonau W (2004) Online-Analyse mit Lab-on-Chip-Systemen. Technisches Messen 71: 613–618
8. Liedberg B, Nylander C, Lundstrom I (1995) Biosensing with surface plasmon resonance – how it all started. Biosens Bioelectron 10: i–ix
9. Lippi G, Favaloro EJ, Plebani M (2011) Direct-to consumer testing: more risks than opportunities. Int J Clin Pract 65:1221–1229
10. Luppa PB, Müller C, Schlichtiger A, Schlebusch H (2011) Point-of-care testing (POCT): Current techniques and future perspectives. Trends Anal Chem 30: 887–898
11. Orth M, Luppa PB (2015) Direct-to-Consumer-Testing – Fluch oder Segen für die Patienten? Dtsch Ärztebl 112: A174–A176
12. Piepenburg O, Williams C, Stemple D, Armes N (2006) DNA detection using recombination proteins. PLoS Biol 4: e204
13. Seme K, Mocilnik T, Komlos KF, Doplihar A, Persing DH, Poljak M (2008) GeneXpert enterovirus assay: one-year experience in a routine laboratory setting and evaluation on three proficiency panels. J Clin Microbiol 46: 1510–1513
14. Spindel S, Sapsford K (2014) Evaluation of optical detection platforms for multiplexed detection of proteins and the need for point-of-care biosensors for clinical use. Sensors (Basel) 14: 22313–22341
15. St John A (2004) Benchtop Instruments for Point-of-care Testing. In: Price CP, St John A, Hicks JM (eds) Point-of-care Testing, 2nd ed. AACC Press, Washington, pp 31–46

Methodology and analytical techniques

Content

Pre- and post-analytical phases

Andreas Bietenbeck

P. B. Luppa, R. Junker (Eds.), *Point-of-Care Testing*
https://doi.org/10.1007/978-3-662-54497-6_4

4.1 Introduction

The objective of all quality assurance measures in laboratory medicine is to produce a proper diagnosis based on conclusive findings. Thereupon rests the foundation of optimal treatment for the patient. Conversely, this means for both laboratories and POCT users a paradigm shift from quality control of analyses to comprehensive quality management. This objective can only be achieved if the overall diagnostic process is calculated into the equation; this process extends from the correct patient-focused ordering of tests for the disease-relevant analytes to the correct interpretation of the analysis results aimed at rendering the proper diagnosis and monitoring the therapy based on that diagnosis.

This process can be divided into a pre-analytical analytical and a post-analytical phase. The pre-analytical phase includes everything from choosing an appropriate test and correctly identifying the patient to taking a sample, transporting it if necessary and preparing for analysis it in the laboratory. Validation and the appropriate reporting of findings follow post analysis. Recently, considerations about non-analytical factors (extra-analytical quality) have become increasingly important [3, 12, 17] and been incorporated into the German Medical Association's Guideline on Quality Assurance in Medical Laboratory Tests (RiliBÄK) [1].

Although fewer errors happen in the pre-analytical phase of POCT than in a central laboratory, this phase is still meaningful. Approximately one third of all errors in POCT occur prior to the actual analyses [14]. POCT pre- and post-analytical phases depend on the specific analytes to be assayed as well as on the devices used. Several overarching principles shall be addressed below.

4.2 Pre-analytical phase

The most important pre-analytical subroutines

1. Choosing a suitable test
2. Preparing the patient (e.g. diet, drugs, position of the body, time of the sample collection)
3. Collecting the sample
4. Transporting and storing the sample
5. Inspecting the sample (hemolyzed, icteric, lipemic appearance)
6. Processing the sample (e.g. centrifugation).

Bullet points 4 to 6 play a less important role within the POCT concept. As POC tests are carried out promptly, possible false readings caused by sample instability are not an issue either. The correct way of sample collection (point 3), however, is particularly important in the pre-analytical process. Many POCT systems use venous, arterial or capillary blood samples. Swabs are increasingly used to detect infectious diseases. POC tests usually require considerably less sample material. Rigorously implemented hygiene measures are particularly important for mobile devices to avoid the spread of pathogens [21].

4.2.1 Choosing a suitable test

Many POCT systems deliver results quicker than conventional central laboratories as no sample transport is required. However it is often not possible to achieve the same accuracy with highly integrated POCT devices as with larger laboratory devices. Therefore, investigations with a POCT device should not be regarded as equivalent to those carried out in a central laboratory even if they measure the same analyte [18]. POCT is, however, well suited for disease monitoring and controlling therapy.

4.2.2 Capillary blood sampling

Capillary blood is a mixture of blood from arterioles, venules and capillaries, sometimes diluted by interstitial or intracellular fluid (hemolysis). The relative composition depends on the blood circulation of the puncture site; heating leads to arterialization of the blood sample.

Differences exist between venous and capillary blood, which can influence hematological tests or oral glucose tolerance tests. In adults, capillary blood is usually taken from the fingertip or ear lobe, while the heel is used in neonates. For therapy monitoring, capillary blood can also be taken from other skin areas (▶ Chapter 12). Good blood circulation around the puncture site is important. To obtain arterialized capillary blood for blood gas analysis, the puncture site needs to be treated with warm water compresses (42° C) or by applying a special cream (e.g. Finalgon) to hyperemize the local tissue.

There is a linear correlation between blood volume and puncture depth. Therefore, puncture aids that can be adjusted to the required sample volumes are recommended. For self-monitoring, an extensive range of puncture aids for capillary blood sampling are available where the puncture depth can be adjusted. In addition, there are multiple lancets to go with some puncture aids, which differ in their bevels and suppression of painful vibrations.

In Germany, the technical regulations for biological materials (TRBA 250) need to be considered when using puncture aids in hospitals, physicians' practices, emergency vehicles etc. In blood sampling, "the use of safe instruments where body fluids are present in amounts that can transmit infections" is mandatory. Among other tasks, the following device safety specifications must be met:

- The safety mechanism must be part of the device and needs to be equipped with an audible or tactile signal. It may be used only once.
- The safety mechanism must be triggered single handedly, immediately after use. Self-activating systems are recommended as they are usually easier to handle.
- Safe work equipment must be compatible with the accessories and other systems used.

When using puncture aids in pediatrics/neonatology, the puncture depth on the child's heel is critical because of the danger of injuring the calcaneus, particularly in pre-term infants. For this reason, lancets with a shorter puncture depth (max. 2.0 mm) are preferred.

After skin disinfection, a single-use lancet is used to puncture. The blood sample is collected by gently pressing onto the tissue (not squeezing) to avoid or minimize hemolysis – one of the most important pre-analytical confounding factors. The first drop of blood is discarded as it is often contaminated with tissue components. In blood glucose monitoring, the first drop can be discarded; although this is not the case in INR self-monitoring where the first drop needs to be analyzed. The next few drops are collected in capillary tubes (e.g. end-to-end capillary tubes) or special capillary blood containers. A wide range is available from different manufacturers. Containers with additives should be inverted 5 times (not shaken), after filling with blood.

The correct technique for capillary blood sampling from the heel of an infant is shown in ◘ Fig. 4.1.

4.2.3 Venous blood sampling

Before phlebotomy, the patient should have been in a seated or prone position for 15 minutes, if possible. After disinfection of the puncture site, venous stasis is achieved with a tourniquet or blood pressure cuff (stasis between systolic and diastolic blood pressure) applied for no longer than a minute. The patient should not be asked to make a fist! The stasis is terminated immediately after successful venipuncture. There are different recommendations regarding the order in which samples are taken [4, 8], not all of which specifically address POC tests.

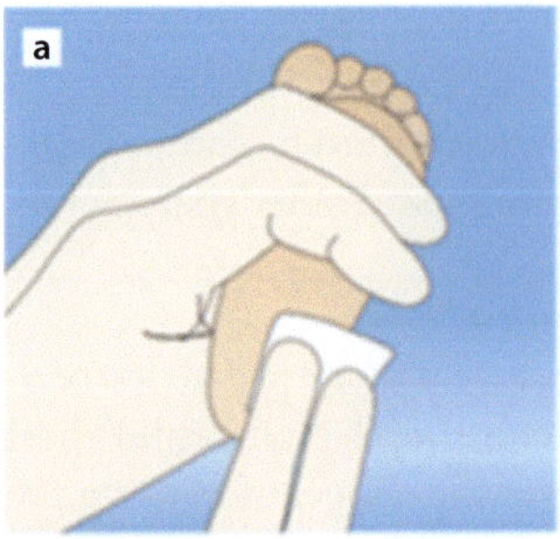

Select and disinfect a suitable puncture site

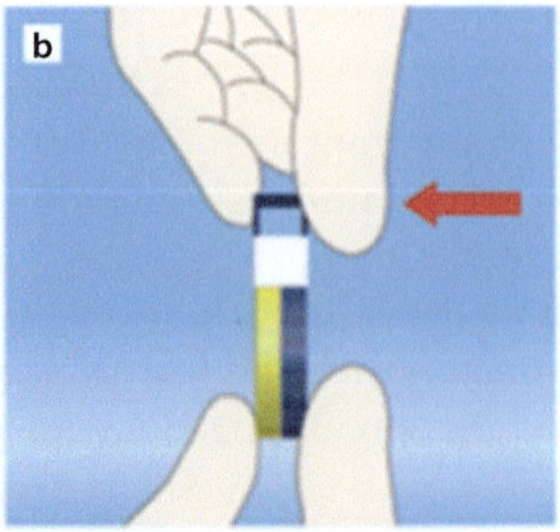

Remove the safety mechanism by pressing on the sides with thumb and forefinger.

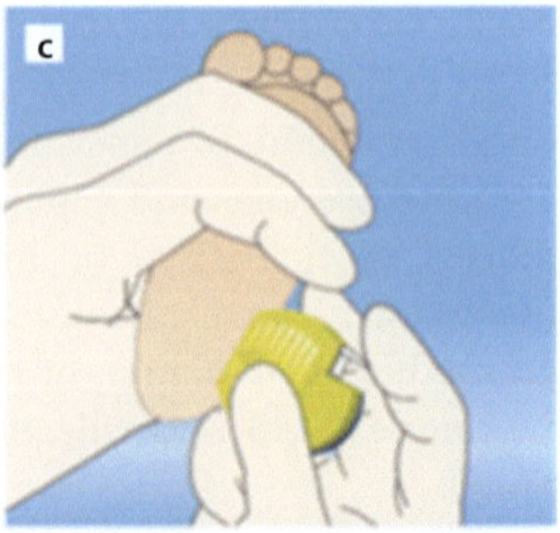

Lift the foot to a suitable position. Place the lancet against the selected, disinfected puncture site. Position the safety incision lancet parallel along the length of the foot and never across the heel. Note that the tip of the triangle points toward the blade exit. Press the firing button.

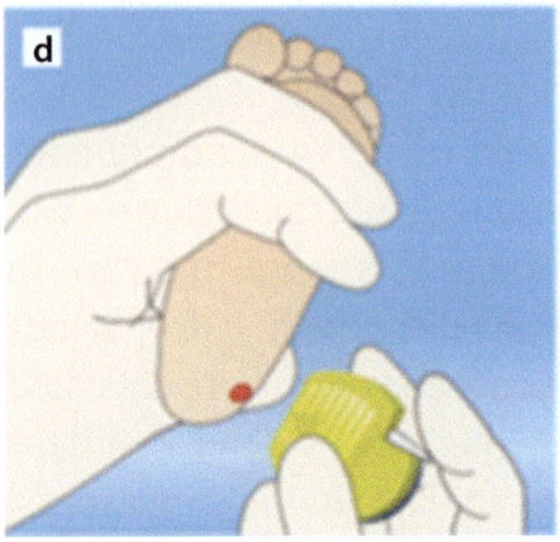

After pressing the firing button, remove the lancet from the heel.

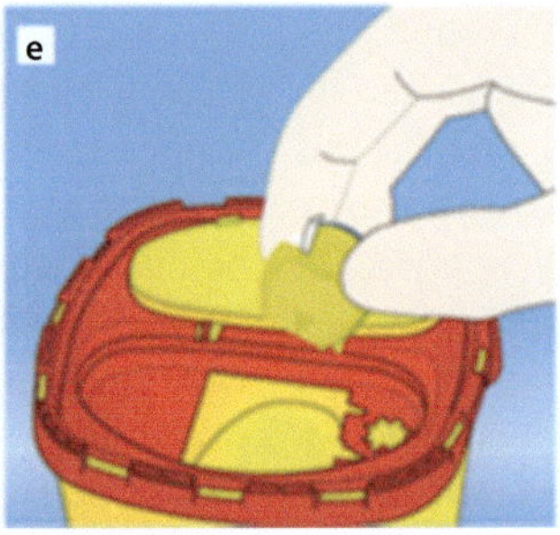

Dispose of the lancet in a suitable sharps container.

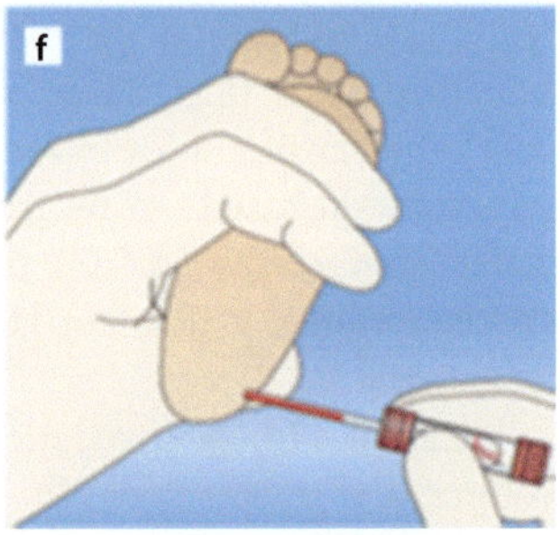

Wipe away the first blood drop. Next, fill the collection tube or capillary.

Fig. 4.1 Correct use of an incision safety lancet for newborns (Courtesy of Sarstedt, Nümbrecht, Germany).

In German speaking countries, for safety reasons, only closed blood collection systems are used. These are based on two different principles: Vacuum and stamp blood collection. The systems are difficult to combine as they are designed differently and have various attachments; with different sizes and accessories.

4.2.4 Arterial blood sampling

Arterial blood sampling is indicated, particularly for blood gas analysis (▶ Chapter 14) [2]. The samples are usually taken by puncturing the femoral, brachial or radial artery or via an arterial line. After the blood collection, firm pressure should be applied to the puncture site. Capillary blood can also be used for blood gas analysis when taken from hyperemized skin areas on the earlobe or fingertip.

Special closed glass or plastic needle-tube systems (see below) allow automatic filling due to the arterial pressure, while minimizing the risk of air bubbles in the sample. Anaerobic blood samples should be ensured; any air bubbles need to be removed immediately. Otherwise, the blood-dissolved gases equilibrate with the respective gas molecules inside the air bubbles. The CO_2 concentration is more affected than O_2 as CO_2 has a higher partial pressure in the blood than in the air. In contrast, the measurement of arterial O_2 partial pressure in the hyperoxic range (>200 mm Hg) is flawed because the diffusion gradient is greater towards room air [7]. After drawing the sample, it should be mixed by careful rolling between both hands and promptly analyzed within 5–10 minutes. If this is not possible, it is recommended to store the sample in ice water for up to 30 minutes [5]. However, it is very important to avoid hemolysis. Mix it again before testing! If sample transport by tube mail is possible, it is important to check the specifications of the system. Pneumatic tube systems, which generate low shearing forces thanks to their controlled acceleration and careful cornering and are approved for blood sample transport, are completely suitable for blood gas samples [7].

4.2.5 Blood sampling systems and anticoagulants for blood gas analysis

Ideally, whole blood samples should be taken with glass syringes, as dissolved gases cannot diffuse through glass. When using plastic syringes, it should be remembered that false results of pO_2 and pCO_2 can occur due to diffusion. Therefore, the sample should be analyzed promptly, as described above. Heparin-coated plastic capillaries are suitable for taking capillary blood samples. Every manufacturer of blood gas devices offers their own blood sampling systems. Examples are Siemens Medical Solutions RAPIDLyte pro, a self-filling system with capacity selection, and Radiometer's safePICO with vented safeTIPCAP and an integrated mixing ball, ensuring fast and thorough mixing of the sample before analysis. Roche offers, for example, the BS2 Blood Sampler and the MICROSAMPLER for its blood gas systems. Other systems are also available on the European market.

Ca^{2+}-balanced heparin is the most frequently used and most suitable anticoagulant, as it does not change the acid-base balance. If lyophilized heparin syringes are not used, it is usually satisfactory to rinse a 2-mL syringe with heparin solution, making sure that only a volume of 0.1 mL of heparin remains in the syringe conus. This way, an excess of heparin solution can be avoided, which would otherwise lower the pH of the sample (heparin is an acid mucopolysaccharide) and would lead to dilution errors. A heparin solution using 500 or 1000 IU/mL to rinse the syringe is recommended. A final concentration of 20–30 IU/mL will remain in the sample.

Blood gas analysis is often combined with a measurement for electrolytes and therefore only lithium heparin is suitable, not sodium or potassium heparin. As all heparins can bind positively charged ions such as Ca^{2+}, syringe systems with electrolyte compensating heparins were developed for the use of combined blood gas and electrolyte analysis. These electrolyte-balanced heparin solutions eliminate the interference from ion binding (e.g. Pico sampler by Drott Medizintechnik).

4.2.6 Blood samples from central lines

The sampling of blood via intravascular needles, cannulas or central lines is very common, particularly in intensive care units. Before the sample is taken, the central line should be flushed with heparin and a minimum of twice the line volume, approximately 2–5 mL blood, should be discarded to avoid sample contamination with infusion solutions or medications [10]. The time between the last infusion and blood sampling should be at least 15 minutes. Particular caution needs to be taken when POCT systems are employed, as they are more sensitive to interfering substances than large central laboratory devices.

4.2.7 Taking swabs

Various sample swabs are used particularly for POCT diagnostics of pathogens. Common sites are nasal, pharyngeal, inguinal, rectal or wounds, depending on the suspected pathogen. Multiple sites may need to be swabbed to identify a specific pathogen [19]. Different swabs are used for different sites [15]. Generally, the swab should be moistened when taking a sample from a dry surface. Avoid contamination.

All contaminated materials need to be disposed of safely. The swab often remains in the single-use cassette of the POCT device. While antibiotic use prior to sample taking must be avoided with conventional microbiological methods based on pathogen culturing, this precaution is not relevant for molecular POCT methods as these tests adopt a nucleic acid amplification technique.

4.2.8 Urine sampling

Urine is usually collected by non-invasive methods, which makes it particularly practical for POC testing by the patients themselves. The time of collection is, however, important as many analytes in the urine show a significant circadian rhythm. The first morning urine is particularly useful for nitrate and protein detection. Urine fractions are not uniform. The first urine fraction is often contaminated with pathogens from the urethra. Mid-stream urine is the preferred sample for many tests, as most analytes are diluted in the last urine fraction.

4.2.9 Inspection of the sample

Routine inspections in the central laboratory identify hemolytic, lipemic and icteric blood samples, which is not possible with POCT, as whole blood samples are mainly used. Hemolytic blood samples are of particular concern and cause by far the most frequent pre-analytical errors in many assay methods. The only solution – albeit unsatisfactory – is the avoidance of hemolysis by employing the proper sampling technique. A similarly difficult problem to solve involves micro-clots present in samples for hemostaseological or blood gas analyses and often not detected by inspection. Many POCT systems carry out internal checks to identify and reject unsuitable samples. To date, blood gas analyzers do not yet feature integrated hemolysis detectors as explained in ▶ Chapter 14. Intra-vascular hemolysis cannot be detected when heparinized blood samples are used.

4.2.10 Reliable identification of patient and sample

Many POCT devices take the sample directly. Therefore, sample containers usually do not need to be labeled, but the patient must be identified properly and the test documented correctly. Where POCT devices are used professionally, a barcode e.g. from the patient's wrist band is scanned directly into the device. This kind of technical option is not available in simpler POC tests, such as lateral flow assays.

Important pre-analytical errors and problems

- Inadequate patient preparation (e.g. diet prior to function tests)
- Inadequate information about a patient's condition (e.g. medication history, body temperature and body position in blood gas analysis)
- Incorrect patient identification
- Incorrect sampling time (for example in oral glucose tolerance test and other function tests, lack of consideration of circadian rhythm)
- Incorrect sampling technique (for example by capillary blood: incorrect site or insufficiently hyperemized skin, hemolysis due to tissue squeezing; contamination when collecting from a central line)
- Transmission of infections from insufficient hygiene measures

4.3 Post-analytical phase

The post-analytical phase of the diagnostic process begins once the parameters have been measured. This phase comprises the following key steps:

- Technical and possibly medical validation of test results
- Reporting the results to the attending clinician
- Entry of the results in the (electronic) medical record

- Documentation of the person who carried out the test
- Recording of the results in an electronic information system (if available)
- Securing documentation over the period of time for medical, legal and organizational reasons.

Post-analytical errors are often less obvious than pre-analytical ones, but equally important for the quality of the results. The following post-analytical errors and problems are not necessarily POCT-specific, but definitely POCT-relevant. Because fewer checks are available with POCT, there is a risk that POCT errors will have a bigger impact on a patient until the findings affect further action by the physician [12]. Post-analytical errors often depend on organizational situations, such as how far devices, wards, outpatients, laboratories and hospital administration are interlinked (▶ Chapter 26). As in the pre-analytical phase, a well-designed device can prevent errors. Clearly visible warning signs displayed on the device can ensure that extremely divergent values are recognized as such.

Important post-analytical errors and problems

- Insufficient validation of results carried out under pressure for short turnaround times
- Incorrect classification of results
- Incorrect or incomplete verbal reporting of results, such as missing or wrong metrological units
- Delayed reporting of alarm limit excursions
- Confusing result reporting, missing notification of results outside the normal reference range
- Errors in data storage in the laboratory or hospital information system
- Incorrect or incomplete documentation, such as verbal result reporting without entry in the medical record. No documentation of person who performed the test

4.4 Avoiding pre- and post-analytical problems

Pre- and post-analytical – as well as analytical – errors are not completely avoidable [11, 13], but can often be significantly reduced by organizational measures. A number of recommendations have been issued [6, 9, 11, 12, 16, 17, 20]; detailed information can also be found in the RiliBÄK and DIN EN ISO 15189 (▶ Chapter 36 and 38). The likelihood of pre- and post-analytical errors can be significantly reduced by employing a well-designed POCT device. For example, if a device only allows a measurement when a patient's barcode has been scanned, this considerably reduces the risk of inadequate patient identification. Automated data transmission into a laboratory information system can eliminate transmission errors, too.

Sources of errors, which cannot be resolved by design changes, need to be minimized carefully when carrying out the test. All pre- and post-analytical steps ought to be described correctly and detailed procedural instructions established (standard operating procedures, SOP), which can be summarized in the POCT quality management manual.

The DGKL working group "Reference Values" has published an exemplary SOP for the pre-analytical phase, which is an important aid for the standardization of pre-analytical conditions [8].

Furthermore, a competent governance committee (e.g. the POCT committee; ▶ Chapter 30) should develop strategies for error prevention as well as track and reduce incidents, including possible changes in the working processes. Intensive staff training on a regular basis and good communications is of vital importance as well.

References

1. Bundesärztekammer (German Medical Association), Instand e.V. (2015) Guidelines of the German Medical Association on quality assurance in medical laboratory testing. GMS Z Forder Qualitatssich Med Lab 6:Doc03
2. Burnett RW, et al. (1995) International Federation of Clinical Chemistry (IFCC). Scientific Division. Committee on pH, Blood Gases and Electrolytes. Approved IFCC recommendations on whole blood sampling, transport and storage for simultaneous determination of pH, blood gases and electrolytes. Eur J Clin Chem Clin Biochem 33(4): 247–253
3. Carraro P, Plebani M (2007) Errors in a stat laboratory: types and frequencies 10 years later. Clin Chem 53(7): 1338–1342
4. CLSI document H3-A6. Edition, A.S.S., Wayne, PA: Clinical and Laboratory Standards Institute, 2007
5. Davis MD, et al. (2013) AARC Clinical Practice Guideline: Blood Gas Analysis and Hemoximetry: 2013. Respiratory Care 58(10): 1694–1703
6. Ehrmeyer SS, Laessig RH (2007) Point-of-care testing, medical error, and patient safety: a 2007 assessment. Clin Chem Lab Med 45(6): 766–773
7. Gruszecki AC, et al. (2003) Utilization, reliability, and clinical impact of point-of-care testing during critical care transport: six years of experience. Clin Chem 49(6 Pt 1): 1017–1019
8. Gurr E, et al. (2011) Musterstandardarbeitsanweisung Präanalytik/Exemplary standard operation procedure pre-examination/Arbeitsgruppe Richtwerte der Deutschen Vereinten Gesellschaft für Klinische Chemie und Laboratoriumsmedizin. LaboratoriumsMedizin 35:55–60
9. Laposata M, Dighe A (2007) "Pre-pre" and "post-post" analytical error: high-incidence patient safety hazards involving the clinical laboratory. Clin Chem Lab Med 45(6): 712–719
10. Lippi G, et al. (2006) Phlebotomy issues and quality improvement in results of laboratory testing. Clin Lab 52:217–230
11. Lippi G, et al. (2006) Preanalytical variability: the dark side of the moon in laboratory testing. Clin Chem Lab Med 44: 358–365
12. Meier, F.A. and B.A. Jones (2005) Point-of-care testing error: sources and amplifiers, taxonomy, prevention strategies, and detection monitors. Arch Pathol Lab Med 129(10): 1262–1267
13. Nichols JH (2005) Medical Errors: Can We Achieve An Error-Free System? Point of Care 4(4): 139–141
14. O'Kane MJ, et al. (2011) Quality Error Rates in Point-of-Care Testing. Clinical Chemistry 57(9): 1267–1271
15. Panpradist N, et al. (2014) Swab Sample Transfer for Point-Of-Care Diagnostics: Characterization of Swab Types and Manual Agitation Methods. PLoS ONE 9(9): e105786
16. Plebani M, Carraro P (1997) Mistakes in a stat laboratory: types and frequency. Clin Chem 43(8 Pt 1): 1348–1351
17. Plebani M (2007) Errors in laboratory medicine and patient safety: the road ahead. Clin Chem Lab Med 45(6): 700–707
18. Schifman RB, Howanitz PJ, Souers RJ (2016) Point-of-Care Glucose Critical Values: A Q-Probes Study Involving 50 Health Care Facilities and 2349 Critical Results. Archives of Pathology & Laboratory Medicine 140(2): 119–124
19. Senn L, et al. (2012) Which anatomical sites should be sampled for screening of methicillin-resistant Staphylococcus aureus carriage by culture or by rapid PCR test? Clinical Microbiology and Infection 18(2): E31-E33
20. Stahl M, Lund ED, Brandslund I (1998) Reasons for a laboratory's inability to report results for requested analytical tests. Clin Chem 44(10): 2195–2197
21. Thompson ND, Schaefer MK (2011) "Never events": hepatitis B outbreaks and patient notifications resulting from unsafe practices during assisted monitoring of blood glucose, 2009–2010. J Diabetes Sci Technol 5(6):1396–1402

Analytical methods, biosensor technology

Peter B. Luppa, Günter Proll, Michael Imhoff, Theodor Koschinsky

P. B. Luppa, R. Junker (Eds.), *Point-of-Care Testing*
https://doi.org/10.1007/978-3-662-54497-6_5

Originating from the simple hand-held blood glucose meters and the first blood gas analyzers of the 1960s, POCT technology has spawned a myriad of analytical methodologies and applications [14]. As the areas of application are so diverse, it is not possible to exhaustively present all analytical methods; nevertheless, general detection principles will be explained in brief. These are in particular:

- Electrochemical methods (e.g. electrophoresis, potentiometry, amperometry)
- Mass change methods (e.g. quartz crystal microbalance)
- Optical methods (e.g. spectrometry, refractometry) and
- Chromatographic methods.

5.1 Biosensor technology

The setup of analytical systems based on biosensor technology provides a helpful overview [13]:

- Surface-immobilized, biologically active sensor
- Transducer unit
- Electronic amplifier
- Sample application/fluidic unit.

POCT systems also evolved from rapid tests (e.g. pregnancy test). In these simple lateral-flow systems, the human eye assumes the role of the transducer in reading out the signal. All other more complex analytical systems are defined as sensors if their signal is read by the transducer and converted into an electrical signal. The underlying principle is referred to as chemosensor and/or biosensor technology [24]. ◘ Fig. 5.1 illustrates and explains a sensor flow cell.

5.1.1 Sensor (bioreceptor)

The sensor can be regarded as a selective signal generator. That is the case when a specific ligand layer is immobilized on the surface of a sensor/transducer system. This ligand can have either biospecific or chemospecific properties. The surfaces are made of various plastic mate-

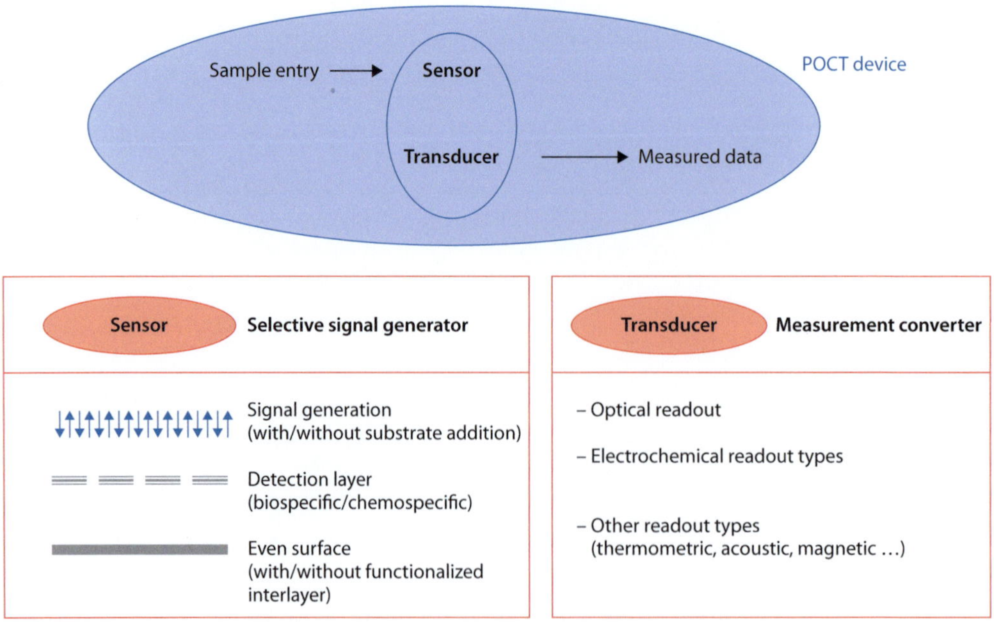

◘ **Fig. 5.1** Basic setup of a biosensor

Tab. 5.1 Properties of a biosensor

Signal generation	**Micro-reflectometry and micro-refractometry** (label-free optical surface analysis)	**Spectroscopy**	**Microgravimetry**	**Electro-chemistry**
	Ellipsometry Reflectometric interference spectroscopy Surface plasmon resonance Diffraction grating Photonic crystals Mach-Zehnder interferometry	Fluorescence (Chemi)luminescence Total internal reflection fluorescence Reflection Absorption Turbidimetry Optical movement analysis	Quartz crystal microbalance (QCM) Surface acoustic wave (SAW)-based systems Microcantilever	Amperometry Potentiometry Conductivity Coulometry
Sensor surface Self-assembled monolayers (SAM), polymers or hydrogels on different carrier materials	**In micro-reflectometric and micro-refractometric methods**	**In spectroscopic methods**	**In electro-chemical methods**	**In mass-sensitive methods**
	Silicon wafer, glass or various plastic materials Nobel metal surfaces Glass or plastic materials with optic gratings Glass, silicon-wafer with integrated optic	Glass or plastic materials (often microtiter plate and microscope slides), in TIRF also in integrated optic or optical gratings Silicon wafer Homogeneous phase	Nobel metal electrodes	Quartz crystal, coated with a metal film Glass or plastic materials in microcantilever sensors
Ligand layer on the sensor surface	**Biospecific**		**Chemospecific**	
	Antibodies Antigens Receptors Ligands Other binding proteins (e.g. anticalins)	Enzymes (substrate conversion, enzyme activation/inhibition) Aptamers DNA/RNA oligomers (Protein) scaffolds Living cell	"Molecular imprints" Indicators Ion-selective membranes (in electro-chemical methods)	

rials, glass, silicone or noble metals. Metal surfaces can be prepared with or without a functionalized interface.

The signal can be generated in different ways, either with or without substrate addition. There are optical surface analytical methods, such as ellipsometry or surface plasmon resonance, which generate specific information from the surface. Other spectroscopic methods can be employed, such as absorption, fluorescence, (chemi)luminescence or even optical motion analysis. Furthermore, electrochemical methods, such as amperometry and potentiometry, or microgravimetry, such as quartz crystal microbalance (QCM), are utilized. The properties of the aforementioned sensors are summarized in Tab. 5.1.

Biospecific ligands are particularly noteworthy [12]. These are often immobilized antibodies or enzymes. Antibodies serve as a biospecific ligand in the reaction with an analyte. Enzymes, by contrast, utilize biospecific catalytic reactions after substrate addition. Examples are glucose sensors (▶ Chapter. 12). Binding proteins can be used as "recognition elements", e.g. in coagulation analysis [19].

Chemospecific ligands include all kinds of ion- and molecule-selective membranes, molecular imprints or indicator molecules that are suitable as ligands because of their selectivity to the analyte.

Electrochemical sensors

Potentiometric sensors

The Nernst equation [3] is the foundation for all potentiometric transducer types. According to this equation, the changes in the potential in a membrane at zero current show a logarithmic proportionality to the specific ionic activity. Potentiometrie transducer electrodes are classified by their methodology:

- **Transmembrane potential.** This transducer is based on the accumulation of a potential along a sensor membrane. Ion-selective electrodes (ISE) utilize ion-selective membranes that generate a charge separation between the sample and the sensor surface.
- **Electrode potential.** This transducer is similar to the transmembrane potential sensor; however, the electrode itself provides the surface for the biochemical detection reaction, whereby the electrode potential changes as a function of analyte concentration
- **Field effect transistor (FET).** As a semiconductor, FET is able to analyze the smallest charge changes on the surface of an electrode ("gate"), which is positioned between the source and drain electrodes. In ion-selective FET (ISFET), which is a combination of ISE and FET, the measuring principle is based on the change in the field effect (development of a space charge zone). This field effect can be observed, depending on the concentration of ions in the sample between source and drain when, instead of an electrical contact at the gate, an ion-selective layer is applied and brought into contact with the solution to be measured. This causes a change in the source-drain current that is directly proportional to the change in the analyte concentration and can be measured very accurately.

The advantage of potentiometric sensors is seen in the simplicity and robustness of their measuring system. Advantageous is also the small sensor size, which is important for POCT systems. All potentiometric sensor methods show problems with sensitivity and non-specific effects, which are reflected in a poor signal-to-noise ratio.

Parameters for potentiometric sensors are pH, pCO_2, Na^+, K^+, Ca^{2+} and Cl^-.

Amperometric sensors

Amperometric sensors measure the current that is generated by an electrochemical reaction at constant voltage. The application of amperometric sensors is only successful if an analyte can also function as redox partner in an electrochemical reaction. Such an analysis system was first described in 1956 by L.C. Clark [4]. He described an oxygen electrode which consisted of an electrolyte-containing chamber, a platinum sensor cathode (polarized at –0.7 V) and an Ag/AgCl reference electrode. The chamber was covered with an O_2-permeable membrane.

The cathode reactions occur as follows:

$$O_2 + 2\,H_2O + 2\,e^- \rightarrow H_2O_2 + 2\,OH^-$$

$$H_2O_2 + 2\,e^- \rightarrow 2\,OH^-$$

The anode reaction is:

$$4\,Ag + 4\,Cl^- \rightarrow 4\,AgCl + 4\,e^-$$

There are a number of enzymes with high catalytic conversion rates ($>10^3\ s^{-1}$), which ensure a substrate transformation in the amperometric systems [7]. In addition to the oxygen generated from H_2O_2 by catalase, there are further amperometrically detectable compounds, such as ferrocene derivatives, In^{2+} salts [1] or the redox polymer PVP-Os(bipyridyl)$_2$Cl that can be co-immobilized with antibodies in immunosensors [11]. Enzymes such as horseradish peroxidase (HRP), glucose oxidase, glucose dehydrogenase and others have been successfully implemented in amperometric sensors [16].

Amperometric sensors have excellent analytical sensitivity. By contrast, the transport rate limitation inherent to the system can have an adverse effect on the redox partners on the electrode surface.

Parameters for amperometric sensors are pO_2, glucose and lactate. ► Chapter 12 (diabetes diagnostics) offers more detailed information, especially for glucose analysis.

▪▪ Conductometric and capacitive sensors

These sensors measure the change in electric conductivity in a solution at constant voltage. The change results from biochemical reactions that produce or reduce ions in a specific way. The capacity changes are measured via an electrochemical structure where the bioactive element is attached to a pair of Au or Pt electrodes. Except for hematocrit measurements, there are only a few clinical applications, as the high ionic strength in whole blood or serum makes it difficult to measure the relatively low conductivity changes in a signal generating reaction [2].

Optical sensors

The methods that are utilized in these sensors are categorized into three groups:

- Optical detection
- Optodes
- Optical surface measurement

▪▪ Optical detection methods

The most frequently used optical sensors are part of the first group and measure light absorption or light reflection of an analyte as long as the parameter is present in at least millimolar concentration. The absorption of light as electromagnetic radiation at a given energy or wave length is subject to the Beer-Lambert law if the analyte is distributed homogeneously in the solution. The use of multi-wavelength photometers in oximetry is presented in ► Chapter 14. If the analyte is expected in a range far below a millimolar concentration, other spectrometric methods such as nephelometry, turbidimetry, fluorescence or (chemi) luminescence are needed as with immunoassays. However, this requires the use of appropriately labeled tracer substances.

▪▪ Optodes

Fiber optic chemosensors, called optodes, represent a further group of optical sensors, which in the future may enable continuous intra-arterial monitoring of various parameters (electrolytes, blood gases) for example. They utilize fluorescence dyes that are bound onto ion-selective (ionophore) or gas-permeable membranes. When these membranes only encase the fiberoptic, determination of electrolytes such as Na^+, K^+ and Ca^{2+} or the measurement of pO_2, pCO_2 and pH is possible. After excitation with a particular excitation wavelength at constant energy, the intensity of the emitted fluorescence is directly proportional to the concentration of a cation present or to the partial pressure of a gas.

▪▪ Optical and surface analytical methods

The use of label-free optical surface techniques in POCT [6] is rarely employed despite their high analytical power. Conversely, optical movement or picture analyses have already been utilized quite frequently in coagulation analyzers to detect the formation of blood clots. As an example, the pattern of movement of paramagnetic ferrous oxide particles under the influence of fibrin formation can be observed in an orthogonal oscillating magnetic field using a photodiode [17].

Total internal reflection spectroscopy (TIRS)

This optical sensor comprises two materials with different reflection indices (RI), where the light strikes the sensor surface entering through the layer with the higher RI under total internal reflection conditions. Thus, an evanescent wave (as electrical light vector) is formed in the material with the lower RI. This wave propagates (to ca. ½ λ, equivalent to a few 100 nm) into the medium at an exponentially decaying amplitude. Hence, biomolecules that are immobilized at the sensor surface can interact with these evanescent waves. This leads to a reduction of the reflected density intensity. By modifying this technique, a total internal refection fluorescence measurement (TIRF) can also be achieved by applying fluorescent substances [6].

Reflectometric Interference spectroscopy (RIfS) RIfS is a detection method based on the interference of white light at thin films (layers). In practice, it is used to investigate molecular interactions. The underlying principle of multiple reflections at thin layers corresponds to that of the Fabry-Pérot interferometer. The shift in the characteristic interference spectrum of the respective layer system allows biochemical binding events to be observed over time [5].

5

Ellipsometry Ellipsometry measures the change in polarization of light upon reflection on a surface with an immobilized sample. In general, linearly or circularly polarized light is used. After reflection, it is elliptically polarized, hence the name of this method. The orientation of the ellipse depends on the incident radiation angle, the orientation of the polarization of the incident light and of the sensor surface reflection properties (RI, density of the biologically modified surface).

Surface plasmon resonance The underlying concept of surface plasmon resonance (SPR) is based on a quantum-physical phenomenon that takes place on the surface of gold. In the gold layer, plasmons (fluctuations in the density of the swinging electrons in the metal, quantum-mechanically treated as quasiparticles) are generated by irradiation with monochromatic and polarized light under total reflection conditions. The plasmons dissipate energy in the form of an evanescent wave (like TIRS) and have a shadow-like effect (loss of intensity) on the reflected light. The reflectance angle of the light changes under otherwise constant optical parameters. This happens because of the interaction of molecules with a biochemical layer along the underside of the chip as the RI of the medium changes slightly due to a different surface composition [10].

The principle of refractometry and reflectometry are presented in ◘ Fig. 5.2.

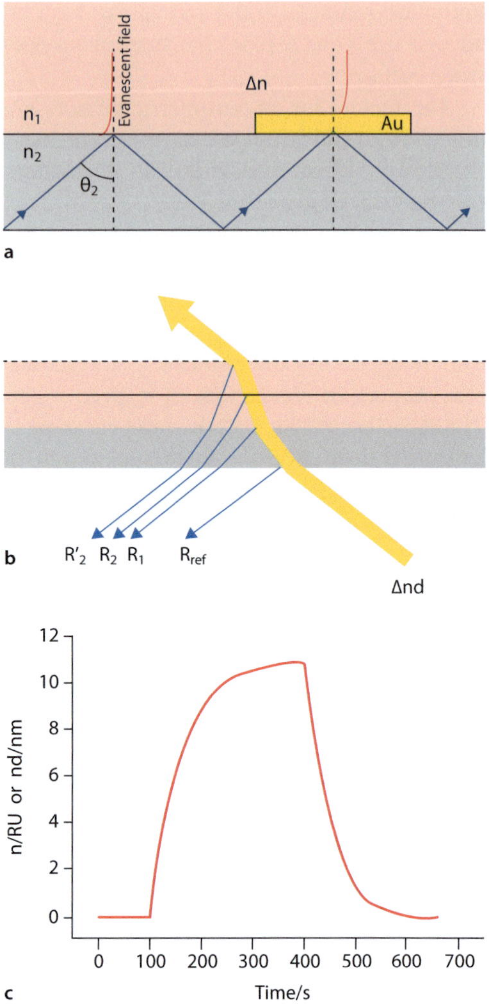

◘ **Fig. 5.2a–c** Label-free optical methods. **a** Refractometry (evanescent field methods) with evanescence field in general (e.g. grid couplers) and surface plasmon resonance (SPR); **b** Reflectometry as reflectometric interference spectroscopy (RIfS); **c** Typical binding curve of a biomolecular reaction as seen in label-free methods

Micro-gravimetric sensors

In mass-sensitive sensors, three different signal transduction methods are adopted: Quartz crystal microbalance, surface acoustic wave-based systems and microcantilevers.

Quartz crystal microbalance (QCM) In the QCM method, the oscillating quartz crystal is covered with a thin metal film (usually Au) and brought into its resonant frequency by alternating voltage. This resonant frequency is indirectly proportional to the mass of the material that is adsorbed on the surface. These label-free sensors react very sensitively to every change in the solution. Therefore, such sensors need a reference system that enables them to distinguish between interactions with the target analyte and non-specific bonds or viscosity changes at the sensor surface [9, 18].

Surface acoustic wave (SAW)-based systems SAW-based sensors consist of micro-electromechanical systems (MEMS), which are also based on vibration. The acoustic waves are restricted to the sensor surface and decay exponentially to the surface distance [22]. Surface waves are excited electrically by interdigital transducers (IDT). One SAW system suitable for POCT is the Love Wave Surface Acoustic Wave (LW-SAW) immunosensor that minimizes the acoustic losses which are affected by the thickness of the aqueous layer above the sensor. The analytic sensitivity of the device can be increased into the nanomolar range [20].

Microcantilevers The microcantilever is also a MEMS device that consists of a miniaturized "trampoline". Here, a piezo-resistive bulk material loaded with a biological layer is only fixed to a platform on one side so that the other end of the elongated carrier can swing freely. The analyte binding to the sensor surface causes the microcantilever to bend. This can be measured either as an electrical (e.g. capacitive), piezoresistive or optical effect [15]. Microcantilever sensors are already used in POCT applications [21].

Sensors for hemostaseological and hematological analyses

For coagulation analyses, known principles that are adopted in mechanized coagulation analyzers are utilized to detect blood clot formation. This is in addition to the above-mentioned optical movement sensors. Electrochemical methods are also employed, for example, to detect electrochemically generated active fibrinogen cleavage products due to the effect of thrombin. For hematological POCT analyses, the principles of particle counting (Coulter principle) analogous to the mechanical hematology analyzers are adopted. These are based on the measurement of electrical conductivity changes when cells pass between two electrodes in a sensitive aperture. Alternatively, there is a quantitative buffy coat method, based on an electro-optical principle. The buffy coat layer is formed by centrifugation of a coated microhematocrit capillary. This layer is analyzed with an optical system.

Further details can be found in ▶ Chapters 15 and 16.

5.1.2 Transducers, electronic amplifiers

The transducer is part of a biosensor and is the equivalent of a transformer that allows optical, electrochemical or other readings (thermometric, acoustic, magnetic etc.) The transducer converts biological/chemical reactions into a physically measurable quantity that can be changed into an analog or digital signal and then be further used in various ways. The signal processing system conditions the sensor signal by electronic smoothing and noise filtration.

For the processing of signals from the transducer, many POCT systems use common computer or smart phone systems (Windows Phone, iOS, Android) or Internet-based apps. Thereby, not only (quantitative) results become visible (display, printer etc.), but also all analyses calculations (calibration etc.) and data management (storage of quality control measurements, patient results etc.) are carried out in the device. Many devices have a built in Wi-Fi-enabled network card, which allows network-compatible communication in the hospital.

5.1.3 Sample application/fluidic unit

The sample application in POCT devices can be very simple in contrast to the complex loading methods used in clinical-chemical analyzers. POCT devices often utilize a fluidic unit that mostly consists of single-use components. The fluidic unit format can be defined as follows:

- Systems with absorbent materials, flow cells, lateral flow or (immuno)chromatography,
- Cassettes, capillary, tube or centrifugal systems.

The flow is maintained using vacuum, pumps or centrifugal force; alternatively, the absorbent properties of many materials as well as capillary force and electroosmosis are utilized. Many fluidic unit systems are cassettes which provide all reagents in segments where the sample material can be easily loaded through an opening.

Simple unit-use POCT devices such as glucose meters utilize test strips that are inserted into the device manually. Then, the sample material is loaded manually onto an application pad and absorbed by capillary force into the strip. In contrast, blood gas analyzers suction the blood sample directly from the syringe or capillary with roller pumps and tube systems.

5.2 Continuous monitoring methods

5.2.1 Continuous monitoring methods

For some years now, the previously described in-vitro determination of metabolic parameters has been supplemented by the principle of continuous monitoring in the interstitial fluid of subcutaneous fatty tissue, mainly in the abdominal area or the upper arm over a limited time period (presently max. 14 days) [8, 23]. Via a minimally invasive microdialysis catheter placed in the subcutaneous tissue (Fig. 5.3), that can also be inserted by trained patients by themselves, continuous measurement of metabolites over a few days is possible. Microdialysis uses an aqueous perfusion solution with low flow rates and therefore highly water soluble metabolites with a low molecular weight are suitable for dialysis. This applies in particular to important parameters such as glucose and lactate but also creatinine. Determination of

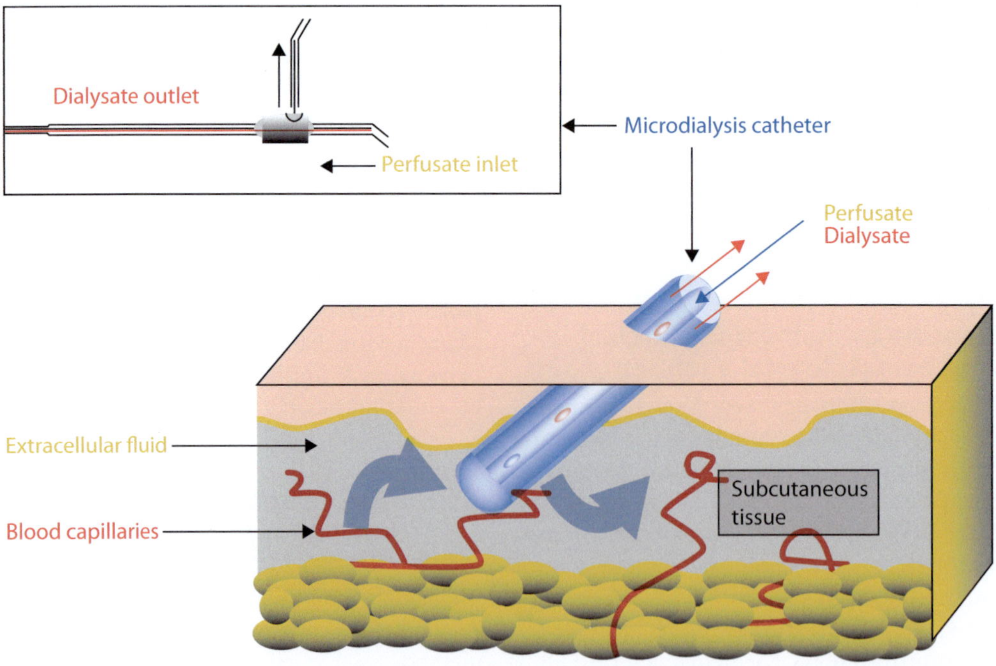

Fig. 5.3 Principle of subcutaneously applied microdialysis

parameters from the dialysate is carried out using the analytical principles described in other chapters. As a matrix for analysis, dialysate is easier to handle than capillary blood, plasma or urine.

5.2.2 Continuous glucose monitoring (CGM)

At present in Germany, various technologies are approved that transmit results to an extracorporeal device to collect, process, store and present data at short intervals (<1 to 5 min). They provide immediate data access (on-line) for a needs-based therapy adjustment including alarm settings for hypo- and hyperglycemia. Examples include s.c. in-vivo needle biosensors (needle electrodes with integrated glucose oxidase monitoring systems) like those featured in the Guardian REAL-Time (Medtronic), FreeStyle Libre (Abbott) (◘ Fig. 5.4) and SEVEN PLUS (DexCom, distributed by Nintamed in Germany) or GlucoDay S (Menarini Diagnostics) as extracorporeal in-vitro-biosensors (electrochemical via glucose oxidase). These features are combined with an s.c. microdialysis fiber whose filtrate volume of interstitial fluid is continually pumped to an ex-vivo biosensor for glucose monitoring.

Clearly, the main advantages of CGM methods primarily lie in the increased number of glucose readings with 500–1,500 per 24 h compared to 2–4 per hour with BGSM.

- This information yield is useful to identify subclinical hypoglycemic states, particularly during sleep,
- postprandial blood glucose profiles,
- overall glucose variability in 24 h for targeted therapy adjustments, including the prospect of connection to an insulin pump ("closed loop").

Secondly, in this way, there is a prognostic potential for trend analyses such as for alarm functions mentioned above.

To date, all CGM systems share a common problem that glucose is measured from or in the

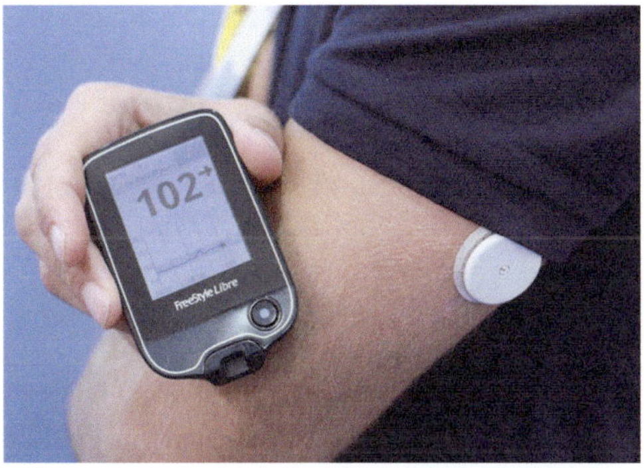

◘ **Fig. 5.4** FreeStyle Libre (Abbott Laboratories, Wiesbaden). The glucose sensor requires the insertion of a sensor filament under the skin. It can be left in situ for up to 14 days. An additional check of glucose values with a blood glucose meter is required when glucose levels change quickly as the levels in tissue fluid may not represent the true blood glucose levels. The same applies in situations when the system indicates hypoglycemia or a developing hypoglycemia or when symptoms do not correspond with the measured values of the system.

s.c. interstitial fluid and cannot be objectified by an independent method. Clinical evidence is based on glucose levels in blood and not in s.c. interstitial tissue. As a result, CGM readings are converted and given as blood glucose equivalents in mg/dL or mmol/L, based on a few parallel capillary blood glucose measurements using different adjustment processes. These are unknown to the operator and allow for physical and physiological time delays caused by glucose flux between blood and interstitial tissue. This approach also has a considerable potential for error due to anatomical and physiological variables that occur inter- and intra-individually over the course of diabetes as well as due to additional system-dependent errors. The resulting requirements for quality assurance of these CGM systems vary considerably from those of the glucose POCT single measurements. However, a regulatory legal framework for this does not yet exist in Germany.

Notwithstanding, the CGM systems user group is growing year on year, not only in the outpatient sector, but also in inpatient settings, in particular in specialized diabetes wards and intensive care driven by the clinical advantages presented above. The use of such CGM systems, particularly in clinical settings, meets the hallmark criteria to count as POCT.

In contrast, **non-invasive glucose monitoring** systems (▶ Chapter 11) do not yet have market readiness and the required approval, despite considerable interest, very high investment in development and various technological approaches. It is currently expected that the development period for **non-invasive glucose monitoring** systems will be between 5–20 years.

References

1. Aizawa M (1994) Immunosensors for clinical analysis. Adv Clin 31: 247–275
2. Berney HC, Alderman J, Lane WA, Collins JK (1998) Development of a capacitive immunosensor: a comparison of monoclonal and polyclonal capture antibodies as the primary layer. J Mol Recognit 11: 175–177
3. Buerk DG (1993) Biosensors. Theory and applications. Technomic Publishing, Lancaster, PA, pp 39–61
4. Clark LC (1956) US Patent a2,913,386
5. Gauglitz G, Brecht A, Kraus G, Mahm W (1993) Chemical and biochemical sensors based on interferometry at thin (multi-) layers. Sens Actuator B: Chemical 11: 1–3 and 21–27
6. Gauglitz G (2010) Direct optical detection in bioanalysis: an update. Anal Bioanal Chem 398:2363–2372
7. Ghindilis AL, Atanasov P, Wilkins M, Wilkins E (1998) Immunosensors: electrochemical sensing and other engineering approaches. Biosens Bioelectron 13:113–131
8. Henrichs HR (2010) continual Glukosemessung (CGM) in der Gewebeflüssigkeit – Wissenschaftliche Bewertung of CGM and medizinische Beurteilung of the Nutzens for the Diabetestherapie (AGDT-DDG-GKV CGM-Positionspapier). Im Internet: www.diabetes-technologie.de/vortraege/vortraege-2010.htm
9. Ittarat W, Chomean S, Sanchomphu C,Wangmaung N, Promptmas C, Ngrenngarmlert W (2013) Biosensor as a molecular malaria differential diagnosis. Clin Chim Acta 419: 47–51
10. Liedberg B, Nylander C, Lundstrom I (1995) Biosensing with surface plasmon resonance – how it all started. Biosens Bioelectron 10: i–ix
11. López MA, Ortega F, Dominguez E, Katakis I (1998) Electrochemical immunosensor for the detection of atrazine. J Mol Recognit 11: 178–181
12. Luppa PB, Sokoll LJ, Chan DW (2001) Immunosensors – Principles and applications to clinical chemistry. Clin Chim Acta 314: 1–26
13. Luppa PB, Müller C, Schlichtiger A, Schlebusch H (2011) Point-of-care testing (POCT): Current techniques and future perspectives. Tr Anal Chem 30:887–898
14. Luppa PB, Bietenbeck A, Beaudoin C, Giannetti A (2016) Clinically relevant analytical techniques, organizational concepts for application and future perspectives of point-of-care testing. Biotechnol Adv. 2016 34: 139–160
15. Mathew R, Ravi Sankar A (2015) Design of a triangular platform piezoresistive affinity microcantilever sensor for biochemical sensing applications. J Phys D Appl Phys 48, 205402 (IOP Publishing)
16. McNeil CJ, Athey D, Renneberg R (1997) Immunosensors for clinical diagnostics. EXS 81: 17–25
17. Oberhardt BJ, Dermott SC, Taylor M, Alkadi ZY, Abruzzini AF, Gresalfi NJ (1991) Dry reagent technology for rapid, convenient measurements of blood coagulation and fibrinolysis. Clin Chem 37: 520–526
18. Prakrankamanant P (2014) Quartz crystal microbalance biosensors: prospects for point of-care diagnostics. J. Med. Assoc. Thail. 97 (Suppl. 4)
19. Price CP, St John A, Hicks JM (2004) Point-of-care testing: what, why, when, and where? In: Price CP, St John A, Hicks JM (eds) Point-of care testing, 2nd edn. AACC Press, Washington, pp 3–12
20. Puiu M, Gurban A-M, Rotariu L, Brajnicov S, Viespe C, Bala C (2015) Enhanced sensitive love wave surface acoustic wave sensor designed for immunoassay formats. Sensors 15: 10511–10525
21. Ricciardi C, Canavese G, Castagna R, Ferrante I, Ricci A, Marasso SL, et al. (2010) Integration of microfluidic and cantilever technology for biosensing application in liquid environment. Biosens. Bioelectron 26: 1565–1570
22. Thalhammer S, Wixforth A (2013) Surface acoustic wave actuated Lab-on-Chip system for single cell analysis. Biosens Bioelectron 4: 1–7
23. Vashist SK (2013) Continuous Glucose Monitoring Systems: A Review. Diagnostics (Basel) 3:385–412
24. Ziegler C, Göpel W (1998) Biosensor development. Curr Opinion Chem Biol 2: 585–591

Laboratory coagulation tests

Michael Spannagl, Dirk Peetz

P. B. Luppa, R. Junker (Eds.), *Point-of-Care Testing*
https://doi.org/10.1007/978-3-662-54497-6_6

6.1 Introduction

Hemostasis can be divided into four categories below to better illustrate the systematics of laboratory methods for coagulation testing and provide a systematic classification into the implications for therapy:

- Primary hemostasis
- Thrombin/fibrin formation
- Clot formation
- Fibrinolysis

6

Some of these sub-steps can be illustrated by hemostaseologic test procedures. The analytic technologies of the POCT systems used for coagulation testing are therefore diverse [3, 22]. Nevertheless, improved outcomes in patient care have been demonstrated for individual patient populations [1]. The increasing experience with POC coagulation testing systems for monitoring therapy is reflected in international recommendations [24].

6.2 Classification of hemostaseologic POCT methods

Substantively, the available methods can be subdivided as follows:

- Analysis of plasmatic coagulation
- Combined recording of plasmatic coagulation, platelet count and fibrinolysis (viscoelastic methods)
- Analysis of platelet function

6.2.1 Analysis of plasmatic coagulation

Analysis of plasmatic coagulation

- Activated clotting time (ACT; various manufacturers)
- Heparin management system (protamine titration; Medtronic)
- POCT systems for measuring prothrombin time (PT, Quick test or international normalized ratio, INR) and activated partial thromboplastin time (aPTT) (various manufacturers)

In general, the ACT is used for peri-interventional treatment monitoring of unfractionated heparin and other anticoagulants. A defining characteristic of these test procedures is, that the coagulation is triggered by thromboplastin (to measure PT/INR) and/or contact activators (to measure aPTT and the ACT). For this purpose, the beginning of coagulation time is detected directly or indirectly via the gelation of the sample or by utilizing fluorogenic or electrochemical substrates. There is also the option to integrate coagulation measurements into multifunctional POCT systems (e.g. i-STAT, Abbott). The principles for quantifying aPTT, ACT, PT/INR and ECT are presented in Fig. 6.1.

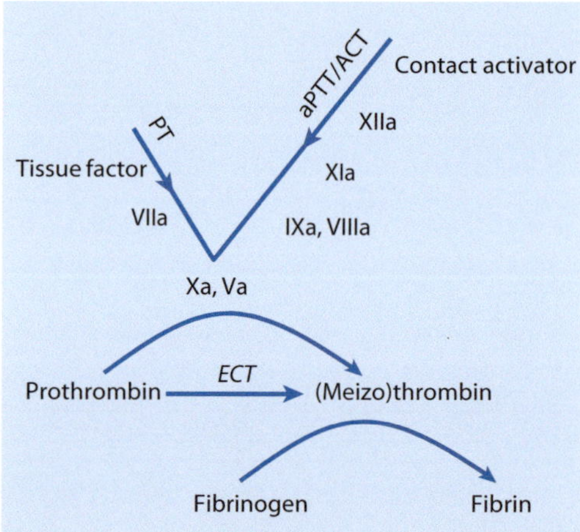

Fig. 6.1 Comparison of the measurement principles of activated partial thromboplastin time (aPTT), activated clotting time (ACT), prothrombin time/international normalized ratio (PT/INR) and ecarin clotting time (ECT). PT/INR, aPTT and ACT each quantify the cascade of coagulation activation in several stages. By contrast, ECT utilizes the snake venom ecarin to convert prothrombin to meizothrombin, an intermediate compound in physiological thrombin activation. Because it activates prothrombin directly, ECT is more specific in measuring direct thrombin inhibitors than ACT, aPTT or PT/INR.

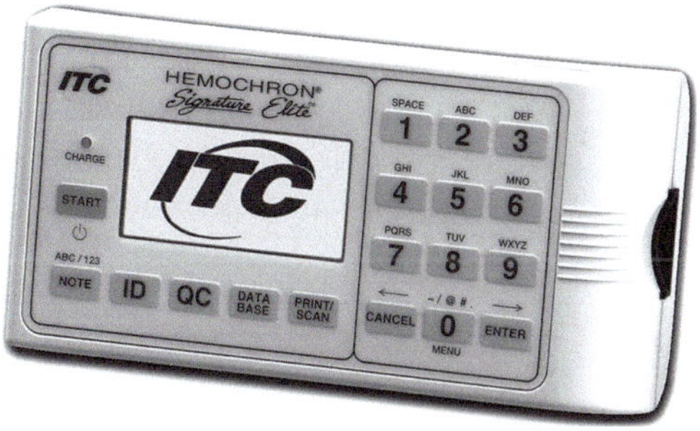

■ **Fig. 6.2** Hemochron Signature Elite. (Courtesy of Keller medical GmbH, Bad Soden, Germany)

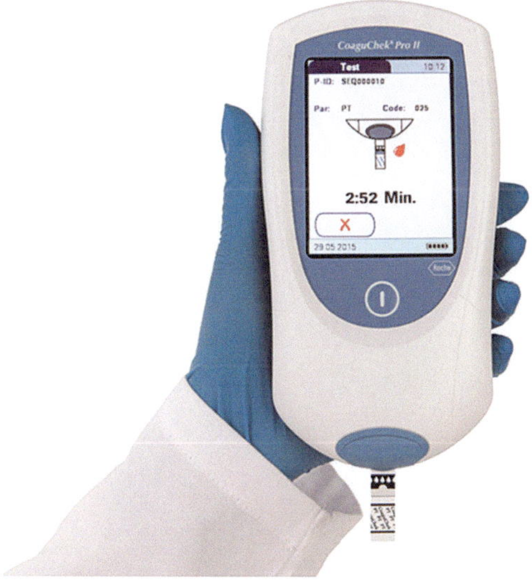

■ **Fig. 6.3** CoaguChek XS plus. (Courtesy of Roche Diagnostics GmbH, Mannheim, Germany)

There is currently a variety of analytical systems on the market. Fundamentally, the analytical principle of ACT corresponds to that of aPTT. By contrast, POCT systems mostly measure in whole blood to avoid the time-consuming process of centrifugation. By varying the type of single-use cartridges for the different anticoagulants and their multiple dosages, most anticoagulant monitoring devices can be used interchangeably. Hemochron systems (Keller Medical; ■ Fig. 6.2) utilize either strips that contain a surface activator alone, or cartridges that contain a surface activator and phospholipid. Other analyzer systems (e.g. GEM PCL Plus, Instrumentation Laboratory or i-STAT, Abbott) feature single-use cartridges, in which whole blood is driven through capillary tubes.

Systems employed to measure ACT in patients receiving high-dosed heparin rely on different contact activators (Celite, silica, kaolin or glass particles). When whole blood methods are used, phospholipid surfaces are presented by blood cells or their fragments. However, there are also tests that use additional phospholipids. Mostly, fresh blood, less often citrate blood is used. Here, it should be noted that the inter-comparability of results is poor due to the different activators and detection methods of the various ACT systems [2, 4, 12, 14].

A variety of test systems are available for INR self-monitoring by patients receiving treatment with vitamin K antagonists (Marcumar). Below is a selection:

- The **CoaguChek XS** (Roche; ■ Fig. 6.3) is based on an electrochemical detection method. The test strip contains recombinant thromboplastin, phospholipids and the peptide substrate Electrocyme TH in dried form. Activation of the coagulation cascade produces thrombin, which enzymatically cleaves the peptide substrate into a residual peptide and the electrochemically active phenylenediamine. The latter generates an electrical signal that is converted into the PT/INR value. The CoaguChek XS is the first analyzer on the market that is not affected by heparin in the sample.
- The **INRatio system** (Alere) likewise consists of an analyzer and single-use test strips. The system measures the change in impedance in a mixture of blood and reagents during the coagulation process and calculates the INR from the impedance signal. The device uses 3-channel technology to analyze the patient sample and 2 controls simultaneously. The measuring chip is depicted in ■ Fig. 6.4.
- The **Xprecia Stride** is a new system supplied by Siemens Healthcare Diagnostics. It is designed to achieve an excellent correlation with INR laboratory methods based on the Dade Innovin reagent (■ Fig. 6.5).

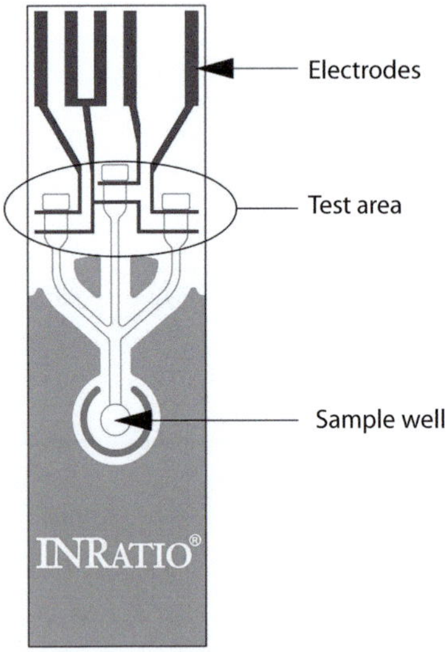

Fig. 6.4 INRatio test chip. (Product withdrawn from the market. Courtesy of Alere GmbH, Cologne, Germany)

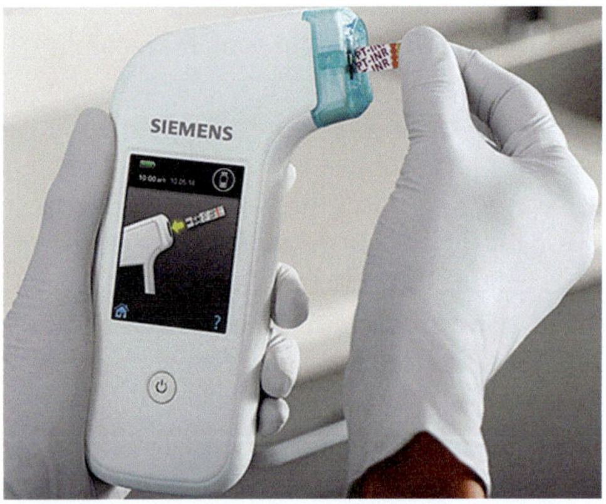

Fig. 6.5 Xprecia Stride. (Courtesy of Siemens Healthcare Diagnostics, Eschborn, Germany)

- The **ProTime microcoagulation system** (ITC) is a portable photometer. Its plastic cuvette features a chamber with dried thromboplastin. The **ProTime** has 5 microchannels for triplicate testing of the patient sample with two levels of controls. The blood is drawn into the channels and pumped back and forth. An array of LED detects when clotting occurs and the motion stops; the INR value can be calculated from this.

Since 2016, test strips for measuring aPTT in capillary whole blood have also been available on the CoaguChek Pro System (Roche) [15]. The detection principle is similar to a PT measurement with electrochemical detection.

6.2.2 Combined recording of plasmatic coagulation, platelet count and fibrinolysis (viscoelastic methods)

Viscoelastic methods
- Rotational thromboelastometry (ROTEM system, Pentapharm) [13]
- Thromboelastography (TEG, Haemoscope, USA) [6]

Thromboelastography was first introduced by H. Hartert in 1948 [6]. The principle consists of a cylindrical pin suspended freely on a thin wire that is submersed into a rotating, blood-filled cuvette. When the sample coagulates, fibrin filaments form between the wall of the cuvette and the pin. The fibrin filaments transfer the motion of the cuvette to the pin as a function of how strong the fibrin filaments are. Irrespective of the technological method employed, the motion of the pin over the time is traced on a thrombelastographic curve. Previously, native whole blood was used as the specimen. Nowadays, analysis is usually performed on citrated blood that is made to coagulate with defined activators (extrinsic, intrinsic).

These viscoelastic methods continuously record the clot strength. This allows the functional evaluation of plasmatic coagulation, fibrinolysis and clot strength. The latter includes the fibrinogen levels as well as the properties of fibrin polymerization and the platelet count (for illustration, Fig. 6.6 and Fig. 6.7).

It was just a few years ago that global tests measured non-activated clots; this led to increased variability, lower specificity and long measuring times. Nowadays, activated meas-

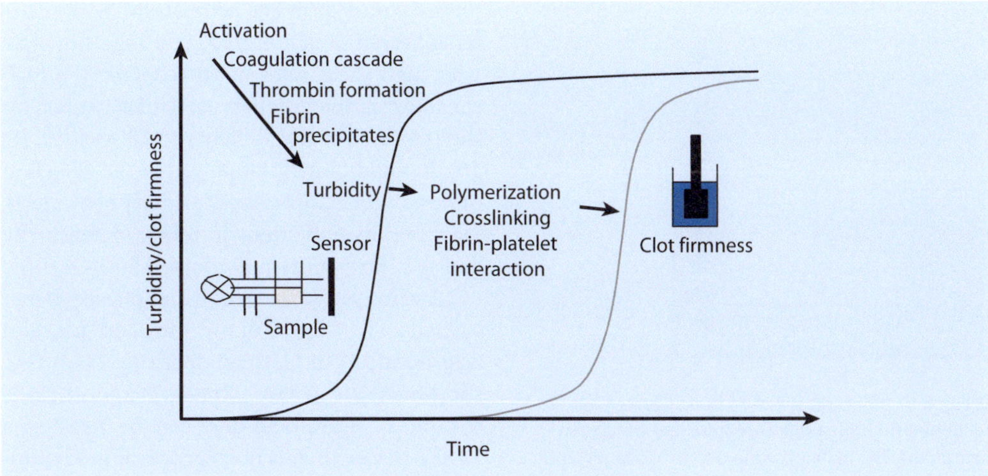

Fig. 6.6 Recording the clot strength in the whole blood platelet aggregator

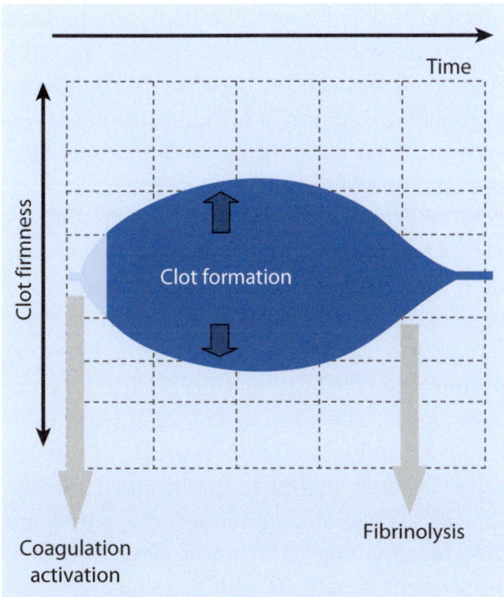

Fig. 6.7 Thrombelastographic curve

urements are commonly used that distinguish between the multiple aspects of hemostasis. This includes, among others, "EXTEM" (coagulation is activated by tissue thromboplastin) or "INTEM" (coagulation is activated via the contact phase).

6.2.3 Analysis of platelet function

Analysis of platelet function [5, 23]

- PFA-100/200 (Siemens Healthcare Diagnostics)
- Whole blood aggregation (Multiplate, Roche)
- VerifyNow (Accumetrics, USA)

Systems to analyze platelet function are used comparably rarely at the POC. However, the increasing employment of platelet inhibitors has led to a growing interest in near-patient platelet diagnostics. Methods such as whole blood aggregation as well as the VerifyNow system analyze platelet function under low shear stress after specific stimulation of platelets by arachidonic acid, ADP or other activators. With the PFA system, primary hemostasis is tested under high shear stresses during concurrent stimulation with collagen/epinephrine or collagen/ADP. The system does not require any added reagents; the performance is nevertheless more complex than "true" POCT methods (pre-analytical phase, storage of the measuring cell at room temperature for several minutes, plausibility check), which is why the FDA clas-

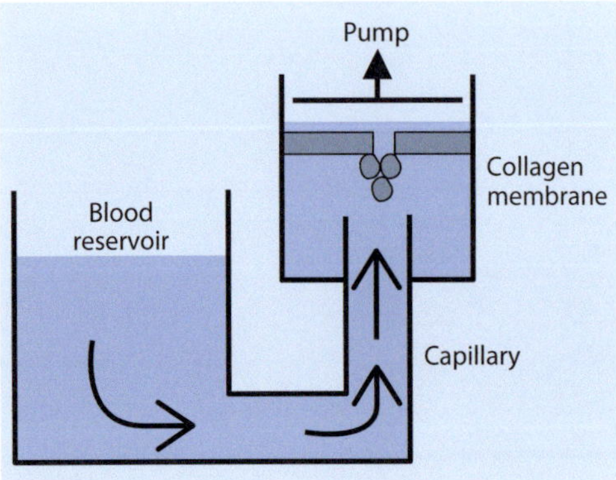

6

Fig. 6.8 Principle of PFA analysis. Citrate blood is suctioned through a small aperture into a collagen membrane. The platelets adhere to the collagen and cause the blood flow to stop. Closure time ("clotting time") is a measure of platelet function. For example, it is prolonged in patients receiving acetylsalicylic acid

sified the PFA in the group of devices with greater operating requirements [7].

Recently, Paniccia et al. [17] presented an excellent comparative overview of the various systems and their clinical applications.

The **measurement principle inherent to platelet function analyzers** (PFA-100/200 systems, Fig. 6.8) consists of a constant negative pressure forcing citrated blood to be pumped through a capillary. In this process, the blood is first passed through a 3 cm long steel capillary with an internal diameter of 0.2 mm, then through an aperture in a collagen membrane with a capillary of 0.15 mm. The membrane is coated with either ADP or adrenaline. The blood passing through activates platelets, which seal the membrane over time until the blood flow stops. The measuring variable "closure time" is longer or shorter depending on the degree of activation of the platelets. Closure time (in seconds) is the time from the beginning of the measurement to the time at which a predetermined flow rate is exceeded. It is indicated separately for the collagen/epinephrine (Col/Epi) and the collagen/ADP cartridges (Col/ADP).

This measuring system is based on the principle of in vitro bleeding time according to Kratzer and Born [11] and attempts to imitate physiological "primary hemostasis" with platelet adhesion or aggregation at a high flow rate with high shear rates in vitro. Given the high shear forces, the system is particularly suited for the screening of von Willebrand's disease [9, 10, 16]. It is additionally employed in the detection of various other disorders of primary hemostasis. However, its limitations in terms of sensitivity and specificity must be accounted for [8, 9, 10].

A variety of whole blood and plasma-based systems are available for **induced platelet aggregation**. In German-speaking countries, the Multiplate device (Roche Diagnostics) is the most widespread and can be used as a POCT device thanks to its practical methodology. Fundamentally, impedance aggregometry – which measures anticoagulated whole blood – can be used as a POCT for platelet function (Fig. 6.9). With specific agonists, personalized monitoring can be carried out, even in patients on combined antiplatelet medication [10]. A marked association between ADP-induced aggregation and previous stent thrombosis was shown for the monitoring of clopidogrel after implantation of drug-eluting stents [19].

The VerifyNow system (Accumetrics) measures platelet aggregation in whole blood based on the agglutination of fibrinogen-coated latex beads [20]. In the USA, the VerifyNow system has mainly become established for the drug monitoring of P2Y12 and GPIIb/IIIa receptor antagonists. The more recent P2Y12 antiplatelet agents (clopidogrel, prasugrel, ticagrelor) achieve nearly complete inhibition. However, the GRAVITAS study did not show any advantage for clopidogrel in monitoring with dose adjustment in patients with high-on-treatment reactivity after drug-eluting stent implantation [18]. The results of the Tropical ACS study should be published soon.

6.2.4 POCT applications with coagulation testing methods

The "true POCT" methods can be differentiated from "POCT-appropriate" methods in terms of performance. Only the "true POCT

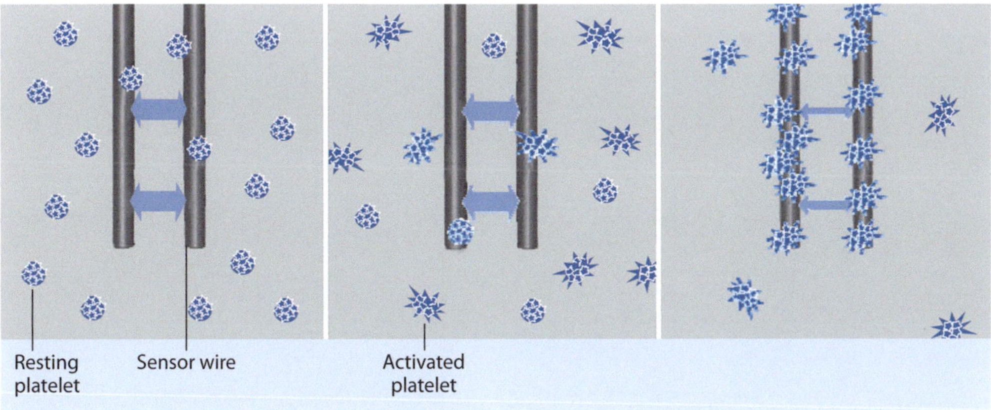

Fig. 6.9 Functional testing of platelets in whole blood (impedance aggregometry). When activated platelets adhere to the surface of two sensor wires, the electrical resistance between them increases significantly. This can be recorded and plotted graphically (impedance vs. time).

methods" that can be carried out by non-specially trained operators using single patient-near measurements without sample preparation.

Classification of various POCT coagulation methods by degree of difficulty

- "True" POCT methods:
  - ACT
  - Heparin management system (protamine titration)
  - POCT systems for INR/aPTT
  - Rapid platelet function analysis (RPFA)
- POCT-appropriate methods (with increasing degree of complexity):
  - PFA-100/200 indicator
  - Rotational thromboelastometry (ROTEM)
  - Thromboelastography (TEG)

True POCT methods are comparable with blood gas analysis or glucose measurement in terms of simplicity of operation. By contrast, the examiner performing POCT-appropriate methods requires effort and skill and accordingly a greater degree of training and motivation. As a rule, precise reagent handling is needed to process the sample. To ensure a 24-hour availability, greater expenditure on logistics and personnel is required than with true POCT methods. For this reason, at many institutions diagnostics are carried out by a mixed team of personnel from the central laboratory and, e.g., from the anesthesiology department. Progress towards a true POCT method arrived in 2015 with the introduction of ROTEM sigma, a closed analytical system. The device is fully automated and obviates pipetting or other test preparations. Moreover, the direct application of the sample test tube into the test cartridge prevents direct contact with the blood specimen.

However, the test cartridge must always be operated with all coagulation activation modes (see above) (cost factor!). More details on the clinical applications of POCT coagulation analyses can be found in ▶ Chapter 15.

6.3 Confounders and influencing variables

When performing blood coagulation analysis by POCT methods, one needs to account for several confounders and influencing variables, linked to both sample and handling factors. Depending on the detection method, different instruments react in varying ways to variations

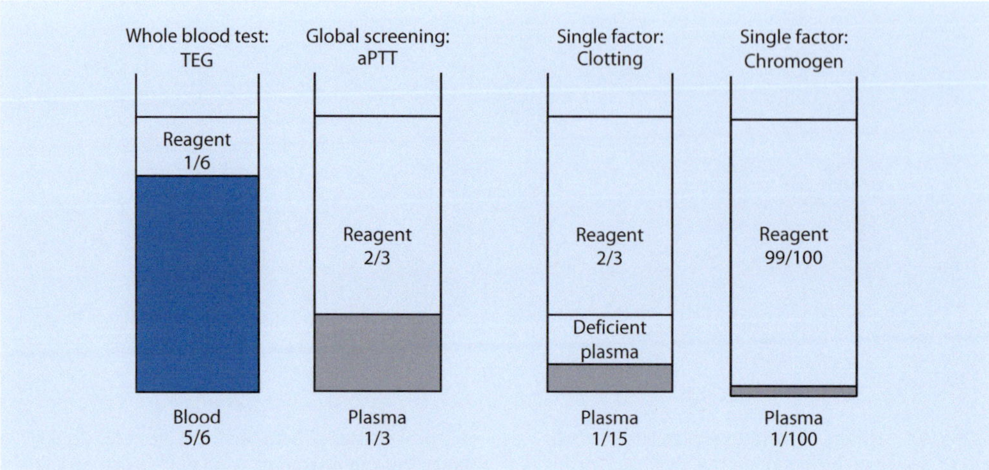

Fig. 6.10 Ratio of plasma to cellular blood components in different hemostaseological analyses. *aPTT* Activated partial thromboplastin time; *TEG* Thromboelastography

6

in hematocrit (Hct), to the influence of colloid and the formation of circulating microaggregates. In addition, near-patient procedures are frequently used in the monitoring of critically ill patients (in the shock room, intensive care unit), in whom marked changes in the ratios between plasma and cellular blood components alongside drug effects can be expected (Fig. 6.10). Oftentimes, the influences of special metabolic conditions (e.g. acidosis) or ambient conditions (e.g. hypothermia) on hemostasis diagnostics have not been systematically studied. Unfavorable blood draw conditions and catheter blood draws can be additional confounding factors. If small volumes or no reagents are added, whole blood methods are sensitive to confounders and influencing factors from the sample matrix. That is why POCT methods intended for hemostasis diagnostics in the respective clinical settings (e.g. cardiac operating room, ICU) should be evaluated scrupulously before clinical conclusions are drawn from the obtained values.

6.4 Quality management

The internal quality control of POCT methods in coagulation [21] is encumbered by the fact that the sample material needed for these procedures (whole blood) is only stable for a short period of time. Lyophilization of the sample causes the blood cells to lyse. Other procedures employed to stabilize blood or blood-like fluids (like those used for hematology controls) fail because the coagulation proteins and platelets in aqueous solution lack stability. Therefore, most of the available control materials consist of lyophilized plasma and can therefore only be used with certain test systems. These controls can be used on instruments which measure plasmatic coagulation and in viscoelastic methods. In some systems, artificial calibration fluids are employed (VerifyNow, Multiplate), for others (e.g. PFA-100) no control materials are available; in these cases, only a plausibility check is carried out, comparing measurement data from patients and healthy persons. Similar problems are associated with the calibration of POCT methods. Calibration of the parameters cannot be performed at regular intervals. For that reason, the manufacturer aims for stabile batch calibration. One way this is done is to define conversion factors for each batch and use these factors to convert the variables measured (e.g. coagulation time) into the result (e.g. INR) on all instruments.

The participation in interlaboratory tests for quantitative analyses mandated by RiliBÄK (▶ Chapter 38) is only possible to a limited extent for hemostaseologic POCT methods due to the complex and hard-to-handle matrix. Nevertheless, the interlaboratory testing organizations do offer some quality control materials, e.g. for INR and ACT. Not all devices available on the market, however, are suitable for participation.

In the case of POCT methods for which analog procedures are available in the central laboratory (aPTT, INR), an external control can be performed indirectly by comparative testing on the POCT system and in the laboratory which is subject to external quality controls, called the split sample technique.

References

1. Avidan MS, Alcock El, Da Fonseca JH et al. (2004) Comparison of structured use of routine laboratory tests or near-patient assessment with clinical judgement in the management of bleeding after cardiac surgery. Br J Anaesth 92: 178–86
2. Bosch YP, Ganushchak YM, de Jong DS (2006) Comparison of ACT point-of-care measurements: repeatability and agreement. Perfusion 1: 27–31
3. Calatzis A, Heesen M, Spannagl M (2003) Patientennahe Sofortdiagnostik von Hämostaseveränderungen in der Anästhesie und Intensivmedizin. Anaesthesist 52: 229–37
4. Ferring M, Reber G, de Moerloose P, Merlani P, Diby M, Ricou B (2001) Point of care and central laboratory determinations of the aPTT are not interchangeable in surgical intensive care patients. Can J Anaesth 48: 1155–60
5. Gibbs NM (2009) Point-of-care assessment of antiplatelet agents in the perioperative period: a review. Anaesth Intensive Care 37:354–69
6. Hartert H (1948) Blutgerinnungsstudien mit der Thrombelastographie, einem neuen Untersuchungsverfahren. Klin Wschr 26: 577–83
7. Haubelt H, Anders C, Vogt A, Hoerdt P, Seyfert UT, Hellstern P (2005) Variables influencing Platelet Function Analyzer-100 closure times in healthy individuals. Br J Haematol 130: 759–67
8. Hayward CP, Harrison P, Cattaneo M, Ortel TL, Rao AK (2006) Platelet function analyzer (PFA)-100 closure time in the evaluation of platelet disorders and platelet function. J Thromb Haemost 4: 312–9
9. Karger R, Donner-Banzhoff N, Müller HH et al. (2007) Diagnostic performance of the platelet function analyzer (PFA-100) for the detection of disorders of primary haemostasis in patients with a bleeding history-a systematic review and meta-analysis. Platelets 18: 249–60
10. Koscielny J, Ziemer S, Radtke H et al. (2004) A practical concept for preoperative identification of patients with impaired primary hemostasis. Clin Appl Thromb Hemost 10: 195–204
11. Kratzer MA, Born GV (1985) Simulation of primary haemostasis in vitro. Haemostasis 15:357–62
12. Leyvi G, Shore-Lesserson L, Harrington D, Vela-Cantos F, Hossain S (2001) An investigation of a new activated clotting time "MAX-ACT" in patients undergoing extracorporeal circulation. Anesth Analg 92: 578–83
13. Luddington RJ (2005) Thrombelastography/thromboelastometry. Clin Lab Haematol 27(2): 81–90
14. Murray DJ, Brosnahan WJ, Pennell B, Kapalanski D, Weiler JM, Olson J (1997) Heparin detection by the activated coagulation time: a comparison of the sensitivity of coagulation tests and heparin assays. J Cardiothorac Vasc Anesth 11:24–8
15. Niederdöckl J, Dempfle CE, Schönherr HR, Bartsch A, Miles G, Laggner A, Pathil A (2016) Point-of-care PT and aPTT in patients with suspected deficiencies of coagulation factors. Int J Lab Hematol 38:426–34
16. Nitu-Whalley IC, Lee CA, Brown SA, Riddell A, Hermans C (2003) The role of the platelet function analyser (PFA-100) in the characterization of patients with von Willebrand's disease and its relationships with von Willebrand factor and the ABO blood group. Haemophilia 9: 298–302
17. Paniccia R, Priora R, Liotta AA, Abbate R (2015) Platelet function tests: a comparative review. Vasc Health Risk Manag 11:133–48
18. Price MJ, Berger PB, Teirstein PS, Tanguay JF, Angiolillo DJ, Spriggs D, Puri S, Robbins M, Garratt KN, Bertrand OF, Stillabower ME, Aragon JR, Kandzari DE, Stinis CT, Lee MS, Manoukian SV, Cannon CP, Schork NJ, Topol EJ; GRAVITAS Investigators (2011) Standard- vs. high-dose clopidogrel based on platelet function testing after percutaneous coronary intervention: the GRAVITAS randomized trial. JAMA 305:1097–105
19. Sibbing D, Braun S, Morath T, Mehilli J, Vogt W, Schömig A, Kastrati A, von Beckerath N (2009) Platelet reactivity after clopidogrel treatment assessed with point-of-care analysis and early drug-eluting stent thrombosis. J Am Coll Cardiol 53:849–56

20. Smith JW, Steinhubl SR, Lincoff AM et al. (1999) Rapid platelet-function assay: an automated and quantitative cartridge-based method. Circulation 99:620–5
21. Spannagl M, Dick A, Junker R (2010) Qualitätssicherung bei POCT-Methoden in der Hämostaseologie. Hämostaseologie 30;2: 82–90
22. Spannagl M, Dick A, Junker R (2010) POCT in coagulation. Hämostaseologie 30:82–90
23. Toth O, Calatzis A, Penz S, Losonczy H, Siess W (2006) Multiplate electrode aggregometry: a new device to measure platelet aggregation in whole blood. Thromb Haemost 96:781–8
24. Zucker ML (2007) The National Academy of Clinical Biochemistry Laboratory Medicine Practice Guidelines for Point of Care Coagulation Testing. Point of Care 6: 223–6

Analysis of cellular blood components

Dorthe Kixmüller, Ralf Junker

P. B. Luppa, R. Junker (Eds.), *Point-of-Care Testing*
https://doi.org/10.1007/978-3-662-54497-6_7

7.1 Introduction

Hematological measurements by POCT range from singular determinations of hemoglobin (Hb) concentration to complete blood counts. The term "blood count" refers to the differential quantification of erythrocytes (or red blood cells, RBC), leukocytes (white blood cells, WBC) – including a differentiation into granulocytes, lymphocytes and monocytes – and platelets (thrombocytes). In addition to counts of these cellular components, the following parameter are evaluated: hematocrit (Hct), Hb, cell characteristics, e.g. erythrocyte indices such as mean corpuscular volume (MCV), mean corpuscular hemoglobin (MCH) and mean corpuscular hemoglobin concentration (MCHC) along with further information about the stages of maturity of single cell lines, e.g. normoblasts, reticulocytes, immature granulocytes, immature thrombocytes.

7.2 Device technology and methods

In principle, there are three main classes of devices used in hematological POCT:

- Full-fledged POCT devices for blood count analysis that are the equivalent of a miniaturized conventional laboratory analyzer encompassing the relevant technical measuring spectrum
- Devices which measure specific parameters, e.g. Hb concentrations, in parallel with other variables, e.g. devices for blood gas analysis (BGA)
- Devices which determine singular hematological parameters, e.g. Hb or total leukocyte count

The required sample volume depends on the device used – on average ranging from a few microliters to 200 µL. Analysis usually takes a few seconds to minutes. All of the aforementioned devices are able to store and/or print out the results after analysis where required or transmit them via a standard interface to an existing electronic data processing system (POCT network, laboratory or hospital information system).

7.2.1 POCT blood count analyzers

Although they use same technology, systems which allow a differential cell count are distinct from standard systems by their degree of miniaturization.

By employing various adapters, it is possible to use different sample tubes, where appropriate, also capillary tubes. Examples of this type of device are shown in ◘ Fig. 7.1 (pocH-100i by Sysmex) and ◘ Fig. 7.2 (Yumizen 500 by Horiba).

Beckman Coulter (DxH 500) and other in vitro diagnostics (IVD) companies also offer fully automated hematology analyzers with manual processing of single samples.

RBC and platelets are usually detected by the electrical impedance method. The impedance measurement principle is based on the fact that cellular components of blood are less conductive compared to plasma. Using hydrodynamic focusing (◘ Fig. 7.3), blood cells in a fluid stream are passed through a measuring

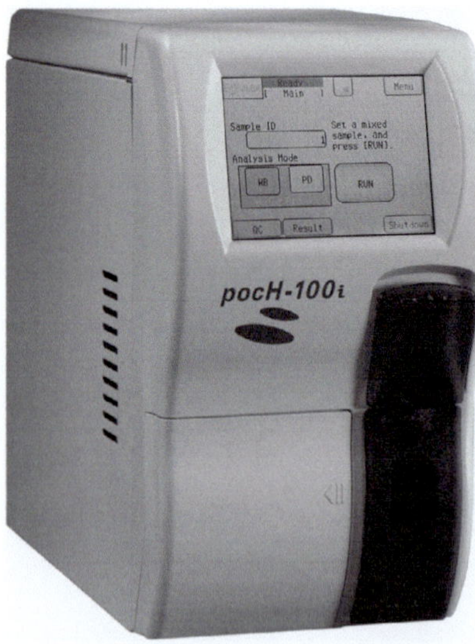

◘ **Fig. 7.1** PocH 100i by Sysmex. (Courtesy of Sysmex Deutschland GmbH, Norderstedt)

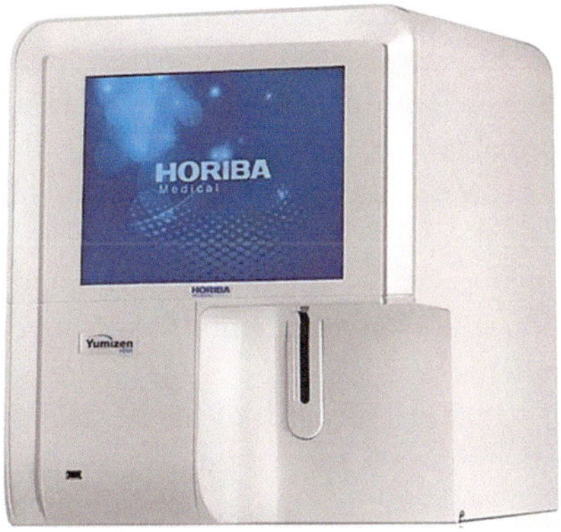

Fig. 7.2 Yumizen H500 by Horiba. (Courtesy of Axon Lab, Reichenbach)

capillary that jackets the cells to allow counting with greater precision. As a cell passes through the measuring aperture, it produces an increase in current resistance due to its poorer conductivity compared to plasma. The charged electrodes located on either side of the aperture detect each individual cell as an electrical resistance signal. This change in impedance is proportional to the cell's volume and size, thereby enabling an identification into platelets (approx. 8–12 fl) and RBC (approx. 80–100 fl). The number of particles is counted per volume or per time unit.

Other methods, such as those used for platelet counts, are based on a cell type-specific pattern of light scatter on the cell surface.

After lysis of the RBC, WBC are likewise mostly detected and identified in a fluid stream by impedance measurement. Based on the strength of the impulse transmitted at impedance measurement, WBC can be differentiated by cell size into lymphocytes, monocytes (including eosinophilic and basophilic granulocytes) and neutrophilic granulocytes.

Direct measurement of Hb is usually carried out by photometry. Using the cyanmethemoglobin method, Hb is oxidized by potassium hexacyanoferrate(III) to metHb (methemoglobin) which is further converted by potassium cyanide to cyanmethemoglobin. Cyanmethemoglobin has a spectrophotometric absorption maximum at 540 nm, which is proportional to the Hb concentration.

Generally, hematology analyzers are not unit-use systems (one measurement using a single reagent pack). Therefore, quality control has to be carried out the same as on conventional laboratory devices. Relevant devices are supplied by several manufacturers. They differ in the technological features employed for the mechanical counting and differentiation of cells and in their handling and degree of automation. The quality of these POCT systems for answering typical clinical questions is mostly adequate and has been verified in several studies [3, 4, 11, 15, 18]. However, users must be trained to use the system and, when analyzing results, the interpreter must be sufficiently knowledgeable.

The various cellular components (RBC, WBC with simple differentiation, platelets) can also be separated by centrifugation in a coated

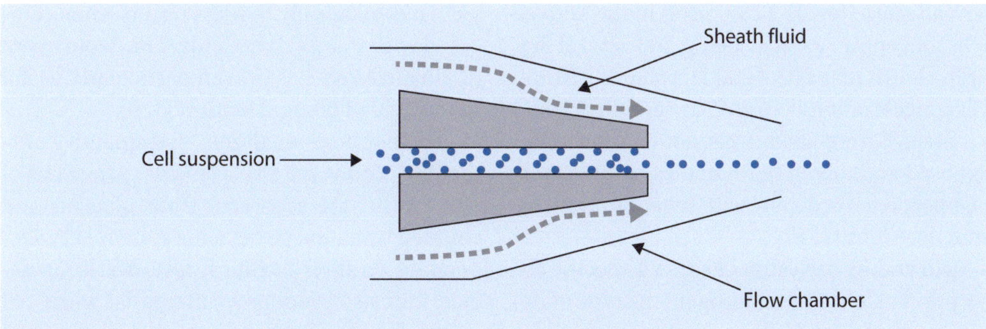

Fig. 7.3 Principle of hydrodynamic focusing

microhematocrit capillary. The buffy coat layer is then analyzed electro-optically for the different cell populations [8, 13]. Hence the name QBC: quantitative buffy coat.

7.2.2 Blood gas analyzers

In addition to electrolytes and blood gases, blood gas analyzers (BGA) usually also measure Hb concentration and Hct (▶ Chapter 14). Here, the principle for analyzing Hct is based on a conductivity method. The volume of non-conducting blood components can be calculated based on the measured conductivity of a calibration solution with known electrolyte concentration, the measured electrolyte concentration and the conductivity in the sample. The quality of method is acceptable at normal Hb concentrations. Due to the fact that plasma components like proteins and not just electrolytes, contribute to the conductivity of the sample, the corresponding shift in conductivity can then lead to false-high or false-low Hct values. This is the case in highly diluted samples (e.g. those taken during operations) when the protein concentration is very low or after the transfusion of blood products. High electrolyte concentrations (Na^+, K^+) can also cause false-low measurements. The use of a correction factor is possible, although complete compensation cannot always be achieved [11]. In some systems, the electrolyte concentration measured in parallel is taken as a direct correction factor.

The Hb concentration can be determined indirectly using the calculated Hct value. A potentially false-low Hct can result in the derived Hb concentration also being too low. It has been shown that calculated Hct values <30 and Hb concentrations <10 g/dL are not sufficiently accurate. Severe anemia cannot be reliably diagnosed and therefore these measurands must be interpreted critically with regard to transfusion decisions [1, 10].

Hb and its derivatives exhibit a specific absorption pattern. Depending on the type of device used, the total Hb concentration can also be measured photometrically at multiple wavelengths of light by CO-oximetry on intact or lysed RBC. Any further diagnostic differentiation (i.e. cell morphology and RBC indices) is not possible.

As several studies have demonstrated, the analytical quality of the different systems is generally good and the results show good correlation with standard methods, even in pediatric patients [6, 7].

If additional analyses to the hematological investigations are carried out, e.g. a complete BGA, some systems requires sample volumes of up to 200 µL, which can cause problems with capillary blood samples. In most cases, however, a few microliters of blood are sufficient.

7.2.3 Single analyses

Systems that exclusively measure **Hb** primarily use a modified wet-chemical azide methemoglobin method. Results are available within 60 seconds using single microcuvettes coated with sodium deoxycholate to lyse the RBC. Furthermore, the microcuvettes contain sodium nitrate which converts Hb to metHb (methemoglobin) and sodium azide for a final staining reaction [16]. The absorption of azide methemoglobin is measured at two wavelengths, 570 nm and additionally 880 nm, in order to compensate for turbidity in the sample [17].

The sample volume needed is no more than a few microliters. This method has produced good results in numerous comparative studies [6, 9, 11]. Other devices which use the same method or measure Hb in blood by photometry yield metrologically flawless results when compared with standard methods. The deployment of Hb analyzers has proven particularly useful in the field of blood donation [3, 5].

Furthermore, analyzers that quantify only the **WBC count** are also available. After lysis of RBC, WBC are separated from platelets and counted by an image detector system [12]. Depending on their technical features, some devices can also produce a differential white cell count, classifying leukocytes into lymphocytes, neutrophils, monocytes, basophils and eosinophils. After initial lyses of RBC and staining of WBC, the cells are identified and classified using a camera. The result is available within a few minutes.

Although the reliability of transcutaneous, non-invasive Hb assessment (SpHb) is still under critical appraisal [2, 14], its benefit lies in continuous real-time patient monitoring (▶ Chapter 11). The pulse CO-oximeter utilizes a spectrophotometric sensor and multiple light wavelengths. Unexpected changes are detected in real time, which can guide decision-making with regard to blood transfusions. The accuracy of this technology, though promising, is not yet sufficiently reliable and should only be used as an adjunct to, not a replacement for invasive Hb analysis.

References

1. Abbott Point of Care. Hematocrit/Hct and Calculated Hb. Article no. 714178-01Q. 15 August 2016
2. Barker SJ, Shander A, Ramsay MA (2016) Continuous noninvasive hemoglobin monitoring: A measured response to a critical review. Anesth Analg 122: 565–572
3. Benjamin JT, Baisden CR (2002) Advances in the physician office laboratory. In: Kost GJ (ed) Principles and practice of point-of-care Testing. Lippincott Williams & Wilkins, Philadelphia, pp 391–406
4. Briggs C, Kunka S, Pennaneach C, Forbes L, Machin SJ (2003) Performance evaluation of a new compact hematology analyzer, the Sysmex pocH-100i. Lab Hematol 9: 225–233
5. Despotis GJ, Saleem R, Bigham M, Barnes P (2000) Clinical evaluation of a new, point-of-care hemocytometer. Crit Care Med 28: 1185–1190
6. Gehring H, Hornberger C, Dibbelt L et al. (2002) Accuracy of point-of-care-testing (POCT) for determining haemoglobin concentrations. Acta Anaesthesiol Scand 46: 980–986
7. Hinds LE, Brown CL, Clark SJ (2007) Point of care estimation of haemoglobin in neonates. Arch Dis Child Fetal Neonatal Ed 92: F378–F380
8. http://www.qbcdiagnostics.com/products/hema/
9. Jahr JS, Lurie F, Driessen B, Davis JA, Gosselin R, Gunther RA (2002) The HemoCue, a point of care B-hemoglobin photometer, measures haemoglobin concentrations accurately when mixed in vitro with canine plasma and three hemoglobin-based oxygen carriers (HBOC). Can J Anaesth 49: 243–248
10. Masimo, Corporation (2009) Technisches Merkblatt, Messungen des Gesamthämoglobins: Genauigkeit von Laborgeräten und Auswirkung physiologischer Variation
11. McNulty SE, Torjman M, Grodecki W, Marr A, Schieren H (1995) A comparison of four bedside methods of hemoglobin assessment during cardiac surgery. Anesth Analg 81: 1197–1202
12. Osei-Bimpong A, Jury C, McLean R, Lewis SM (2009) Point-of-care method for total white cell count: an evaluation of the HemoCue WBC device. Int J Lab Hematol 31: 657–64
13. Paul RI, Badgett JT, Buchino JJ (1994) Evaluation of QBC Autoread performance in an emergency department setting. Pediatr Emerg Care 10: 359–63
14. Rice MJ, Gravenstein N, Morey TE (2013) Noninvasive hemoglobin monitoring: how accurate is enough? Anesth Analg 117:902–907
15. Seguin P, Kleiber A, Chanavaz C, Morcet J, Mallédant Y (2011) Determination of capillary hemoglobin levels using the HemoCue system in intensive care patients. J Crit Care 26:423–427
16. Vanzetti G (1966) An azidemethemoglobin method for hemoglobin determination in blood. J Lab Clin Med 67: 116–26
17. von Schenck H, Falkensson M, Lundberg (1986) Evaluation of „HemoCue," a new device for determining hemoglobin. B Clin Chem 32:526–529
18. Whisler S, Dahlgren C (2005) Performance evaluation of the Sysmex pocH-100i automated hematology analyzer. Lab Hematol 11: 107–117

Clinical chemistry parameters

Ralf Junker, Norbert Gässler

P. B. Luppa, R. Junker (Eds.), *Point-of-Care Testing*
https://doi.org/10.1007/978-3-662-54497-6_8

8.1 Introduction

The wide analytical spectrum offered in clinical chemistry laboratories is equally accessible to near-patient diagnostics, e.g. measurement of enzymes (ALP, ALT, AST, GGT, amylase, CK), electrolytes (Na^+, K^+, Ca^{2+}, Cl^-, Mg^{2+}) and numerous metabolic parameters (bilirubin, HDL, LDL and total cholesterol, triglycerides, glucose, uric acid, creatinine, urea, ammonia and lactate).

The advantage of using POCT in clinical chemistry is the instant availability of the analysis results, particularly when samples would otherwise need to be transported a longer distance prior to analysis or have to undergo special pre-analytical preparation, as is the case for ammonia measurement.

8.2 Device technology and methods

In POCT, a range of sample types can be used including capillary, venous and arterial blood. To avoid the need for centrifugation of the sample, the assay must either use whole blood or integrate an additional process step to eliminate cellular blood components from the whole blood. Usually, a few micro-liters are sufficient for analysis, whereby the actual analysis itself only takes a few minutes. The commonly used testing systems can, to a certain extent, store data and are equipped with an integrated printer or IT interface that allows the results to be exported directly to the laboratory or hospital computer information systems.

In principle, one differentiates between three different device classes (▶ Chapter 3):

- Full-scale clinical biochemistry analyzers are essentially miniaturized versions of conventional laboratory systems, which are able to analyze a wide range of parameters – either singularly or in parallel. Wet and dry biochemistry analyzers fall into this group.
- Devices where biochemical analyses can be made to supplement other measurements, e.g. blood gas analyzers (BGA), which use electrodes or photometry to determine parameters such as electrolytes, glucose, lactate, creatinine, bilirubin and other parameters.
- Test systems designed specifically for singular parameters, e.g. devices for lipid profile, lactate or glucose assessments (▶ Chapter 12).

8.2.1 Dry chemistry

Dry chemistry systems employ single-use test strips for the actual analysis. After the sample (whole blood or plasma) is applied to the test strip, plasma separation occurs and cellular blood components are filtered out. The plasma is then collected in a reservoir on the test strip. The reagents adhere to membranes which are pressed onto the plasma reservoir when the test strip is inserted into the device. The plasma or resulting reaction products then successively permeate the reagent-soaked strip (◘ Fig. 8.1).

The final step consists of a detection reaction, for example, color development where the intensity of the dye is directly related to the analyte concentration. After the reaction chain has run its course, the result is delivered on an optical color spectrum.

One widespread representative of this device class is the Roche Reflotron, which has been available in different versions for a number of years. This analyzer works by measuring light reflectance from a light-emitting diode (LED) within an Ulbricht sphere. The reflection at the test strip area, i.e. at site of the reaction, is different to other reflections within the sphere, giving an indication of the reaction strength and therefore the analyte concentration. All the information about device control, measurement methods and calibration, as well as the mathematical constants for analysis, are stored in a magnetic strip under each test strip, making measurements possible without batch management. Test methods differ from those used in wet chemistry (IFCC-compliant methods). This is because dry chemistry requires other

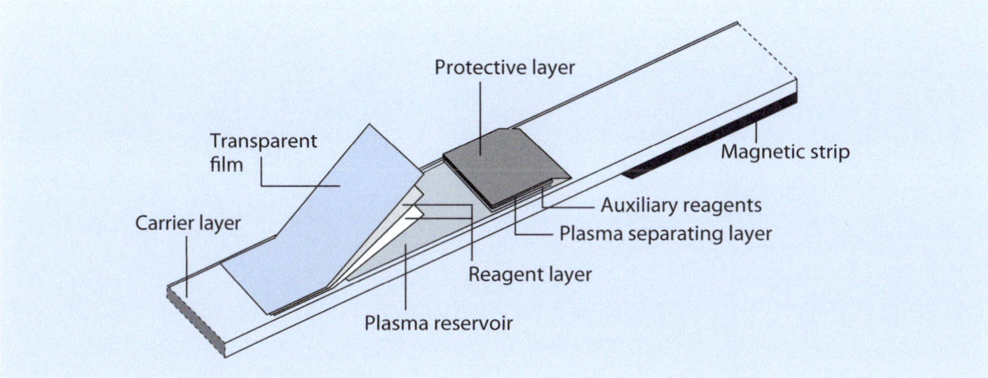

Fig. 8.1 Schematic diagram of a test strip

conditions, for example, fixation on special reagent carriers, which must be highly stable and can be stored at room temperature.

Nevertheless, the methods are usually calibrated, so that their results largely correspond to those of standard methods. Deviations from the reference range may, however, need to be taken into account, particularly in the context of patient follow-up monitoring when POCT and central laboratory methods are used alternately.

Depending on the type of device, analytes can be measured either singularly or in parallel. The analytical quality of manifold systems has been published in several studies [7, 9].

8.2.2 Wet chemistry

Despite their excellent analytical quality, wet chemistry analyzers are less commonly used, partially due to their comparatively high operating costs and the mechanical effort required. In general, established laboratory methods are used in this context, often as IFCC-compliant methods in miniaturized devices, designed for single-sample measurements.

No external centrifugation of the sample is needed as such devices have built-in centrifuges. They are therefore just as suitable for use with whole blood as with serum or plasma. For example, all reagents and diluents are encased in pre-fabricated self-contained discs, supplied in various combinations for a range of different analyses. A sample volume of 100 µL is often sufficient for a complete panel of more than 10 analyses.

The wet chemistry system illustrated in Fig. 8.2 is typical for this class of modern device. After transferring the sample onto the uptake point, the disc is inserted into the analyzer drawer where the sample is moved by centrifugal and capillary force to the appropriate site inside the cuvette where the reaction occurs. In this way, diluents and reagents also flow to the "correct" reaction site. After a few minutes, the analysis is performed photometrically; self-calibration and continuous quality controls are

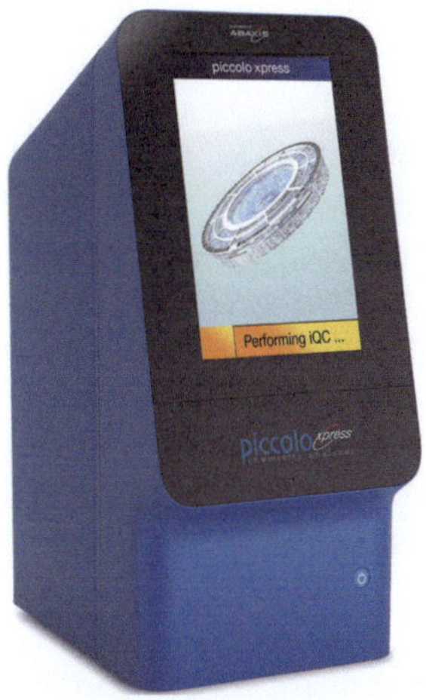

Fig. 8.2 Abaxis Piccolo Xpress with reagent disc. (Courtesy of Abaxis Europe, Darmstadt, Germany)

8

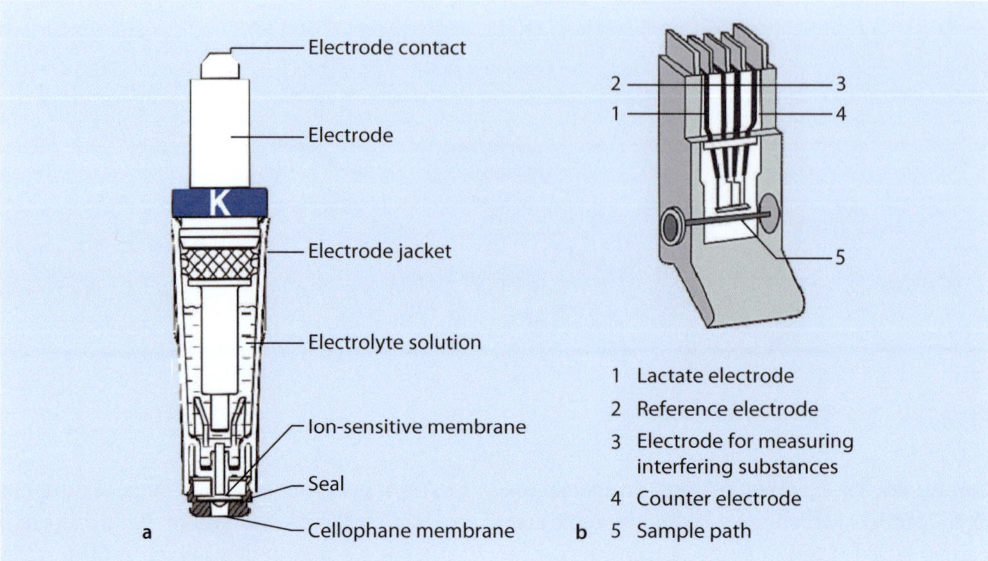

Fig. 8.3a,b Ion-selective electrodes (ISE) in blood gas analyzers. **a** ISE for K+ (Radiometer ABL505). (Courtesy of Radiometer, Willich). **b** ISE for lactate (Siemens Rapidlab 865). (Courtesy of Siemens Medical Solutions Diagnostics, Bad Nauheim, Germany)

integrated in the analyzer. Evaluation data are available that demonstrate sufficient analytical quality and a good correlation with standard methods [4, 6].

Blood gas analyzers may also use wet chemistry methods to perform clinical chemistry analyses. These panels cover parameters like urea, creatinine, bilirubin, glucose and lactate. The electrolytes Na^+, K^+, Ca^{2+}, Mg^{2+} and Cl^- are measured by ion-selective electrodes (Fig. 8.3).

The Abbott i-STAT and its successor are highly miniaturized analyzers, where minimal volumes are sufficient for carrying out multiple methods. This is a system for analyzing blood gases, electrolytes and other parameters in different combinations. Test cartridges of a few centimeters in size contain all the necessary reagents and technologies, including a single-point calibrant (► Chapter 3). The analytical quality of the system has been repeatedly evaluated [1].

Blood gas analyzers often additionally feature measurements for metabolites such as glucose, lactate and creatinine. Validated sensor systems specifically for creatinine have only recently become available. As an example, the amperometric creatinine sensor in the ABL837 FLEX (Radiometer) should be mentioned here, which is based on the enzymatic conversion of creatinine to H_2O_2 [8]. Here, creatinine is converted initially to creatine, which is then metabolized by creatine kinase to sarcosine. Sarcosine is further metabolized by sarcosine oxidase to glycine and H_2O_2. The latter is oxidized amperometrically at the electrodes and the resulting electrolysis current is directly proportional to the creatinine concentration.

8.2.3 Dedicated devices for singular analytes

The assessment of **total bilirubin** in newborns is often performed by direct spectrometry of non-diluted serum or plasma in a "bilimeter". These are mostly filter photometers, where the absorption of plasma is measured at 455 nm, i.e. near the absorption maximum of bilirubin. Bilirubin can be assessed quantitatively in this way, as plasma from newborns is free of lipochromes, such as carotene, which is also absorbed within this wavelength range as long as a spectral interference by hemoglobin is compensated for. This compensation is achieved by an additional reading at 575 nm. Given that the molar extinction coefficient of Hb is identical at 455 nm and 575 nm, the bilirubin concentration can be calculated from the difference

$$\Delta E = E455\ nm - E575\ nm.$$

A hematocrit (Hct) capillary tube is used as a cuvette, which is filled with (capillary) blood and then spun in a special centrifuge. Usually 20–30 µL of blood is sufficient [2, 10]. In the past, a lack of linearity and calibration problems was blamed as the main reason for false results. Newer devices appear to basically have overcome these problems, as recent interlaboratory testing for neonatal bilirubin conducted by the German Society for Clinical Biochemistry Laboratory Medicine (DGKL) has demonstrated. ▶ Chapter 11 and ▶ Chapter 22 contain further explanations on the determination of neonatal bilirubin.

Special systems, e.g. Cholestech LDX by Alere, are used to determine a patient's **lipid status** in specialized outpatient practices, but also in pharmacies. Whole blood is used to analyze triglycerides, total, HDL and LDL cholesterol as well as glucose and AST.

In sports medicine, **lactate measurements** in blood or saliva play an important role (▶ Chapter 22). A number of POCT devices are available for this, like the AccuTrend Plus by Roche, for example. It should also be mentioned here that the intra-partum fetal scalp lactate assessment can give important information about the fetal condition during birth, as an alternative or in addition to pH [3]. Rapid testing can, for instance, be performed with the StatStip by Nova.

A fast **creatinine check** in capillary blood is important in many radiological and interventional settings, when creatinine analysis in a central laboratory is otherwise not possible within an hour. The rapid identification of renal impairment is important for nephroprotection to avoid acute renal failure when intravenous contrast media is given [5]. Here, for example, the StatSensor by Nova can be used.

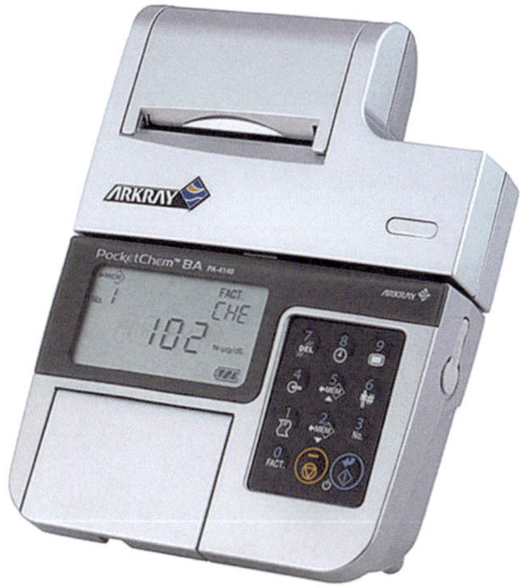

Fig. 8.4 Arkray's POCKETCHEM BA PA-4140. (Courtesy of Arkray Europe, Amstelveen, Netherlands)

In various metabolic disorders, an accurate determination of the **ammonia concentration** as ammonium ions (NH_4^+) is indicated, mostly in patients with cerebral and neuromuscular dysfunctions concurrent with liver cirrhosis. Small POCT devices for measuring blood ammonia are available from different manufacturers. In Fig. 8.4 the POCKETCHEM BA PA-4140 by Arkray is pictured. Confounding factors, primarily hemolysis, but also analyte instability can impact the results of the analysis. Consequently, the time from sample taking to processing in the laboratory should be no longer than 15 minutes. Therefore, POCT technology is often used for measuring blood ammonia.

Furthermore, there are a number of methods for potentiometric measurement of electrolytes and other special applications. The development of other dedicated systems is likely to progress and expand the spectrum of such clinical chemistry methods in the future.

8.3 Applications and indications

Point-of-care biochemical analyses are mainly used in smaller hospitals, where a 24-hour laboratory service is not viable, also in larger GP practices. The main reasons for using POCT for clinical biochemistry analyses in hospitals are as follows:

- For emergencies in outpatient settings and admission units, the waiting times for results from the central laboratory are frequently too long.
- In hospitals without access to a central laboratory, clinical chemistry analyzes can be carried out as POCT for rapid diagnostics in urgent cases, e.g. pre-operative screening of new admissions (even if they are non-emergency, but still require timely analysis).
- In intensive or critical care, POCT methods are an integral part of close patient monitoring, e.g. for blood gas analysis.

In specialized medical practices like dialysis centers or oncology units, rapidly available results help expedite urgently required decision-making (less in emergency situations, as these tend to be rare here). It may also become important before initiating therapy, as therapeutic regimes can be adapted, if necessary, according to the real-time renal retention values, electrolytes and other biochemical parameters.

References

1. Bingham D, Kendall J, Clancy M (1999) The portable laboratory: an evaluation of the accuracy and reproducibility of i-STAT. Ann Clin Biochem 36:66–71
2. Grohmann K, Roser M, Rolinski B et al. (2006) Bilirubin measurement for neonates: comparison of 9 frequently used methods. Pediatr 117:1174–1183
3. Heinis AM, Dinnissen J, Spaanderman ME, Lotgering FK, Gunnewiek JM (2011) Comparison of two point-of-care testing (POCT) devices for fetal lactate during labor. Clin Chem Lab Med 50:89–93
4. Lee EJ, Shin SD, Song KJ, Kim SC, Cho JS, Lee SC et al. (2011) A point-of-care chemistry test for reduction of turnaround and clinical decision time. Am J Emerg Med 29:489–495
5. Meersch M, Schmidt C, Zarbock A (2016) Patient with chronic renal failure undergoing surgery. Curr Opin Anaesthesiol 29:413–420
6. Park H, Ko DH, Kim JQ, Song SH (2009) [Performance evaluation of the Piccolo xpress Point-of-care Chemistry Analyzer]. Korean J Lab Med 29:430–438
7. Price CP, Koller PU (1988) A multicentre study of the new Reflotron system for the measurement of urea, glucose, triacylglycerols, cholesterol, gamma-glutamyltransferase and haemoglobin. J Clin Chem Clin Biochem 26: 233–250
8. Skurup A, Kristensen T, Wennecke G (2008) New creatinine sensor for point-of-care testing of creatinine meets the National Kidney Disease Education Program guidelines. Clin Chem Lab Med 46:3–8
9. St John A, Ritter C (2010) Benchtop instruments for point-of-care testing. In: Price CP, St John A, Kricka LJ (eds) Point-of care testing, 3nd ed. AACC Press, Washington, pp 43–62
10. Thaler M, Luppa PB, Schlebusch H (2008) Die Bilirubinbestimmung – Eine aktuelle Übersicht. J Lab Med 32: 1–9

Immunological methods

Peter B. Luppa, Ralf Junker, Ingolf Schimke, Enno Stürenburg

P. B. Luppa, R. Junker (Eds.), *Point-of-Care Testing*
https://doi.org/10.1007/978-3-662-54497-6_9

9.1 Methods

9.1.1 Immunosensors

The specificity of antibodies for their molecular antigen counterparts is the basis for both immunoassay technologies as well as for immunosensors, where the antibodies are immobilized on a solid phase substrate (▶ Chapter 3 and ▶ Chapter 5). This immunological reaction principle not only allows the qualitative detection of an analyte, but also a quantitative assessment of its concentration. Problematical areas of crucial analytical importance for the selective detection of an antigen-antibody complex are the bioconjugate chemistry applied for immobilization and the orientation of capture antibodies, whose specificity must not be compromised by the binding [6].

9

We distinguish between four different types of **immunosensors** [14, 15]:

- Electrochemical sensors (potentiometric, amperometric or conductometric, capacitive)
- Optical sensors
- Microgravimetric sensors (quartz crystal microbalance)
- Thermometric sensors.

The microgravimetric and thermometric sensor types have not proven their merits in POCT applications. All sensor types can work either as direct (label-free) or indirect (labeled) immunosensors. Direct sensors are able to track physicochemical changes during immune complex formation, whilst indirect sensors mainly utilize fluorescent or luminescent markers, thereby conferring high detection sensitivity. The authors are nevertheless convinced that direct immunosensors have a future in POCT applications and that they will enjoy widespread use due to the fact that their simple reagent concept offers analytical advantages, as long as the relevant analyte concentration range can be achieved.

There are extensive immunolabeling options for both immunosensors and immunoassays, which can only be covered briefly here: The most reliable of these labels are enzymes such as peroxidase, glucose oxidase, alkaline phosphatase, catalase or luciferase. Ferrocene or In^{2+} salts are used as electroactive compounds. Fluorescent labels include rhodamine, fluorescein, Cy5, ruthenium diimine complexes, phosphorescent porphyrin derivatives among others. In particular, laser-induced fluorometric resonance energy transfer between 2 different fluorophores has methodological advantages and can specifically be employed in fiber optic sensors [26].

9.1.2 Homogeneous and heterogeneous immunoassays

Similar to larger laboratory devices, POCT devices can incorporate homogeneous and heterogeneous immunoassays in order to quantify analytes in fluids by specific binding of antibodies to the respective analyte's epitope. Heterogeneous non-competitive **sandwich assays** are the most commonly used type, where bound-from-free separation is achieved by magnetic force. **Fluorescence** and **chemiluminescence techniques** are mainly used for signaling with labeled primary or secondary antibodies or labeled tracers. Homogeneous immunoassays do not require immune complex separation and often utilize **fluorescence polarization techniques** or the **Förster resonance energy transfer (FRET) principle**. Two different fluorescent molecules are used, which overlap in their spectral emission ranges (donor molecule) and excitation (acceptor molecule). Both labeling molecules must be spatially close to one another for the FRET process to take place. This proximity can be achieved by formation of an immune complex. During this process, FRET causes the fluorescence of the donor to decrease and that of the acceptor to increase.

The application of **paramagnetic nanoparticles** coated with specific antibodies has engendered immunoassay procedures with high analytical sensitivity. The example in ◘ Fig. 9.1 shows the reaction principle employed by PATHFAST (LSI Medience Corporation) which

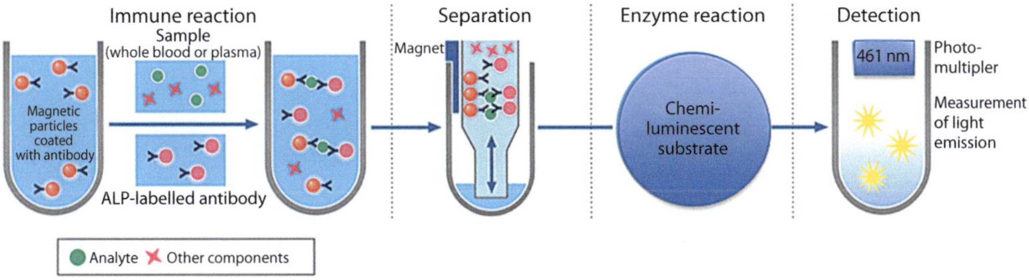

Fig. 9.1 The chemiluminescence immunoassay principle of PATHFAST

can quantify parameters such as troponin I, CK-MB or D-dimer.

More novel detection methods are employed in such immunoassay systems as well. One example is the m16 magnetic immunoassay analyzer (Edan Instruments, China). This biosensor uses magnetic beads to measure immune complex formation by means of **giant magnetoresistance** (GMR). GMR is observed in electrically conductive, alternating magnetic and non-magnetic layers of nanometer thinness. Due to the antiparallel alignment of the magnetization of the two magnetic layers (antiferromagnetic coupling), the electrical resistance in the entire structure is significantly higher than when magnetized. By applying an external magnetic field, the antiferromagnetic coupling is cancelled out and the magnetic moments caused to become orientated in the same direction. The electrical resistance is significantly decreased as a result.

9.1.3 Immunological rapid tests

In POC diagnostics, immunological tests are often used in addition to pure immunosensors. These testing methods are based on the immune complex principle, whereby an antigen binds to a specific antibody to form an immune complex. The test setup of such immunological tests can vary considerably, in particular regarding the generation and evaluation of the test signals. Particle agglutination and immunochromatography have proven their usefulness and are frequently used for rapid tests. Both techniques are applied, for example, in pregnancy tests, but also for microbial antigen detection in bodily fluids. The microbiological applications of these techniques are discussed specifically in ▶ Chapter 20.

Particle agglutination assays

Systems based on identification by reaction (assays) enjoy widespread use for the detection of a broad array of analytes in body fluids. Here, carrier beads with an approx. 0.8 µm diameter – mostly made of latex (polystyrene) – are coated with specific antibodies by adsorption. If microbial antigens, for example, are present in the patient sample, they agglutinate with the latex particles. This agglutination is seen macroscopically as a precipitant in the otherwise homogeneous, milky suspension. The quantitative particle agglutination tests are performed on microscope slides, plates, tubes, capillary tubes or microtiter plates (ELISA plates). Latex reagents have proved useful for diagnosing rotaviruses, but also for antigen detection, particularly for a fast meningitis diagnosis: these detect Streptococcus pneumoniae, Neisseria meningitidis – serotypes A, B, C, Y and W135 – as well as Haemophilus influenzae type b. In addition, numerous other microbial antigens can be detected in this way.

Immunochromatographic tests

The most frequently mentioned names for immunochromatographic strip tests in the medical literature are immunochromatographic tests (ICT), lateral flow assays (LFA), lateral flow devices (LFD), dipstick assays and one-

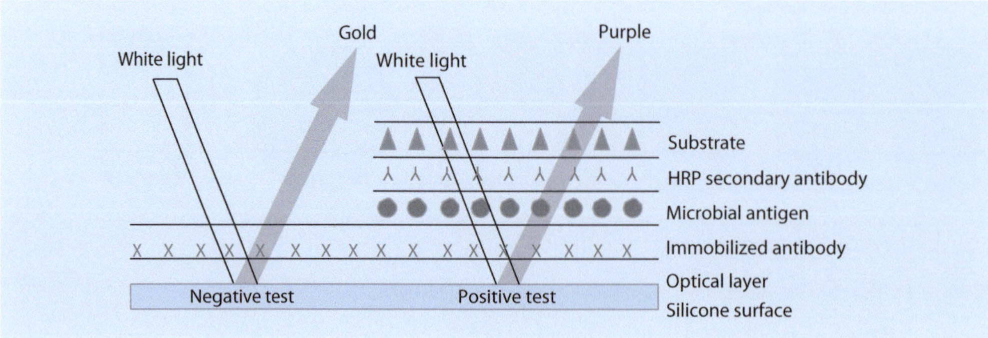

Fig. 9.2 Functional principle of an immunochromatographic test strip. *HRP* Horseradish peroxidase

step tests. These tests are based on the principle of immune complex formation. However, such tests also rely on thin-layer chromatographic separation techniques: The antibodies are immobilized on a membrane and the sample to be analyzed (blood, urine, cerebrospinal fluid) is moved onto the membrane by capillary force [11]. ICT are of particular interest because they are not only fast, but easy to use and easy to analyze. There are qualitative versions, but also tests that allow semi-quantitative or quantitative analysis. The best-known example of a qualitative ICT is the home pregnancy test.

The functional principle of an immunochromatographic test strip is illustrated in Fig. 9.2 [11]. The patient's liquid sample is dropped onto a sample pad. The pad acts as a prefilter and cleans the sample fluid of contaminants. Next, the sample pad releases buffers to adjust to a pH that is optimal for the immunological reaction to take place. The sample fluid then flows through the reagent pad. Here, a labeled antibody (conjugate), usually labeled with colloidal gold, binds to the antigen (if present in the sample) and forms an antigen-antibody complex. Since the conjugate is present in excess, the non-bound conjugates are also transported by capillary action. The sample fluid then migrates onto a nitrocellulose membrane. Here, two different reagent zones are applied in a row, one after another. The first zone (test line) refers to an immobilized secondary antibody directed against a second detection site on the antigen that thereby binds the immune complex. The other zone (control line) contains an anti-species antibody that reacts with the free conjugate. Either one or two lines will appear accordingly. Due to the colloidal gold, both zones and lines appear after a certain time in a reddish color. The excess sample fluid flows further and absorbed onto an absorbent pad to avoid back flow.

In the past years, numerous technical modifications have been developed to enhance ICT sensitivity. This includes using labels with higher specific signal intensity, such as paramagnetic microparticles (magnetic ferric oxide) as well as carbon particles instead of colloidal metal particles or blue latex particles [11]. A different path was chosen in the development of **optic immunoassays** [9]. Since this technique is the basis for many of today's rapid tests, a more detailed description is warranted here (Fig. 9.3). In this type of assay, a color change occurs because the immune reaction causes a change in the reflective properties of a silicone surface. When a protein analyte, extracted from the patient sample or a microbial antigen is placed on the silicone surface, it binds to the immobilized specific antibodies. By adding a secondary antibody, coupled to horseradish peroxidase (HRP), an immune complex sandwich is formed, consisting of the following layers (from bottom to top): immobilized antibody, (microbial) antigen, secondary antibody. This is followed by washing steps before the substrate tetramethylbenzidine (TMB) is added. The whole procedure ultimately alters the

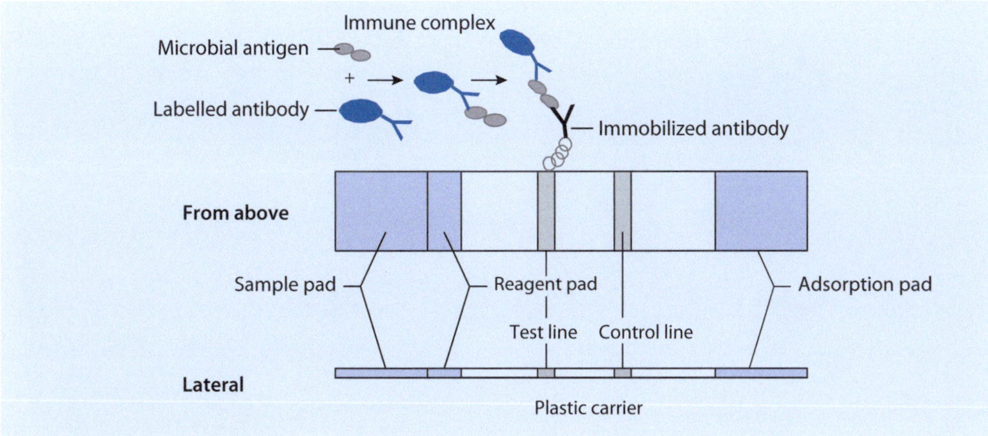

Fig. 9.3 Functional principle of an optical immunoassay

optic properties of the silicone surface and the layering thus created. The reflection of white light through this layer causes a golden hue to change to a color that appears purple to the naked eye. A positive result is therefore seen as a purple colored dot on a golden background.

9.2 Device format and quality

Simple LFA-test strips are available for numerous analytes whose results are read visually. Alternatively, for a restricted analyte spectrum, smaller automated detectors are utilized to read test strips but also give quantitative results. This includes, for example, devices in the size of cobas h 232 (Roche) (Fig. 9.4), the Triage MeterPro (Alere) (Fig. 9.5), the LABGEO IB10 (Samsung/Thermo Fisher) (Fig. 9.6), or the already mentioned m16 magnetic immunoassay analyzer (Edan Instruments).

Also larger benchtop-formats are available, for example the Stratus CS by Siemens, the PATHFAST by LSI Medience Corporation and the AQT90 Flex by Radiometer (Fig. 9.7). In the middle range are the i-STAT (Abbott) and the Minicare I-20 (Philips), which, as handhelds, are also suitable for immunological tests and can therefore be used for true "bedside testing". Automated benchtop devices are mainly used in hospital settings. The methods employed range from fluorescence and chromatographic detectors to enzyme immunoassays. In general, the tests are performed on whole blood, less often on saliva or urine, whereas serum and plasma can also be used [24].

A crucial problem in testing blood samples with test strips lies in the use of capillary blood. Here, a mix with interstitial fluid occurs when blood is taken, causing a change in concentration; the measured concentration of the analyte does not necessarily represent the blood concentration, which is problematic when tests lack sensitivity or the concentration is low, causing analytical limitations in the qualitative detection of the analyte. Furthermore, for most

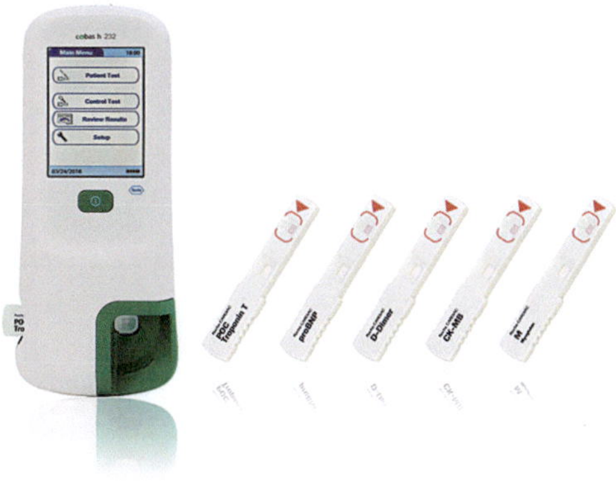

Fig. 9.4 cobas h 232 test strips and barcode scanner. (Courtesy of Roche Diagnostics, Mannheim, Germany)

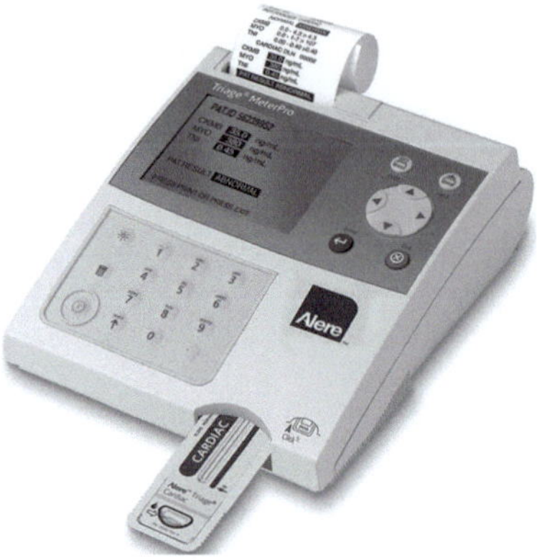

Fig. 9.5 Triage MeterPro. (Courtesy of Alere, Waltham, MA, USA)

analytes measured with test strips, valid reference ranges are missing [25].

Whilst numerous studies have been published about the quality and efficiency of automated detection devices, only limited data exists for the use of test strips, often just mandatory evaluation data or poorly comparable study data. For example, in Germany there are about 10 different Influenza POC tests available. The variation in the test sensitivity is between 50 % and 96 % and the specificity between 72 % and 100 %, which is dependent on the test itself, type of material tested and the age of the patient. Part of that disparity in results is explained by the varied study designs [20]. Specifically for virological diagnostics, the use of molecular biological test will expand in the future [23] (▶ Chapter 10).

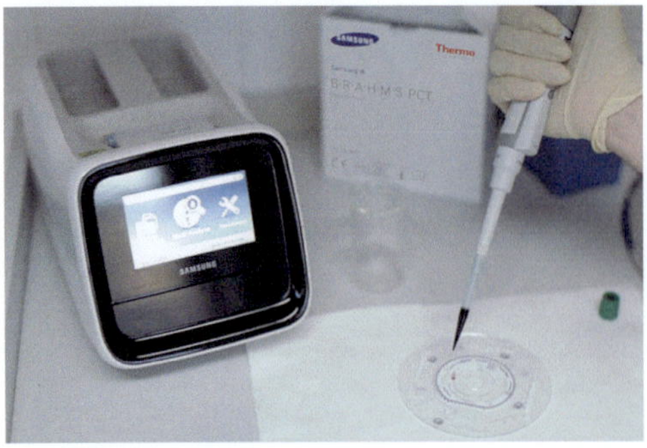

Fig. 9.6 LABGEO IB10. (Courtesy of Samsung Electronics Germany)

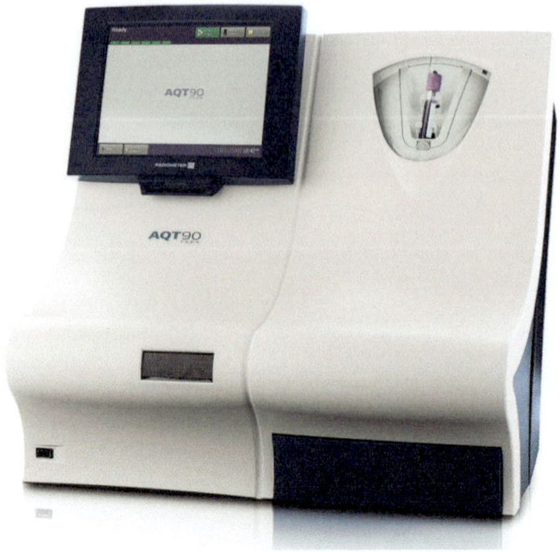

Fig. 9.7 AQT90 Flex. (Courtesy of Radiometer, Willich, Germany)

9.3 Areas of application

Immunological POCT methods are not only used in clinical settings, but also in general practice and for home testing. The increasing demand for this type of test in various markets, particularly the USA, has been documented many times [1, 13]. The limited spectrum of clinical chemical analyses has been established through various procedures in dry- and wet chemistry. This similarly applies to hematological and coagulation tests, where the focus is not on the extension of the analysis spectrum but on further methodological optimization. In contrast, immunological methods have a much higher development potential. Compared to other countries, such as the USA, the adoption in Germany is significantly lower due to licensing procedures and restricted reimbursement options for general practitioners (▶ Chapter 29 and ▶ Chapter 31).

One classic application is the point-of-care determination of **cardiac troponins** in emergency rooms, ICUs and cardiac catheter suites where therapeutic strategies are formulated based on the result of these markers. Presently, there are qualitative and quantitative tests for troponins T and I available as well as for creatinine kinase MB (CK-MB; mass) and myoglobin

[8]. Given the evidence on the superiority of troponins in diagnosing acute myocardial infarction, the latter parameters have, however, become less important [4].

In recent years, the **b-type natriuretic peptides** (BNP) brain natriuretic peptides and N-terminal NT-pro-BNP have been established as characteristic markers for chronic heart failure [17].

Immunological POC tests (rapid intraoperative immunoassays) are used even in surgery. Some examples include **parathyroid hormone** (PTH) measurements to intraoperatively document the outcome of parathyroidectomy by demonstrating a significant fall in PTH plasma concentration or to estimate the likelihood of postoperative hypocalcemia [22].

As mentioned earlier, immunological POCT methods are used in **infectious diseases** to detect pathogens or antibodies. For example, the diagnosis of streptococcal pharyngitis on clinical symptoms alone is not reliable and a conventional microbiological culture takes 1–3 days. Therefore, easy-to-perform streptococcus rapid tests have become increasingly widespread in recent years. Almost all tests have a high specificity (>95 %), while their sensitivity has increased in recent years thanks to technical improvements (>85 %) [18]. For the detection of HIV infections, the sensitivity of POCT methods is largely comparable with standard methods. Algorithms based on simultaneous application of two different tests can be compared with Western blots in terms of specificity.

POCT could help in acute diagnostics of infectious diseases that could affect public health, i.e. the general public, for example after a large-scale disaster like a (bio) terrorist attack by aiding the responsible rescuers onsite in making difficult ethical decisions in triaging the affected persons. Mainly immunological methods would be used. The same applies to diagnostics in the case of natural rare epidemics [7], like the 2014/2015 Ebola outbreak in Africa [12].

Further examples of immunological POCT applications are listed in ◘ Tab. 9.1 but make no claim to completeness. Attention is drawn again to the diagnostic importance of immunological methods in cardiovascular diseases, as will be discussed further in ► Chapter 17.

9.3.1 Hospital setting

The medical benefit conferred by immunological rapid tests depends on the individual clinical circumstances. For smaller hospitals without an emergency laboratory, POCT is often indicated for the assessment of cardiac markers but also CRP, pregnancy tests and several other analytes. If a laboratory is present in the hospital grounds, the need for POCT use would then depend, for instance, on the time taken for sample transport and transmitting the findings. POC tests for detection of antibodies as well as proteins like tumor markers and hormones, can usually be dispensed with in hospital settings and are only helpful in isolated cases, for instance to detect streptococci or for drug screening. The situation is, however, very different in third world countries. Further details are given in ► Chapter 35.

9.3.2 Physician practice setting

In the physician practice setting, POCT is being used less for medical emergencies than to improve organizational processes. This means that the patient does not need to re-schedule an appointment to discuss test results. In some cases, tests performed in the practice can be of immediate medical importance, for example, a CRP measurement prior to antibiotic therapy [27]. The use of immunological POC tests in the physician practice setting is presently restricted by regulations regarding cost reimbursement and invoicing. Usually the costs for material and generation of the test are much higher than what the health insurance covers.

Tab. 9.1 Examples of characteristic variables, which can be assessed at the POC using immunological methods (immunosensors, immunoassays, immunochromatographic tests).

Areas of application	Characteristic variables
Cardiac markers	Cardiac troponins, NT-pro-BNP, creatine kinase muscle-brain (CK-MB), myoglobin, interleukin 1 receptor-like 1 (ST2)
Coagulation	D-dimer
Pregnancy test	Human chorionic gonadotropin (HCG)
Acute phase, inflammation	C-reactive protein (CRP), procalcitonin, IL-6
Infectious diseases Antigen detection	Streptococci (group A and B, Pneumococcus), Clostridium difficile (toxin or antigen), Chlamydia spp., Neisseria gonorrhoeae, Influenza, Respiratory Syncytial Virus (RSV), Plasmodium spp., Legionellae, Rotavirus
Infectious diseases Antibody detection	Hepatitis, human immunodeficiency virus (HIV), Epstein-Barr virus (mostly heterophilic antibodies), Treponema pallidum hemagglutination assay (TPHA), Helicobacter pylori, Mycobacterium tuberculosis
Endocrinology	Follicle-stimulating hormone (FSH), luteinizing hormone (LH), estriol-3-glucuronide (E3G), pregnanediol glucuronide, parathyroid hormone (PTH)
Allergy, immunology	Immunoglobulin E (IgE), single allergen-specific IgEs
Rheumatology	Antistreptolysin, rheumatoid factor, antibodies against mutated citrullinated vimentin
Tumor markers	Prostate-specific antigen (PSA), fecal occult blood; nuclear matrix protein 22 (NMP 22) and bladder tumor antigen (BTA) in urine
Drug screening (mostly group screening with 4–10 single tests)	Amphetamines, barbiturates, benzodiazepines, buprenorphine, cocaine, methamphetamine, ecstasy, morphine, methadone, tricyclic antidepressants, cannabis
Miscellaneous	Transglutaminase/gliadin antibodies, lactoferrin

9.3.3 Home testing

The self-monitoring of blood glucose and INR in anticoagulation therapy has long been established and contributes significantly to the quality of patient care [5, 13, 21]. Fertility monitors in women with a minimum 2-year history of infertility have also proven successful [19]. Identifying the optimum time for conception during each menstrual cycle by measuring estriol-3-glucuronide and LH in urine with POCT improves the chance of pregnancy during the first two cycles in comparison to no fertility monitoring.

By contrast, the benefit conferred by other immunological rapid tests is less clear-cut. One of the main advantages of POCT, namely the rapid availability of test results for immediate therapeutic intervention, does not apply here. While patients who carry out self-testing subjectively like to know the result earlier, there is usually no medical need for this. The legitimate question arises whether tests for the likes of infectious pathogens, tumor markers or hormones are not better carried out in a central laboratory in order to avoid false-positive or false-negative results.

Note

When using POCT, it is to be considered that not only the test itself but also incorrect handling can lead to false results.

On the other hand, such self tests give patients greater health awareness.

9.4 Aptamers as adjuncts or alternatives to antibodies

Aptamers [3, 10, 16, 28, 29] are single-stranded RNA or DNA oligonucleotides (10–80 bases); the name combines Latin ("aptus", meaning fit) and Greek ("meros", meaning part). By virtue of their three-dimensional structure, aptamers bind to a specific target molecule, for example by adaptive binding, electrostatic interaction, hydrogen bond formation and base-stacking; their binding affinity is comparable to that of antibodies (in the pico or nanomolar range). Aptamers are therefore also called "chemical antibodies".

The aptamer with a high affinity and specificity to the target molecule is isolated from an oligonucleotide library of 10^{15}–10^{16} molecules by the SELEX technique (systematic evolution of ligands by exponential enrichment). Initially, the library molecules are incubated with the immobilized target molecule in multiple, repeated cycles, followed by later incubation cycles using the exponentially increasing number of oligonucleotides enriched for the target molecule. Non-bound molecules are removed by washing; bound oligonucleotides are eluted and amplified. Lastly, a few high-affinity aptamers remain, which after cloning are sequenced to determine their chemical composition. In general, aptamers can be generated against any target molecule, including metal ions, small organic molecules (e.g. amino acids, natural products, antibiotics), peptides, proteins, pathogens and even cells. ◘ Fig. 9.8 shows the diversity of aptamer applications.

The key advantages of aptamers over antibodies include:
- Highly reproducible, cost-effective and simple chemical synthesis with the ability for modification, including substitution with marker molecules such as chromophores and fluorophores
- High stability (e.g. thermostability)

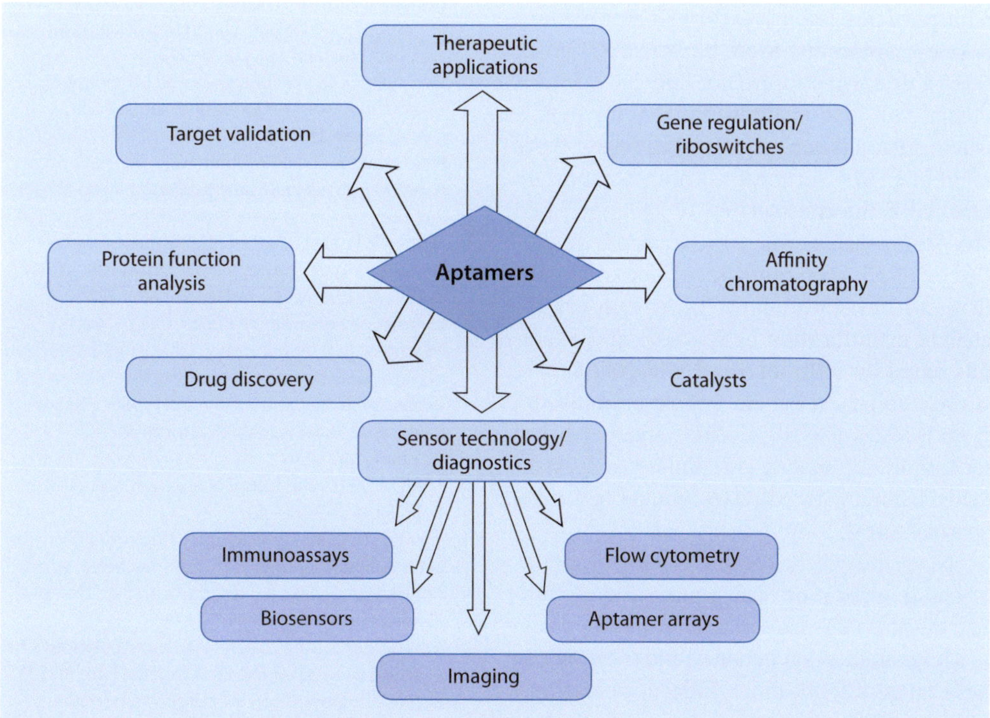

◘ **Fig. 9.8** Applications for aptamers. (Courtesy of A. Wochner)

- Can be reversibly denatured
- Bind to non-immunogenic substances
- Low or no toxicity
- Low or no immunogenicity
- Low metabolism rate in blood (DNA aptamers)
- Highly reproducible, cost effective and simple chemical modification to modulate pharmacokinetics and pharmacodynamics

Particularly the first six advantages enable aptamers to be applied broadly in general analytics but also in clinical laboratory diagnostics, including POCT.

Aptamers can generally be used in immunoassays and FACS analyses to replace antibodies as binding partners or detectors. Assays combining antibodies and aptamers can be used to achieve higher sensitivity. Assays based on aptamers can be developed to analyze substances hitherto inaccessible to antibody-based analysis.

A significant development potential is attributed to the **conjugation of aptamers and nanoparticles.** Intensive effort is being made to develop **aptamer-based biosensors,** which show a promising future not only for environmental and food analysis, but also for POCT, where aptamers can be used in all sensor types mentioned in ▶ Section 9.1. Another important pipeline is the construction of aptamer-based sensors **(aptasors)** where the analyte is quantified by FRET. Even more intensified will be the drive to develop aptamer-based sensors for analyte quantification by spectroscopic methods based on SPR, localized SPR (LSPR) and surface-enhanced Raman spectroscopy (SERS). It may be feasible to prioritize aptamer-based techniques for analyte enrichment over established laboratory medical techniques (e.g. mass spectrometry).

Nevertheless, the commercial use of aptamer-based applications has remained quite limited despite a 25-year aptamer era, recognition of a huge application potential and the resulting wide range of applications. Indeed, increasingly successful "proofs of concept" are being introduced with a view to aptamer use in laboratory diagnostics. The authors feel that aptamers will have a very promising future in the field of immunological diagnostics, particularly when aptamer applications are not exclusively restricted to the use of antibody-based assays, but open up new possibilities for laboratory medicine, especially POCT in order to provide access to new analytes or more test-specific and economically advantageous assays.

References

1. Agarwal R, Heinz T (2001) Bedside hemoglobinometry in hemodialysis patients: lessons from point-of-care testing. ASAIO J 47: 240–243
2. Briggs C, Guthrie D, Hyde K, Mackie I, Parker N, Popek M et al. British Committee for Standards in Haematology (2008) Guidelines for point-of-care testing: haematology. Br J Haematol 142:904–915
3. Bruno JG (2015) Predicting the uncertain future of aptamer-based diagnostics and therapeutics. Molecules 20:6866–6887
4. Collinson P (2015) Detecting cardiac events – state-of-the-art. Ann Clin Biochem 52:702–704
5. Dunn SG, Visnich MR (2002) Home-based point-of-care testing. In: Kost GJ (ed) Principles and practice of point-of-care testing. Lippincott Williams & Wilkins, Philadelphia, pp 376–390
6. Ekins RP (1999) Immunoassay and other ligand assays: from isotopes to luminescence. J Clin Ligand Assay 22: 61–77
7. Friedewald S, Finke EJ, Dobler G (2006) Patientennahe Diagnostik in Krisensituationen. J Lab Med 30: 211–221
8. Hafner G, Peetz D, Dati F (2003) Patientennahe Bestimmung der Troponine zur Diagnostik akuter Koronarsyndrome. J Lab Med 27: 279–287
9. Inverness medical (2006) BioStar OIA GC. An enhanced optical immunoassay for the rapid detection of Neisseria gonorrhoeae from female endocervical swabs and male urine specimens. Product insert, Revision 02. Inverness Medical Innovations, Waltham, MA, USA
10. Khati M (2010) The future of aptamers in medicine. J Clin Pathol 63:480–487
11. Klewitz TM (2005) Entwicklung eines quantitativen Lateral-Flow-Immunoassays zum Nachweis von Analyten in geringsten Konzentrationen. Doctoral dissertation, Hannover University
12. Kost GJ, Ferguson WJ, Hoe J, Truong AT, Banpavichit A, Kongpila S (2015) The Ebola Spatial Care Path™: Accelerating point-of-care diagnosis, decision making, and community resilience in outbreaks. Am J Disaster Med 10:121–143

13. Lehman CA, Giacini JM (2010) Point-of-care testing in the home and community environment: key ingredients for tomorrow´s community health. In: Price CP, St John A, Kricka LJ (eds) Point-of care testing, 3rd edition. AACC Press, Washington, pp 311–321
14. Luppa PB, Müller C, Schlichtiger A, Schlebusch H (2011) Point-of-care testing (POCT): Current techniques and future perspectives. Tr Anal Chem 30:887–898
15. Luppa PB, Bietenbeck A, Beaudoin C, Giannetti A (2016) Clinically relevant analytical techniques, organizational concepts for application and future perspectives of point-of-care testing. Biotechnol Adv. 2016 Jan 22. pii: S0734–9750(16)30004–0. doi: 10.1016/j.biotechadv.2016.01.003
16. Mairal T, Ozalp VC, Lozano Sánchez P, Mir M, Katakis I, O'Sullivan CK (2008) Aptamers: molecular tools for analytical applications.Anal Bioanal Chem 390:989–1007
17. Peetz D, Hafner G, Lackner KJ (2005) Patientennahe Bestimmung natriuretischer Peptide. J Lab Med 29: 219–228
18. Reinert RR (2007) Streptokokken-Schnelltests. J Lab Med 31: 280–293
19. Robinson JE, Wakelin M, Ellis JE (2007) Increased pregnancy rate with use of the Clearblue Easy Fertility Monitor. Fertil Steril 87:329–334
20. Schweiger B (2006) Influenza rapid tests – advantages and limitations. J Lab Med 30: 219–215
21. Siebenhofer A, Jeitler K, Horvath K, Habacher W, Schmidt L, Semlitsch T (2014) Self-management of oral anticoagulation. Dtsch Ärztebl Int 111:83–91
22. Sokoll LJ, Wians FH Jr, Remaley AT (2004) Rapid intraoperative immunoassay of parathyroid hormone and other hormones: a new paradigm for point-of-care testing. Clin Chem 50: 1126–1135
23. Takahashi H, Otsuka Y, Patterson BK (2010) Diagnostic tests for influenza and other respiratory viruses: determining performance specifications based on clinical setting. J Infect Chemother 16:155–161
24. Tang Z, Louie RF, Kost GJ (2002) Principles and performance of point-of-care testing instruments. In: Kost GJ (ed) Principles and practice of point-of-care Testing. Lippincott Williams & Wilkins, Philadelphia, pp 67–92
25. Thomas L (2012) Labor und Diagnose, 8th Edition, TH Books, Frankfurt/Main
26. Wolfbeis OS (2000) Fiber-optic chemical sensors and biosensors. Anal Chem 72: 81R–89R
27. Wood F, Brookes-Howell L, Hood K, Cooper L, Verheij T, Goossens H, Little P, Godycki-Cwirko M, Adriaenssens N, Jakobsen K, Butler CC (2011) A multi-country qualitative study of clinicians' and patients' views on point of care tests for lower respiratory tract infection. Fam Pract 28:661–669
28. Yonzon CR, Zhang X, Zhao J, van Duyne RP (2007) Surface-Enhanced Nanosensors. Spectroscopy 22:42–56
29. Zhou W, Huang PJ, Ding J, Liu J (2014) Aptamer-based biosensors for biomedical diagnostics. Analyst 139:2627–2640

Molecular biological tests

Enno Stürenburg, Norbert Gässler, Aline Schröder, Udo Reischl

P. B. Luppa, R. Junker (Eds.), *Point-of-Care Testing*
https://doi.org/10.1007/978-3-662-54497-6_10

10.1 Introduction

Molecular biological testing has become a mainstay in the repertoire of infectious disease diagnostics like in no other field of medicine. The major ambassador of this technology is the polymerase chain reaction (PCR). Its popularity arose from the speed and sensitivity with which PCR made it possible to ascertain the etiology of an infection. Classic microbiological diagnostics usually takes at least 36–48 hours before the first results from culturing and resistance testing are available. A long delay can be fatal for the patient, so starting empirical antibiotic therapy is generally the preferred strategy. Depending on the severity of the infection, broad-spectrum antibiotics are chosen, without confirmation of the causative pathogen or its antimicrobial resistance. The development of resistance is an undesired consequence of this antibiotic strategy [4].

Direct pathogen detection, not requiring a bacterial culture and only needing time for pure analysis, offers rapid and targeted diagnostics. In recent years, two main methods have established themselves for this direct detection, one of which is PCR. Besides PCR, immunochromatography is the other method used – often in the form of test strips (lateral flow assays) or test cards (▶ Chapter 9). A significant advantage of test strips is their great ease of use and speed coupled with moderate cost. Notwithstanding these selling points, the performance capability of older systems is only moderate. Moreover, specific antibody responses must generally take place (for antibody detection) or specific antibodies (for antigen detection) need to be available in the target organisms before immunological rapid tests can be established. In challenging situations, where the pathogen density is likely to be low, PCR is often a better choice because of its higher sensitivity [4]. PCR has the advantage that, in addition to its main pathogen detection function, resistance or virulence factors can be determined simultaneously. In the age of multidrug-resistant (nosocomial) pathogens, rapid and reliable molecular biological differentiation is becoming increasingly important given the urgency indicated to effectively isolate affected patients at the earliest possible chance.

10.2 Integrated and miniaturized systems

For a long time, PCR was seen purely as a laboratory method requiring lots of manual input. In the meantime, this has undergone a paradigm shift, with the list of molecular biological methods capable of point-of-care use expanding constantly. This expanded range of the POCT applications was made possible by the development of **single-use systems with integrated cartridges,** where analysis occurs in a closed plastic test carrier. With the reagents pre-packaged and ready-to-use in these test cartridges (unit use), hands-on work for the user of such systems is limited to loading the sample and starting the PCR run. First, the sample is mixed with lyophilized reagents, triggering digestion of the sample. Fluid movements cause the reagent mixture to then move through cast plastic arrays or channels into the reaction chambers where the next PCR process steps (DNA amplification and signal detection) occur. By varying the reagents, the test cartridges can easily be adapted to detect the various pathogens.

The more novel systems not only feature single-use cartridges, but are also distinguished by increasing **miniaturization**. This trend is ongoing for both test cartridges as well as for thermocouples and control electronics. Benchtop devices are now available that are no larger than a small coffee machine. If this trend continues, the next generation will be characterized by fully portable hand-held devices that are completely independent of all stationary technology. However, in the context of infectious diagnostics, the trend toward increasing miniaturization is not always seen as progress. Methodologically, tests using smaller aliquots tend to suffer from limits to the pathogen detection sensitivity achievable because of the mostly heterogeneous sample materials collected (swabs, aspirates, biopsy tissue or the likes) and the often very inhomogeneous distribution of the individual pathogens in the sample. This aspect should always be considered when miniaturizing microbiological diagnostic devices [5].

10.3 Selection criteria for POCT systems

In order to achieve significant improvements in medical care, a point-of-care diagnostic system must be integrated as smoothly possible into clinical processes. The main challenge is often that staff without any laboratory-specific training (e.g. in emergency departments or intensive care units) are tasked with carrying out these diagnostic tests. Under these circumstances, the danger is that a lack of time and (laboratory) experience can lead to unintended sample contamination and operational or diagnostic errors. These inherent risks should not be further exacerbated by making POCT technology too complex. Careful selection of diagnostic systems not only protects the patients but also reduces stress and responsibility for staff.

Note

The careful, coordinated selection of POCT systems for molecular biological diagnostics – adapted to the situation at the place of use – not only protects against diagnostic errors but also reduces responsibility and stress for staff.

It is easier to understand the relevance of certain selection criteria if the diagnostic approach is seen as a three-step sequence (pre-analysis, analysis and post-analysis) A positive aggregate result is only achieved if the characteristics of the chosen technology for each step of the process are relevant to the situation at the point of use (Fig. 10.1). With regard to pre-analysis, this means that the type and volume of the sample to be analyzed must conform to the test specification and that the sample can be used directly, without prior preparation. This prerequisite can only be achieved by PCR systems designed with integrated sample preparation; other systems are excluded from the outset (▶ Section 10.2). In medical microbiology settings, practical experience explicitly dictates that nucleic acid-based pathogen detection is only feasible in sample material that is very likely to contain traceable amounts of the target pathogen [5].

From an analytical viewpoint, technical complexity and hands-on time are critical factors alongside test speed (▶ Section 10.4). In the hectic and stressful routine of an ICU or central emergency room, only straightforward and robust technologies are feasible. The best solutions that minimize hands-on time are those systems where manual input is reduced to loading the sample and starting the reaction, which is otherwise fully automated. This walk-away function frees up staff to do other jobs. The output, reading and interpreting of results (post-analysis) should be clear and straight forward, not requiring further interpretation, e.g. "pathogen detected/not detected" or "mutation present/not present". More complex interpreta-

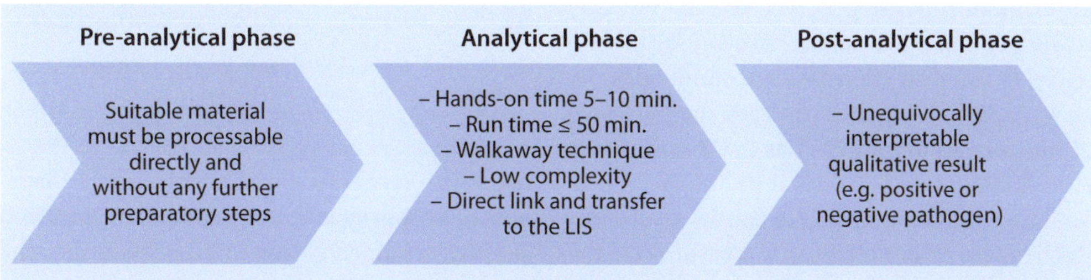

Fig. 10.1 Selection criteria for point-of-care molecular biological diagnostic systems

tion of results (e.g. results from multiplex PCR) should be the responsibility of a physician specialized in microbiology. The result of a PCR analysis not performed in a central laboratory should be transferred immediately and automatically to the laboratory information system to allow cumulative result interpretation (e.g. in microbiology and infectious diseases).

In ◘ Tab. 10.1, some examples of current molecular diagnostic systems are shown that largely meet the above-mentioned selection criteria and are suitable for point-of-care use by virtue of their test speed (▶ Chapter 10.4). There are also other new pipeline products from smaller companies with innovative reagents and/or device designs that are poised to prove their merits on the market. A prime example is the PDQeX 2400 by ZyGEM (Hamilton, New Zealand).

10.4 System concepts to shorten analysis time

Test speed is one of the most important criteria that qualifies a molecular diagnostics system for POCT. PCR, which involves three temperature steps, has been the methodological standard in nucleic acid amplification for many years. GeneXpert by Cepheid was the first point-of-care PCR system to reach market mature and, since its launch, has been the standard among the molecular systems [12, 13]. The range of available GeneXpert test cartridges has steadily increased in recent years to over twenty (◘ Tab. 10.1, ◘ Fig. 10.2). PCR tests take at least 45–60 minutes (even with the GeneXpert system), which makes them slower than test strips in terms of providing POCT results. For that reason, test manufacturers have worked intensively to develop even faster technologies. By cleverly optimizing the processes, the test speed of the conventional PCR has been ramped up even further.

The LIAT system (LIAT, lab in a tube) by Roche is one example of this type of acceleration (◘ Fig. 10.3) [8, 9]. The test cartridge is shaped like a small tube. This tube contains all

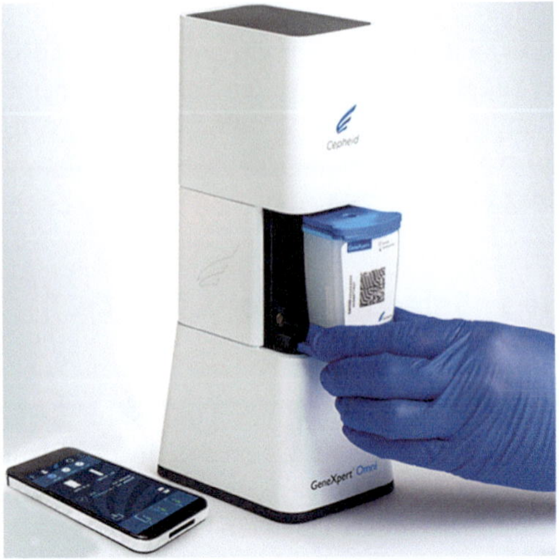

◘ **Fig. 10.2** GeneXpert Omni. (Courtesy of Cepheid, Sunnyvale, CA, USA)

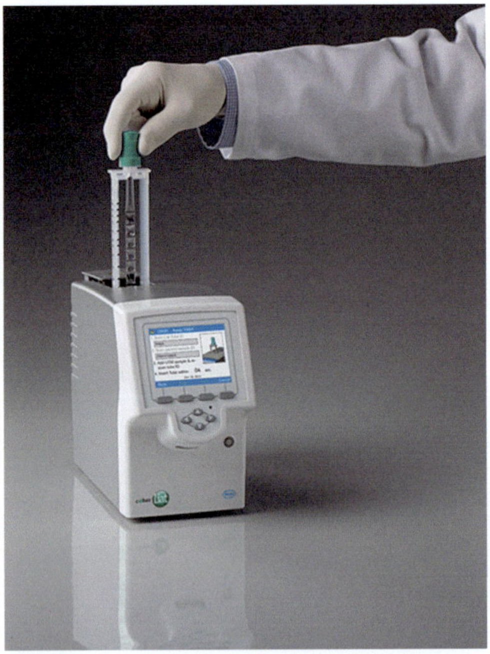

◘ **Fig. 10.3** LIAT. (Courtesy of Roche Diagnostics, Mannheim, Germany)

necessary reagents stored within small chambers of a proprietary plastic compartmentalized system, arranged in rows. After insertion, the sample passes through these chambers step by step. The fluid column of the reagent mix only moves up and down in the analyzer, allowing it to reach the different temperature zones of the PCR. Thanks to this simple principle of up-and-down movement, the PCR requires only a small space, making the reaction significantly faster [8, 9]. Depending on the pathogen-specific application, the LIAT system delivers definitive results within 20 minutes. It therefore more suitably fulfills the above-mentioned POC criteria and assists with rapid treatment decisions on site (◘ Tab. 10.1).

The io system by Atlas Genetics also reflects a PCR process optimization that accelerates the reaction process. Amplification products are measured in the io cartridge by a specific (electrochemical) detection reaction, which is so fast that it produces a PCR result after just 30 minutes [10, 11]. At present, the io system is mainly available for screening of sexually transmitted and nosocomial infections (◘ Tab. 10.1).

In addition to conventional PCR, a number of alternative **nucleic acid amplification techniques** are now on the market [1–3]. Historically, such processes were developed by diagnostics manufacturers who wanted to bypass PCR patent protection in the field of nucleic acid diagnostics in an era of extremely restrictive PCR licensing practices. Of the many PCR-alternative technologies, e.g. branched DNA signal amplification (bDNA), loop-mediated isothermal amplification (LAMP) and recombinase polymerase amplification (RPA), only a few have reached full market maturity to date and many of the assays that have been developed only address some selected niche applications.

The most successful and advanced PCR-alternative technology launched thus far is **isothermal amplification** [3]. Isothermal (Latin: equal heat) means that continuous amplification reactions occur at a constant temperature; this contrasts with conventional PCR, which relies on thermal cycles (three temperature steps). Not only are the whole process and test run times significantly shorter, but the instrumentation is also simpler, since the method only aims at reaching one constant temperature without the need for repeated thermal cycles.

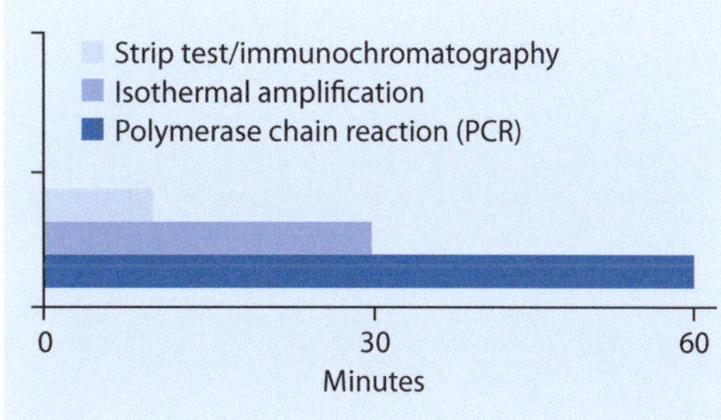

◘ **Fig. 10.4** Comparison of the test run times for conventional PCR vs. isothermal amplification

Compared to PCR, isothermal reactions inherently lose some specificity in the hybridization events. This is compensated, however, by optimizing conditions and through the addition of different enzymatic and biochemical reaction components – a modification which has proven diagnostically satisfactory across a broad range of applications [3]. The first commercial assays for use in POCT settings have now appeared on the diagnostics market. Influenza A and B as well as group A Streptococcus, for example, can be detected with the i system by Alere [6, 7]. Test run times are surprisingly short (less than 20 minutes) because the reaction occurs at a constant temperature (◘ Fig. 10.4).

10.5 Spectrum of molecular biological assays

The recent assays of newer systems are setting a clear trend towards the amplification of multiple pathogens in a single assay (**multiplex PCR**). The big advantage of such a broad approach is the expeditious and sensitive etiological diagnosis of patients' symptoms, which could be caused by various pathogens (e.g. patients with a cough or diarrhea). Further details on this topic are presented in (► Chapter 20). The FilmArray system by bioMérieux is a prime example of testing for "syndromic diagnosis". The panels allow the detection of numerous pathogens, which can cause respiratory, gastrointestinal illnesses and infections alongside the simultaneous detection of resistance markers

Tab. 10.1 Overview of point-of-care molecular biological diagnostic systems (numerical values represent analysis times in minutes)

System/ manufacturer	Method	Single test/duplex test/multiplex test	Sexually transmitted infections									Respiratory infections		
			Human immunodeficiency virus (HIV)a	Hepatitis B virus (HBV)	Hepatitis C virus (HCV)	Human papillomavirus (HPV)	Group B streptococci	Chlamydia trachomatis	Neisseria gonorrhoeae	Trichomonas vaginalis	Mycoplasma genitalium	Influenza A and B	Respiratory Syncytial Virus (RSV)	Group A-Streptococci
Alere i	ITA	ST, DT										15		8
Roche Cobas LIAT	PCR	ET										20		20
		MP										20	20	
Atlas io	PCR	ST						30	30	30	30			
		MP						30	30	30	30			
Cepheid GeneXpert	PCR	ST, DT	90		105	60	52	90	90	40		75		
		MP										40	40	
Spartan RX	ARR													
bioMérieux FilmArray	ARR		**Meningitis/encephalitis panel:** E. coli K1, Haemophilus influenzae, Listeria monocytogenes, Neisseria meningitidis, Streptococcus agalactia											
			Respiratory panel: Adenovirus, Coronavirus (229E, HKU1, OC43, NL63), human Metapneumovirus, human Rhinovirus/Enterovirus,											
			Gastrointestinal panel: Campylobacter (jejuni, coli & upsaliensis), Clostridium difficile (toxin A/B), Plesiomonas shigelloides, Salmonella, Yersinia enterocolitica, Vibrio (parahaemolyticus, vulnificus & cholerae), E. coli O157, enteroaggregative E. coli (EAEC), enteropathogenic E. coli (EPEC),											

ITA Isothermal amplification; PCR Polymerase chain reaction; ARR Microarray with upstream PCR; ST Single test; DT Duplex test; MP Multiplex test.
[a] Also quantitative, as IVD-compliant virus load assessment

	Life-threatening nfections			Nosocomial infections					Non-pathogen-specific detection			
	Ebola virus	Enterovirus in cerebrospinal fluid	M. tuberculosis/rifampicin-resistance	Norovirus	Clostridium difficile	Methicillin-resistant S. aureus (MRSA)	Vancomycin-resistant enterococci (VRE)	Carbapenem-resistant Enterobacteriaceae	Factor II and factor V mutation	BCR-ABL transcription products	Cytochrome P450 allele 2C19	References
												[6, 7]
												[8, 9]
				30	30	30						[10, 11]
	98	150	117	60	45	60	45	50	30	140		[12, 13]
											60	[14]
	and pneumoniae, CMV, Enterovirus, HSV 1 und 2, HHV6, human Parechovirus, VZV, Cryptococcus neoformans											[15, 16]
	Influenza A (A/H1, A/H1–2009, A/H3), Influenza B, Parainfluenza 1–4, RSV, Bordetella pertussis, Chlamydophila pneumoniae, Mycoplasma pneumoniae											
	enterotoxigenic E. coli (ETEC) lt/st, shiga-like toxin-producing E. coli (STEC) stx1/stx2 E. coli O157, Shigella/enteroinvasive E. coli (EIEC), Adenovirus F 40/41, Astrovirus, Norovirus GI/GII, Rotavirus A, Sapovirus (I,II, IV, and V), Cryptosporidium, Cyclospora cayetanensis, Entamoeba histolytica, Giardia lamblia											

(◘ Tab. 10.1) [15, 16]. This parallel detection, however, leads to the greater need for interpretation of results and, for now, is still best suited for laboratory use.

Simpler systems more suitable for deployment near the patient are still dominated by PCR assays for single pathogens or two pathogen variants (e.g. influenza A and B or vancomycin-resistant vanA and vanB with Enterococci genes); such duplex PCR approaches can be advantageous to solve some microbiological problems (◘ Tab. 10.1). Linking the current assays to medical specialties, it is striking that it is the field of molecular biology that has opened up such numerous new areas of application. Five years ago, the assays were used almost exclusively to detect sexually transmitted, nosocomial and respiratory infections. More recently, they have expanded to include infection diagnostics (e.g. detection of hepatitis B and C, some as quantitative and IVD-compliant virus load assessment as well) alongside the development of very promising applications outside of microbiology (◘ Tab. 10.1).

Beyond infectious disease medicine, among the specialties currently benefiting from new-generation products are oncology (BCR-ABL transcription), coagulation centers (detection of factor II and factor V mutations) and cardiology (cytochrome P450 allele 2C19) (◘ Tab. 10.1). The latter-named assay is the first that combines PCR and array hybridization (microarray) for point-of-care use. A cytochrome-P450 2C19 variant allele, which genetically determines the response to clopidogrel treatment, can be detected within 60 minutes [14]. Given its the highly complex nature (the assay was categorized by the FDA as "high complexity" in the American approval), it remains to be seen whether testing for CYP2C19 will become established for direct point-of-care applications.

10.6 Synopsis for infectiology

Molecular biological diagnostic systems for point-of-care testing are now being used for a diverse range of infectious diseases. The operation of devices is becoming increasingly user friendly, the technical workflows more reliable, while devices continue to become more compact. Molecular biological testing is now even evolving into point-of-care applications for non-pathogen-related problems. Isothermal amplification technologies and process optimization in conventional PCR have significantly shortened test run times further yet, where the newest generation of tests can deliver results after just 15–30 minutes. Research and development departments in the diagnostics industry have focused intensively on this dynamic market segment: There is no doubt that new test concepts can be anticipated in the foreseeable future.

References

1. St John A, Price CP (2014) Existing and emerging technologies for point-of-care testing. Clin Biochem Rev 35:155–167
2. Niemz A, Ferguson TM, Boyle DS (2011) Point-of-care nucleic testing for infectious diseases. Trends Biotechnol 240–250
3. Craw P, Balachandran W (2012) Isothermal nucleic acid amplification technologies for point-of-care diagnostics: a critical review. Lab Chip 12:2469–2486
4. Stürenburg E, Junker R (2009) Patientennahe Diagnostik in der Mikrobiologie. Dtsch Ärzteblatt 106:48–54
5. Reischl U, Drosten C, Geißdörfer W, Göbel U, Hoffmann KS, Mauch H, Meyer T, Moter A, von Müller L, Panning M et al. (2011) MiQ 1: Nukleinsäure-Amplifikationstechniken (NAT), 3rd Edition, In: Mikrobiologisch-infektiologische Qualitätsstandards (MiQ) Qualitätsstandards in der mikrobiologisch-infektiologischen Diagnostik. Urban & Fischer, Munich, pp 1–80
6. Nie S, Roth RB, Stiles J, Mikhlina A, Lu X, Tang YW, Babady NE (2014) Evaluation of Alere i Influenza A and B for rapid detection of influenza viruses A and B. J Clin Microbiol 52(9):3339–3344
7. Cohen DM, Russo ME, Jaggi P, Kline J, Gluckman W, Parekh A (2015) Multicenter clinical evaluation of the novel Alere i Strep A isothermal nucleic acid amplification Test. J Clin Microbiol 53(7):2258–2261
8. Scott L, Gous N, Carmona S, Stevens W (2015) Laboratory Evaluation of the Liat HIV Quant (IQuum) whole-blood and plasma HIV-1 viral load assays for Point-of-Care Testing in South Africa. J Clin Microbiol 53(5): 1616–1621

9. Chen L, Tian Y, Chen S, Liesenfeld O (2015) Performance of the Cobas Influenza A/B Assay for rapid PCR-based detection of influenza compared to Prodesse ProFlu+ and viral culture. Eur J Microbiol Immunol (Bp) 5(4): 236–245
10. Pearce DM, Shenton DP, Holden J, Gaydos CA (2011) Evaluation of a novel electrochemical detection method for Chlamydia trachomatis: Application for Point-of-Care diagnostics. IEEE Trans Biomed Eng 58(3): 755–758
11. Pearce DM, Styles DN, Hardick JP, Gaydos CA (2013) A new rapid molecular point-of-care assay for Trichomonas vaginalis: preliminary performance data. Sex Transm Infect 89(6): 495–497
12. Goldenberg SD, Bisnauthsing KN, Patel A, Postulka A, Wyncoll D, Schiff R, French GL (2014) Point-of-Care Testing for Clostridium Difficile infection: A real-world feasibility study of a rapid molecular test in two hospital settings. Infect Dis Ther 3(2):295–306
13. Parcell BJ, Phillips G (2014) Use of Xpert MRSA PCR point-of-care testing beyond the laboratory. J Hosp Infect 87(2):119–121
14. Stimpfle F, Karathanos A, Droppa M, Metzger J, Rath D, Müller K, Tavlaki E, Schäffeler E, Winter S, Schwab M, Gawaz M, Geisler T (2014) Impact of point-of-care testing for CYP2C19 on platelet inhibition in patients with acute coronary syndrome and early dual antiplatelet therapy in the emergency setting. Thromb Res 134(1):105–110
15. Spina A, Kerr KG, Cormican M, Barbut F, Eigentler A, Zerva L, Tassios P, Popescu GA, Rafila A, Eerola E, Batista J, Maass M, Aschbacher R, Olsen KE, Allerberger F (2015) Spectrum of enteropathogens detected by the Film-Array GI Panel in a multicentre study of community-acquired gastroenteritis. Clin Microbiol Infect 21(8):719–728
16. Babady NE (2013) The FilmArray respiratory panel: an automated, broadly multiplexed molecular test for the rapid and accurate detection of respiratory pathogens. Expert Rev Mol Diagn 13(8):779–788

Non-invasive analysis

Peter B. Luppa, Sandeep K. Vashist, John H.T. Luong

P. B. Luppa, R. Junker (Eds.), *Point-of-Care Testing*
https://doi.org/10.1007/978-3-662-54497-6_11

11.1 Glucose monitoring

11.1.1 Introduction

For a long time, there has been a demand for non-invasive analytical methods to achieve effective glucose monitoring in diabetics. A painless analysis would help persons suffering from diabetes to measure their glucose concentration more frequently or even continuously. Fresh impetus in the technical development of non-invasive glucose monitoring (NGM) recently came from the latest trends in data transfer and wearables (e.g. small wristbands equipped with special microchips and a variety of sensors for fitness enthusiasts). In 2014, the introduction of Google's prototype of a smart contact lens drew the public's attention to this topic (see below). Nevertheless, the success of such sensors always depends on the underlying analytical technology. Such advances are currently still causing problems in terms of precision, reliability and calibration. For now, the non-specific interferences observed in individual patients also militate against the use of NGM. Yet, technological advances, especially in the past five years, give grounds to hope for a technological breakthrough in the not too distant future.

11.1.2 Non-invasive optical techniques

Infrared spectroscopy and attenuated total reflectance

The application of optical methods to examine sites on the skin is a very attractive approach to NGM. In this context, attenuated light represents the optical signature of the illuminated tissue. The resulting spectrum strongly depends on the biochemical composition and distribution of the skin, but also on temperature and local blood circulation. That makes it challenging to derive quantified glucose-specific information from such spectra. To date, sensors in the infrared (IR), mid- (MIR) and near-infrared (NIR) range have been developed for glucose detection.

Although the dermis is the target region for spectroscopy, glucose molecules are predominantly found in the interstitial fluid at the junction between epidermis and dermis. The interstitial glucose concentration is proportional to the blood concentration, albeit with a delay of 10 min. In this method, **NIR** and **MIR spectroscopy** are applied on the earlobe, finger pulp, volar forearm or labial mucosa. OrSense NBM 200G (OrSense, Raleigh, NC, USA), a CE-certified wearable device utilizing occlusion red/near-infrared spectroscopy [9], works with a short-term cessation of the blood flow of the finger pulp and projects light at wavelengths of 610 and 810 nm to amplify the red/NIR signal. The measurement takes approximately 1 min and allows for quasi-continuous glucose monitoring for one day without recalibration. Two other NIR spectroscopy systems have been developed as well: SugarTrack (USA) uses light at wavelengths of 650, 880, 940 and 1300 nm, while the Sensys monitor (Sensys Medical, formerly Chandler, AZ, USA) works at 750–2500 nm. Furthermore, a wearable NGM system deserves mention that measures blood glucose in the finger pulp within a second [12].

Attenuated total reflectance (ATR) applies an optical waveguide which conducts the light in total reflectance. For this purpose, a special ATR crystal increases penetration depth [20]. Beyond the reflecting interface, total light reflectance induces evanescent waves within a range of approximately one wavelength. For measurement, the crystal is applied to skin treated with squalane oil [27]. The glucose concentration in the dermal interstitial fluid is derived from the fluid's interaction with the evanescent wave as based on the attenuation of light conducted in the waveguide [4]. On the other hand, MIR-based glucose sensing is strongly affected by the skin's water content.

Reverse iontophoresis

In reverse iontophoresis (RI), uncharged glucose molecules in the interstitial fluid along with ions from dermal tissue are drawn to a cathode where the concentration is determined by a conventional electrochemical glucose sensor [31]. This analytical principle was implemented in the GlucoWatch G2 Biographer (Cygnus, formerly Redwood City, CA, USA), a CE and FDA-approved NGM device worn like a wristwatch. During wear, the skin's interstitial fluid is iontophorized by a current of 300 μA between two electrodes attached to the skin [28]. Skin temperature and moisture are detected by thermal and conductive sensors. The GlucoWatch requires a warm-up phase of 2 to 3 hours and can perform up to six measurements per hour for about 12 hours [18, 29]. Additionally, it only must be calibrated before initial use. The disadvantages of the system are that it can cause skin irritation and also shuts off when the subject sweats excessively, even though increased perspiration might be a sign of hypoglycemia and measurements would certainly be helpful in such situations.

Another non-invasive glucose monitoring device based on laser microporation technology was launched by SpectRx (San Diego, CA, USA). A hand-held laser places micropores in the skin. The leaking interstitial fluid is collected in a special patch and channeled to a glucose sensor. The Symphony CGM (Echo Therapeutics, Iselin, NJ, USA) works in a similar fashion by collecting interstitial fluid through a special transdermal permeation system. After a short warm-up phase, the glucose concentration can be measured every minute.

Bioimpedance and Raman spectroscopy

Bioimpedance spectroscopy measures the tissue's impedance spectrum at different wavelengths within the frequency range of 100 Hz to 100 MHz using alternating currents of known intensity. Changes in the plasma glucose concentration alter the Na^+ and K^+ concentrations, thereby affecting the membrane potential of RBC. The membrane potential can be determined by measuring the impedance spectrum [7, 19]. However, the performance of this non-invasive analysis method is affected by water content and multiple diseases affecting the cells' membrane potential. Pendra, a proprietary POCT device, incorporated such bioimpedance spectroscopy (◘ Fig. 11.1a). It was a CE-certified NGM device worn like a wristwatch (Pendragon Medical Ltd., Switzerland) [32]. Unfortunately, more than 30°% of the patients had to stop using the monitor as their skin types and basic skin impedances were unsuitable. Overall, the measurements also showed poor correlation with conventional glucose meters and, therefore, the product was withdrawn from the market.

Raman spectroscopy utilizes laser radiation in the visible and MIR range to measure scattered light from the illuminated transparent sample. This scattered light has a higher wavelength but lower intensity than the original light [31]. In contrast to IR spectroscopy, there is no specific interference of water with the glucose Raman spectra. Therefore, informative peaks are easy to isolate. However, this method requires long spectral acquisition times and the relative instability of the laser intensity and wavelength is another limitation. The HG1-c (◘ Fig. 11.1b), a compact device by C8 MediSensors (formerly of San Jose, CA, USA) that performed measurement in 3 min. had been CE-certified, but was withdrawn from the market in 2012. Another miniaturized Raman POCT device [2] (◘ Fig. 11.1c) measures glucose concentration in interstitial fluid by radiating the finger pulp with NIR light. The penetration depth in the skin is approximately 0.5 mm. The device also features algorithms to calculate blood glucose from the interstitial concentration based on the glucose diffusion from blood into the interstitium [2].

Photoacoustic and ocular spectroscopy

Energy input into the skin by fast pulsing of laser light leads to continuous alternation between heating and cooling, causing subsequent expansion and contraction of the skin. This vibration is detectable as an acoustic signal by a piezoelectric transducer. This method allows blood glucose to be selectively detected by laser light in a wavelength range from ultraviolet to NIR, without any interference by water [31]. Using conventional methods, the Aprise device

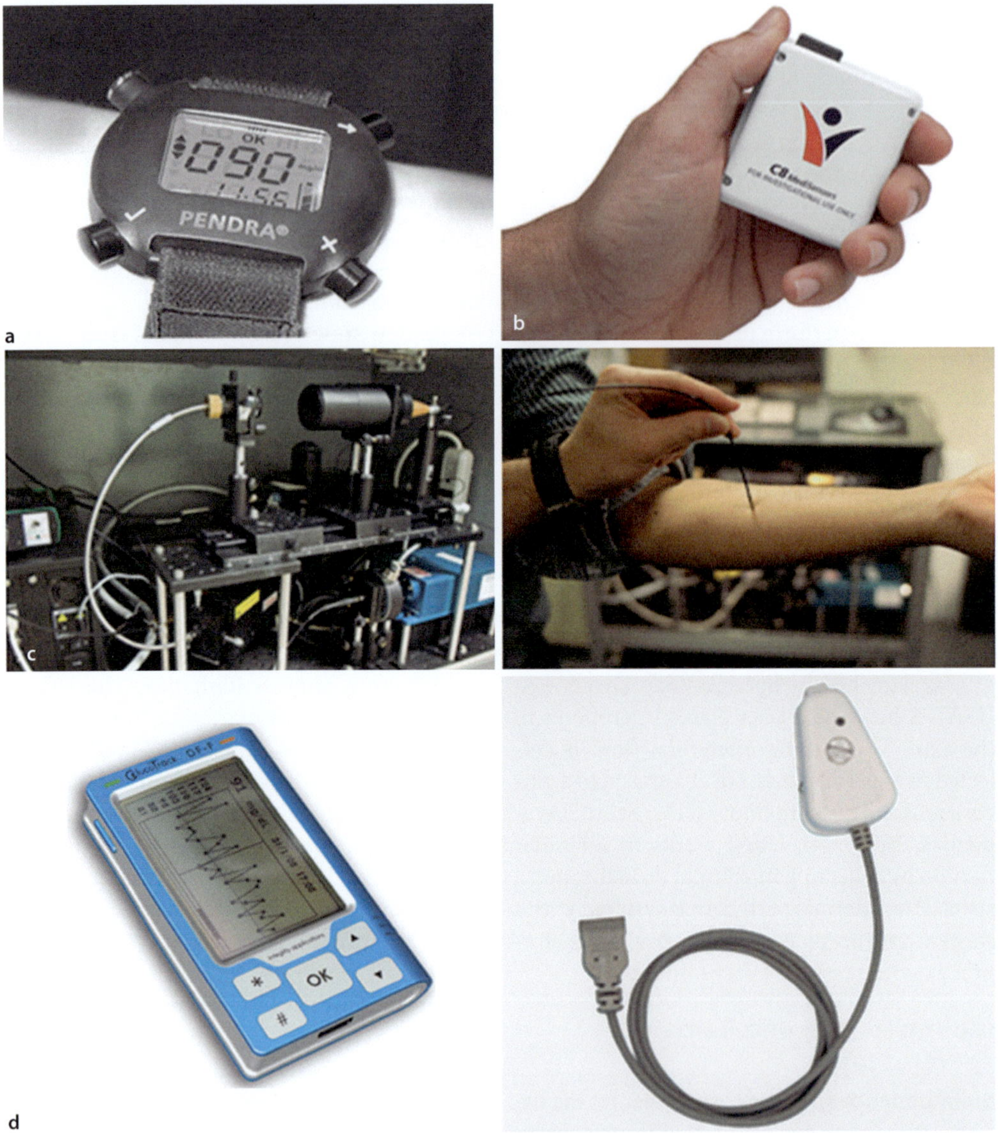

Fig. 11.1a–d Non-invasive analysis methods for POCT devices. **a** Pendra NGM device. **b** HG1-c, Continuous NGM device, manufactured by C8 MediSensors. **c** Prototype of a Raman spectroscopy-based NGM system (*left*) and application in the interstitial fluid (*right*). **d** GlucoTrack NGM analyzer: Main device (*left*) and ear clip for patients (*right*)

by Glucon (Boulder, CO, USA) shows good correlation with measured blood glucose levels, but has mediocre analytical sensitivity, is prone to multiple non-specific interferences and shows a relatively strong temperature dependence.

Spectroscopy of tear fluid can also be regarded as a non-invasive method of glucose monitoring. Google (Mountain View, CA, USA) designed a two-layered contact lens prototype with a sensor and miniature radio chip between the layers. Glucose measurements in the tear fluid can be taken at very short time intervals and the data can be transferred to a special smartphone app [1]. The contact lens is illuminated and the wavelength change in the reflected light attributable to existing glucose is measured spectroscopically. Even though the approach sounds very innovative, the limitations of tear fluid monitoring are evident: Delayed changes of glucose concentration in tears compared to blood, different values in each eye, problems for individuals with insufficient lacrimal fluid production and poor wearing comfort due to relatively rigid lenses.

Ultrasound, thermal emission spectroscopy and electromagnetic sensing

Ultrasound-based technology is based on a piezoelectric transducer creating a 20 kHz ultrasound pulse that, by increasing the dielectric conductivity, transports glucose in the epidermis to the interstitial fluid where the glucose concentration can easily be measured by a conventional electrochemical sensor [31].

By contrast, **thermal emission spectroscopy** is based on the measurement of IR signals naturally emitted from the human body as a function of glucose concentration [31]. Monitoring is done at glucose-specific wavelengths on the forearm, finger pulp or earlobe. Although its performance characteristics are good, this non-invasive technique is disrupted by body temperature fluctuations and body movement.

Electromagnetic sensing is based on the changes in dielectric parameters in the blood in response to different glucose concentrations. An electromagnetic sensor detects eddy currents. These are currents induced in an electric conductor by a magnetic field varying in time [31]. However, the skin reflects most electromagnetic waves. Additionally the measurements are affected by varying temperature and physiological changes to the dielectric parameters in the blood. The successful NGM device GlucoTrack (Integrity Applications Ltd, Israel) (◘ Fig. 11.1d) simultaneously incorporates the described principles of ultrasound, electromagnetic sensing and heat capacity. This combination minimizes interference and promises high precision and correctness in glucose monitoring on the earlobe. The device requires a monthly, personalized, individual calibration against conventional blood-based glucose measurements (basal and postprandial). The calibration and algorithm for data processing still need further technical improvement.

GlucoWise from MediWise (London, UK) utilizes metamaterial thin-film layers which allow deeper skin penetration by high-frequency radio waves (≈65 GHz).

11.1.3 Non-invasive techniques that have not gained acceptance

Many other analysis methods developed for glucose monitoring have not yet gained acceptance on the market due to poor accuracy of the method, inferior correlation between measured glucose values and blood glucose concentration or inferior practicability in patients. Some of these methods are briefly described below.

Temperature-regulated localized reflectance, measurement of metabolic heat conformation and optical coherence tomography

Temperature-regulated localized reflectance analyzes the changes in the refractive index of the skin on the forearm that influences the scattered light at 590 and 935 nm as a function of glucose concentration [31]. However, this method is limited by multiple physiological parameters, but also by fluctuating body temperature.

To monitor glucose concentration, the **metabolic heat conformation** method measure temperature, blood flow and Hb and O_2Hb concentrations [29]. Initially, the temperature at the fingertip, in the ambient room and the background radiation are measured. Then, multi-wavelength spectroscopy is performed to detect glucose. This method is likewise affected by various external influences.

Optical coherence tomography measures the glucose concentration of the skin's interstitial fluid by comparing the delay in scattered light with the light reflected by a mirror [31]. With increasing glucose levels, the refractive index of the interstitial fluid rises, resulting in a change in the scattering coefficient. Here, too, limitations to the analysis are dictated by varying skin temperature and bodily movements.

Polarimetry and fluorescence

In **polarimetry**, polarized light is passed through a solution containing optically active solutes such as glucose. The method is based on the rotation of the linear polarization vector of the passing light as a function of the sample's thickness, temperature and analyte concentration. Nevertheless, skin is a relatively unsuitable medium for such measurements. This is due to the high scattering coefficient that causes complete depolarization of the light beam. The aqueous humor in the anterior chamber of the human eye is more suitable because it is an appropriately clear optical medium. An incident light beam on the cornea traveling into the eyeball is reflected by the retina and returns with information pertaining to the glucose concentration in the aqueous humor [34]. Although this method is not affected by temperature and pH, it fails because of other analytical problems and the inferior practicability of the method.

11

Other non-invasive analysis methods utilize **fluorescence measurements** detected after excitation of the tissue with UV light at specific wavelengths. To measure glucose in tear fluid, polarimetry utilizes polymerized crystalline colloidal arrays that bend visible light as a function of glucose concentration [8]. Current efforts are focused on designing contact lenses that change color in response to the glucose concentration. This would obviate the use of fluorescence detection devices. Another experimental analytical approach relies on an ultraviolet laser to detect the resulting fluorescence signal at 380 nm after excitation of the glucose solution [15]. This fluorescence depends on epidermal thickness, skin pigmentation and other parameters [3].

11.2 Bilirubin testing in newborns

Transcutaneous bilirubin measurement is widely used for screening in both outpatient and hospital settings. There are two such devices on the German market:

- Konica Minolta/Air-Shields JM 103 Jaundice Meter (Dräger Medical)
- Spectrx BiliCheck (Philips Respironics).

Other devices that are not all available in Germany include: Bilimed (Nufer Medical, Guemligen/Bern, Swiss) and Bilitest Technomedica (Lacteromedik, Russia). Detailed information about the devices are not available but two evaluation reports have been published [5, 11].

Although their measuring principle is the same, JM 103 and BiliCheck implement it in different ways. The JM 103 Jaundice Meter uses light at two wavelengths (450/550 nm) that penetrate the skin at different optical paths and depths. There, the light is scattered and partially reflected. The reflected portion is measured with 2 photodiodes. BiliCheck uses white light that is divided after scattering and reflection into multiple spectra which are each measured by individual photodiodes. Both methods try to eliminate the impact of Hb, melanin and the skin's degree of maturity to obtain a "bilirubin-specific" signal.

Detailed evaluation reports have been published on the two devices [14, 17, 21, 33, 35]; their main characteristics are summarized in Tab. 11.1. Starting in the 35th week of gestation, the results of both devices are comparable in newborns of European heritage [14]. The difference to plasma results increases with higher concentrations. Consequently plasma levels of 15 mg/dL can be expected to deviate by 2–3 mg/dL. In premature infants, the deviations are higher. Although both manufacturers state that the results are independent from skin color, BiliCheck appears to produce more reliable results in dark-skinned newborns.

Unlike with the JM 103 Jaundice Meter, manufacturer states that the BiliCheck can also

Tab. 11.1 Meters for transcutaneous bilirubin determination (manufacturer's specifications)

Parameter	JM 103 Jaundice Meter	BiliCheck
Weight	150 g	350 g
Light source	Xenon flashbulb	Tungsten lamp
Optics	Measurement at 2 wavelengths	Measurement at multiple wavelengths
Measuring range	0–20 mg/dL	0–20 mg/dL
Calibration	Non-adjustable; can be tested by testing device in the charging station	By attaching the calibration tip
Measurement	By skin contact; adjustable for 1–6 measurements	By skin contact with calibration tip (single-use material); 5 measurements required
Limitations of use	Only for newborns born after the 35th week of gestation; not for phototherapy	For newborns with a weight >1000 g; also for phototherapy on a covered site (special patch)
Maximal difference to plasma results	±3 mg/dL	±3 mg/dL

be used after phototherapy if the measurement site was previously protected by a photo-opaque patch. However, details are missing on how to evaluate these values; robust investigations on the issue of "transcutaneous bilirubin measurements and phototherapy" have only been rudimentary to date [26]. Starting from a transcutaneously measured value of about 3 mg/dL below the respective limit for phototherapy, a clinical-chemical bilirubin determination is recommended to ensure a safe decision about the use of phototherapy (► Chapter 22).

11.3 Pulse oximetry

On intensive care units, in anesthesiology and emergency medicine, pulse oximetry is part of a patient's standard monitoring. It permits the measurement of pulse frequency and arterial oxygen saturation (SpO_2) [30].

The pulse oximeter, usually attached to the patient's finger or earlobe as clip or adhesive sensor, consists of a photocell and 2 LED that emit light in the red and infrared wavelength range (660/940 nm). Based on the different molar extinction coefficients of O_2Hb and deoxyhemoglobin (deoxy-Hb, HHb) at these wavelengths, the relative proportion of HbO_2 to total Hb, i.e., the oxygen saturation, can be calculated from the absorption values. The measured optical absorption consists of a constant portion, the tissue absorption and a portion fluctuating with the pulse wave rhythm. Only the latter is used to calculate the oxygen saturation. At values >70°%, measured and real arterial O_2 saturation match well. Below that, the measured values are much less accurate.

The relationship between oxygen partial pressure and oxygen saturation is not linear, but follows the known S-shaped dissociation curve. That is why paO_2 changes in the hyperoxic range are hard to detect by SpO_2 measurement.

Important sources of error are motion artifacts, inadequate perfusion at the application site, toxic hypoxia caused by dyshemoglobins (COHb, MetHb) and abnormal Hb variants (e.g. Hb Cologne, Hb Bonn; ► Chapter 14) [36].

Numerous manufacturers supply oxygen saturation meters featuring various designs. As an important development, the company Masimo (Irvine, CA, USA) has been producing the

Rad-57 handheld pulse CO-oximeter to detect increased concentrations of COHb and MetHb since 2007. Based on the rainbow SET technology, the device measures at 7 wavelengths to be able to calculate the percentage portion of COHb and MetHb. The manufacturer offers product experience reports.

11.4 Partial pressures of carbon dioxide and oxygen

For more than 20 years, transcutaneous (tc) pCO_2 and pO_2 measurements have been established methods, especially in neonatology. Today, both blood gases are simultaneously measured with a combination electrode. $tcpCO_2$ is measured using the Stow-Severinghaus sensor, while $tcpO_2$ measurement utilizes dynamic fluorescence quenching. On some devices, the oxygen saturation can be measured instead of or in addition to pO_2 (e.g. OxiVenT by Sentec (Therwill, Swiss) or TCM CombiM Monitor by Radiometer (Willich, Germany). These measurements are used to control ventilators (▶ Chapter 13).

11

The electrode is air-tightly attached to the skin and heated up to a temperature of 42–44° C. Then, it measures the diffusing blood gases, while the electrode's temperature and a minor influence from the skin's metabolism must be accounted for. A more or less distinct erythema will form at the probe site. Therefore, the measurement site should be changed every 4 hours to avoid burns to the skin. In premature infants, this time might need to be shortened and the electrode's temperature lowered (◘ Fig. 11.2).

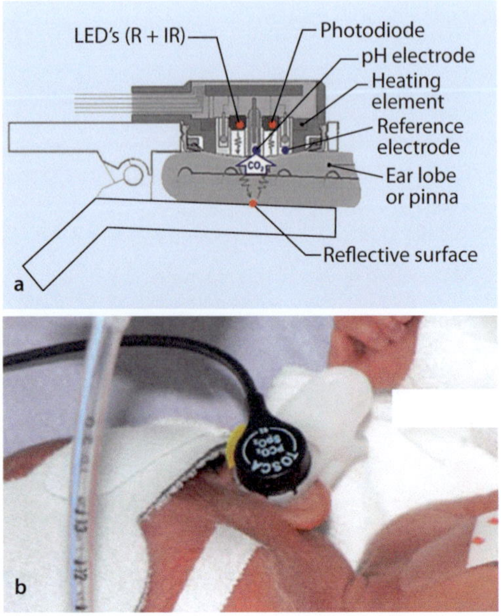

◘ **Fig. 11.2a,b** Measurement of the partial pressures of oxygen carbon and dioxide. **a** The sensor combines an optical sensor for SpO_2 measurement (*red*) with a Stow-Severinghaus CO_2 sensor (*blue*). It is additionally equipped with a heating element (*black*). (Source [6]). **b** Application of the TOSCA sensor at the auricle. (Source [13])

Note

The measurement of $tcpCO_2$ during severe hypercapnia is prone to errors since it underestimates the $paCO_2$ level and is too imprecise at higher concentrations [23].

Compared to neonatology, the application of transcutaneous blood gas determination in adults is currently limited to special clinical questions, e.g. pulmonary function testing or sleep laboratory evaluation. Transcutaneous measurements cannot replace arterial blood gas analyses in blood, but complement it, especially for trending monitoring. Erroneous results must be expected in patients with impaired microcirculation, circulatory centralization or skin edema.

On the German market, measuring devices and the corresponding electrodes are manufactured by Radiometer and Sentec which supply combination electrodes for simultaneous measurement of $tcpCO_2$ and $tcpO_2$ as well [22].

11.5 Measurement of nitric oxide in exhalation air

Current pulmonological guidelines for the treatment of bronchial asthma recommend dosing of inhalable corticosteroids on the basis of symptoms and pulmonary function tests. The analysis of the fractional exhaled nitric oxide (FeNO) is a suitable non-invasive marker and helps in adjusting therapy. This is due to the fact that FeNO correlates well with the degree of airway inflammation. A meta-analysis by Song et al. [25] showed that the consecutive measurements of NO in the exhaled air of patients suffering from asthma is beneficial for optimizing therapy with steroids, while subsequently reduces the frequency of asthmatic attacks. In the differential diagnostics of chronic cough, however, the significance of NO analysis is still debated.

There are several European companies offering devices that measure FeNO, e.g. the CLD 88 sp analyzer (Eco Medics AG, Dürnten, Switzerland), which combines NO analysis with spirometry. The NObreath from Bedfont Scientific (Kent, UK), on the other side, is a handheld device that can be used directly by the patient himself.

11.6 Future trends

In the next years, further non-invasive methods will find acceptance, especially in hospital intensive care environments. For example, this trend will encompass the percutaneous determination of Hb [16] or hematocrit [24]. Several spectroscopic methods have already been developed for both. Fluorescence spectroscopy of zinc protoporphyrin via a flexible optical fiber directly at the bottom lip might be one novel method for diagnosing iron deficiency anemia [10].

11.7 Quality assurance

Quality assurance is still an unsolved problem in non-invasive diagnostics. Its assays are not governed by the Guideline of the German Medical Association on Quality Assurance in Medical Laboratory Examinations (RiliBÄK; ▶ Chapter 38). The device manufacturers' suggestions refer to calibration but not to control procedures. More major errors can only be detected by comparing the findings with invasive measurements.

References

1. Baca JT, Taormina CR, Feingold E, Finegold DN, Grabowski JJ, Asher SA (2007) Mass spectral determination of fasting tear glucose concentrations in nondiabetic volunteers. Clin Chem 53:1370–1372
2. Barman I, Kong CR, Singh GP, Dasari RR, Feld MS (2010) Accurate spectroscopic calibration for non-invasive glucose monitoring by modeling the physiological glucose dynamics. Anal Chem 82:6104–6114
3. Cameron BD, Baba JS, Coté GL (2001) Measurement of the glucose transport time delay between the blood and aqueous humor of the eye for the eventual development of a noninvasive glucose sensor. Diabetes Technol Ther 3:201–207
4. Cohen O, Fine I, Monashkin E, Karasik A (2003) Glucose correlation with light scattering patterns – a novel method for non-invasive glucose measurements. Diabetes Technol Ther 5:11–17
5. De Luca D, Zecca E, Corsello M, Tiberi E, Semeraro C, Romagnoli C (2008) Attempt to improve transcutaneous bilirubinometry: a double-blind study of Medick BiliMed versus Respironics BiliCheck. Arch Dis Child Fetal Neonatal Ed 93:F135–139
6. Eberhard P (2007) The design, use, and results of transcutaneous carbon dioxide analysis: current and future directions. Anesth Analg 105(6 Suppl):S48–52
7. Ermolina I, Polevaya Y, Feldman Y (2000) Analysis of dielectric spectra of eukaryotic cells by computer modeling. Eur Biophys J 29:141–145
8. Fusman R, Rotstein R, Elishkewich K, Zeltser D, Cohen S, Kofler M, Avitzour D, Arber N, Berliner S, Shapira I (2001) Image analysis for the detection of increased erythrocyte, leukocyte and platelet adhesiveness/aggregation in the peripheral blood of patients with diabetes mellitus. Acta Diabetol 38:129–134
9. Heise HM, Marbach R (1998) Human oral mucosa studies with varying blood glucose concentration by non-invasive ATR-FT-IR-spectroscopy. Cell Mol Biol 44:899–912
10. Hennig G, Homann C, Teksan I, Hasbargen U, Hasmüller S, Holdt LM, Khaled N, Sroka R, Stauch T, Stepp H, Vogeser M, Brittenham GM (2016) Non-invasive detection of iron deficiency by fluorescence measurement of erythrocyte zinc protoporphyrin in the lip. Nat Commun 7:10776. doi: 10.1038/ncomms10776

11. Karen T, Bucher FU, Faudere JC (2009) Comparison of a new transcutaneous bilirubinometer (BiliMed) with serum bilirubin measurements in preterm and full-term infants. BMC Pediatrics 9:70
12. Kovatchev BP, Gonder-Frederick LA, Cox DJ, Clarke WL (2004) Evaluating the accuracy of continuous glucose-monitoring sensors: continuous glucose-error grid analysis illustrated by TheraSense Freestyle Navigator data. Diabetes Care 27:1922–1928
13. Lacerenza S, De Carolis MP, Fusco FP, La Torre G, Chiaradia G, Romagnoli C (2008) An evaluation of a new combined Spo2/PtcCO2 sensor in very low birth weight infants. Anesth Analg 107:125–129
14. Maisels MJ, Ostrea EM Jr, Touch S, Clune SE, Cepeda E, Kring E, Gracey K, Jackson C, Talbot D, Huang R (2004) Evaluation of a new transcutaneous bilirubinometer. Pediatr 113: 1628–1635
15. March W, Lazzaro D, Rastogi S (2006) Fluorescent measurement in the non-invasive contact lens glucose sensor. Diabetes Technol Ther 8:312–317
16. McMurdy JW, Jay GD, Suner S, Crawford G (2008) Noninvasive optical, electrical, and acoustic methods of total hemoglobin determination. Clin Chem 54: 264–272
17. Mielsch C, Zimmermann A, Wagner D, Matthes B, Schlebusch H, Luppa P (2010) Point-of-care determination of neonatal bilirubin with the blood gas analyzer RapidLab 1265. Clin Chem Lab Med 48:1455–1461
18. Park HD, Lee KJ, Yoon HR, Nam HH (2005) Design of a portable urine glucose monitoring system for health care. Comput Biol Med 35:275–286
19. Polevaya Y, Ermolina I, Schlesinger M, Ginzburg B, Feldman Y (1999) Time domain dielectric spectroscopy study of human cells. II. Normal and malignant white blood cells. Biochim Biophys Acta 1419:257–271
20. Roychoudhury P, Harvey LM, McNeil B (2006) At-line monitoring of ammonium, glucose, methyl oleate and biomass in a complex antibiotic fermentation process using attenuated total reflectance-mid-infrared (ATR-MIR) spectroscopy. Anal Chim Acta 561:218–224
21. Rubaltelli FF, Gourley GR, Loskamp N, Modi N, Roth-Kleiner M, Sender A, Vert P (2001) Trans-cutaneous bilirubin measurement: a multicenter evaluation of a new device. Pediatr 107:1264–1271
22. Rubortone SA, De Carolis MP, Lacerenza S, Bersani I, Occhipinti F, Romagnoli C (2012) Use of a combined SpO_2/$PtcCO_2$ sensor in the delivery room. Sensors (Basel) 12:10980–10989
23. Ruiz Y, Farrero E, Córdoba A, González N, Dorca J, Prats E (2016) Transcutaneous carbon dioxide monitoring in subjects with acute respiratory failure and severe hypercapnia. Respir Care 61:428–33
24. Sakudo A (2016) Near-infrared spectroscopy for medical applications: Current status and future perspectives. Clin Chim Acta 455:181–188
25. Song WJ, Kim HJ, Shim JS, Won HK, Kang SY, Sohn KH, Kim BK, Jo EJ, Kim MH, Kim SH, Park HW, Kim SS, Chang YS, Morice AH, Lee BJ, Cho SH (2017) Diagnostic accuracy of fractional exhaled nitric oxide measurement in predicting cough-variant asthma and eosinophilic bronchitis in adults with chronic cough: A systematic review and meta-analysis. J Allergy Clin Immunol Pii: S009 1–6749(17)3000 6–4. doi: 10.1016/j.jaci.2016.11.037
26. Tan KL, Dong F (2003) Transcutaneous bilirubinometry during and after phototherapy. Acta Paediatr 92: 327–331
27. Thennadil SN, Rennert JL, Wenzel BJ, Hazen KH, Ruchti TL, Block MB (2001) Comparison of glucose concentration in interstitial fluid, and capillary and venous blood during rapid changes in blood glucose levels. Diabetes Technol Ther 3:357–365
28. Tierney MJ, Tamada JA, Potts RO, Jovanovic L, Garg S (2001) Clinical evaluation of the GlucoWatch biographer: a continual, non-invasive glucose monitor for patients with diabetes. Biosens Bioelectron 16:621–629
29. Tsalikian E for the Diabetes Research in Children Network (DirecNet) Study Group (2004) Accuracy of the GlucoWatch G2 Biographer and the continuous glucose monitoring system during hypoglycemia. Diabetes Care 27:722–726
30. Turnheer R (2004) Pulsoximetrie. Schweiz Med Forum 4: 1218–1223
31. Vashist SK (2012) Non-invasive glucose monitoring technology in diabetes management: a review. Anal Chim Acta 750:16–27
32. Wentholt IM, Hoekstra JB, Zwart A, DeVries JH (2005) Pendra goes Dutch: lessons for the CE mark in Europe. Diabetologia 48:1055–1058
33. Willems WA, van den Berg LM, de Wit H, Molendijk A (2004) Transcutaneous bilirubinometry with the BiliCheck in very premature newborns. J Maternal-Fetal Neonat Med 16: 209–214
34. Yokota M, Sato Y, Yamaguchi I, Kenmochi T, Yoshino T (2004) A compact polarimetric glucose sensor using a high-performance fibre-optic Faraday rotator. Measurement Sci Tech 15:143–147
35. Yasuda S, Itoh S, Isobe K, Yonetani M, Nakamura H, Nakamura M, Yamauchi Y, Yamanishi A (2003) New transcutaneous jaundice device with two optical paths. J Perinat Med 31:81–88
36. Zur B, Hornung A, Breuer J, Doll U, Bernhardt C, Ludwig M, Stoffel-Wagner B (2008) A novel hemoglobin, Bonn, causes falsely decreased oxygen saturation measurements in pulse oxymetry. Clin Chem 54:594–596

Clinical applications

Content

Diabetes diagnostics including analytical methods for glucose monitoring

Hans Günter Wahl, Theodor Koschinsky

P. B. Luppa, R. Junker (Eds.), *Point-of-Care Testing*
https://doi.org/10.1007/978-3-662-54497-6_12

12.1 Introduction

Near-patient blood glucose monitoring is important for optimizing glycemic control in individuals with diabetes and plays an essential role in diabetes management according to the evidence-based guidelines issued by the German Diabetes Association (DDG) [20]. A distinction is made between near-patient quantitative blood glucose monitoring carried out by medical staff that is regulated by RiliBÄK 2014 quality assurance [6] and the self-monitoring of blood glucose (SMBG) (by the patient themselves or with the assistance of relatives) for which no legally regulated quality control is required. The clinical requirements for SMBG systems and their successful use in daily life by non-professionals have importantly shaped the technological development of POCT systems for SMBG. Devices today are smaller, faster, easier to use, less prone to interference and need less capillary blood volume in comparison with those in use approximately 15 years ago.

12.2 Glucose measurement

POCT blood glucose monitors almost exclusively adopt enzymatic analysis methods based on glucose oxidase and glucose dehydrogenase. The reaction products are detected photometrically or electrochemically.

12.2.1 Enzymatic assay reactions

Glucose is oxidized by **glucose oxidase** to gluconic acid in the presence of water and oxygen (◘ Fig. 12.1). The enzyme co-factor flavin adenine dinucleotide (FAD) acts as a first electron acceptor, which is reduced to FADH. Afterwards, FADH is oxidized by oxygen (O_2), the final electron acceptor, forming H_2O_2. The O_2 consumption or formation of H_2O_2 can be measured by electrochemical or chromogenic methods. O-dianisidine, aminophenazone/aminopyrine, as well as iodide/molybdate are used as chromogens. The chromogen is oxidized by the resulting H_2O_2 and measured by reflectometry (e.g. GlucoTouch, LifeScan).

In conventional blood gas analyzers, modified Clark electrodes (oxygen electrodes) are employed for blood glucose determination without sample dilution (direct) (▶ Chapter 8), for example by i-STAT (Abbott) and the GlucometerPro system (BST BioSensor Technology GmbH) or after blood sample dilution (indirect method) by glucose meters like those from Yellow Springs Instruments (YSI Glucose Analyzer). When mediators such as ferrocene (Precision PCx) or hexacyanoferrate (Ascensia Elite, Nova StatStrip) are used instead of molecular oxygen as final electron acceptor, platinum electrodes (modified Clark electrodes) can be replaced by simple, cost-effective single-use electrodes (sensor test strips).

The assay reaction is highly specific, the indicator reaction can, however, be disrupted to a variable degree by substrates such as ascorbic acid or acetaminophen. The results are dependent on the amount of oxygen in the sample and this may lead to analytical problems. A variety of glucose oxidase methods are considered more closely here. Methods which involve oxygen as a final electron acceptor (blood gas analyzers, YSI or GlucoTouch) are not sensitive to changing oxygen concentrations, as long as there is sufficient oxygen in the sample for the reaction to take place. The converse is true for methods which use ferrocene (Precision PCx) or hexacyanoferrate (Ascensia Elite) as a final electron acceptor: Here, oxygen in the sample competes with these mediators as electron acceptors, with the result that glucose measure-

$$\beta\text{-D-Glucose} + H_2O + O_2 \xrightarrow{\text{Glucose oxidase}} \text{D-gluconolactone} + H_2O_2$$

◘ **Fig. 12.1** Oxidation of glucose by glucose oxidase

$$\beta\text{-D-Glucose} \xrightarrow{\text{Glucose dehydrogenase}} \text{D-gluconolactone} + H^+ + e^-$$

Fig. 12.2 Oxidation of glucose by glucose dehydrogenase

ment falsely reads low when the oxygen concentration in the sample is high.

In the oxidation reaction involving **glucose dehydrogenase** (GDH), nicotinamide adenine dinucleotide (NAD; HemoCue as well as Precision PCx, Xceed and Xceed Pro), pyrroloquinoline quinone (PQQ; Accu-Chek Inform, Aviva and Compact Plus) and flavin adenine dinucleotide (FAD, Ascensia CONTOUR) are used as first electron acceptors and are reduced during the reaction to NADH, PQQH and FADH (Fig. 12.2).

The detection of a dye as reaction product is measured photometrically (HemoCue) or reflectometrically (Accu-Chek Compact Plus). HemoCue glucose 201 is the only POCT system that measures glucose by photometry, using single-use cuvettes. After the lysis of red blood cells and oxidization of glucose by glucose dehydrogenase, the resulting NADH reacts with tetrazolium salts which are reduced to formazan by diaphorase activity and measured at 660/840 nm. Their detection time of 40–240 s is concentration-dependent. Calibration is carried out by the manufacturer.

The detection reaction involving hexacyanoferrate (Ascensia CONTOUR) or nitrosoaniline (Accu-Chek Inform and Aviva) as a final electron acceptor is performed electrochemically, using sensor test strips.

> **The advantage of methods using glucose dehydrogenase systems compared to glucose oxidase systems lies in their much lower susceptibility to cross-react with reducing therapeutics or variable oxygen content of the sample.**

12.2.2 Detection methods

Enzymatic reactions in whole blood can be detected photometrically or electrochemically. Many years after the introduction of the first glucose test strips, reflectometers remained the only devices able to quantitatively measure a dye compound as an end product of the reaction. The introduction of non-wipe technology, which made it obsolete to have to wipe off the blood from the strip, as well as the automated reading on the device were significant advancements. This required the separation of red blood cells and other cellular components from the plasma to avoid the color reaction being falsified by the color of whole blood. Although whole blood is used, the measurement occurs in the plasma. The technical challenge is to achieve a quantitative separation of red blood cells while ensuring a relatively fast and sufficient plasma flow to the yreaction carrier for an undisturbed reaction. The manufacturers supply different, partially multi-membrane films with varied properties. It is not clear to what extent lipids are retained in the different systems and therefore how far it is plasma in the strict sense. In contrast, the HemoCue System measures the absorption for a color reaction in a disposable microcuvette using a photometer, directly in the hemolyzed whole blood (Fig. 12.3).

Further progress was made predominantly by introducing sensor technology, which allowed a smaller blood sample volume (0.3–5 µL), shorter measuring times (3–30 s), integrated quality assurance, as well as avoiding interference and direct contact of blood with the device. This consequently improved hygiene and reduced infection risks, particularly in POCT settings. The first sensor test strips were constructed similarly to the photometric

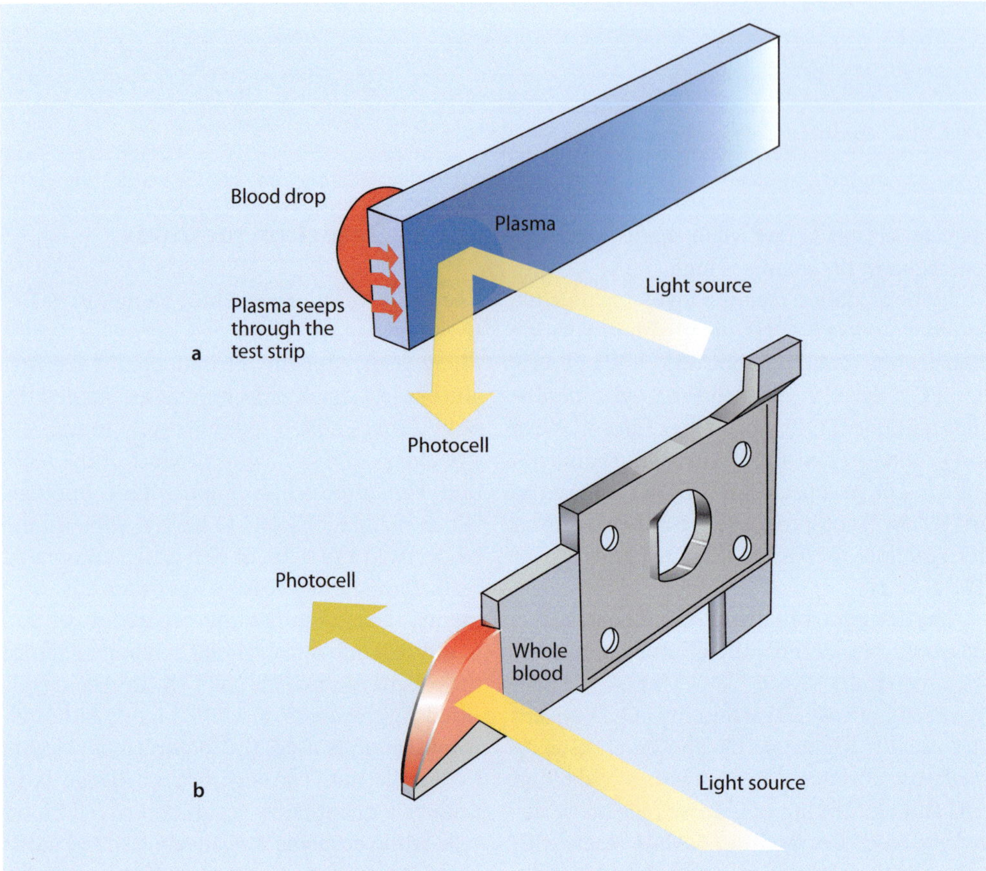

■ **Fig. 12.3a,b** Photometric measurement in hemolyzed whole blood. **a** Reflection method with test strips. **b** HemoCue absorption method. (Courtesy of HemoCue GmbH, Grossostheim, Germany)

test strips to achieve the separation of red blood cells. The introduction of a third electrode, positioned near the end of the test strip, was designed to check that blood volume is sufficient and compensates for interference.

Whereas amperometric measuring techniques had originally been applied to all sensor test strips, Abbott then introduced FreeStyle, the first system on the market to utilize coulometric techniques. The advantage of this technique is the lower impact of hematocrit (Hct) on the result. With amperometric techniques, only a relatively small amount of the glucose present in the sample is measured. The electric current is directly proportional to the glucose concentration. In coulometry, the relationship of electrode surface to blood volume is higher, therefore the total glucose can be measured; the charge ($Q = +\,_0\int^t I\,dt$) is directly proportional to the glucose concentration.

The fact that glucose measuring devices are dependent on Hct makes their use in neonatology and also in intensive care more difficult; new technologies have brought a significant improvement The Nova StatStrip system is the first device to adopt an additional Hct measurement (multi-well, 4 electrodes) and subsequent correction of the glucose reading (■ Fig. 12.4). The Accu-Chek Aviva test strip incorporates a total of 8 electrodes (■ Fig. 12.5). Their function is to check the reagents, humidity, test strip integrity and the sufficiency of blood volume

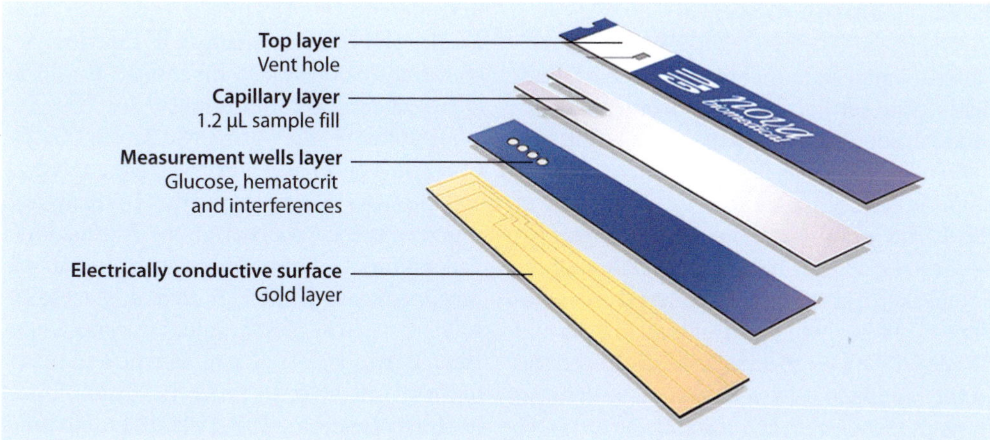

Fig. 12.4 Nova StatStrip, structure of a test strip. (Courtesy of Nova Biomedical GmbH, Rödermark, Germany)

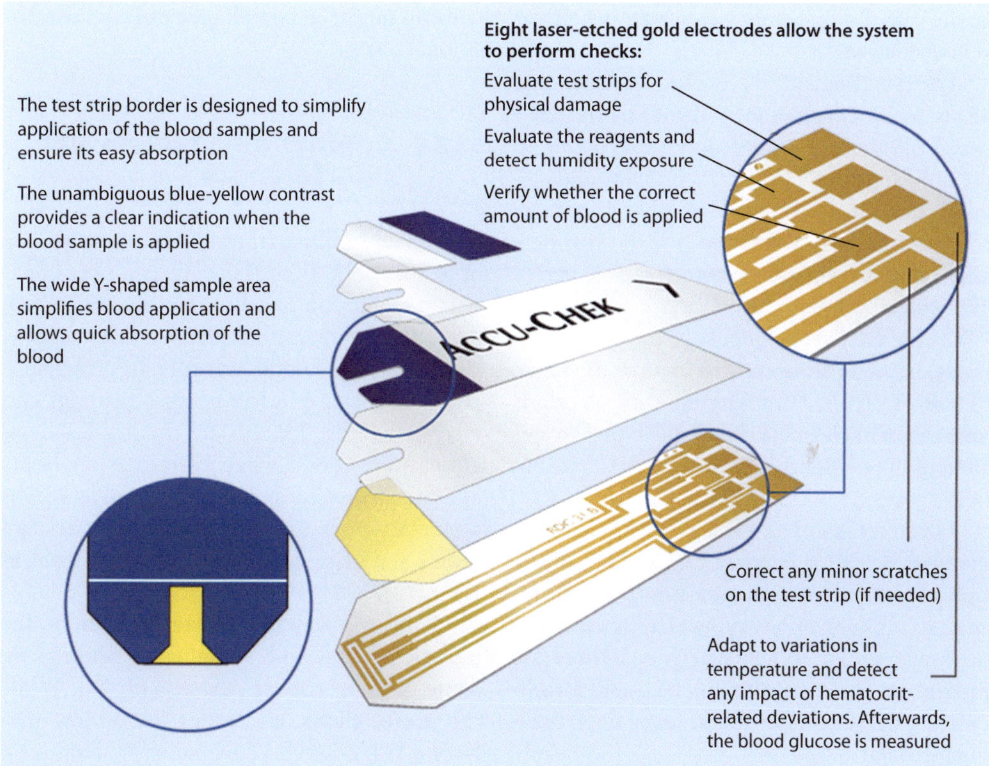

Fig. 12.5 Accu-Chek Aviva, structure of a test strip. (Courtesy of Roche Diagnostics GmbH, Mannheim, Germany)

while checking for temperature and Hct-dependent fluctuations. Temperature and Hct corrections occur as a result of an additional alternating current on the relevant electrodes (impedance measurement), while the blood glucose is measured by conventional direct current methods.

12.2.3 Sample material

Sample material includes whole blood, hemolysate, serum/plasma (with or without deproteinization), urine, cerebrospinal fluid and interstitial fluid or its dialysate, whereby whole blood is generally used for POCT. The site of the blood draw (arterial, capillary, venous) needs to be considered when reporting results because of an arteriovenous concentration gradient of glucose, which can amount to 5–45 mg/dL, depending on metabolic status. Glycolysis in the red blood cells does not matter here, as measurements are taken instantly at the POC. According to the DDG's 2010 practice guidelines [9], the diagnosis of diabetes was based on plasma glucose in a venous sample, as the diagnostic criteria had only been established for venous plasma.

For the clinician, the biggest variable that needs to be considered is the difference between plasma and whole blood; this is due to the differing content of lipids, proteins and cellular components in both sample types. Glucose passes easily through the erythrocyte membrane by passive transport and disperses like water between plasma and erythrocytes.
12
While the molality of glucose (= mg glucose/kg water) in the aqueous compartment is the same, its molarity (concentration) in red blood cells is lower than in plasma, because the typical water content of red blood cells is 0.71 kg H_2O/L, but in the plasma it is 0.93 kg H_2O/L.

Given a mean Hct value of 43 %, this produces a calculation factor of 1.11 for the conversion of whole blood glucose into plasma glucose. In samples with varying Hct, this factor is no longer exact. When extreme outliers are present, an Hct correction can be useful for calculating the conversion factor using the following formula:

$$f = 0.84/(0.93 - 0.22 \times \text{Hct}),$$

where Hct is entered as a fraction. This correction factor calculated in this way is then multiplied by the conversion factor of 1.11.

A distinction should be made between the physiological Hct-dependent whole blood-plasma difference and Hct interference. Such interference plays a role in many POCT devices (see above) and can only be compensated, if at all, by technical design modifications.

In 2009 [11] the German Society of Clinical Chemistry and Laboratory Medicine (DGKL) and the DDG recommended that Germany also implement the proposal of the International Federation of Clinical Chemistry and Laboratory Medicine's (IFCC) [5] that all glucose results be stated as plasma values (plasma equivalents), irrespective of sample type and measurement methods in order to avoid clinical misinterpretation. The IVD manufacturers have followed these recommendations in the designs of more recent glucose POCT systems, meaning that mostly older glucometer models will still not be in compliance during the transitional phase.

12.2.4 Confounding factors and interferences

In hospital settings, especially on intensive care or neonatology units, there are a number of patient-related confounding factors [6], which can alter the measured glucose values. Here, **Hct** plays a very important role. In principle, a rise in Hct causes lower glucose readings and vice versa [8, 18, 20]. To date, the smallest deviations were found when blood gas analyzers and the HemoCue system were used as well as systems which measure and adjust for Hct (Nova StatStrip, Accu-Chek Aviva and Inform II) [8, 18].

A **higher oxygen concentration** in the blood can result in lower glucose readings on some glucose oxidase devices [13, 21]. While the use of classic electrodes (blood gas analyzers, YSI, i-STAT, Glucometer^{Pro}) requires a certain oxygen concentration, in other systems using different mediators (e.g. Ferrocene), this causes a competing reaction, leading to falsely low glucose readings (see above).

Drugs with reducing properties (vitamin C, acetaminophen, dopamine and others) and endogenous metabolites can cause further interference. This refers mainly to procedures based on glucose oxidase, that adopt a peroxide reaction as a detection method [12, 19] Procedures using glucose dehydrogenase as an enzyme are much less confounded by such substances (see above). In addition to these general confounders, there are some factors that cause procedure- or system-related interference. These include interference by methemoglobin (>10 %) or Intralipid in the HemoCue system [1, 17].

Mannose, maltose, xylose, galactose alongside glucosamine and 2-deoxyglucose are metabolized by glucose dehydrogenase (GDH), leading to higher glucose values. Clinically relevant problems, however, only occur in the GDH-PQQ combination. In order to minimize maltose and xylose interference, in 2010, a genetically modified form of GDH was developed that is utilized in the test strips Accu-Check Performa for Accu-Chek Inform II [7, 10]. This option is particularly suitable for peritoneal dialysis patients treated with icodextrin (e.g. Extraneal Baxter), which is metabolized to maltose, or given intravenous maltose-containing solutions such as certain human immunoglobulin products. Devices using a combination of GDH-NAD or GDH-FAD as well as glucose oxidase-based systems are not affected.

12.2.5 Evaluation and validation

Only evaluated and validated devices and procedures should be used at the POC. Manufacturers exclude the use of devices elsewhere that have been developed solely for the purpose of blood glucose self-monitoring. Evaluations of blood glucose monitors have been published extensively. As expected, there are often pronounced differences in the results with different devices but also for the same model, and there can be significant differences across performance data even for the same model. This can be partially explained by the known quality variations in devices themselves and, even more so, in test strips (particularly due to batch-to-batch fluctuations). Methodical shortcomings in evaluating and comparing methods also play a significant part. Indeed, several proposals made by the likes of the Clinical Laboratory Standards Institute (CLSI; previously known as: National Committee on Clinical Laboratory Standards, NCCLS) [16] or the STARD Initiative [2, 3], have been implemented in many studies, albeit inadequately [14]. In 2007, Mahoney and Ellison [15] used these and other recommendations (IFCC, FDA, TNO, SKUP, ISO 15179) to design a practical 14-point checklist, which can be used as an evaluation protocol guide to simplify critical literature reviews. Moreover, in 2008, CLSI published recommendations (POCT 6-P) aimed at standardizing glucose monitoring methods for different sample types [4].

It should be remembered that the conclusions regarding the accuracy of blood glucose measuring devices cannot be derived from interlaboratory test results, as long as no reference method values can be used, due to a lack of suitable whole blood control samples. Rather, device-dependent target values obtained in interlaboratory trials only provide information on the devices' precision and proper handling.

12.3 Glucose POCT monitoring systems

In hospitals, the most efficient way to measure blood glucose is with a POCT system, which features a comprehensive measurand, quality control and electronic data management with access to both laboratory and hospital information systems. It is either used exclusively for blood glucose analysis (◘ Tab. 12.1 and ◘ Tab. 12.2) or combined with other POCT-relevant test systems such as those used on intensive care units for blood gas analysis (▶ Chapter 14).

Technologically, there is a lot of overlap to SMBG systems, which are also used by medical personnel for POCT glucose monitoring, but without adequate quality assurance. In the context of this book, only examples of SMBG systems are presented that fulfill the RiliBÄK quality assurance and meet the much stricter requirements of the ISO 15197:2013 norm rather

Tab. 12.1 Glucose POCT systems with reagents or biosensors for repeated use (for example in specialized diabetes clinics)

Measurement system (manufacturer)	Sample volume [µL]	Measuring time [s]	Measuring range [mg/dL, mmol/L]	Calibrated to	Measurement method (enzyme)	Sample carrier/sensor
Hitado Super GL 2 (Hitado)	10	40	11–900 (0.6–50)	Whole blood or plasma (+Hct correction)	Electrochemical (Glucose dehydrogenase)	Glucose sensor (for measurements of up to 90 days)
GLUCOMETER^{PRO} (BST Bio Sensor technology)	6	5–10	10–600 (0.6–33.3)	Plasma	Electrochemical (glucose oxidase)	Capillary/chip (for measurements of up to 30 days)

Tab. 12.2 Glucose POCT systems (unit-use reagents) featuring comprehensive measurand, quality control and electronic data management (examples)

Measurement system (manufacturer)	Sample volume [µL]	Measuring time [s]	Measuring range [mg/dL, mmol/L]	Calibrated to	Measurement method (enzyme)	Sample carrier/sensor
Accu-Chek® Inform II (Roche Diagnostics)	0.6	5	10–600 (0.6–33.3)	Plasma	Electrochemical (mutated variant of quinoprotein glucose dehydrogenase (GDH)	AC® Inform II test strip
HemoCue® Glucose 201 DM system (Radiometer)	5	40–240	0–400 (0–22.2)	Whole blood or plasma	Photometric (GDH)	Microcuvette
Nova StatStrip® Glucose (Nova Biomedical)	1.2	6	10–600 (0.6–33.3)	Plasma	Electrochemical (glucose oxidase)	Nova StatStrip® Glucose Analyses System
FreeStyle® Precision Pro (Abbott Diabetes Care)	0.6	5	20–500 (1.1–27.8)	Plasma	Electrochemical (nicotinamide adenine dinucleotide-GDH)	FreeStyle® Precision Pro test strips

12

than the previous ISO 15197:2003 [39] (Tab. 12.4). Since 2016, the 2013 version represents the only valid new POCT minimum requirements [40] (Tab. 12.3).

Blood glucose measuring devices have different minimal requirements for quality and quality assurance, therefore three different **areas of application** should be considered:

- the initial diagnosis of different types of diabetes mellitus (DM).
- blood glucose monitoring in the treatment of diagnosed diabetes mellitus and,

Tab. 12.3 "In vitro diagnostic test systems – Requirements for blood-glucose monitoring systems for self-testing in managing diabetes mellitus" (ISO standard 15197:2013)

ISO 15197:2003 (valid up to May 2016)		ISO 15197:2013 (from June 2016 solely valid)	
Blood glucose concentration	Tolerance range*	Blood glucose concentration	Tolerance range*
≥75 mg/dL (4.2 mmol/L)	±20 %	≥ 100 mg/dL (5.6 mmol/L)	±15 %
<75 mg/dL (4.2 mmol/L)	±15 mg/dL (0.83 mmol/L)	<100 mg/dL (5.6 mmol/L)	±15 mg/dL (0.83 mmol/L)

* 95 % of all values must be within the tolerance range

Tab. 12.4 SMBG devices which meet the new POCT minimum standards from 2016 (ISO standard 15197:2013) (examples)

Measurement system (manufacturer)	Sample volume [μL]	Measuring time [s]	Measuring range [mg/dL, mmol/L]	Calibrated to	Measurement method (enzyme)	Sample carrier/sensor
Accu-Chek Aviva (Roche Diagnostics)	0.6	5	10–600 (0.6–33.3)	Plasma	Electrochemical (mutant variant of quinoprotein GDH) (MutQ-GDH)	AC Aviva test strip
Contour XT (Ascensia Diabetes Care)	0.6	5	10–600 (0.6–33.3)	Plasma	Electrochemical (flavin adenine dinucleotide) (FAD-GDH)	Contour Next Sensor
OneTouch Verio (Johnson & Johnson/ LifeScan)	0.4	5	20–600 (1.1–33.3)	Plasma	Electrochemical (FAD-GDH)	OneTouch Verio test strips
FreeStyle Precision Neo (Abbott Diabetes Care)	0.6	5	20–500 (1.1–27.8)	Plasma	Electrochemical (nicotinamide adenine dinucleo-tide-GDH)	FreeStyle Precision Neo test strips

- under certain conditions, to check devices used for the self-monitoring of blood glucose.

Basically, every glucose measurement method – regardless of volume and near-patient use – that employs reagents for multiple uses and is approved for use in a clinical chemistry laboratory can be considered for all three of these areas of application. The slimmed-down versions used in private diabetology practices have been found to be particularly efficient and cost effective as they deal with fewer blood glucose samples per day than satellite laboratories. In practice, internal and external quality assurance of these measurement methods is legally regulated and described in detail in the RiliBÄK.

In contrast, previous POCT glucose measurement systems using unit-use reagents pursuant to RiliBÄK are only licensed for the 2nd and 3rd but not for 1st area of application (initial diabetes diagnosis). The only exceptions to this group of blood glucose meters are the HemoCue Glucose 201+ and HemoCue Glucose 201 RT devices that are licensed for areas 1, 2 and 3.

In physicians' practices, only the internal quality assurance of POCT measurement methods is legally required and described in detail in the respective provisions of RiliBÄK. Irrespective thereof, the DDG guidelines for gestational diabetes (GDM) recommend voluntary external quality assurance as per RiliBÄK for these methods in area of application 1.

Taking account of the quality requirements of ISO standard 15197:2013, further differentiation into the **measurement quality** of blood glucose systems (they all need to be CE-marked) is not yet compulsory. It is, however, technically possible with the existing devices and would enable more meaningful differentiation according to the areas of application defined above. An objective, generally applicable and independent information data base has therefore still been lacking to make recommendations about the suitability of such monitoring systems for all three areas of application.

The wide choice of **sample material** in glucose analysis poses a general problem. Venous, capillary and arterial whole blood as well as plasma and serum can be used. However, the choice of sample material is a significant confounding factor and often leads to misinterpretation of results. In one study, for example, the majority of staff at a diabetes center did not know whether whole blood or plasma was used for laboratory blood glucose testing [8].

Although glucose POCT systems normally use whole blood (but in various preparation forms, e.g. as hemolysate or filtered plasma), the results can either be given as whole blood or plasma glucose, using different manufacturer- or device-specific calibration methods, whose details are treated as company secrets. In Germany, conventional clinical chemistry laboratory methods for measuring glucose in whole blood can be reported either as whole blood or plasma glucose.

Note

In order to avoid misinterpretation of results and the potentially ensuing consequences, detailed specialist knowledge is essential: this applies to the comparative studies of glucose POCT systems versus conventional reference methods in the laboratory as well as to clinical interpretation and therapy recommendations, e.g. glucose-dependent insulin dosing or carbohydrate portion algorithms.

To stop further confusion, in 2005 the International Federation of Clinical Chemistry and Laboratory Medicine (IFCC) proposed that all glucose results be given as plasma values irrespective of sample type or measurement method [10]. Already implemented in the USA and in many European countries, this proposal has also been supported jointly by the DGKL and the DDG in Germany since 2009 [27] and meanwhile become increasingly accepted in clinical day-to-day practice.

In hospitals, particularly on intensive care units, many **confounding and influencing factors** can lead to altered glucose values. These are either dependent on the measurement system used (oxygen) or independent of it (tissue perfusion). It is therefore crucial to elucidate any study results found on the devices to be used and any related confounding factors. Hypotension can result in reduced tissue perfusion which leads to increased glucose utilization and therefore a significantly greater difference between capillary whole blood and venous plasma values [14, 25, 30]. Here, two counter-opposing effects occur: Capillary glucose values are higher than the equivalent venous values, whereas the whole blood samples (capillary) are lower than the equivalent plasma values. When combining capillary whole blood and venous plasma, measurements can show the same values in both. In the case of plasma-correlated capillary whole blood, values may even be higher as compared with venous plasma [12, 26]. These – like nutrition-related – effects must likewise be accounted for in all blood glucose measurement devices, regardless of the measuring method.

There is a difference between blood glucose concentrations in arterial and venous blood, mainly depending on the extent and speed of glucose absorption or the metabolism in the relevant tissue. Known as the "arteriovenous difference", it decisively depends on the nutritional status. According to Bürgi [7], the difference in healthy individuals after fasting is 9 mg/dL (0.5 mmol/L) and rises 30 min after a 50 g oral glucose load up to 45 mg/dL (2.5 mmol/L). It then falls after 120 min to 14 mg/dL (0.8 mmol/L). The difference is lower in patients with peripheral insulin resistance.

Note

Of note is that blood samples from which serum is obtained for measuring other biochemical parameters do not contain any additive that inhibits glycolysis in erythrocytes. According to DDG guidelines [21, 23], blood samples used for measuring glucose under non-POCT conditions, unless centrifuged immediately, should contain an inhibitor of glycolysis in red blood cells. When using glycolysis inhibitors such as sodium fluoride, complete inhibition of glycolysis only occurs after 4 hours. However, it is important to remember that a decrease of up to 6 % or 10 mg/dL is possible after just one hour, compared to the baseline concentration [28]. To achieve immediate and complete glycolysis inhibition in whole blood, a combination of sodium fluoride and a citrate buffer is recommended to achieve a stable pH reduction to pH 5.5 [5, 13].
In general, using glucose POCT measurement systems, no pre-analytical preparations are needed, as the measurement is carried out immediately after blood is taken.

12.4 Initial diagnosis

Blood glucose monitoring by POCT has only been used to a limited extent in the initial diagnosis of impaired glucose metabolism. Generally, it is employed in outpatient settings for clarification in acute emergencies – in patients with suspected hypo- or hyperglycemic coma or, for example, gestational diabetes. This enables immediate discussion with the pregnant woman regarding further treatment and spares her the psychological stress that otherwise would ensue due to unnecessarily waiting for an external laboratory analysis [23]. Otherwise, the initial diagnosis of impaired glucose metabolism remains the domain of clinical chemistry laboratories as immediate treatment is usually not required. According to evidence-based practice guidelines from the DDG, only venous plasma glucose values are suitable for diagnosing diabetes (◘ Tab. 12.5).

According to guidelines, a diagnosis of diabetes mellitus can only be made using a quality-assured laboratory method, i.e. not with devices designed for SMBG. Even when a very precise laboratory method is used, which has a variation coefficient of "only" 2 %, it has to be presumed that a "real" glucose value of, for example, 126 mg/dL (the cut-off for a diagnosis of diabetes), the 95 % confidence interval will range from 121 to 131 mg/dL. Therefore, depending on the clinical situation, glucose values in the cut-off area should be repeated at longer intervals or an oral glucose tolerance test should be performed.

Special conditions apply to the quality assurance of POCT glucose measurements, as per RiliBÄK 2014 [6], when unit-use reagents are applied and the devices need to meet special requirements. Measuring a control sample once a week is sufficient for internal quality control. External quality control has not been required so far for the physician's private practice; neither is it required in the hospital setting, provided the central laboratory is responsible for quality assurance (▶ Chapter 37). According to the evidence-based guideline issued by the DDG in 2011, "Gestational Diabetes Mellitus (GDM)

Tab. 12.5 Diagnostic criteria of glucose metabolism (venous plasma) [21]

Diagnostic criteria	Plasma glucose (venous) mg/dL	Plasma glucose (venous) mmol/L
Normal glucose tolerance		
Fasting glucose	<100	<5.6
2-h-oGTT	<140	<7.8
Abnormal fasting glucose		
Fasting glucose	100–125	5.6–6.9
Impaired glucose tolerance		
Fasting glucose	<126	<7.0
2-h-oGTT	140–199	7.8–11.0
Diabetes mellitus (only by HbA_{1c} values of 5.7 % to <6.5 % or 39 to <48 mmol/mol)		
Fasting glucose	≥126	≥7.0
and/or 2-h-oGTT	≥200	≥11.1
Gestational diabetes mellitus (at least 1/3 of the values)		
Fasting glucose	≥92	≥5.1
1-h-oGTT	≥180	≥10.0
2-h-oGTT	≥153	≥8.5

oGTT Oral glucose tolerance test with 75 g glucose (according to WHO)

Diagnosis, Therapy and Follow-Up Care", the diagnosis of gestational diabetes using glucose POCT methods in the physician's private practice should also undergo external quality control according to RiliBÄK rules [23].

12.5 Monitoring of blood glucose

POCT measurement of blood glucose in blood is mainly performed (up to approx. 90 %) for the following the clinical course of ongoing diabetes therapy in physicians' private practices representing the various sub-specialties, but also in hospitals as well as in clinical chemistry central laboratories. The use of glucose POCT systems has significantly increased over recent years, although the costs are 2 to 4 times higher than for glucose measurement in central laboratories. In private practice, this is due to changes in therapy requirements, as presented in the DDG guidelines [20]. Based on long-term studies for the prevention and reduction of diabetic sequelae, these guidelines make recommendations to achieve long-term near-normal glycemic control while avoiding hypoglycemia and considering the individual targets set.

In blood glucose control, these individual targets need to be regularly monitored and adjusted. This has been neglected for a long time in hospitals primarily as a result of cost optimization in the clinical chemistry central laboratory and secondly due to the relatively short inpatient stay. It was thought that the effect of transient suboptimal blood glucose control occurs frequently, for example, during an acute illness, but has negligible impact on the development of diabetic sequelae (objective: Therapy to prevent diabetic coma).

This changed because of the increasing demand for intensive insulin therapy, guided by frequent near-patient blood glucose monitoring, particularly when preprandial insulin regimes were used. Achieving near normal blood glucose levels, even just for a few days, using intensive insulin therapy and insulin dose adjustments with frequent blood glucose monitoring can lead to significant postoperative improvements in surgical patients in intensive care. This has been shown in ground-breaking publications by van den Berghe since 2001 [35]. In subsequent years, these treatment outcomes were reviewed, modified and critically grouped by disease stage in the different patient cohorts [3, 17, 29, 31, 33, 34, 36]. It had now become medically necessary to use glucose POCT systems in acute illnesses in diabetic patients and those with secondary impaired glucose tolerance. The latter can clinically manifest itself in conditions such as trauma, hemorrhages, burns, oxygen deficiencies, severe infection, sepsis and shock. The stress caused by such critical conditions induces the release of counter-regulatory hormones of glucose metabolism, which cause various changes in carbohydrate metabolism with transient peripheral insulin resistance and a relative lack of insulin. Additionally other treatments such as high-dose corticosteroids and vasopressors or parenteral feeding with carbohydrates can trigger abnormal glucose metabolism.

POCT is mainly used in the context of treatment with rapidly adjusting insulin doses for fast changing blood glucose levels, including emergency monitoring in hypoglycemia. Furthermore, in this clinical situation the severity of an illness can change relatively quickly. Food intake by the patient will also vary. The glucose analysis with POCT needs to be in the immediate vicinity of the patient and performed timely with immediate interpretation of blood glucose levels and their resulting (post-analytical) therapeutic consequences. The established distinction of analytical and post-analytical expertise in diabetes care is not applicable here. The result is that in all clinically relevant areas (hospital, physician's practice, emergency vehicle, "home care" and ambulatory care), new competence, cooperation and logistics models need to be established between all disciplines caring for patients with different types of diabetes and related complications (▶ Chapter 31).

The conduct and interpretation of the above mentioned studies [3, 17, 29, 31, 33, 34, 35, 36], partly from intensive care [12, 14, 25], partly from other clinical settings [1, 22], show a number of methodological problems, which need to be considered when the result are implemented in clinical day-to-day practice. In these studies, different measurement systems (ABL 700 by Radiometer, HemoCue B Glucose analyzer) with different sample materials (arterial and capillary whole blood) were used; different laboratory methods to "reference methods" (YSI Glucose analyzer, Dimension RXL, Glucose analyzer Beckman, Hitachi 747 analyzer) were submitted. The time from taking blood to analysis in the central laboratory was not standardized, or was accepted with an average of 45 minutes, without the addition of a glycolysis inhibitor. Finally, different "sampling sites" were allowed. Nevertheless, in clinical practice, the same glucose laboratory values were used for blood glucose therapeutic targets (80–110 mg/dL as well as for the definition of hypoglycemia (<40 mg/dL). Although this problem has been raised repeatedly [4, 11], a lack of clarity still remains.

12.6 Blood collection from alternative sites

Though not always sufficiently appreciated, there is a clinically relevant characteristic in blood glucose measurement in the pre-analytical setting, which is irrelevant for other POCT applications, namely blood collection from other capillary sites than the fingertip or ear lobe, called alternative/alternate site testing (AST). The reason for this was the desire to find a less painful alternative to the abused "classic" fingertip capillary for blood glucose self-monitoring. The development of SMBG systems (≤3 µL) allowed the collection of capillary blood (skin depth: ≤2 mm) from the arm, leg and abdominal skin, which is generally less painful. There is no difference in blood glucose levels between the above mentioned AST sites and the fingertip in a fasting sample taken at rest. In contrast, when blood glucose levels change quickly (>2 mg/dL/min) in both directions (e.g. after a carbohydrate rich meal or an inadequate preprandial insulin dose, injection of a too high insulin dose or after intense physical exertion/sport) a clinically relevant delay of some 30 minutes in the capillary blood glucose adjustment can be detected in the AST site and the fingertip even after the skin cools down.

In acute settings, misjudgment of these situations can cause hypoglycemia and endanger the patient. Long-term, an inadequate therapy adjustment can hinder a near normal glycemic control [2, 15, 18, 19, 24]. The reasons for this are mainly anatomical differences: There are significantly fewer surface end capillaries and arteriovenous shunts in the AST sites than on the fingertip, which cause a faster blood glucose change in parallel with systemic changes. When capillary blood is taken, the condition of a patient needs to be considered and, if in doubt, AST sites should be avoided.

Alternative sites are also used with continuous glucose monitoring devices [16]. Further Information is given in ▶ Chapter 13.

12.7 HbA_{1c} POCT monitoring systems

The HbA_{1c} value reflects the median plasma glucose levels during the preceding 2 to 3 months and has been seen for a long time as the standard for assessing blood glucose control over longer periods.

This parameter has gained additional importance in diabetes diagnosis [21, 32]: Values of <5.7 % rule out diabetes mellitus, whereas values of ≥6.5 % confirm the diagnosis. In Germany, the DDG guidelines recommend that, in patients with values between 5.7 % and 6.4 %, the diagnosis should be made based on conventional criteria [21]. This is only applicable when irregular hemoglobin variants are absent and the life cycle of erythrocytes is normal.

The HbA_{1c} value can only be considered for diagnosis when standardized and specific assessment methods and sufficiently performed quality controls are used.

HbA_{1c} can be measured using different laboratory devices by various measurement methods: Immunological (Abbott, Beckmann, Roche Diagnostics, Siemens), affinity chromatographic (Alere, Axis-Shield), capillary electrophoresis (Sebia) and enzymatic (Abbott) that allow the RiliBÄK 2014 requirements to be met [6]. Examples of HbA1c-POCT measurement systems using unit use reagents are summarized in ◘ Tab. 12.6. They are all registered and licensed by the manufacturer for use in the follow up monitoring of diabetes. In addition, the DCA Vantage Analyzer is also labeled and licensed by the manufacturer for use in the initial diagnosis of diabetes (◘ Fig. 12.6).

The above-mentioned analytical methods differ significantly from one another, more so than the glucose measurement methods available. This means that the sources of error or problems, which can occur in different measuring techniques vary significantly from method to method. The results of HbA_{1c} measurements can vary significantly and depend on the method used, despite standardization, and therefore the use of HbA_{1c} for diabetes diagnosis is limited [21]. Current comparisons in measurement quality between HbA_{1c} POCT systems in general practice and clinical chemistry laboratories in the Netherlands have shown some consistency, but also significant differences [37]. In Germany current inter-laboratory ring trial results, as per RiliBÄK, using control samples from the Reference Institute for Bioanalytics (RfB) and from INSTAND e.V. have also demonstrated significant differences between HbA_{1c} measurement methods [38]. RiliBÄK requirements for external quality control allow a generous ±18 %: meaning that a real HbA_{1c} value of 6.5 % may vary between 5.3 % and 7.8 %! It is unclear why these requirements for external quality control are not ±6 % as in the USA or ±10 % as in Switzerland. The advisory board of the German Medical Association is indeed aware of this situation.

Tab. 12.6 HbA_{1c} POCT system using unit-use reagents (examples) for capillary or anticoagulated venous whole blood.

Measurement system (Producer/Distributer)	Sample volume [µL]	Measuring time [min]	Referenced to	Measurement method	Sample carrier
DCA Vantage (Siemens Healthcare Diagnostics)	1	6	IFCC/NGSP	Immunological	DCA HbA_{1c}-test
Super ID (Hitado)	10	7	IFCC/NGSP	Immunological	Super-ID HbA_{1c}-test kit
Afinion AS 100 Analyzer (Alere)	1.5	3	IFCC/NGSP	Affinity chromatography	Afinion HbA_{1c}-test
NycoCard Reader II (Alere)	5	3	IFCC/NGSP	Affinity chromatography	NycoCard HbA1c Single test system
Quo-Test Analyzer (EKF Diagnostic)	4	4	IFCC/NGSP	Affinity chromatography	Quo-Test HbA_{1c}

All manufacturers have standardized their methods to the reference method of the International Federation of Clinical Chemistry (IFCC). Therefore the IFCC values can be converted into values (in % HbA_{1c}) according to the National Glycohemoglobin Standardization Program (NGSP) using the following formula and vice versa:

$$\text{IFCC (mmol/mol)} = [10.93 \times \text{NGSP (\%)}] - 23.50 \text{ or}$$

$$\text{NGSP (\% HbA}_{1c}\text{)} = [0.09148 \times \text{IFCC (mmol/mol)}] + 2.152.$$

The new IFCC reference standard (results in mmol/mol Hb) has not yet been established in clinical daily routine. Laboratories, however, often report the results in both units (% and mmol/mol). In summary, the DDG and DGKL both recommend the following for HbA_{1c} [9]: In quality assurance (calibration of devices and reagents), the new HbA_{1c} values are used as per IFCC with mandatory reporting as per RiliBÄK. In clinical practice, that means that the "old" HbA_{1c} units as per NGSP can be used further in addition to the new ones (within the guidelines).

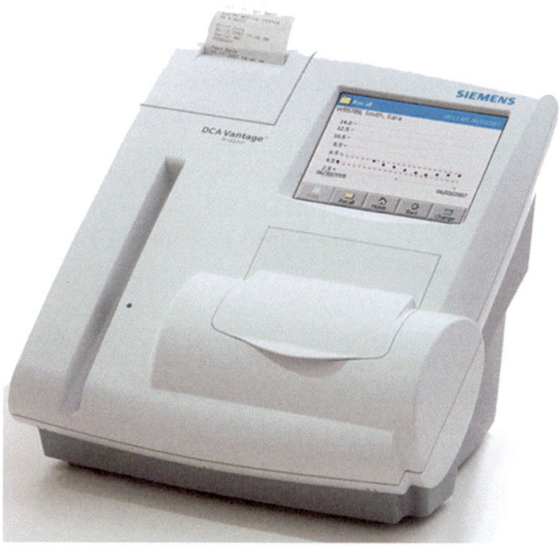

Fig. 12.6 DCA Vantage HbA_{1c} Analyzer. (Courtesy of Siemens Healthcare Diagnostics)

References

Cited in Sections 12.1 and 12.2

1. Asworth L, Gibb I, Alberti KG (1992) HemoCue: evaluation of a portable photometric system for determining glucose in whole blood. Clin Chem 38: 1479–1482
2. Bossuyt PM, Reitsma JB, Bruns DE et al. (2003) The STARD statement for reporting studies of diagnostic accuracy: explanation and elaboration. Clin Chem 49: 7–18
3. Bossuyt PM, Reitsma JB, Bruns DE et al. (2003) Towards complete and accurate reporting of studies of diagnostic accuracy: the STARD initiative. Clin Chem 49: 1–6
4. CLSI (2008) Guidelines for comparison of glucose methodologies that use different sample types; Proposed Guideline. CLSI document POCT 6-P.Clinical and Laboratory Standards Institute, Wayne, PA
5. D'Orazio P, Burnett RW, Fogh-Andersen N et al. (2005) The International Federation of Clinical Chemistry Scientific Division Working Group on Selective Electrodes and Point of Care Testing: Approved IFCC recommendation on reporting results for blood glucose (abbreviated). Clin Chem 51: 1573–1576
6. Dungan K, Chapman J, Braithwaite SS, Buse J (2007) Glucose measurements: Confounding issues in setting targets for inpatient management. Diabetes Care 30: 403–409
7. Janssen W, Harff G, Caers M, Schellekens A (1998) Positive interference of icodextrin metabolites in some enzymatic glucose methods. Clin Chem 44: 2379–2380
8. Karon BS, Griesmann L, Scott R et al. (2008) Evaluation of the impact of hematocrit and other interference on the accuracy of hospital-based glucose meters. Diabetes Technol Ther 10: 111–120
9. Kellerer M, Danne T (Ed.) (2010) Praxis-Leitlinien der Deutschen Diabetes-Gesellschaft; Diabetologie und Stoffwechsel 5 (Suppl 2); Evidenzbasierte Leitlinien DDG, aktualisierte Version auf den Webseiten der DDG: www.deutsche-diabetes-gesellschaft.de/Evidenzbasierte Leitlinien/
10. Koschinsky T (2011) Genauigkeit der Blutglukosemessung – aktuelle Anforderungen und Interferenzen. Diabetologie Stoffw 6: 43–47
11. Koschinsky T, Junker R, Luppa PB, Schlebusch H (2009) Improvement of therapeutic safety through standardized plasma calibration of blood glucose test systems at the point-of-care. J. Lab Med 33: 349–52
12. Kost GJ, Nguyen TH, Tang Z (2000) Wholeblood glucose and lactate: trilayer biosensors, drug interference, metabolism, and practice guidelines. Arch Pathol Lab Med 124: 1128–1134
13. Kost GJ, Vu HT, Inn M et al. (2000) Multicenter study of whole-blood creatinine, total carbon dioxide content, and chemistry profiling for laboratory and point-of-care testing in critical care in the United States. Crit Care Med 28: 2379–2389
14. Mahoney J, Ellison J (2007) Assessing the quality of glucose monitor studies: a critical evaluation of published reports. Clin Chem 53: 1122–1128
15. Mahoney JJ, Ellison JM (2007) Assessing glucose monitor performance – a standardized approach. Diabetes Technology & Therapeutics 9: 545–552
16. NCCLS (2002) Point-of-care blood glucose testing in acute and chronic care facilities; Approved Guideline. NCCLS document C30–A2
17. Patrick L, Lynch M, O'Kane MJ (2002) Methemoglobin interferes with the HemoCue B-glucose Analyzer. Clin Chem 48: 581–583
18. Rao LV, Jakubiak F, Sidwell JS, Winkelmann JW, Snyder ML (2005) Accuracy evaluation of a new glucometer with automated hematocrit measurement and correction. Clin Chim Acta 356: 178–183
19. Tang Z, Du X, Loie RF, Kost GJ (2000) Effects of drugs on glucose measurements with handheld glucose meters and a portable glucose analyzer. Am J Clin Pathol 113: 75–86
20. Tang Z, Lee JH, Louie RF, Kost GJ (2000) Effects of different hematocrit levels on glucose measurements with handheld meters for point-of-care testing. Arch Pathol Lab Med 124: 1135–1140
21. Tang Z, Louie RF, Payes M, Chang KC, Kost GJ (2000) Oxygen effects on glucose measurements with a reference analyzer and three handheld meters. Diabetes Technol Ther 2: 349–362

Cited in Sections 12.3 to 12.7

1. Arabadjief D, Nicholas JH (2006) Assessing glucose meter accuracy. Curr Med Res Opin 22: 2167–2174
2. Bina DM, Anderson RL, Johnson ML, Bergenstal RM, Kendall DM (2003) Clinical impact of prandial state, exercise, and site preparation on the equivalence of alternative-site blood glucose testing. Diabetes Care 26: 981–985
3. Brunkhorst FM, Engel C, Bloos F et al. (2008) German Competence Network Sepsis (SepNet). Intensive insulin therapy and pentastarch resuscitation in severe sepsis. N Engl J Med 358: 125–139
4. Brunkhorst FM, Wahl HG (2006) Blood glucose measurements in the critically ill: more than just a blood draw. Crit Care 10: 178
5. Bruns DE, Knowler WC (2009) Stabilization of glucose in blood samples: why it matters. Clin Chem 55: 850–852
6. Bundesärztekammer (2014) Richtlinie der Bundesärztekammer zur Qualitätssicherung laboratoriumsmedizinischer Untersuchungen (RiliBÄK 2014). Dtsch Ärztebl 111:A1583–A1618

12

7. Bürgi W (1974) Oraler Glukosetoleranztest: unterschiedlicher Verlauf der kapillären und venösen Belastungskurven. Schweiz Med Wochenschr 104: 1698–1699
8. Burrin JM, Alberti KG (1990) What is blood glucose: can it be measured? Diab Med 7: 199–206
9. Danne T, Mueller-Wieland D, Lackner K, Schleicher E (2009) Gemeinsame Empfehlungen der DDG und DGKL zur Qualitätssicherung der HbA1c-Messung. Dtsch Ärzteblatt 106 (33)
10. D'Orazio P, Burnett RW, Fogh-Andersen N et al. (2005) Approved IFCC recommendation on reporting results for blood glucose (abbreviated). Clin Chem 51(9): 1573–1576
11. Dungan K, Chapman J, Braithwaite SS, Buse J (2007) Glucose measurements: Confounding issues in setting targets for inpatient management. Diabetes Care 30: 403–409
12. Finkielmann JD, Oyen LJ, Afessa B (2005) Agreement between bedside blood and plasma glucose measurement in the ICU setting. Chest 127: 1749–1751
13. Gambino R, Piscitelli J, Ackattupathil TA et al. (2009) Acidification of blood is superior to sodium fluoride alone as an inhibitor of glycolysis. Clin Chem 55: 1019–1021
14. Harrison JG (1995) Accuracy of fingerstick glucose values in shock patients. Am J Crit Care 4: 44–48
15. Haupt A, Berg B, Paschen P et al. (2005) The effects of skin temperature and testing site on blood glucose measurements taken by a modern blood glucose monitoring device. Diabetes Technol Ther 7: 597–601
16. Henrichs HR (2010) Kontinuierliche Glukosemessung (CGM) in der Gewebeflüssigkeit – Wissenschaftliche Bewertung von CGM und medizinische Beurteilung des Nutzens für die Diabetestherapie (AGDT-DDG-GKV CGM-Positionspapier). Im Internet: www.diabetes-technologie.de/vortraege/vortraege-2010.htm
17. Ingels C, Debaveye Y, Milants I et al. (2006) Strict blood glucose control with insulin during intensive care after cardiac surgery: impact on 4-years survival, dependency on medical care, and quality-of-life. Eur Heart J 27: 2716–2724
18. Jungheim K, Koschinsky T (2002) Glucose monitoring at the arm: risky delays of hypoglycemia and hyperglycemia detection. Diabetes Care 25: 956–960
19. Jungheim K, Koschinsky T (2002) Glucose monitoring at the thenar: evaluation of upper dermal blood glucose kinetics during rapid systemic blood glucose changes. Horm Metab Res 34: 325–329
20. Kellerer M, Danne T (Ed.) (2015) Praxis-Leitlinien der Deutschen Diabetes-Gesellschaft; Diabetologie und Stoffwechsel 10 (Suppl 2); Evidenzbasierte Leitlinien DDG, current version available on the DDG websites: www.deutsche-diabetes-gesellschaft.de/Evidenzbasierte Leitlinien
21. Kerner W, Brückel J (2015) Definition, Klassifikation und Diagnostik des Diabetes mellitus. Diabetologie und Stoffwechsel 10 (Suppl 2) S 98–S101; current version available on the DDG websites: www.deutsche-diabetes-gesellschaft.de/Evidenzbasierte Leitlinien/Praxisempfehlungen der Deutschen Diabetes Gesellschaft/ Definition
22. Khan AI, Vasquez Y, Gray J, Wians FH Jr, Kroll MH (2006) The variability of results between point-of-care testing glucose meters and the central laboratory analyzer. Arch Pathol Lab Med 130: 1527–1532
23. Kleinwechter H, Schaefer-Graf U (2011) Gestationsdiabetes mellitus; Diabetologie und Stoffwechsel 5 (Suppl); Evidenzbasierte Leitlinien DDG, current version available on the DDG website: www.deutsche-diabetes-gesellschaft.de/Evidenzbasierte Leitlinien/Gestationsdiabetes mellitus
24. Koschinsky T, Jungheim K, Heinemann L (2003) Glucose sensors and the alternate site testing-like phenomenon: relationship between rapid blood glucose changes and glucose sensor signals. Diabetes Technol Ther 5: 829–842
25. Kulkarni A, Saxena M, Price G, O'Leary MJ, Jaques T, Myburgh JA (2005) Analysis of blood glucose measurements using capillary and arterial blood samples in intensive care patients. Intensive Care Med 31: 142–145
26. Kuwa K, Nakayama T, Hoshino T, Tominaga M (2001) Relationships of glucose concentrations in capillary whole blood, venous whole blood and venous plasma. Clin Chim Acta 307: 187–192
27. Lackner K, Luppa PB, Koschinsky T, Danne T (2009) Ein einheitlicher Kalibrationsbezug (Plasma statt Vollblut) bei der patientennahen Glukosebestimmung verbessert die Therapiesicherheit beim Einsatz von Glukosekonzentrationswert-abhängigen Therapiealgorithmen: Eine gemeinsame Initiative der DGKL, diabetesDE und der DDG. Internet: http://www.deutsche-diabetes-gesellschaft.de/redaktiion/news/POCT-Glukose_Plasmastandard_DGKL_diabetesDE_1209.pdf
28. Mikesh LM, Bruns DE (2008) Stabilization of glucose in blood specimen: mechanism of delay in fluoride inhibition of glycolysis. Clin Chem 54: 930–932
29. NICE-SUGAR Study Investigators, Finfer et al. (2009) Intensive versus Conventional Glucose Control in Critically Ill Patients N Engl J Med 360: 1283–1297
30. Rao LV, Jakubiak F, Sidwell JS, Winkelmann JW, Snyder ML (2005) Accuracy evaluation of a new glucometer with automated hematocrit measurement and correction. Clin Chim Acta 356: 178–183
31. Schetz M, Vanhorebeek I, Wouters PJ, Wilmer A, Van den Berghe G (2008) Tight blood glucose control Is renoprotective in critically ill patients. J Am Soc Nephrol 19: 571–578

32. The International Expert Committee (2009). International Expert Committee report on the role of the A1c assay in the diagnosis of diabetes. Diabetes-Care 32: 1327–1334
33. Van den Berghe G (2008) Insulin therapy in the intensive care unit should be targeted to maintain blood glucose between 4.4 mmol/l and 6.1 mmol/l. Diabetologia 51: 911–915
34. Van den Berghe G, Wilmer A, Hermans G et al. (2006) Intensive insulin therapy in the medical ICU. N Engl J Med 354: 449–461
35. Van den Berghe G, Wouters P, Weekers F et al. (2001) Intensive insulin therapy in the critically ill patients. N Engl J Med 345: 1359–1367
36. Vanhorebeek I, Langouche L, Van den Berghe G (2007) Tight blood glucose control: what is the evidence? Crit Care Med 35 (9 Suppl): S496–S502
37. Lenters-Westra E, Slingerland RJ (2014) Three of 7 hemoglobin A1c point-of-care instruments do not meet generally accepted analytical performance criteria. Clin Chem 60:1062–1072
38. Kerner W, Freckmann G, Müller UA, Roth J, Schleicher E, Niederau C, Müller-Wieland D, Landgraf R, Heinemann L (2015) Positionspapier der Kommission für Labordiagnostik in der Diabetologie der DGKL und der DDG zur HbA1c-Messung. Diabetologie und Stoffwechsel 10: 329–333
39. International Organization for Standardization (2003) In Vitro Diagnostic Test Systems – Requirements for Blood-Glucose Monitoring Systems for Self-Testing in Managing Diabetes Mellitus. EN ISO 15197:2003 (E). Geneva
40. International Organization for Standardization (2013) In Vitro Diagnostic Test Systems – Requirements for Blood-Glucose Monitoring Systems for Self-Testing in Managing Diabetes Mellitus. ISO 15197:2013 (E). Geneva

Continuous monitoring of metabolic parameters

Michael Imhoff, Theodor Koschinsky

P. B. Luppa, R. Junker (Eds.), *Point-of-Care Testing*
https://doi.org/10.1007/978-3-662-54497-6_13

13.1 Definition of monitoring

In the narrower sense, patient monitoring is the systematic, scheduled and repeated collection and presentation of parameters on bodily and organic functions, biochemical and other processes based on registered biosignals with the primary objective of making information about a patient's current state available. Disorders of the observed functions and processes should be detected as early as possible [3]. More comprehensively, the measuring values collected form the basis for decision-supporting systems which reflect the impact of the detected disorders on the clinical objective; from this, possible diagnostic, preventive and therapeutic measures can be derived and reproducibly presented.

A subgroup of patient surveillance is continuous patient monitoring, which hereinafter will simply be referred to as monitoring alone. However, monitoring is more narrowly described and can be defined as follows:

- Measurement of one or more physiological variables
- Continuous or automatically activated discontinuous function
- Alarm option
- Recording and displaying changes over time
- Up-to-date, clinically relevant measurements

13

In the broader sense of term "patient monitoring", the collected values are preferably analyzed according to pre-set rules, with any deviations being reported to the user by appropriate means.

The **objectives of monitoring** include, but are not limited to

- Creating or improving patient safety
- Supporting therapy and diagnostics
- Optimizing quality and ensuring quality assurance of medical processes
- Expanding the "safe free space" of a monitored patient

Particularly with a view to the collection of laboratory values, two fundamental characteristics of monitoring should be considered: First, automatic measurements are carried out repeatedly or continuously. By definition, no direct interaction takes place between user and patient or user and device. Second, further processing of the collected laboratory findings takes place rapidly without validation by a specialist in laboratory medicine (e.g. for generating automatic alarms), the findings are presented to the end user. In physiologic closed-loop controllers, the findings are directly converted into therapeutic decisions. It is particularly important to take these characteristics into account in both the technical implementation and the technical and clinical risk management.

13.2 Monitoring and therapeutic implications

Depending on each patient's disease severity or nature and invasiveness of the intervention performed, the bodily and organ functions as well as biochemical processes may not reliably lie within sufficiently specific physiological limits. That is why supportive interventions are called for to prevent or avert harm. The use of monitoring helps identify such situations.

In order to enable interventions, to detect their necessity in a timely manner, to monitor their efficacy and possible side effects and to be able to reliably and safely attend to chronically ill patients over the clinical course of their disease, it is essential to measure multiple bodily and organ functions and the biochemical processes they represent based on characteristic data and, finally to display and perform a workup on these findings. Depending on their clinical importance, the characteristic data are collected and analyzed by monitoring methods in varyingly time-critical and regular fashions in order to expose any changes early and react to them appropriately.

As with any other diagnostic measure, monitoring is only clinically feasible when a time-critical interpretation of the findings leads to a therapeutic decision or further diagnostic measures. A therapeutic decision may also mean doing nothing because the measured values are within the reference range.

13.2.1 Warning against danger

Biochemical variables can be monitored to obtain warnings against emerging or existing dangers. This can then lead to immediate therapeutic measures or give occasion to further diagnostics. The key aspect is that the monitoring is carried out automatically and continuously or the measurements are repeated at close intervals. Thereby, warnings or alarms can also be triggered without the clinician previously having a suspicion or a clinical reason to order the corresponding laboratory test. Naturally, it is important that the warning or alarm can also be transmitted to the clinician in a time-critical manner; this can be done by integration into a patient monitoring system.

Capnography is one example where warnings are generated by biochemical monitoring on anesthesia devices – an alarm is triggered if the respiratory system is disconnected or the calcium is depleted on the absorber. In this example, a continuous measurement with a correspondingly fast response is required because there is the danger that a critical situation might develop within seconds or minutes [1]. Another example is sepsis monitoring which enables the staff to react early to emerging parametric symptoms. Early administration of antibiotics to treat manifest sepsis often dictates the clinical outcome [2]. In such cases, intermittent measurement of sepsis markers in blood can facilitate the appropriate monitoring. The same parameters my also guide determination of the proper time point to stop antibiotic therapy.

13.2.2 Titration of therapy

Titration of therapy is another common implementation for monitoring to guide therapeutic decisions. This involves adjusting a drug's dose rate according to changes in one or more measured variables. This is typically how the settings on the dosage dispenser (anesthetic vaporizer/humidifier) for a volatile anesthetic agent are made based on the concentration of the anesthetic agent measured in the patient's exhaled air. The expiratory breathing gas concentration of volatile anesthetic agents in the steady state is closely correlated with the concentration at the site of action in the brain and is thereby a strong surrogate for clinical effectiveness, i.e. for the depth of anesthesia.

Dosing insulin guided by blood glucose levels in diabetics is certainly the most frequently used therapy titration based on a laboratory parameter worldwide (▶ Section 13.6 and ▶ Chapter 12).

In these applications, therapy is generally titrated by the user after interpretation of the measured values. This type of control is also known as an "open-loop controller" because the physician is still always "in the loop" of the circuit and makes the final decision.

13.2.3 Physiologic closed-loop controllers

In closed-loop systems, the current measured value is returned to the controller, which continuously prevents any deviation from the target value (negative feedback). Physiologic closed-loop controllers represent a special case where the controlled loop lies with the individual person, i.e. with the patient in our context here. In these closed-loop systems, the regulation or titration of therapy takes place fully automatically, without continuous monitoring or intervention by the healthcare staff or patient. Examples of this include regulation of inspiratory oxygen concentration on ventilators for neonates and pre-term infants based on the pulse oximetric monitoring of arterial oxygen concentration (▶ Chapter 11), automatic monitoring of fresh gas concentration in anesthetic gases based on expiratory concentrations or the fully automated control of insulin and glucose infusion based on continuously measured blood concentrations as part of blood glucose clamp studies. Particularly in physiologic closed-loop systems, risk management is paramount [4].

13.3 Systematics of monitoring methods

Monitoring methods can be classified according to different criteria. Fundamentally, all interventions that injure or penetrate the body's surface are referred to as invasive. This also includes any insertion through natural body orifices. Interaction with the body means that through the monitoring methods energy or substances are caused to enter into the body. The nature, intensity and duration thereof are decisively important for risk assessments.

One possible classification might be as follows (without any claim to completeness):

- **Invasiveness**
  - Non-invasive
    - Direct contact with the body, e.g. through electrodes, sensors.
    - Non-contact
  - Invasive
    - Implants
    - Penetration of the skin: intravascular catheters, catheters/probes/electrodes in subcutaneous tissue)
    - Insertion through natural body orifices
    - Minimally invasive: Transdermal method
- **Sampling**
  - No drawing of blood, bodily fluids or gases
    - In-line sensors are inserted into a blood vessel, for example. That makes them invasive. The measurement is taken in the flow of the blood stream. That means that no sampling takes place.
  - Repeated or continuous draws of blood, bodily fluids or gases
    - Without return (the collected samples are discarded)
    - With return (collected samples are returned back into the body after measurement)
- **Matrix**
  - Gases
  - Blood
    - Whole blood
    - Serum
    - Plasma
  - Tissue fluid
    - Interstitial fluid
  - Other bodily fluids
- **Interaction with the body**
  - No interaction
    - Passive measurement
    - Extracorporeal analysis without backflow of the substrate
  - Mechanical interaction
    - Ultrasound
    - Pneumatics
  - Electrical, electromagnetic, thermal interaction
    - Electrical flows (e.g. bioimpedance)
    - Magnetic fields (e.g. MRI)
    - Heat (e.g. transcutaneous O_2/CO_2 measurement)
  - Chemical interaction
    - Indicators (e.g. lithium, indocyanine green (ICG) fluorescence angiography)
    - Change in chemical characteristics of the blood etc., when returned back into the body
- **Clinical laboratory test methods** (please refer to the chapters in Section II).

13.4 Application examples

Out of the many applications possible for monitoring by POCT, only several typical examples shall be mentioned here.

The monitoring of individual **blood glucose levels** in diabetics or the critically ill can probably be said, by far, to be most frequently used biochemical monitoring method worldwide (▶ Chapter 12). Moreover, the past few years has seen a rise in the use of continuous subcutaneous glucose monitoring both as an

13

independent measure as well as in combination with continuous subcutaneous insulin infusion systems (insulin pumps), sometimes even included as a closed-loop in the application (▶ Section 13.6).

Another area of monitoring highly relevant in intensive care, emergency medicine and anesthesiology is the monitoring of **blood gases,** including oxygen saturation of the blood. While conventional blood gas analysis depends on the user to draw blood and bring the sample to the appropriate analyzing device, monitoring methods enable automatic user-independent measurement of pO_2, pCO_2 or oxygen saturation or a combination of these variables. A variety of these types of methods are in clinical use:

- Pulse oximetry for continuous, non-invasive measurement of arterial sO_2.
- In-line sensors for directly monitoring pO_2 and pCO_2 in the blood stream of an extracorporeal circulation. These are standard sensors used for heart-lung machine monitoring during heart surgeries.
- Transcutaneous measurement of pO_2 and pCO_2 in neonates and pre-term infants (▶ Chapter 11). This measurement is only reliable for pO2 in neonates and pre-term infants. Due to the delicate dermal structure in these tiny patients, the partial pressures of O_2 and CO_2 very closely approximate arterial levels in skin locally heated to over 40° C.
- Transcutaneous measurement of pCO_2 and pulse oximetric measurement of oxygen saturation in adults. Unlike O_2, the diffusion of CO_2 is also so good in adults that the CO_2 partial pressures measured after warming the skin correlate very closely with arterial levels. In combination with pulse oximetry, this enables continuous, non-invasive monitoring of blood gases in adults as well.
- Since the end-tidal pCO_2 value in the exhaled air of persons with healthy lungs is only approx. 1–5 mmHg below the arterial value, capnometry can also be used as a surrogate in $paCO_2$ monitoring.

The exhaled air can also be used to continuously monitor trace gases. By measuring propofol in the breath (ppb), conclusions can be drawn about the serum concentrations of this hypnotic in blood. Numerous detection methods can be used for this, such as electrochemical sensors or gas chromatography mass spectrometry (GS-MS) [5, 6]. These methods, however, have not found their way into clinical routine yet.

By contrast, the standard in clinical practice is to continuously monitor the inspiratory and expiratory concentrations of the anesthetic gases (O_2, N_2O, volatile anesthetics) during anesthesia. For this purpose, the monitoring devices employ various methods like paramagnetic oxygen sensors, photoacoustic or infrared absorption methods for measuring volatile anesthetic agents, N_2O and CO_2. Usually, these devices for monitoring all anesthetic gases are integrated into the anesthesia machine or patient monitor and measure the gases in the side stream.

13.5 Methodology of clinical studies on monitoring

The study of the performance capability and clinical benefits of modern medical technologies is growing in importance. Beside the performance capability and safety demonstrated within the scope of the respective marketing approval, clinical users are increasing demanding that outcome studies be conducted before a new technology can be considered for clinical application. Outcome studies are particularly challenging when non-therapeutic methods like monitoring or POCT are being investigated.

Typically, laboratory methods and monitoring methods are investigated exhaustively with regard to the accuracy and precision of their measurements and the sensitivity and specificity of their diagnostic conclusions. Nevertheless, almost all possible outcome impacts are dictated by the therapeutic interventions that have been triggered or guided by the monitoring methods in question. Hence, such studies require that the monitoring methods be embedded in a therapeutic protocol. The treatments described therein must have unequivocal effectiveness and potentially affect the endpoints investigated. The chronological behavior of monitoring and therapy must also be adequate. Whenever the aim is to investigate a new monitoring method, it is not advisable to investigate a new therapeutic intervention in one and the same study, since this usually makes is difficult to distinguish with certainty the effects of the each individual method on the target variables. And, unless the validity of the monitoring method has already been demonstrated elsewhere, the target variables should moreover be distinguishable from the variables that are measured by the each of the respective monitoring methods. As in every other clinical or preclinical study, it is imperative that monitoring methods studies investigate clinically relevant endpoints in adequately defined populations.

13.6 Continuous glucose monitoring (CGM)

In ▶ Chapter 12, a form of in vitro testing of blood glucose is presented that is based on the principle of continuously monitoring glucose in the interstitial fluid of the subcutaneous fatty tissue. This monitoring is done via needle electrodes integrated in a glucose oxidase measuring system. The main target group is the population of diabetics receiving intensified insulin therapy. For many years now, this type of continuous glucose (CGM) has been enhanced and combined in various ways with continuous subcutaneous insulin infusion systems (CSII). The test is mainly done in the areas of the abdomen and the arms. At present in Germany, various systems are market-approved that transmit the current measured values to an extracorporeal device; at brief intervals (<1–5 min), this device continuously collects, processes, stores and reads out the data with immediate availability (online operation/real-time CGM) to achieve needs-based therapy adjustments. These techniques incorporate alarm functions, i.e. they warn about hypo- and hyperglycemia (Tab. 13.1) [7, 8, 9, 10, 12].

Clearly, the main advantages of CGM procedures primarily lie in the increased number of glucose readings with 500–1,500 per 24 h compared to 2–4 per hour by SMBG 24 h and in the resultant information gain, e.g. regarding

- subclinical hypoglycemic episodes, particularly during sleep,
- postprandial blood glucose profiles and
- overall 24-h glucose variability

The data can be used

- for targeted therapy adjustments by a properly trained diabetic (open-control system) or
- by teaching special algorithms for the targeted control of an s.c. insulin pump without intervention by the patient (closed-loop circuit).

Positive evidence from clinical studies relating to a significant improvement in metabolic control as measured by HbA_{1c}, a decrease in the frequency and extent of hypoglycemia and a lower variability of the 24-hour blood glucose levels was obtained for

- the sensor-supported treatment decision (SSTD) where the interruption and adjustment of insulin therapy is made by the diabetic alone (open-loop controller) and
- sensor-supported pump therapy (SSPT) in 2 forms:
  - as a hypoglycemia alarm with patient-independent, automatic threshold-suspend of the insulin pump, with its re-start and dosage adjustment left up to the patient, and
  - as predictive hypoglycemia management, e.g. during sleep, linked to the continuous sensor glucose values with a patented, algorithm-based, hypoglycemia-preventive, automatic insulin dose adjustment over several hours without patient intervention (time-limited phase of a closed-loop controller) [8, 10, 12].

13

Tab. 13.1 Examples s.c. CGM sensor systems for insulin therapy without/with s.c. insulin pumps

Glucose sensor	Sensor device system (manufacturer/sales)	Sensor period of use (days)	Calibration frequency (n/day)	Coupled to an insulin pump (no/yes)	Insulin pump manufacturer
FreeStyle Libre	Flash glucose monitoring system (Abbott Diabetes Care)	14	0	No	n/a
DexComG5	Mobile CGM System (DexCom/ninta med)	7	2	No	n/a
DexComG4	Platinum CGM System (DexCom/ninta med)	7	2	Animas Vibe Platinum for SSPT*	Johnson & Johnson Animas
Enlite	Glucose sensor (Medtronic/MiniMed)	6	2	MiniMed Paradigm-Veo for SSPT*+ LGS**	Medtronic
Enlite II	Optimized glucose sensor (Medtronic/MiniMed)	6	2	MiniMed 640 G + SmartGuard for PLGM***	Medtronic

* SSPT: Sensor supported pump therapy; ** LGS: Low-glucose suspend; *** PLGM: Predictive low glucose management

With this incremental progress in the use of increasingly automated SSPT, it can be justifiably predicted for the foreseeable future that the evolution of such closed-loop systems will deliver even greater alleviation to patients requiring this therapy. The relief will come in their everyday metabolism monitoring and insulin-dependent control.

To date, all CGM systems share a common problem that glucose has been "measured" from or in the s.c. interstitial fluid and could not be objectified by an independent method. Clinical experience is based on data from glucose levels in the blood (and not in the subcutaneously accessible interstitium). As a result, CGM readings are converted and displayed as **blood glucose equivalents** in mg/dL or mmol/L, based on a few parallel capillary blood glucose measurements using different adjustment algorithms. These procedures take place unknown to the user and allow for physical and physiologic time delays caused by glucose flux between blood and interstitial tissue. Alongside a breadth of system-dependent errors, this approach also has a considerable potential for errors arising from anatomical and physiological variables that fluctuate inter- and intra-individually over the course of diabetes [11]. The resulting quality assurance requirements for such CGM systems vary considerably from those of the single measurements by in-vitro glucose POCT. However, a regulatory legal framework for this does not yet exist in Germany.

Moreover, the previous CGM systems only featured relatively short periods of use of 6–14 days at most. Afterwards, a new sensor must be re-inserted and connected to the matching insulin pump. It would be ideal if these invasive glucose sensors could be replaced by analytical systems for non-invasive glucose monitoring. Despite the large fundamental interest of the last 40 years and more, the very high investments put into development and a diversity of technological approaches, these systems have not yet reached the maturity for marketing authorization. Currently, the necessity for further development is expected to extend over the next 5–20 years. Nevertheless, its clinical advantages have enabled the circle of users of s.c.

CGM systems to keep growing from year to year – not only in outpatient, but also in clinical settings: Above all, this presently applies to internistic diabetes wards and intensive care. The use of such CGM systems, particularly in clinical settings, complies with the key attributes for POCT.

References

1. American Society of Anesthesiologists (2015) Standard for Basic Anesthetic Monitoring (http://asahq.org/~/media/Sites/ASAHQ/Files/Public/Resources/standards-guidelines/standards-for-basic-anesthetic-monitoring.pdf [last accessed: 21 April 2016])
2. Dellinger RP, Levy MM, Rhodes A, Annane D, Gerlach H, Opal SM, Sevransky JE, Sprung CL, Douglas IS, Jaeschke R, Osborn TM, Nunnally ME, Townsend SR, Reinhart K, Kleinpell RM, Angus DC, Deutschman CS, Machado FR, Rubenfeld GD, Webb S, Beale RJ, Vincent JL, Moreno R; Surviving Sepsis Campaign Guidelines Committee including The Pediatric Subgroup (2013) Surviving Sepsis Campaign: international guidelines for management of severe sepsis and septic shock, 2012. Intensive Care Med 39 (2): 165–228
3. DGBMT- Fachausschuss Methodik der Patientenüberwachung (2009) Definition: Patient monitoring (http://www.vde.com/de/fg/dgbmt/ueber %20uns/vorstand/documents/definition %20patientenüberwachung.pdf [last accessed: 21 April 2016])
4. DIN EN 60601– 1–10:2008 – Medical electrical equipment – Part 1–10: General requirements for basic safety and essential performance – Collateral Standard: Requirements for the development of physiologic closed-loop controllers
5. Grossherr M, Varadarajan B, Dibbelt L, Schmucker P, Gehring H, Hengstenberg A (2011) Time course of ethanol and propofol exhalation after bolus injection using ion molecule reaction-mass spectrometry. Anal Bioanal Chem 401 (7): 2063–2067
6. Hornuss C, Wiepcke D, Praun S, Dolch ME, Apfel CC, Schelling G (2012) Time course of expiratory propofol after bolus injection as measured by ion molecule reaction mass spectrometry. Anal Bioanal Chem 403 (2): 555–561
7. Henrichs HR (2010) continuous Glukosemessung (CGM) in der Gewebeflüssigkeit – Wissenschaftliche Bewertung of CGM and medizinische Beurteilung of the Nutzens for the Diabetestherapie (AGDT-DDG-GKV CGM White Paper). On the Internet: www.diabetes-technologie.de/vortraege/vortraege-2010.htm
8. Kropff J, DeVries J H (2016) Continuous glucose monitoring, future products, and update on worldwide artificial pancreas projects. Diabetes Technol Ther 18 (Suppl 2): 53–63
9. Statement issued by the Diabetes & Technology (AGDT) Working Group of the German Diabetes Society on substituting blood glucose monitoring with measurement systems for continuous glucose monitoring (CGM) or flash glucose monitoring (FGM) ("replacement claim"): 2016; on the Internet: www.AGDT_Stellungnahme_Replacement_20160130-L.doc
10. Rodbard D (2016) Continuous Glucose Monitoring: A review of successes, challenges, and opportunities. Diabetes Technol Ther 18 (Suppl 2) 3–13
11. Schmelzeisen-Redeker G, Schoemaker M, Kirchsteiger H, Freckmann G, Heinemann L, del Re L (2015) Time delay of CGM sensors: Relevance, causes, and countermeasures. J Diabetes Sci Tech 9: 1–10
12. Siegmund T, Thomas A (2015) Aktueller Stand der Insulinpumpentherapie. Diabetologie 10: R67–R84

Blood gas analysis and disorders of acid-base balance – including analytical methods

Peter B. Luppa, Jan Martin, Philipp Deetjen

P. B. Luppa, R. Junker (Eds.), *Point-of-Care Testing*
https://doi.org/10.1007/978-3-662-54497-6_14

14.1 Introduction

For decades, blood gas analysis (BGA) has been established in intensive care, anesthesiology and emergency medicine as well as in lung function diagnostics and is therefore primarily conducted at the POC [1, 2, 10]. BGA helps objectify a patient's oxygenation status, ventilation and acid-base balance. In light of the increased prevalence of chronic obstructive pulmonary disease, BGA is becoming increasingly important, particularly in preclinical settings.

In hospitals, the central laboratory is usually involved in quality assuring the various parameters, measured by BGA systems. In different versions, these devices analyze the following parameters:

- The basal variables – partial oxygen pressure (pO_2), partial CO_2 pressure (pCO_2), and pH – are measured in arterial, capillary or venous blood; whereas a series of other variables in BGA (e.g. standard bicarbonate) and of the metabolic acid-base status (e.g. base excess) are calculated. Blood gases can be corrected to the patient's current body temperature.
- Blood oxygenation (O_2 saturation, sO_2), total hemoglobin (Hb) concentration, hematocrit (Hct) and Hb fractions (e.g. FO_2Hb, MetHb) are measured by CO-oximetry.
- The measurement of the electrolytes Na^+, K^+, ionized Ca^{2+}, ionized Mg^{2+} and Cl^- supplement the determination of the metabolic acid-base status.
- Metabolites such as glucose, lactate, creatinine, urea, bilirubin and others can optionally be measured as well.

Fundamental principles in sensor technology have not changed since blood gas analysis was in its nascent stage. Only the principles of metabolite measurements are new and differ across the various systems. Validated sensor systems to determine creatinine levels have only recently become available [17] (▶ Chapter 8).

Inconsistent nomenclature used by various manufacturers of blood gas analyzers to measure and calculate blood gas parameters is still an unsolved problem in clinical practice.

14.2 Monitoring methods for pH, pO_2 and pCO_2

Two different biosensor technologies are used in modern blood gas analyzers to measure pO_2, pCO_2 and pH in heparinized whole blood.

- Electrochemical methods
- Optical methods

Automated functional checks of the single sensors, calibration at set time intervals and cleaning procedures are common features in all of these device types. A range of conventional systems is shown in ◘ Tab. 14.1.

The devices i-STAT, Epoc (◘ Fig. 14.1) and OPTI CCA-TS2 are unit-use systems. The i-STAT and the Epoc use electrochemical sensors in cartridges, whereas the OPTI cassette binds analytes with the fluorescent sensor molecule to optically measure blood gas parameters.

The novel Proxima device by Sphere Medical, Cambridge, UK allows a quasi-continuous measurement of blood gas parameters via an arterial line operating as a closed system. The device, which has a touchscreen interface, is employed directly at the bedside in intensive care units.

Devices by Alere (Epoc), Werfen (GEM Premier 4000), Nova (Critical Care Xpress) and Radiometer (ABL800 FLEX) are presented are depicted in ◘ Fig. 14.1, ◘ Fig. 14.2, ◘ Fig. 14.3 and ◘ Fig. 14.4.

14.2.1 Electrochemical sensor

The electrochemical BGA sensor consists of a pH glass electrode, a Stow-Severinghaus pCO_2 electrode and a Clark pO_2 electrode. These sensors use gas-selective membranes that act as chemospecific detection layers. The measuring principle is based on electrochemical reactions that occur between electrode surface and blood. The electric current (pO_2) or tension

Tab. 14.1 Blood gas systems established on the German market

Manufacturer	Device Names
Abbott GmbH	i-STAT
Werfen, Germany	GEM Premier 3000, 3500, 4000, 5000, GEM OPL
Alere GmbH	Epoc
Keller Medical GmbH	Irma TRUpoint, AVOXimeter 1000E and 4000
Nova Biomedical GmbH	Stat Profile pHOx/pHOx Ulltra, Stat Profile Prime Critical Care System, Bioprofile Analyzer Series 100/300/400
OPTI MEDICAL Systems GmbH	OPTI CCA, OPTI CCA-TS2
Radiometer GmbH	ABL80 FLEX, ABL80 FLEX CO-OX, ABL90 FLEX, ABL90 FLEX PLUS, ABL800 FLEX
Roche Diagnostics Germany GmbH	cobas b 123, cobas b 221
Siemens Medical Solutions Diagnostics GmbH	Rapidlab 248/348(EX), Rapidlab 1200; Rapidpoint 400/405/500

The method spectrum varies considerably: Blood gases with or without CO-oximetry; blood gases with or without CO-oximetry plus electrolytes; blood gases with or without CO-oximetry plus electrolytes and substrate/metabolites.

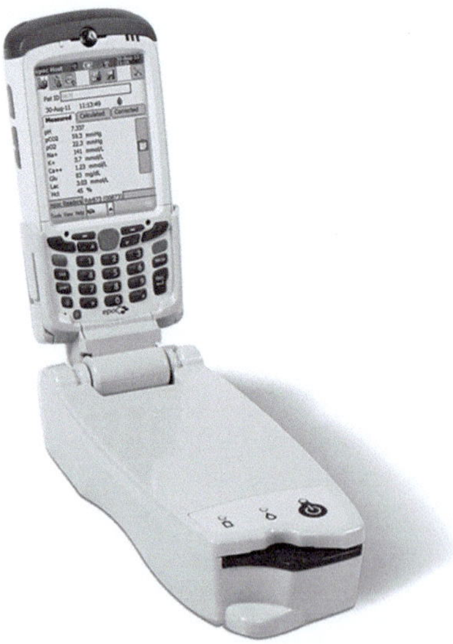

Fig. 14.1 Epocal Epoc. (Courtesy of Alere GmbH, Cologne, Germany)

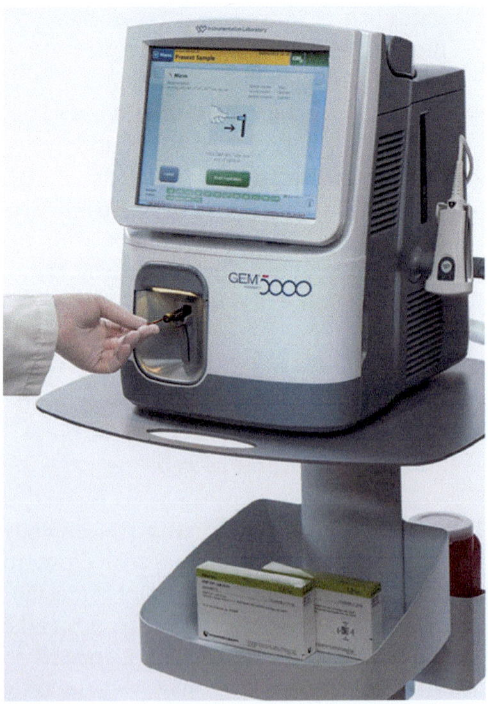

Fig. 14.2 IL GEM Premier 5000 blood gas analyzer. (Courtesy of Instrumentation Laboratory GmbH, Kirchheim, Germany)

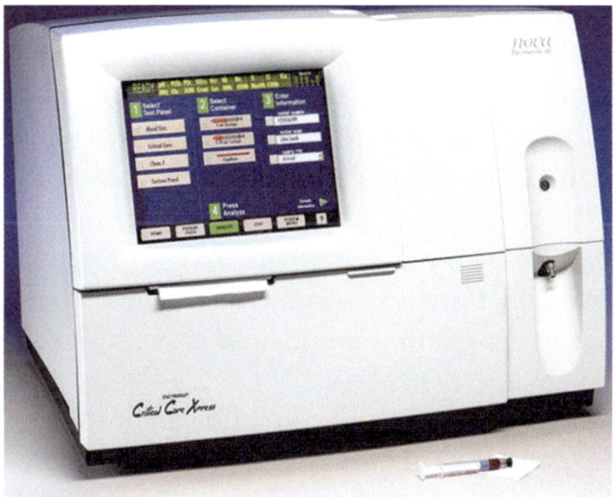

Fig. 14.3 Nova Critical Care Xpress. (Courtesy of Nova Biomedical GmbH, Rödermark, Germany)

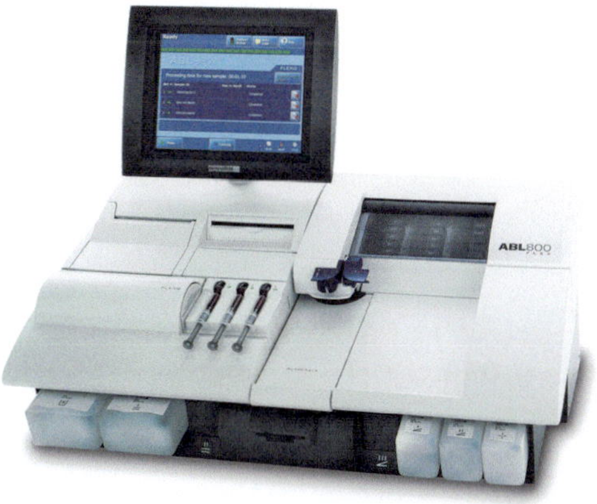

Fig. 14.4 Radiometer ABL800 FLEX. (Courtesy of Radiometer GmbH, Willich, Germany)

(pH and pCO_2) are proportional to the H^+ concentration or the partial O_2 and CO_2 pressures. Electrodes referred to here are combination electrodes, i.e. a single cylinder contains both the membrane (measuring) and reference electrodes. The use of silicone membranes in pCO_2 sensors accelerates the diffusion of CO_2 gas molecules and therefore shortens the total measurement time significantly. In pO_2 sensors, oxidative disproportionation of the resulting H_2O_2 is the speed-determining factor. The reaction is optimized using catalytically active platinum black (a particular powder form of platinum) that is fixed to the inner layer of an O_2-permeable Teflon membrane. This allows a total analysis time of less than 60 s at a sample volume of <400 μL (including CO-oximetry).

14.2.2 Optical sensor

In optical BGA sensors that generate spectroscopic signals (optodes), pO_2 is determined by an excitable fluorescent dye, whose luminescence lifetime is modulated by the O_2 concentration in the blood sample. This change can be correlated to a pO_2 value using a standard series [10]. The **Stern-Volmer equation** quantifies the relationship between the emitted fluorescence and pO_2. The emission intensity of the fluorescence (luminescence) is inversely proportional to the pO_2 as quencher. In contrast to the electrochemical Clark electrode, optodes do not consume any O_2 molecules during measurement.

In contrast, the principle of pCO_2 measurement is based on measuring absorption in the infrared range. Dissolved CO_2 absorbs infrared light. Photometers measure absorption at three different wavelengths and calculate the dissolved CO_2 concentration and the pCO_2 according to the Lambert-Beer law. The determination of pH in whole blood is based on the spectroscopic detection of specific light absorptions by a color indicator in the visible spectral range by means of a diode array photometer. These characteristic absorptions can be assigned to the actual pH. The indirect measurement of pCO_2 is historically interesting. According to the method by Astrup and Siggaard-Andersen, presented in the late 1950s, the endogenous pCO_2 can be calculated after equilibration of blood with a gas mix of a known CO_2 concentration followed by measurement of changes in the pH.

In order to calibrate older blood gas analyzers, ultra-pure CO_2 and O_2 from gas bottles are required to guarantee reproducible partial O_2 and CO_2 pressures in solutions. Buffer solutions with constant partial gas pressures stored in plastic containers are difficult to produce as these containers are gas permeable to a lesser degree. In the most modern blood gas analyzers, room air and calibration solutions are used, which are mixed from 2 base solutions immediately before their calibration. One solution is subsequently O_2-depleted.

14.3 Oximetry

CO-oximeters are multi-wavelength photometers that measure total Hb concentration, O_2 saturation (sO_2) and the Hb fractions O_2Hb, HHb (deoxy-Hb), COHb, met-Hb etc. in a blood sample in the range of 400 to 1000 nm. Oximetry allows the diagnosis of CO intoxication (smoke intoxication) or poisoning with substances that produce MetHb (nitrite etc.). This method compares absorption spectra of Hb fractions in relation to wavelength. It allows their concentration to be measured using matrix equations. The following relationship applies to each fraction for absorption A:

$$A = \lambda \times \varepsilon \times c$$

Here, ε is: The extinction coefficient of the particular Hb species, λ: Wavelength and c: concentration.

The total absorption is the sum of the individual absorptions of the Hb fractions analyzed. Measurements need to be taken at a minimum of at least two different wavelengths to determine single concentrations. Each Hb fraction is determined by its characteristic absorption maximum. Interference of bilirubin, for example, can thereby be eliminated.

Devices from the market-leading manufacturers Radiometer, Siemens, Roche, Nova and Werfen (Instrumentation Laboratory) carry out between 128 and 512 different measurements at a measuring range of 460–680 nm (e.g. Radiometer ABL at 128 wavelengths in the range of 478–672 nm; Roche cobas b 221 at 512 wavelengths in the range of 460–660 nm; Siemens Rapidlab 1200 at 256 wavelengths in the range of 500–680 nm). Here, the individual extinction maxima are detected automatically. The devices from Roche and Radiometer use ultrasound to hemolyze blood samples, whereas the Nova and Werfen analyzers use lysis reagents. Only the Siemens analyzers (400, 500, 1200) do not rely on hemolysis. Intact red blood cells (RBC) cause a scattered light glare in the optical unit that is eliminated after complex data analysis by least square matrix programs. Avoiding the use of hemolyzed samples has the advantage of lower mechanical costs and a simpler oximetry process.

The total Hb concentration can be directly measured by photometry using the hemiglobincyanide method or indirectly via Hct as an alternative to the multi-wavelength photometry described above. The hemiglobincyanide method quantitatively measures the stable Hb derivative A by detecting its characteristic absorption maximum at 540–546 nm. Hb derivative A is produced when Hb is converted with potassium hexacyanoferrate III:

$$Hb(Fe^{2+}) + [Fe(CN)_6]^{3-} \rightarrow Hb(Fe^{3+}) + [Fe(CN)_6]^{4-}$$

$$Hb(Fe^{3+}) + CN^- \rightarrow Hb(Fe^{3+})CN \text{ (Derivative A)}$$

Hct, which is the ratio of RBCs to whole blood volume, can be determined by centrifugation or – as in most blood gas analyzers – by means of conductivity sensors (▶ Chapter 5). The conductivity of blood is indirectly proportional to the number and size of RBCs and is measured by the difference in the potential at two sensor points. Conductivity measurements are, however, prone to faults; interference occurs in patients on anticoagulation therapy, with significant leukocytosis and those receiving infusions of plasma expanders or crystalloid fluids in larger volumes. This may be relevant in cardiopulmonary bypass operations. Opti-Medical offers an alternative with its OPTI-CCA blood gas analyzer that determines Hb directly: Laser diode light is directed onto an oxygen sensor positioned at a window in the cassette. It is partially absorbed and scattered by the RBCs. The light is directed via an optical wave guide to a photo diode. This measures the light intensity, which can be used to calculate the total Hb concentration.

The spectrophotometric measurement of **total bilirubin** in whole blood is possible using the oximeter modules of the following devices:
- ABL 830 FLEX, 835 FLEX and 837 FLEX (Radiometer)
- Cobas b 123 and b 221 (Roche)
- GEM Premier 5000 (Werfen)
- Rapidlab 1265 (Siemens)
- Stat Profile pHOx Ultra (Nova)

Whereas the devices from Roche and Siemens are only suitable for measuring bilirubin in newborns, the Radiometer devices and GEM Premier 4000 can also analyze adult samples. For this purpose, the calculation module on ABL devices can be changed with a correction button. The Nova device has only been evaluated so far for adult blood samples.

The measurement principle is identical for all devices: In the CO-oximetry module, Hb fractions as well as bilirubin are determined by multi-wavelength measurements in hemolyzed (ABL, cobas, pHOx Ultra) and non-hemolyzed (Rapidlab) blood samples. Although the spectra of bilirubin and Hb differ greatly, high analytical demands are made as there is a large difference between the bilirubin and the "interfering" Hb concentrations. The bilirubin concentration is calculated from the absorption measurements using multicomponent analysis. The applied algorithm – in addition to an exacting wavelength calibration – is crucial to the accuracy of the results [22].

Depending on the device, 35–100 µL blood is required for measuring bilirubin. The measurement range for most devices lies between 3 and 30 mg/dL. Manufacturers' own control samples allow quality controls. Evaluation studies have been published showing that the ABL devices, cobas b 221 and Rapidlab 1265 are comparable in terms of their performance data (precision, accuracy, interference). Method comparisons with the Jendrassik-Grof reference method show that deviations of up to ±2 mg/dL are to be expected for all devices in a concentration range of 3–20 mg/dL [3, 6, 11, 14, 21].

14.4 Calculated variables describing oxygen status

14.4.1 Oxygen saturation – sO_2 and psO_2

Modern multi-wavelengths CO-oximeters are able to measure all Hb fractions (F) (O_2Hb, HHb, MetHb, COHb and others) present in a blood sample, and thereby calculate the real oxygen saturation ("**fractional saturation**", sO_2); sO_2 is thus the ratio of O_2Hb to total Hb.

This is different than **partial saturation** psO_2, which is also called "functional saturation" and refers only to the O_2Hb fraction of the total binding capacity of Hb with bivalent heme iron (O_2Hb + HHb). At elevated COHb concentrations or when MetHb fractions are present (e.g. in smokers), psO_2 is measured falsely high. This parameter is read off older blood gas analyzers, but also pulse oximeters where transmission spectrometry is only measured at 2 wavelengths (▶ Chapter 11).

Note

Although the physiologic principles described are undisputed, the terminology has unfortunately not yet been unified and may therefore be confusing for users. Frequently, the "fractional saturation sO_2" is called "FO_2Hb", whereas "partial saturation psO_2" is referred to as "oxygen saturation sO_2".

14.4.2 Maximum oxygen-binding capacity, BO_2

The maximum oxygen-binding capacity is the amount of oxygen in milliliters that can be maximally transported in a blood volume of 1 dl. The value (in mL/dL) gives the total amount of oxygen that is bound to Hb. BO_2 is a helpful variable for assessing the effectiveness of O_2 ventilation therapy.

14.4.3 Oxygen concentration, cO_2

The oxygen concentration (cO_2), also referred to as "arterial total oxygen concentration", is usually measured by CO-oximetry, as is sO_2. The O_2 concentration in mL O_2/dL blood is the total of the Hb-bound and physically dissolved O_2 in the blood:

$$cO_2 = FO_2Hb \times cHb \times 1.34 + 0.0031 \times pO_2$$

Included in the formula is the Hüfner number, which refers to the amount of O_2 of 1 g Hb in vivo that can be maximally bound (1.34 mL O_2/g Hb). The solubility coefficient for O_2 is 0.0031 mL O_2/dL blood × mmHg.

All changes in the pO_2, Hb concentration and the O_2 binding behavior are captured by cO_2. This characteristic parameter is vital for assessing a patient's global oxygenation status. However, the above applies only in conjunction with cardiac output (volume per minute), as the body's oxygen (DO_2) is the product of cO_2 and cardiac output.

14.4.4 Arteriovenous oxygen difference, $avDO_2$

Together with cO_2, the arteriovenous oxygen difference ($avDO_2$) is an important characteristic parameter for evaluating the peripheral oxygen supply. It is the product of the difference between oxygen concentrations in arterial ($cO_2(a)$) and in mixed venous blood ($cO_2(v)$) that can be sampled via a pulmonary vein catheter (Swan-Ganz catheter). High $avDO_2$ value indicates increased O_2 extraction/utilization from the blood and therefore an inadequate oxygen supply to vital organs. This can also occur from a drop in cardiac output at sufficient cO_2 levels. Venous oxygen levels $cO_2(v)$ at constant $cO_2(a)$ and cardiac output are also important for assessing the oxygen supply in critically ill patients.

14.4.5 p50

On the sigmoid-shaped oxyhemoglobin binding curve, p50 is the partial O_2 pressure in the blood at which the hemoglobin is 50 % saturated (at 37° C and a pH of 7.4). p50 gives information about the O_2 release in the peripheral tissue and is read out with the results on some blood gas analyzers. Its value can also be calculated from the pH, pO_2 and sO_2.

14.4.6 Oxygen partial pressure in the alveolar gas mix, $pO_2(A)$

The O_2 partial pressure in the alveolar gas mix $pO_2(A)$ is a primary component of the gas exchange indices. This value is important for calculating the alveolar-to-arterial partial pressure gradient $pO_2(A–a)$ and the alveolar-to-arterial oxygenation index.

The **alveolar gas equation** is formulated for $pO_2(A)$:

$$pO_2(A) = F_iO_2\,(p_B - pH_2O) - (pCO_2(A)/RQ)$$
(simplified version)

The following parameters should be defined:

$[F_iO_2\,(p_B - pH_2O)]$ = inspiratory O_2 partial pressure with F_iO_2 = inspiratory oxygen fraction (at room air = 0.21), **p_B = barometric pressure** (is measured by the blood gas analyzer, mostly = 760 mmHg) and pH_2O = water vapor pressure (= 47 mmHg, 37° C);

$pCO_2(A)$ = alveolar CO_2 partial pressure (under physiological conditions, the $pCO_2(A)$ is approximately equal to the arterial $pCO_2(a)$);

RQ = respiratory quotient = CO_2 release/O_2 absorption per time unit (nutrition-dependent, for a mixed diet, an RQ of 0.85 can be assumed, for extracorporeal CO_2 elimination frequently falls to 0.4).

14.4.7 Alveolar-to-arterial oxygen partial pressure gradient – $pO_2(A–a)$

The alveolar-to-arterial O_2 partial pressure gradient $pO_2(A–a)$ can be calculated using the alveolar gas equation (see above):

$$pO_2(A–a) = pO_2(A) – pO_2(a) = F_iO_2\,(p_B – pH_2O) – (pCO_2(a)/RQ) – pO_2(a)$$

The alveolar-to-arterial (or alveolar-arterial) O_2 partial pressure gradient provides information about the efficiency of the oxygenation process in the lungs and is therefore important for evaluating pulmonary gas exchange in intensive care medicine.

14.4.8 Respiratory Index – RI

The respiratory index (RI) is calculated from the $pO_2(A–a)$ difference and pO_2 in the arterial blood at patient temperature. It is frequently used instead of the $pO_2(A–a)$ pressure difference:

$$RI = pO_2(A–a)/pO_2(a).$$

14.4.9 Oxygenation index (Horovitz index), OI

Sometimes referred to as the Horovitz index, the oxygenation index, OI, is calculated to evaluate the oxygenation function of the (damaged) lungs. The formula includes the arterial oxygen partial pressure pO_2 (a) and the concentration of oxygen in the inhaled air (F_iO_2).

$$OI = pO_2(a)/F_iO_2.$$

14.5 Calculated parameters describing the metabolic acid-base balance

14.5.1 Bicarbonate and base excess

Using pH, pCO_2, pO_2 and cHb, measured by BGA, the following acid-base parameters are calculated.

Bicarbonate concentration in plasma, $cHCO_3^-$

The $cHCO_3^-$ can be calculated from the pH and pCO_2, using the Henderson-Hasselbalch equation. However, $cHCO_3^-$ is dependent on the current pCO_2. In acute respiratory acidosis, a small rise in $cHCO_3^-$ levels occurs, while a fall occurs in an acute respiratory alkalosis. On its own, $cHCO_3^-$ is only of limited use when assessing the metabolic component of the acid-base balance as both respiratory and metabolic disorders change $cHCO_3^-$ levels.

Base excess

Siggaard-Andersen devised the concept of base deviation by introducing the term "base excess" (BE) into blood gas analysis [18]. In the case of positive or negative base deviation, the BE indicates the amount of a strong acid or base in mmol required to return the pH of vitro blood or another body fluid to a normal pH of 7.4 under standard conditions ($pCO_2 = 40$ mmHg, 37° C). Siggaard-Andersen's acid-base curve nomogram makes it possible to determine the BE using the known variables pH, cHb and pCO_2. Modern blood gas analyzers calculate the BE from pH, pCO_2, cHb and sO_2. The metabolic component of an acid-base balance disorder can be calculated, as the BE is not influenced by acute changes in pCO_2. Incorporating O_2 saturation has the advantage that the same BE is found in all arterial, mixed venous and venous blood samples [23].

Standard bicarbonate concentration, $cHCO_3^-$ (std)

Similar to the BE, the $cHCO3^-$ (std) is a calculated value that is independent of acute respiratory changes in pCO_2. It is defined as the $cHCO_3^-$ concentration that would be present in fully oxygenated plasma at a pCO_2 of 40 mmHg (37° C). A deviation from the reference value corresponds to the metabolic, not the respiratory, component of a metabolic acid-base balance disorder.

Base excess of extracellular fluid, BE (Ecf) or standard base excess, SBE

BE refers to the in-vitro ratios in heparinized whole blood. The concept of SBE or BE (ecf) was developed to be able to correctly quantify metabolic disorders in vivo. It is calculated by on modern blood gas analyzers with the following formula from the Clinical Laboratory Standards Institute (CLSI) [13]:

$$\text{BE (Ecf)} = cHCO_3^- - 24.8 + 16.2 \times (\text{pH} - 7.40)$$

The numerical term 24.8 corresponds here to the buffer capacity of extracellular fluids incorporating the RBCs. Hb – the most important buffer base in the erythrocytes – not only exchanges with the plasma but also with the total protein-poor extracellular space. When peripherally produced CO_2 is taken up by RBCs, carbonic acid is formed by the enzyme carbonic anhydrase and then broken down intracellularly into H^+ and HCO_3^-. The proton is bound by Hb, while the HCO_3^- diffuses back into the plasma and by 2/3 into the interstitial space. The buffer capacity of Hb is notionally translated to the extracellular space to take into account the interstitial HCO_3^- diffusion under in vivo conditions. The BE (Ecf) is therefore calculated as 1/3 of an assumed Hb value of 15 g/dL, which equals 5 g/dL. BE (Ecf) is also stated in variants that incorporate the oxygen saturation.

14

Graphic methods for diagnosing disorders of metabolic acid-base status

The widely known acid-base diagnostic nomogram by Müller-Plathe [12] is used for interpreting the significance areas for compensated acid-base disorders. The points where the coordinates of the relevant pCO_2 (x-axis) and $cHCO_3^-$ (y-axis) values intersect allow classification into pure or mixed disorders. There are other graphics available, such as by Driscoll et al. [4] which can help diagnose primary or combined disorders of the metabolic acid-base balance. Some blood gas analyzers (e.g. ABL 800 of Radiometer) can print these graphics in addition to their numerical results.

14.5.2 Anion gap

The determination of the anion gap (AG) allows differentiation of the causes of metabolic acidosis [10]. The AG reflects the presence of unmeasured anions in the plasma. By reason of electroneutrality, the total concentration of anions in all fluid compartments is equal to the total concentration of cations:

$$[Na^+] + [\text{unmeasured cations}^+] = [Cl^-] + [HCO_3^-] + [\text{unmeasured anions}^-]$$

This produces:

$$[\text{unmeasured anions}^-] - [\text{unmeasured cations}^+] = [Na^+] - ([Cl^-] + [HCO_3^-]) \text{ \{mmol/L\}}$$

The AG therefore equals the difference between unmeasured anions (polyanionic albumin, phosphate, sulfate, lactate, acetoacetate, β-hydroxybutyrate) minus unmeasured cations (e.g. K^+, Mg^{2+}, Ca^{2+}):

$$AG = [Na^+] - ([Cl^-] + [HCO_3^-]) \text{ \{mmol/L\}}$$

The normal value in healthy persons is 3–11 mmol/L [2].

Most blood gas analyzers that use ion selective electrodes to measure the electrolytes cNa^+, cK^+, cCl^-, and cCa^{2+} calculate the AG by the above formula (also without cK^+ and cCa^{2+}).

Metabolic acidosis with high anion gap

The cause for metabolic acidosis is either the exogenous supply of fixed acids (not HCl) or the increased accumulation of acids in the metabolism. The acid valency of non-chloride acids is neutralized by HCO_3^- to H_2O and CO_2. CO_2 is exhaled. The HCO_3^- concentration in the blood decreases and a sodium salt of the unmeasured acid anions is formed. Thereby, the expression ($[Cl^-] + [HCO_3^-]$) in the above equation is lowered. As the Na^+ concentration remains unchanged, the resulting AG is higher.

Metabolic acidosis without a high anion gap

This state occurs when either HCl or hydrochloride compounds are supplied exogenously or infusion solutions with higher chloride concentration are used or HCO_3^- is lost. The expression ($[Cl^-] + [HCO_3^-]$) in the above formula nevertheless remains unchanged, as the reduction in the HCO_3^- concentration is compensated for by a similar increase in the Cl^- concentration (hyperchloremic metabolic acidosis). Therefore, a normal AG occurs either in gastrointestinal or renal HCO_3^- loss (e.g. In renal tubular acidosis or after administration of carbonic anhydrase inhibitors) or from the delivery of HCl or solutions with high Cl^- concentrations such as NaCl 0.9 %.

When the serum albumin concentration is very low (albumin is negatively charged at pH 7.4), a low AG occurs (approx. 2.5 mmol/L per 1 g/dL albumin-reduction). The AG is also lowered when the concentration of unmeasured cations increases, e.g. in multiple myeloma (accumulation of polycationic monoclonal immunoglobulins) or in hypercalcemia.

14.6 Interpreting acid-base balance disorders

Nowadays, the clinician has three different approaches available for interpreting acid-base status. In addition to the two traditional methods of the Copenhagen [19] and Boston Schools [5] propagated by Siggaard-Andersen, there is the 'modern' Stewart method, developed by Peter Stewart [20]. Although the traditional and modern models conceptualizing the processes involved in acid-base balance differ, both approaches have features that can be usefully combined in clinical practice.

14.6.1 The physiologic method of the Boston School

This approach focuses on the variables pH and pCO_2, as well as on $cHCO_3^-$, calculated with the Henderson-Hasselbalch equation. A differentiation is made between primary respiratory disorders that present with changes in the pCO_2 concentration and primary metabolic disorders that show typical changes in $cHCO_3^-$. The strength of this approach lies in the incorporation of in vivo compensatory responses of the human body to a disturbance. The Boston School offers a system, developed by in vivo titration experiments (Tab. 14.2) that can predict compensatory mechanisms and detect concomitant disorders [5, 9]. The simple rules of the Boston School allow changes in $cHCO_3^-$ to be detected, although $cHCO_3^-$ is not independent of pCO_2. In patients with acute hypoventilation, elevated pCO_2 and acute respiratory acidosis (pH = 7.21, $paCO_2$ = 70 mmHg, $cHCO_3^-$ = 27.9 mmol/L, SBE = –0.2 mmol/L), for example, a reduction in pH is not only observed but also a small increase in $cHCO_3^-$. This increase in $cHCO_3^-$ is not a manifestation of a concomitant metabolic disorder or a compensation effort by the body, but can be explained by the elevated total concentration of carbon dioxide. In addition to an elevated pCO_2, carbon dioxide in its dissociated form is increased as HCO_3^-. The resulting H^+ ions are

Tab. 14.2 Guidelines for the interpretation of acid base balance developed from patient monitoring studies

Acute respiratory	
Acidosis	$\Delta[HCO_3^-] = 0.1\ \Delta pCO_2$
Alkalosis	$\Delta[HCO_3^-] = 0.2\ \Delta pCO_2$
Chronic respiratory	
Acidosis	$\Delta[HCO_3^-] = 0.35\ \Delta pCO_2$
Alkalosis	$\Delta[HCO_3^-] = 0.5\ \Delta pCO_2$
Chronic metabolic	
Acidosis	$pCO_2 = 1.5\ [HCO_3^-] + 8$
Alkalosis	$pCO_2 = 0.9\ [HCO_3^-] + 15$

resorbed by non-bicarbonate buffers. The Boston rule for acute respiratory acidosis (Tab. 14.2) predicts a corresponding increase

$$\Delta[HCO_3^-]/\Delta pCO_2 = 0.1 \text{ mEq/L per mmHg.}$$

According to the rule, a metabolic compensation occurs in chronic hypoventilation with a consequent additional increase in $cHCO_3^-$.

$$\Delta[HCO_3^-]/\Delta pCO_2 = 0.3 \text{ mEq/L per mmHg.}$$

14

In chronic respiratory disorders, the change in $cHCO_3^-$ is attributable to the human body's renal compensatory mechanisms. If $cHCO_3^-$ deviates from the expected value, a concomitant metabolic disorder is likely. Conversely, corresponding respiratory compensatory mechanisms can also be detected in metabolic disorders. If the respiratory response deviates from the prediction, this is evidence of an additional underlying respiratory disorder. This approach uses the AG to further differentiate metabolic disorders. The AG is best explained by the Steward approach, as mentioned below.

14.6.2 Base excess approach of the Copenhagen School

This concept of BE is to have a quantitative variable for estimating the metabolic component of acid-base balance, irrespective of respiratory changes. On its own, the $cHCO_3^-$ concentration, calculated from pH and pCO_2, is not a good measure for the extent of an existing metabolic disorder. The variable BE aims to eliminate respiratory influences. What remains is the deviation in the $cHCO_3^-$ concentration, which can only be caused by a metabolic disorder. The BE thereby simulates the situation reflecting how much the $cHCO_3^-$ deviates from the normal value at a presumed pCO_2 of 40 mmHg.

There are a number of formulas to calculate BE. A calculation that incorporates oxygen saturation is beneficial, because of the ability of Hb to buffer CO_2 as a function of oxygen saturation. The calculated BE value is therefore the same for either arterial or venous blood. The Copenhagen School was criticized for the concept of BE as its prediction is only related to in vitro blood samples and does not reflect in vivo situations correctly [16]. The in vivo situation, however, is, that the plasma (where the pH is measured) interacts not only with the intracellular space of the erythrocytes but also with the interstitial space. Therefore, the buffer capacity of Hb is slightly reduced. The concept of SBE or BE of extracellular fluids was developed to reflect the in-vivo situation more accurately. As already mentioned in Section 14.5.1, the impact of Hb is reduced to 1/3 at a presumed normal Hb value of 15 g/dL so that the standard BE or actual BE of the extracellular fluid is adjusted to an Hb value of 5 g/dL. This simplification may not be correct in patients with an expanded extracellular space, such as patients with edema or anemia.

Irrespective of these deviations, the BE has the advantage of being assessed quickly and simply in clinical practice. However, seen in isolation, it does not give a total picture of the acid-base disturbance, as is possible with the Boston School approach. Looking at chronic respiratory acidosis, for example, renal metabolic compensation usually occurs, reflected by a positive BE. The BE value does not indicate whether this metabolic component reflects a normal compensation mechanism or the presence of a concomitant metabolic disorder.

14.6.3 Stewart approach

The two traditional approaches use the pCO_2/$cHCO_3^-$ system to quantify a disorder and to differentiate between respiratory and metabolic changes. The strength of the Stewart approach lies in further differentiation of a metabolic disorder. In the traditional approach, the actors in the acid-base balance are the weak acid carbon dioxide and the non-volatile weak acid albumin in the plasma, with Hb in the erythrocyte as the main protagonist. They interact with the H^+ ions and determine the pH due to their buffer ability. The Stewart approach has a different model. For example, all strong and weak acids and bases present in a compartment such as plasma (where the pH is measured), determine the actual pH. In addition to weak acids in the plasma such as carbon dioxide and albumin, there is also water with its acid-base properties alongside strong acids and bases. HCl and NaOH feature as the strong acids and bases in the plasma. They are present in combination with weak acids as salt solutions in the form of **strong ions**. The strong anion Cl^- represents the effect of the strong acid HCl, where the strong cation Na^+ represents the strong base NaOH. In his model, Stewart showed that, in a compartment such as plasma, the effect of concentrations of different strong and weak acids/bases can be condensed down to three calculated variables.

- The first variable is the **strong ion difference (SID)** that is composed of all strong cations and all strong anions.
- The second factor is the concentration of all weak non-volatile acids. Albumin dominates in the plasma.
- The third factor is pCO_2, which reflects the effect of the weak acid carbon dioxide.

This simplification is, of course, only a mathematical shortcut, albeit one that depicts the driving forces of the relevant concentrations of weak and strong acids and bases correctly.

The interactions of the plasma with the compartment of the intracellular space of RBCs and the interstitial space are neglected when plasma is seen in isolation. They can, however, be depicted within the Steward approach, for example, in a multi-compartmental model. The principles of the Steward model understandably reproduce the function of the BE, which particularly incorporates the prediction of the compartment of erythrocytes with the dominant Hb. The same applies to the differentiation between primary and secondary reactions of the acid-base balance. For the clinician, however, the Steward approach does not offer any suitable tools yet.

In clinical practice, the changes of SID can be used. A reduction in SID in hyperchloremia or lactatemia is manifest as acidosis. An increase in SID, like in hypochloremia after loss of gastric juices, presents as alkalosis. With reference to the example of gastric juice loss, the Stewart approach is not dictated by the loss of protons but by the lower concentration of HCl, the effect of which is represented by the lower concentration of the strong ion Cl^-. Among the changes to non-volatile weak acids, hypoalbuminemia is one of the clinically more relevant conditions that shifts the pH toward alkalosis. A lower concentration of a weak acid like albumin shifts the pH to the alkaline range. This disturbance often masks other disorders. The "**strong-ion-gap**" (SIG) is helpful in identifying undetected anions as cause of a metabolic acidosis [9].

$$\begin{aligned} SIG = {} & cNa^+ + cK^+ + cCa^{2+} + cMg^{2+} \\ & - cLactate^- - cCl^- - cHCO_3^- \\ & - cAlbumin\ \{g/L\} \times (0.123 \times pH - 0.631) \\ & - cPhosphate\ \{mmol/L\} \times (0.309 \times pH \\ & - 0.469) \end{aligned}$$

Note

The SIG is nothing more than an extended AG. The incorporation of albumin is particularly helpful here. A normal albumin is presumed when calculating the AG. In hypoalbuminemia, the portion of the albumin present as anions is smaller. Undetermined anions can theoretically be present and take the place of albumin without increasing the AG. It is not yet possible to determine the SIG on blood gas analyzers, as they cannot determine phosphate or albumin, for example.

14.7 Temperature correction and pre-analysis

14.7.1 Temperature correction of blood gas analysis results

Blood gas analyzers perform measurements at 37° C as a default setting. Since the physical solubility of gases in blood is temperature-dependent, empirical correction formulas were introduced to adjust blood gas values to a patient's body temperature if different from normal (37° C). Here, O_2 and CO_2 are corrected differently. It should be noted that for CO_2, the solubility coefficient αCO_2 increases at lower temperatures, i.e. the gas dissolves better. However, as the CO_2 concentration cCO_2 remains constant in the sample, the pCO_2 must decline accordingly. This occurs according to **Henry's law**:

$$cCO_2 = \alpha CO_2 \times pCO_2$$

The solubility of O_2 in hypothermia is higher. There is also an increase in the affinity of Hb to O_2 (left shift of the O_2Hb binding curve), thereby resulting in a significant reduction in pO_2.

14.7.2 Pre-analytical phase

The many specific pre-analytical problems in (arterial) blood gas analysis should only be sketched out briefly at this juncture. Further important information is given in ▶ Chapter 4. In general, blood taken to measure volatile blood gas parameters should be obtained with a plastic syringe and analyzed within 30 min at the most. After 10 min, the homogeneity of the sample is no longer guaranteed, which requires careful resuspension of the blood sample. The test tube should be rolled for more than 1 min between the palms of the hands [8].

When using ion-selective electrodes (ISE), various confounders and influencing factors interfere with the determination of the electrolytes (Na^+, K^+, ionized Ca^{2+} and ionized Mg^{2+}). **Intravascular hemolysis**, in particular, can cause falsely high K^+ values in plasma. Without hemolysis detectors built into the devices, hemolysis cannot be detected in heparinized blood samples for BGA. Unfortunately, current blood gas analyzers do not yet feature such detectors.

Ionized calcium in the blood can show falsely low concentrations in the presence of sodium perchlorate (Irenat). This is given when thyroid function needs to be blocked prior to administering iodine contrast agents [7]. On the other hand, different blood gas analyzers can show a discrepancy in the lower range of 0.2–0.5 mmol/L when ionized Ca^{2+} is measured in samples taken and analyzed immediately after hemodialysis from the hemofiltration giving set [15].

14.8 Transcutaneous measurement of pO_2 and pCO_2

As already described in ▶ Chapter 11, the blood gas parameters pCO_2 and pO_2 can also be measured transcutaneously (tc). This form of analysis is primarily used in neonatology. In addition to $tcpO_2$, the oxygen saturation can be measured frequently (e.g. OxiVenT from Sentec, Therwill, Switzerland or TCM CombiM Monitor from Radiometer, Willich, Germany).

References

1. Boemke W, Krebs MO, Rossaint R (2004) Blutgasanalyse. Anaesthesist 53: 471–492
2. Boemke W, Francis RC, Reinnhardt HW (2008) Blutgasanalyse und Säure-Basen-Haushalt. In: Rossaint R, Werner C, Zwissler B. Anästhesiologie. 2. Edition Springer, Berlin Heidelberg, S 140–157
3. Chen J, Gorman M, O'Reilly B, Chen Y (2016) Linearity analysis and comparison study on the epoc(®) point-of-care blood analysis system in cardiopulmonary bypass patients. Data Brief 6:847–852
4. Driscoll P, Brown T, Gwinnutt C, Wardle T (1997) A simple guide to blood gas analysis. BMJ Publishing Group, London
5. DuBose TD (2001) Acidosis and alkalosis. In: Braunwald E, Fauci AS, Kasper DL, Hauser SL, Longo DL, Jameson JL (Ed.) Harrison's principles of internal medicine. McGraw-Hill, Philadelphia, S 283–291
6. Grohmann K, Roser M, Rolinski B et al. (2006) bilirubin measurement for neonates: comparison of 9 frequently used methods. Pediatr 117:1174–1183
7. Gruber M, Nehring C, Creutzenberg M, Graf B, Hopf S (2011) Perchlorate (Irenat®) may falsely lower measured ionised calcium. Clin Chem Lab Med 49:1019–1024
8. Gruber MA, Felbermeir S, Lindner R, Kieninger M (2016) Preanalytics: The (in-)stability of volatile POCT parameters and the homogeneity of blood in syringes at the market. Clin Chim Acta 457: 18–23
9. Kellum JA (2007) Disorders of acid-base balance. Crit Care Med 162:2246–2251
10. Mikulcik P (2005) Rapidanalyse – Blutgase und mehr, 2. Edition Bayer HealthCare, Fernwald
11. Mielsch C, Zimmermann A, Wagner D, Matthes B, Schlebusch H, Luppa P (2010) Point-of-care determination of neonatal bilirubin with the blood gas analyzer RapidLab 1265. Clin Chem Lab Med 48:1455–1461
12. Müller-Plathe O (1987) A nomogram for the interpretation of acid-base data. J Clin Chem Clin Biochem 25: 795–798
13. NCCLS (2001) Blood gas and pH analysis and related measurements; Approved Guideline. NCCLS document C46-A (ISBN 1–56238–444–9). NCCLS, Wayne, PA
14. Rolinsky B, Okorodudu AO, Kost G et al. (2005) Evaluation of total bilirubin determination in neonatal whole-blood samples by multiwavelength photometry on the ROCHE OMNI S point-of-care analyzer. Point of Care 4: 3–8
15. Schwarzer P, Kuhn SO, Stracke S, Gründling M, Knigge S, Selleng S, Helm M, Friesecke S, Abel P, Kallner A, Nauck M, Petersmann A (2015) Discrepant post filter ionized calcium concentrations by common blood gas analyzers in CRRT using regional citrate anticoagulation. Crit Care 19:321
16. Severinghaus JW (1993) Siggaard-Andersen and the "Great Trans-Atlantic Acid-Base De-bate". Scand J Clin Lab Invest 53(Suppl 214):99–104
17. Skurup A, Kristensen T, Wennecke G (2008) New creatinine sensor for point-of-care testing of creatinine meets the National Kidney Disease Education Program guidelines. Clin Chem Lab Med 46: 3–8
18. Siggaard-Andersen O, Engel K, Jørgensen K, Astrup P (1960) A micro method for determination of pH, carbon dioxide tension, base excess and standard bicarbonate in capillary blood. Scand J Clin Lab Invest 12: 172–176
19. Siggaard-Andersen O (1974) The acid-base status of the blood. Munksgaard, Copenhagen
20. Stewart PA (1981) How to understand acid-base. A quantitative acid-base primer for biology and medicine. Elsevier, New York (Internet: http://www.acidbase.org)
21. Suen WW, Ridley B, Blakney G, Higgins TN (2003) Comparison of lactate, bilirubin and hemoglobin F concentrations obtained by ABL 700 series blood gas analyzers with laboratory methods. Clin Biochem 36: 103–107
22. Thaler M, Luppa PB, Schlebusch H (2008) Die Bilirubinbestimmung – Eine aktuelle Übersicht. J Lab Med 32: 1–10
23. Zander R (1995) Die klassische Blutgasanalyse (Säure-Basen-Status): Interpretation and errors. In: Deutsche Akademie für Anästhesiologische Fortbildung (Ed.) Refresher Kurs: Aktuelles Wissen für Anästhesisten. Springer, Berlin Heidelberg New York, pp. 27–35

Coagulation diagnostics

Dirk Peetz, Jürgen Koscielny, Michael Spannagl

P. B. Luppa, R. Junker (Eds.), *Point-of-Care Testing*
https://doi.org/10.1007/978-3-662-54497-6_15

15.1 Introduction

A large number of global and special tests for plasmatic coagulation are applied in routine clinical practice. For this purpose, the clinical monitoring of therapy with anticoagulants and the evaluation of the hemostasis status prior to surgeries/interventions are the two areas of greatest demand. Specific diagnostics of plasmatic coagulation disorders or thrombotic disturbances is mostly carried out in special laboratories, although the results are usually not available until days later. This is not primarily due to the time these analyses take, but to logistical aspects like scheduling and the fact that the specialized analyses are sometimes performed batchwise, e.g. once a week. The key arguments for this include cost effectiveness and improved analytical quality associated with the measurement of larger sample series under controlled conditions.

In recent years, complex analytical methods are increasingly being employed in the management of surgical and intensive care patients, not least in coagulation diagnostics, which can be conducted at the bedside with anticoagulated or sometimes even untreated whole blood. POCT for monitoring unfractionated heparin has long been used in catheterization labs and on intensive care and dialysis units. The decisive advantage of such bedside analysis is the fast availability of results that allows the option of targeted therapeutic intervention [3]. For example, timely adjustment of the anticoagulant dose is imperative during extracorporeal therapeutic procedures. Coagulation tests are conducted near the patient for the diagnosis and control of therapy with procoagulant drugs, antifibrinolytics and blood products [2].

These coagulation POCT methods encompass easy-to-use tests for the previously mentioned monitoring of anticoagulants and for assessing plasmatic coagulation alongside more complex, viscoelastic analytical methods for the simultaneous measurement of plasmatic and cellular hemostasis as well as for conducting tests of platelet function.

The currently available methods for POC coagulation diagnostics cover many, although not all, processes involved in physiological hemostasis in vitro. Indeed, the different reactions in the hemostasis cascade are weighted variously and not always represented by the individual POCT methods in terms of their physiologic balance. Ultimately, all methods for detecting clotting in whole blood require artificial components in their test systems. Therefore, training and knowledge about reaction conditions, confounders and influencing factors coupled with a certain amount of clinical and analytical experience are required by the user to be able to interpret the findings.

15.2 Documentation of primary hemostasis and/or platelet function

Isolated disorders of primary hemostasis are usually pre-existing, i.e. drug-induced, illness-related or congenital, making them relevant to intra- and postoperative bleeding episodes etc. For instance, the Platelet Function Analyzer PFA-100/200 by Siemens is used as a screening test to establish preoperative bleeding tendencies in patients with a positive bleeding history. One large prospective, preoperative screening study on patients with a positive bleeding history found that 97.7 % showed pathologically prolonged PFA results, which could be attributed by >99 % to acquired thrombocytopathy, von-Willebrand syndrome or liver cirrhosis. By contrast, secondary (plasmatic) disorders were less common. PFA screening was not conducted in patients with a negative bleeding history, meaning that this study provided no data on sensitivity or specificity [15]. The meta-analysis determined pooled sensitivities of 82.5 % and 66.9 % and specificities of 88.7 % and 85.5 % for the PFA collagen/epinephrine cell and the collagen/ADP cell, respectively [15]. The Austrian Anesthesia Society recommends the PFA system as a diagnostic cornerstone for establishing defects of primary hemostasis as part of preoperative history-based

15

coagulation evaluation (evidence level D, expert opinion) [21, 21a].

Preoperative, therapeutic staged concepts have been developed for the management of disorders of primary hemostasis. The efficacy of hemostatic drugs such as DDAVP (desmopressin) can be shown based on PFA readings. In patients with disorders of (primary) hemostasis, the number of bleeding episodes requiring transfusions can be reduced by a preoperative regimen and the therapeutic follow-up procedure can be supported by POC coagulation testing [16, 16a, 17]. For example, current guidelines on thoracic surgery recommend a POCT-supported transfusion regimen [11].

However, it should be emphasized here that platelet function analysis in the whole blood alone is not sufficient to comprehensively diagnose a preoperative bleeding tendency and that further clarification is indicated which gives consideration to the bleeding history and corresponding staged diagnostic concepts [11, 21].

When used with suitable measuring cells, the PFA system is sensitive to acetylsalicylic acid and ADP receptor antagonists. Traditionally, **induced platelet aggregation** is used to monitor therapy with acetylsalicylic acid, clopidogrel and other anti-platelet agents; meanwhile, multiple POCT systems have been introduced for this as well [22, 23, 24]. The PFA-200 (PFA-100 after a software update) can be used for monitoring clopidogrel as well as for other P2Y12 receptor-mediated platelet aggregation inhibitors (prasugrel, ticagrelor) with the Innovance PFA P2Y measuring cell.

15.3 Thrombin/fibrin formation

In routine clinical practice, POCT methods that detect thrombin/fibrin formation are mostly employed for monitoring therapy with parenteral anticoagulants. Another area of application is for patient self-measurement of the international normalized ratio (INR) and/or patient self-management of oral therapy with vitamin K antagonists (see below).

In operating rooms and resuscitation units, but also when applying extracorporeal methods like hemodialysis or extracorporeal membrane oxygenation, coagulation inhibition is an unconditional prerequisite for the administration of the appropriate therapeutic modalities. The necessity for individualized dosing of heparin or other parenteral anticoagulants has led to an increase in the application of POCT analyses that are based on whole blood.

15.3.1 Heparin monitoring and activated clotting time (ACT)

In general, the activated clotting time (ACT) is used for the peri-interventional treatment monitoring of unfractionated heparin and other anticoagulants. Important criteria for selecting a method to measure ACT include the volume of the whole blood used and the linearity of the dose-response relationship (desired measuring range).

When interpreting ACT values, it must be considered that a coagulation factor deficiency – either in isolation or in combination with heparin – can prolong ACT.

The ACT test does not aid in the differential diagnosis of a factor deficiency or a heparin effect, whereas the use of additional, modified test approaches utilizing heparinase can. This enzyme cleaves the heparin chain into short fragments that lack anticoagulant efficacy, thereby cancelling out the heparin effect. By the heparinase approach, prolonged coagulation times are thus generally to be interpreted as a factor deficiency. The ACT PLUS from Medtronic (Minneapolis, MN, USA) utilizes this aspect, for example: The sample is added to a test tube containing an intrinsic activator. A pin is suspended in the test tube and raised upwards by a mechanism. Gravity then causes the pin to drop downwards. If coagulation now takes place in the sample, the lowering of the pin takes place more slowly – a process captured by an optical system. A second test tube contains the activator in addition to heparinase. An aliquot of the blood sample is pipetted into the heparinase test tube and the measurement re-started. By comparing the two coagulation times (with and without heparinase), a factor deficiency can be distinguished from a heparin effect. If protamine is used instead of heparinase, the heparin concentration can also be estimated via protamine titration.

The generation of heparin calibration curves using on-site substances along with the activated partial thromboplastin time (aPTT)/ ACT method is complicated and time-consuming. For this purpose, the anticoagulants to be tested must be added to normal plasma in vitro. Defined amounts of unfractionated heparin are added to the normal plasma in vitro. Prefabricated calibrators can be used alternatively. However, these are not available for all anticoagulants across all dose ranges. Depending on the coagulation test employed (prothrombin time (PT) or aPTT), both direct and indirect thrombin and factor Xa inhibitors can be measured during this process. A precise assessment of these dose-response curves is crucial for therapy control. For instance, aPTT-based measuring methods show a low sensitivity to thrombin inhibitors in the high-dose range.

15.3.2 Oral anticoagulation – patient self-management

More than 800,000 individuals in Germany are receiving long-term therapy with oral anticoagulants (vitamin K antagonists). This includes patients with artificial heart valves, cardiac arrhythmias (e.g. atrial fibrillation), heart failure, recurrent thromboses of the leg and pelvic veins, lung embolisms or other cardiovascular diseases. A successful outcome of oral anticoagulation is not least guided by close monitoring of the International Normalized Ratio (INR) and an individualized adjustment in medication.

Coagulation self-monitoring was introduced in 1986, markedly enhancing the reliability and safety of anticoagulant therapy. The patients themselves can measure their current INR in capillary blood and adjust the individual dose of their oral anticoagulant independently. The treating physician and the team at a qualified training center attend to patients learning coagulation self-monitoring and counsel them in issues relating to their therapy. Prospective studies – systematically reviewed in several meta-analyses – clearly show that better therapeutic outcomes are achieved by self-monitoring. The cost effectiveness gained by the significantly lower thromboembolism rate and bleeding incidences was emphasized [6, 14, 20]. A more recent study by Matchar et al. [19], contrarily showed no superiority for the self-testing in terms of stroke or bleeding episodes. Likewise, a recent Cochrane review on the target criteria "time in therapeutic range" was unable to determine any significant gain in benefit of patient self-monitoring in combination with educational and behavioral interventions over INR control and care by the attending physician [7].

Further explanations on the importance of INR self-monitoring are given in ▶ Chapter 33.

15.3.3 POCT in non-vitamin K antagonist oral anticoagulants (NOAC)

In acute situations in patients known or presumed to be receiving non-vitamin K antagonist oral anticoagulants (NOAC), e.g. for acute bleeding or verifying the indication for a lysis to treat acute stroke, the determination of INR by POCT may supply initial information about the efficacy of oral anti-Xa inhibitors, e.g. rivaroxaban, apixaban, edoxaban, whenever no specific anti-Xa tests happen to be available [9, 10]. However, these measuring results can be only accessed qualitatively because of the varying sensitivity of the different PT tests. Serial measurements can moreover allow certain conclusions to be drawn about the pharmacokinetics as well. For interpretation purposes, it is additionally necessary to consider, among other variables, the time of the last drug administration alongside the renal and hepatic organ function. Currently, there are no specific POC tests available for the drug monitoring of NOAC.

The question remains open as to whether testing the efficacy of NOAC as POCT method, e.g. in acute hemorrhaging, could establish itself in clinical routine, for instance, by measuring a change in color in the urine or serum [13].

15.4 Determining clot formation by viscoelastic methods

In Central Europe, the conventional thromboelastography (used in the Haemoscope TEG [5]) has been mostly superseded by modern procedures of rotational thromboelastometry (ROTEM, ROTEM sigma; TEM International, Munich, Germany) (◘ Fig. 15.1). Diagnostically, this would primarily encompass acute bleeds in polytraumatized and surgical patients. A targeted, individually adapted therapy is not only clinically challenging, but also requires sophisticated diagnostics. Therapy is encumbered by the fact that coagulopathy is mostly a multi-causal event (loss/consumption of plasmatic and cellular components, possibly concurrent with hyperfibrinolysis) and each coagulation situation is subject to the dynamics of time as well. Additionally, the effects of transfused colloidal solutions – predominantly hydroxyethyl starch – come into play which can promote impaired polymerization of the fibrin clot. The actions of coagulation-activating agents like heparin, any acquired thrombocytopathy, effects emanating from foreign surfaces (e.g. in an extracorporeal circulation) alongside causes of surgical bleeding are to be weighed into equation during differential diagnostic and therapeutic deliberations.

In near-patient settings (resuscitation unit, operating room), the thromboelastogram can be continuously monitored and leads to direct therapeutic implications. Alternately, analysis by thrombelastography can be performed in the central laboratory and the thromboelastogram can then be transmitted **online** to the acute treatment setting [8]. One prerequisite is a fast sample transport to the central laboratory, e.g. by a pneumatic tubing system, that takes place gently without abrupt acceleration and braking

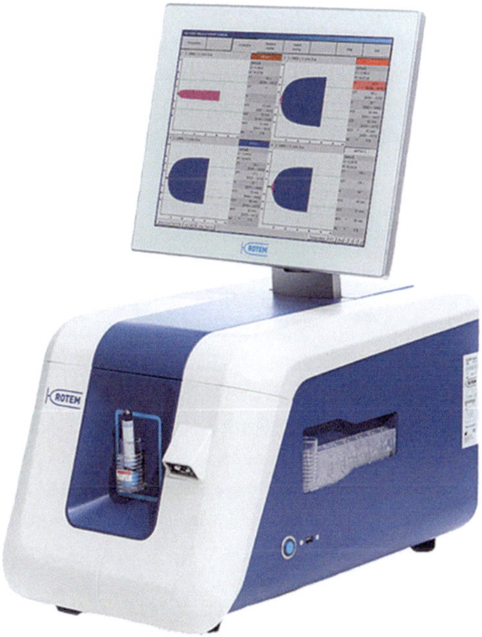

◘ **Fig. 15.1** ROTEM sigma. (Courtesy of Tem International GmbH, Munich)

which would prematurely activate coagulation cascades in the blood sample. Particularly at larger hospitals with central laboratories where several devices are operated simultaneously, precautions must be additionally undertaken to ensure that the analytical findings are transmitted accurately to those physicians who are directly treating the patients. When these conditions are met, ROTEM analysis will lead to a high availability for all affiliated clinical areas. When laboratory personnel are also trained accordingly, a high analytical quality of the tests is guaranteed.

The advantage of thrombelastography over conventional coagulation testing methods lies in the fact that clot quality, clot stability and the effect of platelets on both can be assessed alongside clotting time.

In clinical settings, thrombelastography is primarily used to analyze complex disturbances of hemostasis perioperatively. Further indications include the specific detection of hyperfibrinolysis and the assessment of thrombocytopenia or dilutional coagulopathy.

Patient-near thromboelastography carried out in the operating rooms and shock units to directly stop bleeds has become indispensable these days and an integral part of the modern **patient blood management**. Many studies [18] have shown that its judicious diagnostic application significantly improve patient management while simultaneously reducing the number of administered blood products. In this context, Goodnough et al. [12] and Ak et al. [1] showed that the use of ROTEM in cardiac surgery and hemorrhaging trauma patients significantly decreased the number of administered units of packed red blood cells, pooled platelets and fresh frozen plasma. A prospective, randomized study by Weber et al. [25] arrived at the same conclusion; they moreover showed that the significant reduction in blood transfusions may prevent complications in patients and moreover can reduce the costs for coagulation therapy (blood products, fibrinogen and other individual coagulation factors) in cardiac surgery patients from € 3,100 to € 1,500.

For differentiating between coagulation disorders, ROTEM analyses are furthermore conducted with the additive of the heparin-cleaving enzyme heparinase or of a fibrinolysis inhibitor (e.g. aprotinin, tranexamic acid), but also using the in-vitro blockade of platelets by cytochalasin D or fibrinogen receptor antagonists for the qualitative assessment of the fibrin clot.

Clot strength is plotted against time. For historical reasons, a two-legged curve is generated and parametrized in millimeters. A large number of parameters have been described for evaluating thromboelastograms.

15

The most important are:

- Clotting time (CT or reaction time r)
- Clot formation time (CFT or coagulation time k)
- Maximum clot firmness (MCF or maximum amplitude MA)
- Maximum lysis (ML, percent reduction of clot firmness after MCF)

It is recommendable to enter in the findings report the measured values (mostly CT, CFT, MCF and ML) and reference range along with a semi-quantitative assessment (e.g. normal coagulation activation, abnormal clot formation, lacking evidence of hyperfibrinolysis etc.). In the case of borderline findings, the tests should be repeated and/or the findings reported as borderline. It is recommended to establish institutional reference ranges by examining patients or subjects with normal hemostaseology and/or at least review the correlation between the institutionally determined values with external reference ranges.

▪▪ Evaluating fibrinolysis

When the clot dissolves in vitro, it is an indication of fibrinolysis. This is essentially never observed in healthy individuals because fibrinolysis inhibition in the peripheral blood usually clearly predominates over the fibrinolysis activators. Fibrinolysis observed intraoperatively is frequently associated with massive bleeding episodes. The faster fibrinolysis is observed in vitro, the more pronounced the clinical bleeding symptoms usually tends to be (fulminant lysis with clot dissolution in less than 30 min). A late emergent lysis (after at least 45 min) frequently stops spontaneously.

▪▪ Evaluation of clot formation

A normal-range clot formation is used as a functional marker of intact whole blood coagulation; diminished clot strength can be corrected by administering pooled platelets or fibrinogen (in the form of fresh frozen plasma or fibrinogen concentrate).

▪▪ Assessment of coagulation activation

The diagnostic value of clotting times is basically equivalent to that of aPTT or INR, albeit with lower specificity. Indeed, this is due to the stronger influence caused by clot polymerization in thrombelastography, as compared to the determination clotting time by conventional methods. The CT measurement is much less specific for a factor deficiency (e.g. on oral anticoagulation) than the INR (especially when the sample has a high fibrinogen content).

Overall, thrombelastography is a functional global test that can aid in the assessment of the

Tab. 15.1 Conventional POCT methods in hemostaseology by system type

Procedural sequence	Test tube systems, heparin management systems	Test strip systems	RPFA (Accumetrics)	PFA-100/200 (Siemens)	Viscoelastic methods (thrombelastography, ROTEM*, Sonoclot, Multiplate)
Prepare the measuring cell	None	Pre-heat the measuring cell at room temperature (partially)			None
Insert measuring cell	Insert the test strip and/or measuring cell in the analyzer				
Insert blood sample	Inject and/or pipette the blood sample	Place a blood drop on the sample card	Attach the closed blood test tube to the measuring cell	Pipette the whole blood sample into the measuring cell	
Reagent handling	No addition of reagents				Addition of 1–3 various reagents
Uses control material	Some plasma-based controls		Artificial control fluid	No control available	Plasma-based controls

* On the ROTEM sigma, simplified handling by closed analysis system. Details ▶ Chapter 6.

coagulation situation, particularly in complex hemostasis disorders [5]. The method is not sensitive to platelet aggregation inhibitors, low molecular heparins and pentasaccharide (Fondaparinux), nor for the von Willebrand syndrome, while it is relatively insensitive to the effects of fibrinogen receptor (P2Y12) antagonists and oral anticoagulants. A factor XIII deficiency is more easily detected in the plasma than in whole blood; particularly at high fibrinogen levels however, it is only detectable at very low factor XIII levels.

Tab. 15.1 summarizes the conduct of the conventional POCT methods for coagulation analyses. As discussed in ▶ Chapter 6, a distinction is made between "true" and "POCT-appropriate" methods. Mostly, POCT-appropriate methods require considerably greater effort and skill on the part of the testing person and, therefore, that person's appropriate training and motivation.

15.5 Benefits, risks and cost effectiveness

The benefits and risks of POCT methods are chiefly impacted by the urgency of the analysis, the reaction times under the respective conditions "on-site", the available personnel and logistic resources as well as the spectrum of coagulation changes to be expected (special patient populations). Despite its usually higher costs, POCT diagnostics makes sense when it produces better process quality of hemostasis management (e.g. time advantage, personalized control of the anticoagulant dose, targeted hemostasis management instead of a polypragmatic application of various therapeutic options).

Coagulation analysis systems deployed for POCT feature a great range of different designs. The original claim of simple measurement of a sample without special preparation (ACT) is not fulfilled for the complex methods required for documenting primary hemostasis or for measuring thrombus elasticity. In this context, several measuring channels are provided; sometimes, the manual addition of reagents is required. Nevertheless, a near-patient deployment of these methods frequently makes sense, particularly under intensive care and/or perioperative conditions, because it allows targeted and rapid treatment.

References

1. Ak K, Isbir CS, Tetik S, Atalan N, Tekeli A, Aljodi M, Civelek A, Arsan S (2009) Thromboelastography-based transfusion algorithm reduces blood product use after elective CABG: a prospective randomized study. J Card Surg 24:404–10
2. Avidan MS, Alcock EI, Da Fonseca JH et al. (2004) Comparison of structured use of routine laboratory tests or near-patient assessment with clinical judgement in the management of bleeding after cardiac surgery. Br J Anaesth 92: 178–86
3. Boldt J, Walz G, Triem J, Suttner S, Kumle B (1998) Point-of-care (POC) measurement of coagulation after cardiac surgery. Intensive Care Med 24: 1187–93
4. Calatzis A, Heesen M, Spannagl M (2003) Patientennahe Sofortdiagnostik von Hämostaseveränderungen in der Anästhesie und Intensivmedizin. Anaesthesist 52: 229–37
5. Cammerer U, Dietrich W, Rampf T, Braun SL, Richter JA (2003) The predictive value of modified computerized thromboelastography and platelet function analysis for postoperative blood loss in routine cardiac surgery. Anesth Analg 96: 51–7
6. Christensen TD, Johnsen SP, Hjortdal VE, Hasenkam JM (2007) Self-management of oral anticoagulant therapy: A systematic review and meta-analysis. Int J Cardiol 118: 54–61
7. Clarkesmith DE, Pattison HM, Lane DA (2013) Educational and behavioural interventions for anticoagulant therapy in patients with atrial fibrillation. Cochrane Database of Systematic Reviews, Issue 6. Art. No.: CD008600
8. Colucci G, Giabbani E, Barizzi G, Urwyler N, Alberio L (2011) Laboratory-based ROTEM analysis: implementing pneumatic tube transport and real-time graphic transmission. Int J Lab Hematol 33:441–6
9. Cuker A, Siegal D, Crowther M, Garcia D (2014) Laboratory measurement of the anticoagulant activity of the non–vitamin K oral anticoagulants. J Am Coll Cardiol 64: 1128–39
10. Ebner M, Peter A, Spencer C, Härtig F et al. (2015) Point-of-care testing of coagulation in patients treated with non-vitamin K antagonist oral anticoagulants. Stroke 46 (10): 2741–7
11. Ferraris VA, Ferraris SP et al. (2011). Update to The Society of Thoracic Surgeons and The Society of Cardiovascular Anesthesiologists Blood Conservation Clinical Practice Guidelines. Ann Thorac Surg 91: 944–82
12. Goodnough LT, Hill CC (2012) Use of point-of-care testing for plasma therapy. Transfusion 52 Suppl 1:56S-64S. Erratum in: Transfusion 52:2496
13. Harenberg J, Du S, Wehling M, Zolfaghari S et al. (2016) Measurement of dabigatran, rivaroxaban and apixaban in samples of plasma, serum and urine, under real life conditions. An international study. Clin Chem Lab Med. 54: 275–83
14. Heneghan C, Alonso-Coello P, Garcia-Alamino JM, Perera R, Meats E, Glasziou P (2006) Self-monitoring of oral anticoagulation: a systematic review and meta-analysis. Lancet 367: 404–11
15. Karger R, Donner-Banzhoff N, Müller HH et al. (2007) Diagnostic performance of the platelet function analyzer (PFA-100) for the detection of disorders of primary haemostasis in patients with a bleeding history–a systematic review and meta-analysis. Platelets 18: 249–60
16. Koscielny J, Tempelhoff GF, Ziemer S et al. (2004) A practical concept for preoperative management in patients with impaired primary hemostasis. Clinical and Applied Thrombosis/Hemostasis 10: 55–166 and [16a] 195–204
17. Koscielny J, Kiesewetter H, Tempelhoff GF (2006) Platelet Function Analyzer (PFA)-100 closure time in the evaluation of platelet disorders and platelet function. Journal of Thrombosis and Haemostasis 4:1426–7
18. Levi M, Hunt BJ (2015) A critical appraisal of point-of-care coagulation testing in critically ill patients. J Thromb Haemost 13:1960–7
19. Matchar DB, Jacobson A, Dolor R, Edson R, Uyeda L, Phibbs CS et al. (2010) Effect of home testing of international normalized ratio on clinical events. N Engl J Med 363:1608–20. Erratum in: N Engl J Med 364: 93
20. Mc Cahon D, Murray ET, Jowett S et al. (2007) Patient self management of oral anticoagulation in routine care in the UK. J Clin Pathol 60: 1263–7
21. Pfanner, G, Koscielny J, Pernerstorfer T et al. (2007) Empfehlungen der Arbeitsgruppe perioperative Gerinnung (AGPG) der Österreichischen Gesellschaft für Anästhesiologie, Reanimation und Intensivmedizin (ÖGARI). Die präoperative Blutungsanamnese. Anästhesist, 56(6):604–11 und [21a] Präoperative Anamnese-basierte Gerinnungsevaluierung. Update June 2013. http://www.oegari.at/web_files/dateiarchiv/editor/anamnese-basierte_gerinnungsabklaerung_vor_op_2013_1.pdf (Status 19 May 2016)
22. Simon DI, Liu CB, Ganz P et al. (2001) A comparative study of light transmission aggregometry and auto-

15

mated bedside platelet function assays in patients undergoing percutaneous coronary intervention and receiving abciximab, eptifibatide, or tirofiban. Catheter Cardiovasc Interv 52:425–32
23. Smith JW, Steinhubl SR, Lincoff AM et al. (1999) Rapid platelet-function assay: an automated and quantitative cartridge-based method. Circulation 99:620–5
24. Steinhubl SR, Talley JD, Braden GA et al. (2001) Point-of-care measured platelet inhibition correlates with a reduced risk of an adverse cardiac event after percutaneous coronary intervention: results of the GOLD (AU-Assessing Ultegra) multicenter study. Circulation 103:2572–8
25. Weber CF, Görlinger K, Meininger D, Herrmann E, Bingold T, Moritz A, Cohn LH, Zacharowski K (2012) Point-of-care testing: a prospective, randomized clinical trial of efficacy in coagulopathic cardiac surgery patients. Anesthesiology 117:531–47

Hematological diagnostics

Dorthe Kixmüller, Norbert Gässler, Ralf Junker

P. B. Luppa, R. Junker (Eds.), *Point-of-Care Testing*
https://doi.org/10.1007/978-3-662-54497-6_16

16.1 Introduction

In hematology, POCT is clinically important for near-patient rapid diagnostics. The isolated determination of Hct and Hb, for example as part of blood gas analysis (BGA), often provides adequate patient monitoring and is thus an integrated component used during surgeries and on intensive care units. By contrast, the overall determination of blood count without or with the differentiation of leukocytes (small and/or complete blood count) is primarily the task of central laboratories at hospitals or physician's offices specialized in laboratory medicine. In general, a need for POCT hematology diagnostics here is obvious in only few scenarios.

16.2 Indications and areas of application

In principle, there are two conceivable reasons for using POCT in hematological examinations:

- In emergencies, where an immediately available finding leads to a rapid medical decision, and
- To improve organizational workflows, which can secondarily result in advantages for patient care.

Potential areas of application are thus hospital outpatient departments, surgical settings, intensive care units, pediatric hospitals and occasionally even physician's offices [8]. The Guidelines of the British Committee for Standards in Haematology (BCSH) can be useful when introducing hematology methods at the POC [4].

With the known conventional systems on the market, the analysis generally only takes seconds to a few minutes, making their use for emergency analyses justified. Unlike other laboratory sectors, where POCT shortens the pre-analytical phase by obviating centrifugation, the aspect of time savings in hematology is less appreciable because centrifugation of the sample material is not required.

16.2.1 Hospitals

The use of POCT systems for hematologic tests can help enhance the efficiency of patient management in hospital settings [1, 9]. The time saved by POCT is of paramount importance in hospitals, especially when it addresses the question of the need for transfusions in patients with acute blood loss. The assessment of Hb in surgical settings or in emergency rooms is therefore particularly pertinent in the face of a severe blood loss.

Note

Nevertheless, it should be considered that the Hb value can remain unchanged for the first 30 min of a significant hemorrhage. This is because the volume deficiency is not compensated by fluids until up to 36 h later, which then leads to a measurable drop in Hb.

By contrast, POCT is generally not required to determine the leukocyte count or their differentiation [11]. In pediatrics, e.g. in patients with abdominal pain or fever of unknown origin, the leukocyte count is occasionally helpful, at best by means of a simple differentiation [3, 7, 8]. It is questionable whether the use of POCT is indispensable given that decisions derived from these results are usually not time-critical or based on the concurrent results of other tests, e.g. acute-phase parameters such as CRP. Moreover, automated hematology is subject to special requirements particularly in the intensive care of premature infants/newborns and children given that many leukocyte precursor cells that are difficult to allocate can be found in the circulation of these patients [16, 17]. Furthermore, existing erythrocyte precursors are difficult to lyse, a fact which impairs leukocyte analysis [10, 17]. Hematologic tests are therefore prone to erroneous measurements, if the measuring devices are not suitable to answer the question at hand. Furthermore, specialized knowledge is required for interpretation of the findings; this in turn tends to put POCT diagnostics in question since POCT is generally performed by personnel who lack experience in the area of laboratory diagnostics.

The purposeful importance of POCT in determining platelet count in conjunction with coagulation tests for estimating the transfusion needs has already been demonstrated [5, 6] as was its positive value in the clinical course in disseminated intravascular coagulation (DIC), sometimes also referred to as consumptive coagulopathy, provided that the turn-around time (TAT) at the central laboratory is not sufficient for time-critical decision-making.

A general advantage can be found in the small sample volumes used for POCT: Studies have demonstrated that the transfusion requirement can be lowered in intensive neonatal care [2, 15].

16.2.2 Physician's offices

In physician's offices and at outpatient centers outside of emergency care settings, the use of POCT devices to determinate hematologic parameters only makes sense in a few indications. This includes oncological practices among circumstances, where the current treatment decision may depend on the blood count (Hb, leukocytes, platelets). However, the use is thereby limited by the fact that specifically leukocytes can be strongly altered in appearance by disease or therapy and therefore cannot always be correctly allocated morphologically by the device [8]. As in the clinical setting, it is therefore imperative that the measuring device be suitable to answer the medical question and that users possess the specialized knowledge and skills to successfully use such devices.

An Hb assessment is generally sufficient to screen for the presence of anemia, but is not time-critical to answer this question. Given that the costs for an Hb assessment by POCT are generally higher than the costs for the whole blood count in a large laboratory, the advantages of Hb screenings in practice are questionable. In regions, where laboratories – unlike in most regions of Europe – are difficult to achieve, this use may be justified. In several fields, the successful use of POCT for Hb assessment has already been demonstrated for anemia diagnostics in rural populations [12, 13, 14].

References

1. Agarwal R, Heinz T (2001) Bedside hemoglobinometry in hemodialysis patients: lessons from point-of-care testing. ASAIO J 47: 240–243
2. Alex CP, Manto JC, Garland JS (1998) Clinical utility of a bedside blood analyzer for measuring blood chemistry values in neonates. J Perinatol 18: 45–48
3. Benjamin JT, Baisden CR (2002) Advances in the physician office laboratory. In: Kost GJ (Ed.) Principles and practice of point-of-care Testing. Lippincott Williams & Wilkins, Philadelphia, pp 391–406
4. Briggs C, Guthrie D, Hyde K, Mackie I, Parker N, Popek M et al. British Committee for Standards in Haematology (2008) Guidelines for point-of-care testing: haematology. Br J Haematol. 142: 904–915
5. Brohi K, Cohen MJ, Davenport RA (2007) Acute coagulopathy of trauma: mechanism, identification and effect. Curr Opin Crit Care 13: 680–685
6. Despotis GJ, Joist JH, Goodnough LT (1997) Monitoring of hemostasis in cardiac surgical patients: impact of point-of-care testing on blood loss and transfusion outcomes. Clin Chem 43: 1684–1696
7. Despotis GJ, Saleem R, Bigham M, Barnes P (2000) Clinical evaluation of a new, point-of-care hemocytometer. Crit Care Med 28: 1185–1190
8. Duguid JKM, Newland AC (2010) Point-of-care testing in hematology. In: Price CP, St John A, Kricka LJ (eds) Point-of care testing, 3rd ed. AACC Press, Washington, pp 535–543
9. Gehring H, Hornberger C, Dibbelt L et al. (2002) Accuracy of point-of-care testing (POCT) for determining haemoglobin concentrations. Acta Anaesthesiol Scand 46: 980–986
10. Hermansen MC (2001) Nucleated red blood cells in the fetus and newborn. Arch Dis Child Fetal Neonatal Ed 84: F211–F215
11. Holloway PAH (2002) Point-of-care testing in intensive care. In: Kost GJ (Ed.) Principles and practice of point-of-care Testing. Lippincott Williams & Wilkins, Philadelphia, pp 133–156
12. Lewis SM, Osei-Bimpong A, Bradshaw A (2004) Measurement of haemoglobin as a screening test in general practice. J Med Screen 11: 103–105
13. McNulty SE, Torjman M, Grodecki W, Marr A, Schieren H (1995) A comparison of four bedside methods of hemoglobin assessment during cardiac surgery. Anesth Analg 81: 1197–1202

14. Weiss M, Dullenkopf A, Moehrlen U (2004) Evaluation of an improved blood-conserving POCT sampling system. Clin Biochem 37: 977–984
15. Widness JA, Kulhavy JC, Johnson KJ et al. (2000) Clinical performance of an in-line point-of-care monitor in neonates. Pediatrics 106: 497–504
16. Xanthou M (1970) Leucocyte blood picture in healthy full-term and premature babies during neonatal period. Archives of Disease in Childhood 45: 242–249
17. Zandecki M, Genevieve F, Gerard J, Godon A (2012) Spurious counts and spurious results on hematology analyzers: white blood cells, red blood cells, hemoglobin, red cell indices, and reticulocytes. In: Kottke-Marchant K, Davis B (eds) Laboratory Hematology Practice. John Wiley & Sons, pp 79–95

Diagnosing cardiovascular diseases

Evangelos Giannitsis, Ingolf Schimke, Peter B. Luppa, Dirk Peetz

P. B. Luppa, R. Junker (Eds.), *Point-of-Care Testing*
https://doi.org/10.1007/978-3-662-54497-6_17

17.1 Introduction

In Germany, cardiovascular disease is considered the most common cause of death with a rate of almost 50 %. In cardiovascular diseases, more than 40 % of deaths are related to coronary artery disease (CAD); 20 % of these are attributed to acute coronary syndrome (ACS) and, specifically, to acute myocardial infarctions (MI). MI account for approximately 75,000 deaths per year. After suffering an MI, 50 % of survivors develop chronic heart failure, which, combined with chronic heart failure of other etiologies (e.g. chronic CAD, cardiomyopathy, hypertension, myocarditis, valvular heart disease, arrhythmias, anemia), has become the principle cause of death with advancing age among the populations of western countries. In the total population, the prevalence rises from <1 % (in persons aged 45 to 55) to 2–5 % (aged 65–75) and further to almost 10 % in those over 80. Acute decompensated heart failure is the most frequent cause of hospitalization in the over 65 age group and is the most costly cardiovascular disease in western countries [17].

Considering the importance of CAD with regard to healthcare policy and economics, it is not surprising – thanks to their favorable cost-benefit ratio – that clinical laboratory tests are of paramount importance in diagnostics, therapy monitoring and judging the prognosis in patients with CAD. Shaped by current developments ("natriuretic peptides" are the catchphrase here), the role of clinical laboratory tests in patients with heart failure is growing in significance.

17

A decision must be made as to whether tests for CAD and heart failure should be centralized or available as POCT, depending on time constraints, local circumstances and the specific indication. For the diagnosis of MI, qualitative but also semi-quantitative and quantitative test systems are available for POCT. Quantitative test systems using biomarker concentrations in serum or whole blood enable diagnostic decision-making to confirm or refute the presence of disease by using definable cutoff values, while the trend in marker levels allows an estimation of risk, clinical course and therapy.

The deployment of POCT is recommended in hospitals, where central laboratories (or comparable clinical laboratory units providing acute care for patients with ACS) cannot achieve a laboratory turn-around time (TAT) for cardiac biomarker testing that is <60 min. TAT encompasses time for sample transport, preparation, analysis and reporting of results. A shorter laboratory TAT with POCT can be advantageous, not only in ACS but also in clinical scenarios where an MI or a re-infarction might occur as a complication of cardiac surgery or other interventions. The advantage of clinical laboratory diagnostics by POCT, focusing on cardiology, is only relevant if a shorter laboratory TAT ultimately translates into earlier availability of results with more timely therapeutic decisions. The argument that POCT delivers a shorter laboratory TAT is only valid if to the POC test's clinical implementation shortens the overall TAT, i.e. the time taken from requesting the biomarker by the clinician (laboratory TAT) to his acting (or not acting) on the test results.

Another fundamental issue impacting the use of POC versus central laboratory testing in cardiology-focused laboratory diagnostics is the fact that concentration-time curves for cardiac biomarkers can provide important diagnostic information that enables early detection of any re-infarction and allows the outcome of fibrinolytic therapies to be judged. The latter area of application is particularly relevant in geographical regions outside Europe where fibrinolytic therapy is still widely used. In evaluation of the clinical course, cardiac biomarkers are often measured by a central laboratory during the day with the use of POCT at night or over the weekend. Therefore, the National Academy of Clinical Biochemistry (NACB) recommends that the basic assay principles should be the same for both test methods. In the reality of day-to-day practice, however, completely different assays are used for POCT than in central laboratory techniques. Even when the same testing methods are used, assay technologies can differ significantly and consequently deliver different measurements, making monitoring impossible. This practice clearly makes the fundamental problem of standardizing assays more difficult.

Due to its tissue specificity, **cardiac troponin** (cTn) is the preferred biomarker for diagnosing myocardial infarction and can be measured quantitatively in the serum. High-sensitivity (hs) troponin assays have now largely been displaced other biomarkers such as myoglobin, CK-MB mass and FABP. These usurped biomarkers are no longer recommended as they contribute no additional diagnostic value. One exception is the renewed increase in CK-MB, which may have diagnostic relevance in the recognition of early re-infarction when the cTn is still rising.

Cardiac troponins (cTnT and cTnI) are the established biochemical standard assays for diagnosing ACS. An increase in the concentration of these cardiac-specific biomarkers indicates myocardial necrosis. With their higher cutoffs, conventional, not hs, cTn assays show do not show a rise in blood levels until after a few hours. Therefore, the additional measurement of biomarkers that indicate early myocardial ischemia or myocardial necrosis is recommended in these cases and particularly in the early phase of ACS.

POCT can complement or possibly even replace central laboratory diagnostics under specific local conditions or time requirements. In general, the cost-benefit ratio in this indication favors diagnostics in a central laboratory (▶ Chapter 30 and 31).

In chronic heart failure, which can become life threatening in the context of acute decompensation, serum markers such as **BNP** (brain natriuretic peptide) and the N-terminal fragment of pro-BNP (NT-pro-BNP) can be measured. Indeed, their use is recommended by the European Society of Cardiology (ESC), the American College of Cardiology (ACC) and the American Heart Association (AHA) for the screening and diagnosis of heart failure [11, 20]. When investigating the differential causes of chest pain and dyspnea ("shortness of breath"), the combined measurement of BNP/NT-pro-BNP, cTn and D-dimer is recommended. For BNP/NT-pro-BNP as well as D-dimer, systems are available as POC tests. In contrast to most other parameters developed only as secondary POCT-compatible systems, the clinical validation of BNP as a marker started with a test system developed for POCT (e.g. Triage, Biosite, now Alere).

POCT systems can deliver BNP results within approximately 15 min. To date, no recommendations have been given for which indication and above which laboratory TAT POCT should be preferred over determination in a central laboratory.

17.2 Requirements for the POCT of cardiac biomarkers

Clear improvements in the cost-benefit ratio can be observed when cardiac biomarkers are used in compliance with the guidelines for the diagnosis, therapy monitoring and prognostic evaluation of CAD and heart failure. Yet, general conclusions about the additional benefits of using POCT instead of central laboratory services cannot be made. The common assumption that shortening the laboratory TAT by using POCT (if this can in fact lead to a shorter overall TAT) will further improve the cost-benefit ratio of cardiac marker assays, was not confirmed by one prospective multicenter study (RATPAC trial) [9, 10].

The following still applies to immunoassay-based techniques:

- The antibodies used should be well characterized and targeted against stable analyte epitopes that are not subsequently modified in the blood (proteolysis, oxidation, phosphorylation, complexation).
- If a cardiac marker occurs in different variants (monomeric, complexed) in the blood, the antibody must bind them in a comparable way.
- If no generally accepted standard exists, as is the case for cTnI (see below), then separate reference values and decision-making

Table 17.1 Serum profile after STEMI and heart specificity of myocardial markers

Marker	Heart specificity	Start of concentration increase [h]	Maximum increase in concentration [h]	Factor increase	Time to normalization [days]
hsTn	+++	1–3	12–48	>100	>20
Myoglobin	–	2	6–12	Up to 20-fold	0.5–1
GPBB	++	2–4	7–9	Up to 20-fold	1–2
cFABP	++	2–3	5–10	Up to 35-fold (up to 125-fold)	0.5–1
CK-MB mass	+	3–4	12–18	Up to 30-fold	2–3
cTnT	+++	4–10	12–48	Up to 40 to 60-fold (up to 300-fold)	7–20
cTnI	+++	4–6	12–48	Up to 40-fold	7–14

cFABP Cardiac fatty acid binding protein; *CK-MB* Creatine kinase muscle-brain fraction; *cTnI, cTnT* Troponin proteins; *GPBB* Glycogen phosphorylase BB, *STEMI* ST-segment myocardial infarction

cutoff values should be set for the different assays.

- If at all possible, hospitals should choose tests for POCT that have similar analytical properties as the tests used in central laboratories to achieve harmonized test results.

17.3 Acute coronary syndrome

17.3.1 Laboratory diagnostics

The term "acute coronary syndrome" (ACS) subsumes the acute life-threatening phases of CAD: unstable angina, non-ST-segment elevation myocardial infarction (NSTEMI), formerly known as non-transmural infarct/non-Q-wave infarct/infarct without ST-segment elevation, and ST-segment elevation myocardial infarction (STEMI), formerly known as transmural infarction/Q-wave infarction. This insult is initially reversible, but later leads to irreversible ischemic myocardial damage. Myocardial structural proteins (cardiac markers) are released into the blood particularly as a result of irreversible damage (necrosis). These markers can be measured quantitatively to detect ischemic damage to the myocardium. The ideal marker should be mostly stable against pre-analytical influences, highly specific to cardiac injury detection, open a window for initial diagnosis and monitoring the clinical course and therapy, help detect complications such as re-infarction, while also allowing it to be measured by simple, rapid and cost-effective methods.

In contrast to all other currently known cardiac markers, the myofibril-bound and partially (3–6 %) cytosolic localized cardiac proteins cTnI and cTnT meet the aforementioned requirements and are therefore regarded as the "gold standard" for the diagnosis of irreversible myocardial damage.

Table 17.1 shows the key indicators for the diagnostic window of cTn compared to other biomarkers. Nowadays, the use of sensitive or highly sensitive cTn assays is strongly recommended as they are more accurate in detecting MI earlier than conventional cTn tests.

17

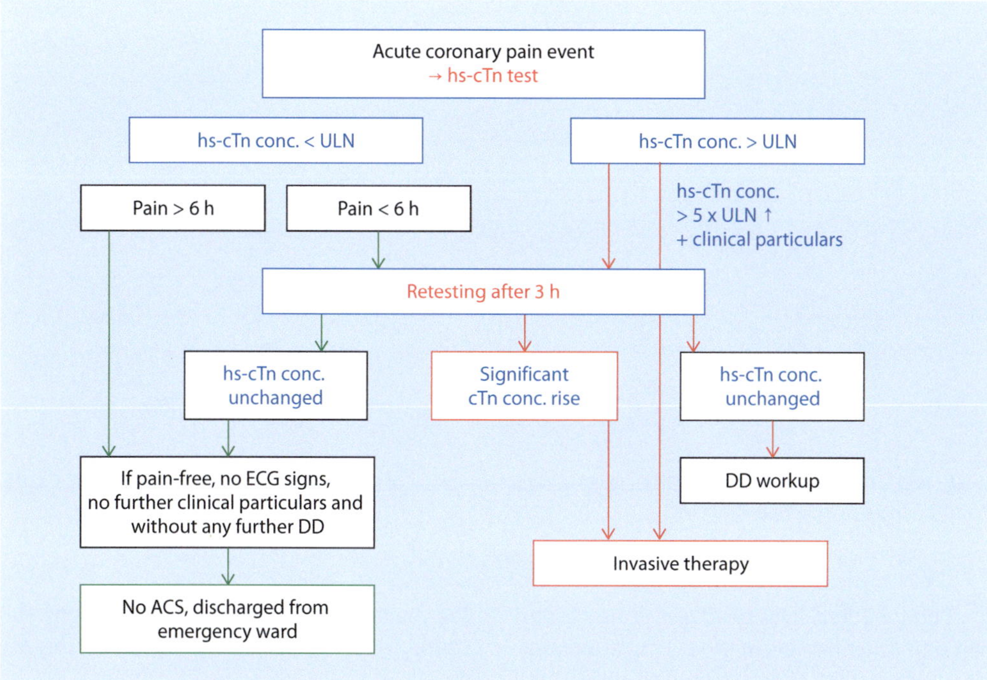

Fig. 17.1 The 0 h/3 h rule-out algorithm for NSTEMI using hs cTn detection methods. *ULN* Upper limit of normal, 99th percentile of healthy controls

Note

Based on these characteristics, current guidelines (e.g. from the ESC [22]) have recommended shortening the repeat interval from 6–9 hours to 1–3 hours after initial blood sampling. When highly sensitive cTnT or cTnI detection methods are used, no other markers are recommended by these guidelines (Fig. 17.1).

Nevertheless, it is important to point out that the cTn concentration, when measured with conventional, **low-sensitivity cTn assays**, might not have reached the upper cutoff point for disease, especially in patients who present early to the hospital after the onset of their chest pain. A second test should therefore be performed in all patients presenting with a last episode of chest pain less than 6 hours ago, but can be omitted in the case of a reported episode > 6 hours. The time between the onset of heart muscle necrosis and the detectability of markers in the blood is called the "troponin-blind interval". Biomarkers which give an early indication for the presence of myocardial ischemia can be a useful addition to cTn measurement when performed within this time interval.

Candidate markers include:
- Myoglobin (one of the longest established markers in use)
- Copeptin (stable fragment of the hormone preprovasopressin)
- Cardiac fatty acid-binding protein (c-FABP)
- Ischemic-modified albumin (IMA)
- Glycogen phosphorylase BB (GPBB)

The available evidence on the very early ischemic phase (<4 hours after symptom onset) is inconsistent; with no additional benefit from the use of other markers compared to **highly-sensitive troponin testing** being confirmed thus far.

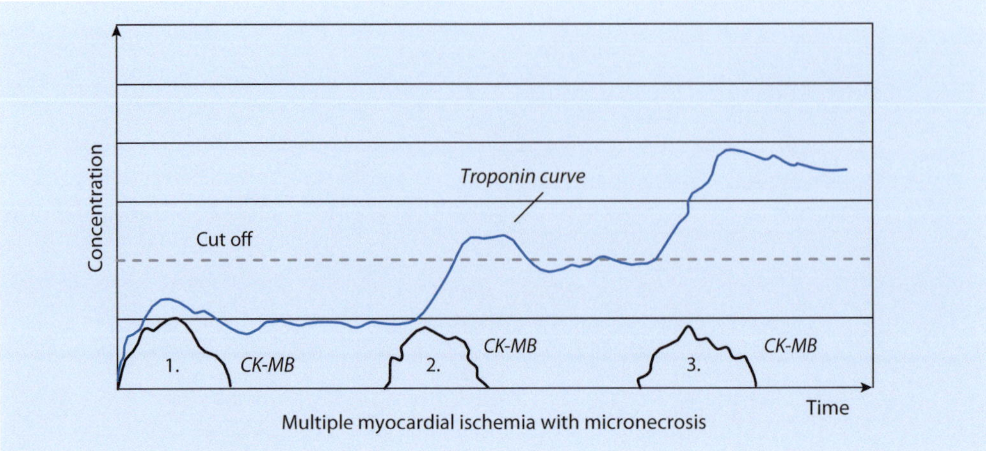

Fig. 17.2 Cumulative model for cTn increase in serum as a result of multiple myocardial ischemic events. *CK-MB* Creatine kinase muscle-brain fraction

The long half-lives of cTn isoforms (compared to other cardiac markers) cause a cumulative effect that allows assays to detect cTn elevations in the serum even in the case of microinfarctions (Fig. 17.2). The currently available tests for cTnT and cTnI can detect myocardial necrosis in the range of ≤1 g.

Comparable biphasic serum profiles post-MI have been described for both cTn parameters. When monitored frequently, an initial rise can be seen caused by the release of cTn from the cytosolic pool (peaks after 1–2 days) followed by the release of the structural (bound) portion (peaks after 3–5 days). In terms of organ specificity, cardiac troponins moreover show decisive advantages over the other cardiac markers (Table 17.1).

The risk profiles among patients with ACS vary according to the time point and number of fatal events. The earliest possible distinction between patients with unstable angina and those with NSTEMI is important because the indicated treatment strategies differ. The diagnosis of STEMI is made using ECG tracings where biomarker measurement is only used for retrospective confirmation and determination of infarction size. This differentiation is based on recommendations by the ESC or on the third universal definition of myocardial infarction [25]. According to these recommendations, the criteria for an acute or recently occurring MI are:

- A typical rise and gradual decline (cTnT, cTnI) with at least one of the following criteria:
  - Ischemic symptoms
  - Pathological Q-wave development on ECG
  - Ischemic ECG changes (ST-segment elevation or reduction)
  - New myocardial wall motion abnormalities
  - Evidence of a myocardial scar
  - Evidence of a coronary thrombus

These recommendations explicitly highlight the key role markers of myocardial necrosis play, especially cTn, and emphasize the importance of marker kinetics. Clear requirements for cardiac marker analyses are also specified. Previously, it was suggested that every cTn elevation in the blood is pathological and indicative of myocardial necrosis. Therefore, a cTn concentration above the 99th percentile of a healthy reference cohort was set as a decision-making cutoff. The imprecision (coefficient of variation; CV) of assays is higher, the lower the marker concentration is. Therefore, cTn assays

17

for diagnosing ACS have been recommended that have a ≤10 % CV at the 99th percentile concentration. Until the introduction of more sensitive assays, cutoffs for the cTn concentration that reflected an imprecision of 10 % (CV ≤10 %) were employed as a convention by many because the conventional assays did not meet the above-mentioned criteria. For example: if the cTnT 99th percentile cutoff is 0.01 µg/L, the CV is >10 %; a CV of 10 % would lead to a cutoff = 0.03 µg/L. In order to meet the precision criteria, assays were developed that have a ≤10 % CV at the 99th percentile. Additionally the lower detection limit of such assays, which lies significantly below the 99th percentile concentration, allows the measurement of cTn concentrations within the range between the lower detection limit and the 99th percentile without losing the cardiac specificity of cTn.

Four groups of tests are identified, based on the sensitivity of the analytical test methods. The first group (sensitivity grade 1; lowest analytical sensitivity) includes assays that can measure cTn concentrations in less than 25 % of a reference population. The group with the highest analytical sensitivity (grade 4) includes assays that can measure cTn concentrations in >95 % of healthy subjects.

Using current assays from group 4 in the central laboratory, measurable cTn blood values can be found in almost all "healthy" subjects (>95 %). By definition, only 1 % of the values measured lie above the 99th percentile. It is more problematic that the improved analytical sensitivity leads to an increased detection of heart muscle damage not caused by myocardial infarction. The improvement in analytical sensitivity is linked to a loss of diagnostic specificity in terms of "MI" diagnosis. Unless the initial values are very high, interpretation of the cTnT or cTnl results thereby requires serial cTn measurements to show release kinetics typical for MI, regardless of the clinical context. Not rarely, cTnI and cTnT measurements have been discredited by the indiscriminate use of such cTn values as an indication for heart catheter examinations when normal coronary arteries were found with no correlation to the elevated concentrations. On the other hand, it has been shown that raised cTn values in those patients who are not suffering from ACS had an increased risk of mortality. Therefore, elevated levels should trigger an early search for the underlying pathology (i.e. pulmonary embolism, chronic pulmonary hypertension, aortic dissection, decompensated heart failure, decompensated valvular disease) in order to initiate disease-specific treatment as early as possible.

In patients with ACS, the advantages of using high sensitivity assays generally outweigh the disadvantages by far. On the one hand, the detection of microinfarctions is now possible. On the other, it has been shown that the majority of infarctions are fully diagnosed at a time of the patient's admission or after 3 hours at the latest. In addition to standard diagnostics of repeated testing after 3 hours, the ESC guidelines recommend an accelerated diagnostic protocol using validated hsTn assays after 1 hour, repeated testing after 2 hours, but also the option of immediate diagnostics to exclude any infarction. The latter is possible if the hsTn test results lie below the limit of detection (LoD) or, if no high-sensitivity troponin test is available, by using combined measurements of copeptin, the stable C-terminal fragment of preprovasopressin. If copeptin and cTn concentrations lie below the relevant cutoffs at the time of the patient's admission, an MI is excluded [12, 14, 19, 26]. Moreover, taking the clinical risks into consideration, patients with a normal copeptin (<95th percentile) and normal cTn (<99th percentile) can be discharged from hospital safely [5]. Presently, implementation is limited by the fact that copeptin can only be measured with a separate analyzer (Kryptor, Thermo Fisher Brahms). Such a POCT device is undergoing clinical testing.

However, when hs cTn assays are used, no additional early biomarkers for ischemia need to be measured [13].

With these hs cTn assays, unstable angina is less frequently diagnosed while the diagnosis of infarction is increased.

17.3.2 POCT in acute coronary syndrome

Compared to the development of assays for central laboratory use, technical developments of POCT systems have unfortunately not advanced at the same speed. Some systems still use cutoff values that are significantly above the 99th percentile concentration [5, 18] and therefore only enable a diagnosis of infarction to be made "too late" or many infarctions are not even detected [15, 16, 23, 27]. Various suppliers have introduced POCT systems that enable the quantitative measurement of cTn with a precision comparable that of central laboratories (e.g. Radiometer AQT90, Roche h 232). In addition, there are two POCT systems on the market to date which make measurements possible using a high-sensitivity cTnI assay (Siemens' Stratus CS Acute Care, Mitsubishi's PATHFAST). An overview of available quantitative POCT devices (e.g. Triage MeterPro from Alere, AQT90 FLEX, i-STAT from Abbott, Meritas from Trinity Biotech, PATHFAST, Stratus CS, h 232, VIDAS from Biomérieux, RAMP from Response Biomedical) together with analytical specifications and performance data was compiled by Amundson and Apple in 2015 [2].

Diagnostic and therapeutic strategies have been developed in various studies that tested POCT-measured cardiac markers with regard to utilization of resources, risk stratification, therapeutic management and clinical outcomes. The ACC/AHA 2002 Guideline update relating to patients with unstable angina and NSTEMI already anticipated that the diagnosis and management of patients will improve thanks to the transition from qualitative or semi-quantitative to quantitative measurements within POCT [7]. In the 2004 ACC/AHA Guidelines for the management of patients with STEMI, it was proposed that qualitative tests could be used to detect elevated cardiac markers, whereas follow-up measurements should be carried out using quantitative tests. If these guideline recommendations are implemented consistently, then only high-sensitivity POCT assays should be used [3].

As previously described in ▶ Section 17.3, the initial diagnosis using cTn measurements as part of a multi-marker strategy can be supplemented with markers that detect an earlier rise in the blood concentration (e.g. copeptin, myoglobin, c-FABP, GPBB). Improved risk stratification or better control of invasive diagnostics can additionally be expected when multi-marker strategies are applied, including cTn and others like cystatin C, natriuretic peptide, C-reactive protein, mid-regional pro-adrenomedullin (MR-proADM) and growth-differentiation factor-15 (GDF-15). In order to measure such marker combinations with POCT, quantitative, in part, highly sensitive test systems are required and already being developed. Practicability and cost-benefit ratios will finally determine the implementation of POCT in the diagnosis of ACS. The postulated positive effects of POCT on patient outcomes, in particular, must be able to stand up to critical validation. Unfortunately, the results from clinical studies so far are contradictory [21]. The decision to implement a POCT concept for the diagnosis of ACS should be guided by local circumstances, logistical and organizational options, the implementation of effective quality management systems and by cost structures. General advantages for POCT in the diagnosis of ACS have not yet been proven in clinical studies. Nevertheless, at hospitals without a central laboratory or in developing countries [1], the use of POCT for measuring cardiac markers can be viewed differently.

17.4 Heart failure

17.4.1 Laboratory diagnostics

ECG, imaging and invasive procedures have so far been the core tools of diagnostic studies for chronic and acute heart failure as well as in the delineation of these pathologies against chronic obstructive lung disease and pulmonary embolism etc. Until recently, laboratory medicine had not played a significant role in the diagnosis, monitoring, therapeutic evaluation or in the differential diagnosis of acute chest pain and dyspnea. The discovery of the natriuretic peptides (especially BNP) and D-dimer heralded a future trend towards a greater role that laboratory diagnostics will play in the care of patients with heart failure, particularly from a cost-benefit perspective. These two characteristic variables will be discussed in greater detail below.

BNP/NT-pro-BNP/NT-pro-ANP

Brain natriuretic peptide (BNP) is produced mainly in the atrium of a healthy heart. In ventricular diseases, particularly in heart failure, BNP gene expression in ventricular myocytes increases, mainly due to the response of the heart to greater myocardial wall tension, dilatation and/or intracardiac pressure. BNP is released into the blood without any prior storage. BNP causes increased diuresis and natriuresis as well as relaxation of vascular smooth muscle. It has an inhibitory effect on mitogenesis and myocardial remodeling.

BNP is the product of proteolytic processing of the precursor molecule prepro-BNP (134 amino acids) which is synthesized in cardiomyocytes. Prepro-BNP is cleaved in the cytoplasm to pro-BNP (108 amino acids) and an N-terminal signal peptide (26 amino acids). After its release into the blood, pro-BNP is cleaved into the active hormone BNP (amino acids 77–108) and the inactive N-terminal fragment NT-pro-BNP (amino acids 1–76). Additionally, low concentrations of pro-BNP are also found in the blood. This mostly has analytical implications for the specificity of BNP and NT-pro-BNP tests. As an active hormone, BNP is proteolytically degraded via binding to NP receptors by internalization into the target cell and by a membrane-bound neutral endopeptidase. The latter is present mainly in the liver, lung and kidneys. Plasma levels of BNP can potentially be affected by inhibitors of the neutral endopeptidase (e.g. candoxatril for the treatment of heart failure) or vasopeptidase (e.g. omapatrilat for the treatment of hypertension). By contrast, NT-pro-BNP undergoes mainly renal excretion and could therefore be more suitable for monitoring of therapy in patients treated with these drug classes.

Meanwhile, numerous studies have become available and recommendations made in national and international guidelines (new 70; ESC Guideline) on BNP as well as NT-pro-BNP and NT-pro-ANP, attesting to their suitability as clinical chemistry markers to exclude acute or chronic heart failure and for the assessment of prognosis in heart failure patients. Currently, it is still unclear if factors such as age, gender and body mass index need to be included when setting decision-making limits. The initial discussion on the principle advantages and disadvantages of BNP vs. NT-pro-BNP in terms of the renal elimination of NT-pro-BNP or the sensitivity of BNP to pre-analytical influences finally failed to establish the superiority of either of the two parameters. The degree of renal impairment that affects NT-pro-BNP blood levels and the resulting implications for decision-making limits are well known [4]. Once the required conditions for BNP in the pre-analytic phase are met (EDTA plasma), it exhibits a comparable stability to NT-pro-BNP for a minimum of 4–24 h (depending on the test antibodies used). It is possible that the different clearance times for BNP (20 min) and NT-pro-BNP (2 h) could deliver future approaches to the targeted use of one or the other marker, for example, in the monitoring of patients with chronic heart failure or for recompensation after decompensated heart failure.

In addition to the main area of applications for BNP/NT-pro-BNP/NT-pro-ANP in the diagnosis of left ventricular systolic dysfunction in patients with chronic heart failure, further potential applications that have emerged include left ventricular hypertrophy, left ventricular diastolic dysfunction, atrial fibrillation, valvular diseases and ACS. In combination with cTnT/cTnI and D-dimer, BNP/NT-pro-BNP offers a possible aid in the differential diagnoses of chest pain and dyspnea. Here, the measurement of BNP/NT-pro-BNP is seen as a screening tool that leads to further diagnostic procedures, (particularly echocardiography) if the result is positive. A negative result excludes cardiac involvement with high probability. The following **areas of application** for measuring BNP/NT-pro-BNP derive from the National Academy of Clinical Biochemistry (NACB) [24] and the ACC or the AHA recommendations [25]:

- Initial measurement in patients in the emergency department (or similar settings) where the diagnosis of heart failure is unclear
- Differential diagnoses in chest pain and dyspnea combined with D-dimer assays
- Patients with left ventricular dysfunction post acute MI
- Presence of risk markers for heart failure such as a medical history of ACS or diabetes mellitus
- Risk stratification and prognosis assessment in patients with ACS, decompensated heart failure, stable chronic heart failure, a heart transplantation or non-cardiac diseases such as pulmonary embolism
- Therapy monitoring for heart failure
- Monitoring of patients treated with potentially cardiotoxic medications (e.g. anthracycline or trastuzumab)

However, for many of the indications listed, only insufficient data have been evaluated to date, particularly regarding decision-making limits. In many cases, no conclusions are available about the cost-benefit ratio.

BNP-/NT-pro-BNP measurement is **not** indicated:

- In patients with obvious clinical signs of heart failure. If this is the case, it is not a replacement for conventional methods like echocardiography and invasive hemodynamic examinations.
- For screening asymptomatic patient cohorts for left ventricular dysfunction.

17

D-dimer

D-dimer is a characteristic product of fibrin cleavage mediated by plasmin and found in plasma at higher concentrations during intravascular activation of coagulation and secondary fibrinolysis.

A normal D-dimer practically excludes pulmonary embolism (PE) when the pre-test probability is low to moderate. If the test is positive, imaging procedures like spiral computed tomography and magnetic resonance imaging, occasionally, scintigraphy and pulmonary angiography are still also employed to confirm the diagnosis. The measurement of D-dimer can be useful in the differential diagnosis of pulmonary embolism, on the one hand, and of pneumonia, pleurisy, pneumothorax and asthma attacks, on the other.

Under the pretest probabilities described, a normal D-dimer is grounds to practically rule out any deep vein thrombosis. At this juncture, it is important to stress that a negative test result at high pre-test probability is still associated with a high post-test probability (typically 20 % and 50 %); even when high-sensitivity tests are used. Only at a (theoretical) test sensitivity of 100 % can a diagnosis be excluded with certainty (independent of the pre-test probability) [8].

A positive test requires further diagnostic work up to confirm the diagnosis seeing as D-dimer is not specific for PE and deep vein thrombosis (DVT). A diverse range of other pathophysiological conditions that are linked with coagulation activation needs to be weighed into the differential diagnoses if the D-dimer test is positive.

The diagnosis of DVT and PE can follow an evidence-based procedure as formulated in the guidelines issued by the Association of the Scientific Medical Societies in Germany (Arbeitsgemeinschaft der Wissenschaftlichen Medizinischen Fachgesellschaften e.V., AWMF) [6]: In suspected DVT, the clinical probability of the presence of the disease is initially evaluated. If the pre-test probability is high, compression ultrasound of the leg veins is immediately requested without laboratory diagnostics. In case of the slightest clinical suspicion, the immediate test for D-dimer is exceptionally important. If the test result is negative, a DVT is excluded because of the very high negative predictive value. If the test turns out positive, a compression ultrasound of the leg veins should be ordered.

If PE is suspected in a clinically stable patient, the diagnostic algorithm is analogous to the procedure for DVT. If there is a high or moderate clinical suspicion, immediate imaging procedures are undertaken (compression ultrasound of the leg veins, spiral computed tomography, scintigraphy). D-dimer is only measured as a first-line diagnostic test in the context of low clinical suspicion. If the result is negative, a PE is excluded because of the test's very high negative predictive value. Positive test results are followed by the relevant imaging studies.

17.4.2 POCT in heart failure

The ability to determine natriuretic peptides alone or in combination with D-dimer has brought significant progress in the diagnosis of and differential diagnostics for patients with suspected heart failure. It is anticipated that natriuretic peptides will also gain importance in the management of patients with heart failure and other cardiac diseases. The significance of POCT in this context, however, still remains unclear: No systematic studies have been carried out to date aimed at comparing the cost-benefit ratio of central laboratory diagnostics versus POCT.

If such a concept is further developed consistently, continuous self-monitoring of patients with chronic heart failure – possibly supported by telemedicine – might not only lead to quicker optimization of therapy, but thereby to lower re-hospitalization rates for decompensation of heart failure as well.

References

1. Altintas Z, Fakanya W, Tothill I (2014) Cardiovascular disease detection using bio-sensing techniques. Talanta 128: 177–186
2. Amundson B, Apple F (2015) Cardiac troponin assays: a review of quantitative point-of-care devices and their efficacy in the diagnosis of myocardial infarction. Clin Chem Lab Med 53: 665–676
3. Antman EM, Anbe DT, Armstrong PW et al. (2004) ACC/AHA guidelines for the management of patients with ST-elevation myocardial infarction – executive summary. A report of the American College of Cardiology/American Heart Association Task Force on Practice Guidelines. J Am Coll Cardiol 44: 671–719; erratum in: J Am Coll Cardiol 2005; 45: 1376
4. Anwaruddin S, Lloyd-Jones DM, Baggish A et al. (2006) Renal function, congestive heart failure, and amino-terminal pro-brain natriuretic peptide measurement: results from the ProBNP Investigation of Dyspnea in the Emergency Department (PRIDE) Study. J Am Coll Cardiol 47: 91–97
5. Apple FS, Wu AHB, Jaffe AS (2002) European Society of Cardiology and American College of Cardiology guidelines for redefinition of myocardial infarction: How to use existing assays clinically and for clinical trials. J Am Heart 144: 981–986
6. AWMF online (2015) Diagnostik und Therapie der Venenthrombose und der Lungenembolie. Leitlinien Angiologie 2015; www.leitlinien.net. Last revised: 10/2015
7. Braunwald E, Antman EM, Beasley JW et al. (2002) ACC/AHA 2002 guideline update for the management of patients with unstable angina and non-ST-segment elevation myocardial infarction – summary article: a report of the American College of Cardiology/American Heart Association task force on practice guidelines (Committee on the Management of Patients With Unstable Angina). J Am Coll Cardiol 40:1366–1374
8. Christenson RH on behalf of the Committee on Evidence Based Laboratory Medicine of the International Federation for Clinical Chemistry Laboratory Medicine (2007) Evidence-based laboratory medicine – a guide for critical evaluation of in vitro laboratory testing. Ann Clin Biochem 44: 111–130
9. Collinson P, Goodacre S, Gaze D, Gray A (2012) Very early diagnosis of chest pain by point-of-care testing: comparison of the diagnostic efficiency of a panel of cardiac biomarkers compared with troponin measurement alone in the RATPAC trial. Heart 98: 312–318
10. Goodacre S, Bradburn M, Fitzgerald P, Cross E, Collinson P, Gray A, Hall AS (2011) The RATPAC (Randomised Assessment of Treatment using Panel Assay of Cardiac markers) trial: a randomised controlled trial of point-of-care cardiac markers in the emergency department. Health Technol Assess 15 (23): 1–102
11. Hunt SA, Abraham WT, Chin MH et al. (2005) American College of Cardiology; American Heart Association Task Force on Practice Guidelines; American College of Chest Physicians; International Society for Heart and Lung Transplantation; Heart Rhythm Society. ACC/AHA 2005 guideline update for the diagnosis and management of chronic heart failure in the adult: a report of the American College of

Cardiology/American Heart Association Task Force on practice guidelines (Writing committee to update the 2001 guidelines for the evaluation and management of heart failure), developed in collaboration with the American College of Chest Physicians and the International Society for Heart and Lung Transplantation: endorsed by the Heart Rhythm Society. Circulation 112: 154–235
12. Karakas M, Januzzi JL Jr, Meyer J, Lee H, Schlett CL, Truong QA et al. (2011) Copeptin does not add diagnostic information to high-sensitivity troponin T in low- to intermediate-risk patients with acute chest pain: Results from the rule out myocardial infarction by computed tomography (ROMICAT) study. Clin Chem 57:1137–1145
13. Keller T, Zeller T, Peetz D, Tzikas S, Roth A, Czyz E et al. (2009) Sensitive troponin I assay in early diagnosis of acute myocardial infarction. N Engl J Med 361:868–877
14. Keller T, Tzikas S, Zeller T, Czyz E, Lillpopp L, Ojeda FM et al. (2010) Copeptin improves early diagnosis of acute myocardial infarction. J Am Coll Cardiol 55:2096–2106
15. Korley FK, Jaffe AS (2013) Preparing the United States for high-sensitivity cardiac troponin assays. J Am Coll Cardiol 61: 1753–1758
16. Lee-Lewandroski E, Januzzi JJ, Grisson R, Mohammed A, Lewandrowski G, Lewandrowski K (2011) Evaluation of first-draw whole blood, point-of-care cardiac markers in the context of the universal definition of myocardial infarction. Arch Pathol Lab Med 135: 459–463
17. McMurray JJ, Stewart S (2000) Epidemiology, aetiology, and prognosis of heart failure. Heart 83: 596–602
18. Panteghini M, Pagani F, Yeo KT et al. (2004) Committee on Standardization of Markers of Cardiac Damage of the IFCC. Evaluation of imprecision for cardiac troponin assays at low-range concentrations. Clin Chem 50: 327–332
19. Reichlin T, Hochholzer W, Stelzig C, Laule K, Freidank H, Morgenthaler NG et al. (2009) Incremental value of copeptin for rapid rule out of acute myocardial infarction. J Am Coll Cardiol 54:60–68
20. Remme WJ, Swedberg K (2001) Guidelines for the diagnosis and treatment of chronic heart failure. Eur Heart J 22: 1527–1560
21. Ryan RJ, Lindsell CJ, Hollander JE, O'Neil B, Jackson R, Schreiber D, Christenson R, Gibler WB (2009) A multicenter randomized controlled trial comparing central laboratory and point-of-care cardiac marker testing strategies: the Disposition Impacted by Serial Point of Care Markers in Acute Coronary Syndromes (DISPO-ACS) trial. Ann Emerg Med 53:321–328
22. Roffi M et al., for the Task Force of the European Society of Cardiology (ESC) (2016) 2015 ESC Guidelines for the management of acute coronary syndromes in patients presenting without persistent ST-segment elevation. European Heart Journal) 37: 267–315
23. Sandoval Y, Apple FS (2014) The global need to define normality: the 99th percentile value of cardiac troponin. Clin Chem 60: 455–462
24. Tang WH, Francis GS, Morrow DA et al. (2007) National Academy of Clinical Biochemistry Laboratory Medicine practice guidelines: Clinical utilization of cardiac biomarker testing in heart failure. Circulation 116: 99–109
25. Thygesen K, Alpert JS, Jaffe AS, Simoons ML, Chaitman BR, White HD; Writing Group on behalf of the Joint ESC/ACCF/AHA/WHF Task Force for the Universal Definition of Myocardial Infarction (2012) Third universal definition of myocardial infarction. Glob Heart 7(4):275–295
26. Vafaie M, Biener M, Mueller M, Abu Sharar H, Hartmann O, Hertel S, Katus HA, Giannitsis E (2015) Addition of copeptin improves diagnostic performance of point-of-care testing (POCT) for cardiac troponin T in early rule-out of myocardial infarction – A pilot study. Int J Cardiol 198:26–30
27. Venge P, Öhberg C, Flodin M, Lindahl B (2010) Early and late outcome prediction of death in the emergency room setting by point-of-care and laboratory assays of cardiac troponin I. Am Heart J 160: 835–841

POCT methods for screening in addiction medicine

Lars Wilhelm

P. B. Luppa, R. Junker (Eds.), *Point-of-Care Testing*
https://doi.org/10.1007/978-3-662-54497-6_18

18.1 Introduction

In drug of abuse (DOA) screening, immunological test strip methods are frequently employed to obtain a general suspicion of drug or substance abuse. The area of application extends from forensic investigations through emergency medicine to uncovering the concomitant use of drugs during addiction treatment.

One important area of application for DOA screening is therapy monitoring of patients being treated for addiction. In this context, differentiations can be made between substitution-supported treatment with methadone, buprenorphine or, since 2015, sustained-release morphine and treatments without maintenance therapy (clean therapy). In 1998, a chapter entitled "Narcotic Drugs" was added to Section 24a of the German Road Traffic Act. This resulted in greater crackdown on drivers operating vehicles under the influence of these drugs. From 1999 to 2004, the proportion of traffic offenses committed by drivers on drugs rose from 1.1 % to 10.1 % [5]. POCT is frequently used in emergency medicine in the hopes that intoxication or poisoning can be diagnosed as quickly as possible. Within the broadest sense of workplace drug testing, employees working in safety-relevant areas can moreover be tested legally defensible for abstinence. At criminal justice facilities, tests must be conducted regularly in order to rule out narcotics abuse. As part of fitness-to-drive diagnostics based on a medical-psychological assessment report, proof of abstinence is required for driver's license re-applicants [9].

Considering the more than 9.5 million persons drinking risky amounts of alcohol, 1.3 million alcoholics and annually between 42,000 to 74,000 alcohol-related deaths, alcohol indeed plays a major role in addiction medicine in Germany. Alcohol abuse in adolescents has risen in recent years. Over 20,000 children and adolescents annually receive inpatient treatment for alcohol poisoning [1]. Alcohol monitoring is a part of preventive measures (for example to ensure road safety) or it serves as proof of abstinence during treatments for addictive diseases. Alcohol is contraindicated in substitution treatment. To screen drivers with suspended driver's license for driving under the influence of alcohol, ignition interlock devices have been in use for several years now, where alcohol breathalyzers control the motor's ignition [4]. In many areas, alcohol breath analysis is carried out to detect acute alcohol abuse.

Meanwhile, test systems for sweat, saliva or solids have come onto the market alongside strips for urine sample testing. Nowadays, the spectrum of detectable substances extends far beyond the classic analytes amphetamine, methamphetamine, cocaine metabolite (benzoylecgonine), opiates (morphine) and cannabis metabolite (THC-COOH); currently, over 200 substances from around 20 substance groups with a diverse range of pharmacokinetic and pharmacodynamic properties can be detected [2]. Numerous (pre-)analytical factors need to be accounted for in the selection of the optimal test system, as well as for analyzing and interpreting the results.

18.2 Analysis planning and optimal test systems

Prior to selecting the test system, considerations should be given as to which sample matrix best achieves the desired objective. The most common practice is to collect **spontaneously voided urine**, this method offers innumerable advantages: The sample is easy to obtain without medical staff. Because of the physiological function of the renal tubuli (reabsorption, secretion), very high concentrations of most narcotic drugs and pharmaceutical agents are normally found in the urine. This allows the analytes to be detectable for a relatively long period of time.

Note

It should be noted that any conclusions about pharmacologically relevant blood concentrations cannot be drawn from urine analysis results.

Tab. 18.1 Monitoring parameters for urine sample tampering [10]

Parameters	Normal findings
Temperature	32–39° C (directly after taking the sample)
pH	4.0–9.0
Creatinine in the urine	<20 mg/dL
Relative density	<1.003 kg/L
Chemicals, metabolites	e.g. glutaraldehyde, chromates, bleaching agents, nitrite, glucose, fructose

One problem is sample tampering. Laboratories constantly discover that water, tea, juice or chemicals have been added artificially to urine samples. Even during supervised sample taking, tampering – including the addition of foreign urine – cannot be ruled out completely. It should also be noted that excessive intake of fluids can dilute the sample considerably. Appropriate concurrent analyses must be carried out to check for sample tampering (Tab. 18.1). Today, a broad spectrum of parameters extending beyond analysis for traditional narcotics is available for urine samples.

Since the end of the 1990s, **saliva** (oral fluid) has played an increasingly important role in narcotics and DOA screening. This school of thought was borne by the prevailing hypothesis in the past that the concentration of an analyte in the saliva would allow pharmacologically relevant conclusions to be drawn about corresponding levels in the blood. Moreover, saliva analysis offers an option that lets the patient avoid the indiscretion of providing a urine sample. Particularly for police officers, taking saliva samples thus appears to be a good replacement for urine samples [12].

On the whole, however, it should be stated that saliva analysis cannot meet all expectations. The primary reason is that not all narcotic drugs and medications are excreted the same way by the saliva glands. Lipophilic and acidic compounds such as the active metabolite in cannabis, Δ^9-THC, are hardly detectable in saliva samples. Detection of this substance is generally attributable to consumption residues. Furthermore, it was determined that secretion from the saliva glands is pH-dependent. Thus, any correlation between serum levels and saliva concentration can only rarely be assumed [5].

The collection of saliva is fraught with problems. Several analytes are adsorbed on the surface of the carrier (e.g. cotton wool). Other systems rely on a rinsing solution for the oral cavity, thereby circumventing adsorption problems. A further difficulty associated with the collection of saliva samples is verification of the sample volume. If a dye is added to the mouth-rinsing solution for obtaining a sample, it can be used to photometrically determine the saliva volume harvested [8]. Despite the aforementioned limitations, saliva analysis offers an interesting supplement to classic POCT techniques, albeit, still dependent on more comprehensive analysis by chromatographic methods.

Yet another analytical option is offered by testing **sweat**. Medicinal agents and narcotic drugs are secreted through the saliva glands by the same physiological mechanisms as through the sweat glands. In turn, the chemical characteristics of a substance have an impact on the excretion rate. It is problematical to delineate any contamination of the skin with analytes which can lead to false-positive results. There are two techniques available for sampling of sweat:

- The wipe test used in Germany is performed by wiping a sampling pad attached to a collector system over the subject's skin. However, the harvested sample volume can vary considerably from case to case and impair the test result.
- Samples are taken on "sweat patches" through a non-woven layer. They are applied to the body and left there for several days. These technique offers the advantage that the corresponding skin area can be cleaned properly, which largely rules out any contamination. Furthermore,

the observation period is extended as well. By contrast, the possibility of sample tampering poses problems.

Because of the aforementioned disadvantages, POCT analyses in sweat samples only play a minor role in testing for narcotic drugs and pharmaceuticals.

Volatile substances like ethanol can be detected in the **breath**. The substances are diffused from the air stream in the lungs through the alveolar barrier walls into the exhaled air, where they can be quantitatively demonstrated. Breath alcohol analysis is an indirect method for determining the level of systemic alcohol intoxication. The ratio between breath alcohol and blood alcohol is 1:2100 on average. The concentration can, however, be affected by the manner of exhalation and any possible residual alcohol in the oral cavity. Thus, the ratio of breath alcohol to blood alcohol fluctuates inter- and intra-individually. A blood sample must be drawn for quantitative verification of the measured results. The length of time available to determine alcohol in the breath is only several hours after the person has stopped drinking; this method can thus only detect an acute alcohol intoxication of the patient [5]. Parameters with a broader detection window such as carbohydrate-deficient transferrin (CDT) or the phase-II-metabolite ethyl glucuronide should be employed for abstinence monitoring.

18.3 Test principle and conduct

The test principle on which the majority of rapid drug tests are based is **GLORIA technology** (GLORIA: gold-labeled optical-read rapid immunoassay). This technology is also incorporated into many other test strips, like those used for albumin detection in the urine, for example (▶ Chapter 19). The test strip consists of a carrier film applied to multiple layers of non-woven material. When the test strips are immersed in the sample, the lower non-woven layer absorbs the sample material. The mark delineating the maximum immersion depth should be observed. This triggers a chromatographic process during which the sample is absorbed upwards along the test swab by virtue of capillary action. This causes the sample to initially flow through a non-woven layer that contains gold-labeled monoclonal antibodies. The analytes contained in the sample form mobile, red-stained complexes with the test-specific monoclonal antibodies. In this process, a portion of the antibodies are complexed; in the next non-woven layer, the excess antibodies are bound to immobilized analyte analogs (haptens) and produce an optically visible control band. The red-stained analyte complex flows through this fleece and accumulates in the detection zone. A specific change in color is a positive indication that the analyte has been detected. Thus, a control line is only visible in negative samples. Above this control line, however, a second detection line can be seen in positive samples [10].

The main sources of error associated with the analysis of test strips results are:

- Misinterpretation of the displayed lines
- Read-off inaccuracies
- Erroneous interpretation of color changes, particularly at cutoff concentrations

The inexperienced examiner will thus inevitably be hard-pressed to interpret weakly stained bands. In the field, insufficient or unfavorable lighting will frequently encumber interpretation. Another problem is posed by the strongly varying intrinsic color of urine samples, which can further complicate the reading of the test strips. The analysis can be objectified with a reflectance photometer or scanner.

One important handling aspect, particularly with urine samples, is the danger of **sample contamination**. Taking into account that the concentrations of some narcotic drugs and pharmaceuticals are in the range of 100 mg/L, mere traces – meaning a fraction of a droplet – can contaminate a second subsequent sample. Such analytical errors can no longer be cleared up by any further comprehensive laboratory analysis. Therefore, it is urgently recommended the principles of trace analysis always be applied to sample handling in order to prevent any contamination.

Several important measures for preventing contamination

- Never process multiple samples simultaneously.
- Only employ single-use materials.
- Always wear single-use gloves.
- Samples should be aliquoted and the original samples kept as retention samples for any follow-up analyses.
- Benchtops should be cleaned scrupulously after the end of each work step.

Note

The use of firmly sealable urine beakers with integrated test strips can eliminate many contamination pathways.

The quality assurance conventionally used in laboratory diagnostics is contravened by limitations with test strips; in practice, only the control band is used for quality control. Control materials are generally not offered by the test suppliers. Given that the test strips are a closed test system, the transferability of quality control results from one test to another must be questioned. One example of the reasons for this is that errors in storage will not impact all tests equally. Not only test suppliers should participate in interlaboratory tests; this is recommendable for every test user.

A variety of test principles are used for breath analysis. In the **Alcotest test tubes,** ethanol is oxidized with potassium dichromate to acetaldehyde; yellow potassium dichromate (VI) is reduced to green chromium(III) oxide. The color change serves visual detection. The more modern analytical systems determine the alcohol concentration by infrared sensors or electrochemical means. In the case of infrared optical sensors, infrared light is guided through a measuring chamber. In the presence of ethanol, a portion of the infrared light is absorbed in a concentration-dependent manner. Absorption is proportional to the ethanol concentration and is used to calculate breath alcohol concentration after calibration of the measuring system [3].

For electrochemical detection, an electrical piston pump is used to transfer an air sample of a pre-set volume (approx. 1 cm^3) into a measuring chamber. The alcohol is oxidized to aldehyde on the catalytic lining of the measuring electrode. The electrons released by this generate a measurable electrical signal that enables determination of the alcohol concentration after calibration [3].

18.4 Decision-making limits (cut-off concentrations)

Decision limits are stated by the manufacturer as cut-offs. These are concentration levels generally given in µg/L and in the context of a reference substance. The test reacts positively when the concentration of the reference substance is above the stated cut-off. The cut-offs commonly refer to the Mandatory Guidelines drafted by the Substance Abuse and Mental Health Services Administration (SAMHSA) in the USA. However, the SAMHSA Mandatory Guidelines [13] only cover a very limited spectrum of controlled substance for which the relatively high cut-off concentrations have been proposed (◘ Tab. 18.2). Given that a majority of the test strips are produced for the US market, tests are frequently supplied with cut-offs that are unconventional for Europe. The European Workplace Drug Testing Society (EWDTS) has published cut-offs for urine and saliva analysis [11]. In 2013, the third edition of the expert review guidelines for fitness-to-drive diagnostics published new cut-offs for chromatographic tests for narcotic drugs in the urine [9].

Tab. 18.2 Cut-offs (stated in µg/L) [9, 13, 11]

Analytes	Medical-psychological assessment of fitness-to-drive (chromatography)	SAMHSA	EWDTS	
			Immuno-chemistry	Confirmatory analysis
Amphetamines		1000	500	–
Amphetamine	25	–	–	200
Methamphetamine	25	–	–	200
MDMA/MDA/MDE(A)	25	–	–	200
Other amphetamines		–	–	200
Barbiturates		–	200	150
Benzodiazepines	50**	–	200	100
Buprenorphine/ norbuprenorphine		–	5	2
Cannabinoids		50	50	–
Δ^9-THC-COOH	10*	–	–	15
Cocaine		300	150	–
Benzoylecgonine	30	–	–	100
Methadone/methadone metabolite EDDP		–	100 (300)	–
Methadone	50	–	–	250
Methadone metabolite EDDP	50	–	–	75
Opiates (total)		2000	300	–
Morphine (total)	25*	–	–	300
6-Acetylmorphine	25*	–	–	10
Codeine (total)	25*	–	–	300
Dihydrocodeine (total)	25*	–	–	300

SAMHSA Substance Abuse and Mental Health Services Administration
* After hydrolysis, ** Selected benzodiazepines

18.5 Cross reactions

Cross reaction is defined as a property of any substance to cause a reaction in a specific test strip product. The cross reactions are indicated either as the minimal concentration that led to a positive test result, or in percentage reactivity as compared to the test-specific reference substance. Unfortunately, the presence of compounds that do not count among the group of target analytes can lead to a positive result. This is referred to as an "adverse cross reaction". As a rule, only tests should be used for which extensive lists of proven cross reactions are available. These lists should also indicate the sensitivity for target analyte collection. Particularly in the case of group tests (e.g. with the benzodiazepines), test strips do not show a sufficient cross-reactivity for all analytes. On the urine tests, it is imperative that the relevant metabolites and not only the active constituents be listed.

In the presence of detected ethanol, cross reactions with other alcohols or ketones can occur. At physiological concentrations, however, these are mostly irrelevant. Medication, hygiene or disinfectants can contain various alcohols in relevant concentrations.

18.6 Hydrolysis

Phase II metabolites like glucuronides generally do not show sufficient cross-reactivity. A high percentage of pharmaceuticals, however, are renally excreted as glucuronide conjugate. Enzymatic hydrolysis is a simple method for detecting these analytes. After pH adjustment (e.g. Helix pomatia: 4.5), a hydrolase for cleaving the conjugates is added to the sample. After incubation in a water bath, the glucuronide conjugates are present in the sample as phase I metabolites and available for test strip analysis. If a test is repeated to confirm uncertain or borderline analyses, these methods can also lead to more unequivocal results.

18.7 Confirmatory analysis

Positive findings for DOA and pharmaceutics on test strips should always be verified. For this purpose, a clear statement by the patient suffices. However, if the information is contradictory, a confirmatory analysis is indispensable – the sample must be tested in the laboratory and analyzed by thin-layer chromatography, high-performance liquid chromatography (HPLC), gas chromatography or HPLC tandem mass spectrometry. A confirmatory analysis with another immunochemical test is not allowed. The analysis report provides information about the constituents detected, e.g. which benzodiazepine is involved [6, 7].

Note

To answer forensic questions, test strips should only be used to obtain a tentative diagnosis. They are required for treating addiction, to discover a relapse and to initiate therapeutic measures in a timely manner. Otherwise, the analyte spectrum of individual tests and their sensitivity is too low for the regular use of test strips in treating addiction. In the event of false-positive findings, consideration should be given to informing the test suppliers, for example, so that they can commission a test analysis for any cross-reacting substances. These facts are then documented in the list of cross reactions for this test.

18.8 Documentation

In practice, documenting test strip results may be troublesome, but it is indispensable in terms of quality assurance in order to obtain reproducible analytical results. The test strips themselves cannot be stored. They are potentially infectious, and the band staining cannot be permanently preserved. Records should be taken down on a pre-printed form containing the date, patient, analyzing technician, batch number of the test, analytical findings, sampling method, check for sample tampering, creatinine concentration, pH value, temperature and any appropriate remarks, such as special medications. The test strips can additionally be documented photographically. Generally, documentation must be archived and retained for a minimum of 10 years. Electronic data processing-supported documentation with special scanners can facilitate interpretation, particularly of cumulative findings and improve data security.

18.9 Pre-analytical phase

Sampling should occur spontaneously and at best immediately in the case of suspected misuse, given that the analytes are partially decayed and excreted very rapidly in vivo. Until analysis, the samples should be stored in a dark, cool place, even though most analytes are temperature- and light-stable. Given that repeat sampling does not produce the same results and frequently throws up new questions later, samples should be kept in long-term storage. Patients should be informed about how important it is to disclose any medicines they may be taking. Consideration should also be given to possible consumption of foods that contain poppy seeds. These can contain morphine and trigger a positive opiate test.

18.10 Brief information on important analytes and analyte groups

18.10.1 Alcohol

Important target analytes: Ethanol and ethyl glucuronide.

Since acute alcohol consumption can be detected easily and reliably with the breathalyzer, additional methods only play a minor role. Respiratory behavior and contamination of the oral cavity can affect the measurement to the extent that conclusions about the blood alcohol concentration cannot be drawn with certainty in every case. Cross reactions with other alcohols can likewise be confounders. In isolated cases, test strips pose a cost-effective alternative to the comparatively expensive alcohol breathalyzers. However, it should be noted that saliva tests are purely qualitative methods. Test strips based on ethyl glucuronide offer the option of being able to detect alcohol intake over a longer time span (up to approx. 80 hours).

18.10.2 Amphetamine-like designer drugs

Important target analytes: Amphetamine, methamphetamine, MDA, MDMA, MDE(A), BDB, MBDB, Ritalin, cathinone and a variety of other designer amphetamines [7].

Amphetamines are generally detected on the test strips as individual analytes. Frequently, only tests for amphetamines and methamphetamines are offered (see SAMHSA Mandatory Guidelines). Solely a few tests offer a sufficient cross-reactivity with other designer amphetamines – MDA, MDMA, MDE(A), BDB and MBDB. Because of the relatively low molecular mass, the informative value of the molecule structures is low, thereby characterizing true group tests by a broad spectrum of adverse cross reactions. Certain pharmaceuticals decompose into amphetamine (e.g. selegiline). Meanwhile, test strips for special questions like Ritalin and the cathinone MDPV have become available as well. The appropriate chromatographic analysis must be carried out to discriminate these false-positive findings. Most amphetamines are not excreted in the urine as conjugates. Detection in the saliva is possible.

18.10.3 Barbiturates

Important target analytes: Phenobarbital, pentobarbital, allobarbital, alphenal, amobarbital, apobarbital, barbital, butobarbital, cyclopentobarbital, secobarbital, vinyl barbital, thiopental [7].

Only a few barbiturates are still approved in Germany. Therefore these drugs are being abused less and less. Conventional test strips detect the commercially available drugs with sufficient sensitivity. Analysis in the saliva is not practicable.

18.10.4 Benzodiazepines

Important target analytes: Diazepam, nordiazepam, temazepam, oxazepam, nitrazepam, clonazepam, flunitrazepam, alprazolam, flurazepam, tetrazepam, lorazepam, lormetazepam, medazepam, midazolam, bromazepam, brotizolam, clobazam, clorazepate, chlordiazepoxide, prazepam, triazolam, phenazepam, etizolam, z-substances (zopiclone, zolpidem), designer benzodiazepines [7].

The group of benzodiazepines includes diverse pharmaceutical agents with many metabolites. This list is compounded by the great pharmacodynamic breadth of drugs that encumbers analysis by group testing. Whereas most tests rate well in the detection of 1,4-benzodiazepines like diazepam, the sensitivity is insufficient for detecting pharmacodynamically highly active 7-nitro-benzodiazepines such as clonazepam as well as their phase-I and phase-II metabolites. Frequently, the term "group test" is not an adequate one. Enzymatic hydrolysis can be conducted to enhance sensitivity. Another major group is also represented by the z-substances. However, these are generally not detected by the antibodies. Meanwhile, a large variety of designer benzodiazepines has appeared on the black market.

It appears questionable, whether the sensitivity of the test strips for saliva can be evaluated as sufficient.

18.10.5 Cannabinoids

Important target analytes: Δ^9-THC, Δ^9-THC-COOH, synthetic cannabinoids [7].

Test strips for cannabinoids contain a monoclonal antibody that targets the main metabolite Δ^9-THC-COOH and shows a high sensitivity and selectivity for these analytes. Different cut-offs are applied, depending on the question at issue. For several years now, synthetic cannabinoids have been gaining importance. For example, cannabinoid receptor agonists (e.g. JHW-018, MDMB-CHMICA or 5F-ADB) are added to herbal mixtures and solutions for e-cigarettes and e-hookahs. The market is starting to offer the first test strips for synthetic cannabinoids. However, the spectrum of detectable substances is limited and the cut-offs are relatively high.

Test strips for saliva currently do not appear suitable.

18.10.6 Cocaine

Important target analytes: Cocaine, benzoylecgonine [6].

As with cannabinoids, the monoclonal antibodies of cocaine tests react to the primary metabolites (here: benzoylecgonine) and are likewise characterized by high sensitivity and specificity. Unlike the cannabinoids, the cocaine metabolite is also detectable in many other bodily fluids (including saliva).

18.10.7 Methadone

Important target analytes: Methadone and methadone metabolite (EDDP) [6].

Besides the test for the parent compound, screening for the primary metabolite EDDP is increasingly being offered to find evidence of the substitute drug methadone. The two tests are characterized by high sensitivity and specificity. The detection of metabolites gives the analyst additional information as to whether the substitute drug was taken, and thereby rules out any substance subsequently added by tampering. Unlike the EDDP, methadone is detectable in measurable concentrations in the saliva.

18.10.8 Opiates

Important target analytes: Morphine, codeine, dihydrocodeine, 6-acetylmorphine [6].

Most test strips detect the opiates morphine, 6-acetylmorphine, codeine and dihydrocodeine. Additional opioids like tramadol, tilidine or fentanyl do not exhibit any relevant cross reactions in the tests. That means these opiate test strips cannot be used as a selective test for heroin abuse or in broader screening for opioids. Heroin abuse can only be reliably demonstrated by evidence of 6-acetylmorphine in a confirmatory analysis. This aspect has gained particular importance since the introduction of sustained-release morphine for substitution-supported treatment. Meanwhile, test strips with a monoclonal antibody for 6-acetylmorphine have also become available. The list of cross-reacting substances should be accounted for, both in the case of positive as well as negative findings. Furthermore, it should be noted that the test cut-offs vary between 300 and 2,000 µg/L. Opiates are detectable in the saliva.

18.10.9 Buprenorphine

> **Important target analytes: Buprenorphine, norbuprenorphine [6].**

For years, buprenorphine has been approved as an opioid replacement drug. It is a partial opiate antagonist with high pharmacodynamic potency. Commercial tests frequently do not exhibit cross reactions with the main metabolite norbuprenorphine, even though it takes up a major portion of the renally excreted active ingredient. Hence, most buprenorphine tests are characterized by good specificity, but low sensitivity. There are no commercial tests available for determination in saliva.

18.11 Sample tampering

18

Test strips should always be used in routine analysis to unmask any sample tampering. Combination test strips are supplied that cover pH, creatinine concentration and specific density, including the following parameters:

- Oxidants
- Nitrite
- Glutaraldehyde
- Chromate
- Peroxidase
- Bleaching agents

These compounds are capable of disrupting the test or eliminating the analytes from the sample by chemical reactions [10].

References

1. Bundesministerium für Gesundheit (2009) Drogen- und Suchtbericht 2009. www.bmg.bund.de, www.drogenbeauftragte.de
2. Külpmann W-R (2003) Nachweis von Drogen und Medikamenten im Urin mittels Schnelltest. Dtsch Ärztebl 100: A1138–A1140, B956, C898
3. Lagois J (2000) Dräger Alcotest 7110 Evidential – das Messgerät zur gerichtsverwertbaren Atemalkoholanalyse in Deutschland. Blutalkohol 37: 77–91
4. Lagois J, Sohoge J (2003) Interlock – ein Beitrag zur Erhöhung der Sicherheit im Straßenverkehr. Blutalkohol 40: 199–207
5. Madea B, Musshoff F, Berghaus MG. (Eds.) (2012) Verkehrsmedizin – Fahreignung, Fahrsicherheit, Unfallrekonstruktion. Deutscher Ärzte-Verlag, Cologne
6. Baselt RC (2011) Disposition of Toxic Drugs and Chemicals in Man. Bimedical Publications, Seal Beach, California
7. Pfleger K, Maurer HH, Weber A (2007) Mass spectral and GC data of drugs, poisons, pesticides, pollutants and their metabolites, 3rd ed. Wiley-VCH, Weinheim
8. Raggam RB, Santner BI, Kollroser M, Gössler W, Schmied B, Schmitt U et al. (2008) Evaluation of a novel standardized system for collection and quantification of oral fluid. Clin Chem Lab Med 45: 287–291
9. Schubert W, Dittmann V, Brenner-Hartmann J (2013) Urteilsbildung in der Fahreignungsbegutachtung – Beurteilungskriterien. 3rd Edition. Kirschbaum, Bonn
10. Schütz H, Paine A, Erdmann F, Weiler G, Verhoff MA (2006) Immunoassays for drug screening in urine. Forensic Sci Med Pathol 2:2: 75–83
11. EWDTS (2015) European Guidelines for Workplace Drug Testing in Urine
12. www.rosita.org
13. www.samhsa.gov

Urine and stool analyses

Norbert Gässler, Harald Schlebusch, Peter B. Luppa

P. B. Luppa, R. Junker (Eds.), *Point-of-Care Testing*
https://doi.org/10.1007/978-3-662-54497-6_19

19.1 Urine analyses

19.1.1 Introduction

In the context of timely primary prevention, the exclusion or early diagnosis of incipient nephropathy and urinary tract infections (UTI) can be rendered using laboratory tests. Here, urine test strips are nearly always used as the first diagnostic step. These test strips, which simultaneously detect a number of parameters, are regarded as classic POC tests. They are used as screening tests in the central laboratory or by patients for self-testing. Their use allows an early diagnosis of diseases affecting kidney function and carbohydrate metabolism as well as the liver and hemolytic functions. Further explanations are provided by the review by Boege et al. [2] and the guidelines of the European Confederation of Laboratory Medicine [6].

The specific gravity of urine has been the subject of physical testing since ancient times. The first chemical method for detecting protein was introduced by the Viennese chemist, J.F. Heller (1813–1871), Heller's ring test. Concentrated nitric acid is added to the urine in a liqueur glass to form two layers. If albumin is present in the urine, a white ring of precipitated protein appears. Since this test was usually carried out directly by the physician in the presence of the patient, this method can be said to be the first POC test described in the literature.

The introduction of test strips in the 20th century revolutionized urine diagnostics because of their practicality and specificity. The reaction area on the test strip is made of absorbent material and contains all the necessary detection reagents. The urine acts as a solubilizer, dissolving the stabilized reagents and triggering the relevant chemical detection reactions. The test strip test includes the chemical detection of corpuscular components, such as erythrocytes and leukocytes, as well as measurements of pH and specific gravity.

The collection of urine samples and the point of diagnostic testing should, whenever possible, be nearby, as urine needs to be analyzed within 2 hours of collection. Spontaneously voided urine is suitable for many chemical and microscopic tests; nevertheless, midstream urine should be collected whenever possible, i.e. the first portion of the urine stream is discarded [13]. A first morning urine sample is preferable as it contains the target substances in higher concentrations than further portions voided during the day. In urology, a **3-portion sample** is examined in males. The first stream and the midstream are collected in separate tubes; the third tube is filled after a gentle prostate massage. This method allows a rough localization of substances or cells such as red blood cells (RBC).

Nitrites and proteins are best detected in the first morning urine. A urine sample taken 2 hours after a carbohydrate-rich meal is suitable for glucose detection. A 24-hour urine collection sample is often helpful for assessing renal function (clearance tests) and detecting other metabolic products. It is important to keep the urine cool and, depending on the parameters to be assessed, a preservative may need to be added. For some specific diagnostics, catheter urine is collected, including by suprapubic puncture.

The test strips should be dipped in the urine for no longer than 1 second. Excess urine should be cast off. Depending on the manufacturer and test field, required incubation times (generally 1 minute) should be followed exactly. Once the reaction is complete, the colored reaction fields are compared against a color scale on the test strip cassette and the results are read off qualitatively or semi-quantitatively. Test strip assessment with a reading device is also possible; usually, such devices contain storage memory for patient and user data and even interfaces for data transfer. Various manufacturers offer such POCT-capable systems. The urine diagnostic system by Roche, the Urisys 1100, is presented in ◘ Fig. 19.1. Similar devices by Arkray (PocketChem UA PU-4010), Siemens Healthcare (Clinitek Status+), Hitado (Doc U Reader 2 Pro) and others are also available on the market.

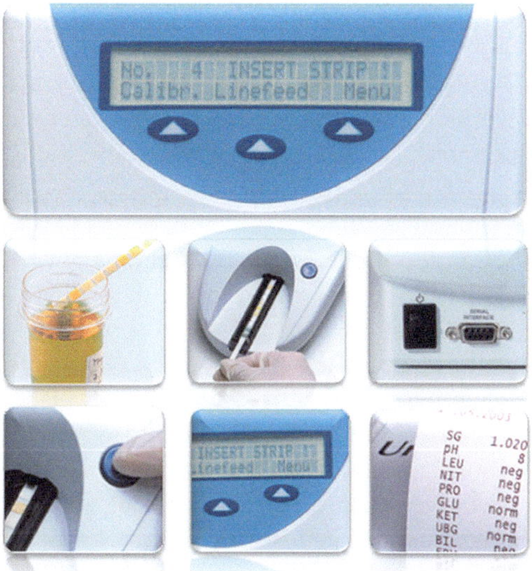

Fig. 19.1 Urisys 1100. (With kind permission from Roche Diagnostics GmbH, Mannheim)

19.1.2 Protein

Proteinuria is a frequent, albeit non-specific symptom. In renal disease, a distinction is made between glomerular and tubular proteinuria occurring transiently, intermittently or continuously. Its detection is strongly indicated in the case of infections of the renal parenchyma, renal pelvis and urinary tract as well as for pregnancy monitoring. Proteinuria due to extrarenal factors can occur concurrently with many acute illnesses. Harmless proteinuria can become manifest after physical exercise (sport) or agitation (stress). Furthermore, it is known that orthostatic control and circadian rhythms can affect proteinuria. Differential diagnosis with specific marker proteins or urine electrophoresis should follow when proteinuria has been positively detected.

The test principle is based on what is called the **protein error** of pH indicators. Here, free protein amino groups in the urine pick up protons from the indicator (tetrabromophenol blue) at a constant pH. This causes the indicator to change color. The sensitivity for albumin is significantly higher than for other globulins, with Bence-Jones protein and other small molecular proteins being difficult to detect. If the urine is strongly alkaline (e.g. in urinary tract infections) or contaminated with cleaning products or disinfectants, the test strip may indicate a false-positive result. Detection sensitivity is at a level of 15–20 mg albumin/dL urine. Hereby, macroalbuminuria, but not microalbuminuria can be detected [9].

19.1.3 Microalbumin

Microalbuminuria is often the first sign of an incipient non-inflammatory nephropathy. Particular risk groups are patients with diabetes and hypertension and therefore regular screening for microalbuminuria is recommended. Threshold values are:

- 30 mg/24 hours for a 24-hour total urine collection,
- 20 mg/L or 30 mg/g creatinine for a spontaneously voided urine sample (first morning urine).

In persistent proteinuria, a quantitative measurement of protein excretion and creatinine clearance are indicated, among other assays. Transiently elevated albumin excretion is generally seen in acute febrile illnesses, urinary tract infections and metabolic disorders as well as in pregnancy [10].

The semi-quantitative test-strip-based measurement of "microalbumin", meaning albumin concentrations <20 mg/dL urine, is possible by means of two different principles: either by immunochemical techniques or an improved "indicator error". In the Micral-Test II (Roche), urine passes through a test strip zone that contains a soluble red antibody-gold conjugate that binds to albumin. Excess conjugate is held back in a capture zone with immobilized albumin so that only the albumin loaded with the gold conjugate migrates to the detection zone. A color shade between white and red then becomes visible. By contrast, the Microalbustix test strip (Siemens Healthcare), is based on the **indicator error principle** (see above), whereby sulfophthalein (DIDNTB) is used as an indicator that turns a blue color. Both tests have a similar sensitivity (approx. 2 mg/dL urine), although the specificity of the immunological test is significantly better [7].

When using a spontaneous urine sample, the validity of the result can be improved when applied to the creatinine concentration in the urine as the effect of diuresis on the result is eliminated. The Microalbustix and the Clinitek microalbumin test strips (Siemens Healthcare) therefore have an additional creatinine test field.

A quantitative microalbumin measurement is feasible, using POC methods. The HemoCue Albumin 201 is a turbidimetric immunoassay, with single-use microcuvettes. The DCA Vantage (Siemens Healthcare) measures albumin and creatinine (immunological/color reaction) and calculates the albumin-creatinine ratio.

19.1.4 Glucose

The detection of glucose in urine as a screening tool is of significant diagnostic importance for diabetes mellitus as well as in the follow up and self-management of this disease. Glucosuria occurs when the blood glucose level exceeds the renal threshold of 150–180 mg/dL. This not only occurs in diabetes mellitus, but also after a very carbohydrate-rich meal (alimentary glucosuria). Liver and pancreatic diseases as well as various hormones and medications can cause glucosuria. When the renal threshold is lowered, which can occur during a febrile illness or pregnancy, glucose is excreted into the urine even at normal blood glucose levels [10].

The glucose measurement is based on oxidation of glucose by the enzyme glucose oxidase. In a second step, hydrogen peroxide is converted to a colorant. The sensitivity is 50–100 mg glucose/dL urine. High concentrations of ketones and ascorbic acid (>50 mg/dL) can lead to a reaction reduction [1], whilst oxidative cleaning products such as H_2O_2 or hypochlorite solution can cause false negative results.

19.1.5 Ketones

The ketone bodies **β-hydroxybutyrate** (β-HBA), **acetoacetate** (AcAc) and **acetone** are metabolic products of free fatty acids, whereby β-HBA and AcAc are normally present in equimolar amounts. The proportion of acetone is less than 5 %. The balance between β-HBA and AcAc is, however, shifted in hypoxia, fasting and ketoacidosis in favor of β-HBA. Ketone bodies in the urine are detected when a higher amount of fat is catabolized. In diabetes, particularly in type 1, the detection of ketones is always a sign of metabolic imbalance and therefore an important warning sign for impending ketoacidosis. Urine test strips are not sufficiently reliable for monitoring diabetic ketoacidosis because only AcAc and acetone are measured, whereas β-HBA is not detected (▶ Chapter 12). Quantitative β-HBA measurements in the blood are necessary to be able to detect the major portion prevailing in metabolic disorders. Test strips are available for use with the relevant blood glucose meters (e.g. Abbott's Precision Xceed or Nova's StatStrip Glucose/Ketone) (◻ Fig. 19.2 and ◻ Fig. 19.3).

Ketones in urine can occur in states of starvation, e.g. slimming diets with reduced or absent carbohydrate intake (zero diet), also in febrile conditions and hyperemesis gravidarum and acetonemic vomiting in infants.

Acetic acid and acetone react with sodium nitroprusside to form a color complex (Legal's test). The sensitivity is 5–10 mg acetic acid/dL urine; the test is insensitive for acetone, β-hydroxybutyric acid is not detected [8]. Ketone bodies should be measured in a fresh urine sample. If a fresh urine sample is stored for too long, the acetic acid becomes unstable or is degraded by bacteria; this causes false-negative results similar to acidic urine, for example after consumption of large amounts of ascorbic acid. Also strongly discolored urine can falsify the test results.

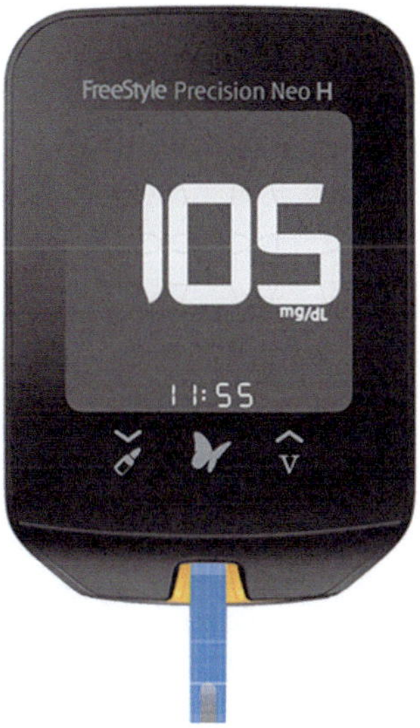

Fig. 19.2 FreeStyle Precision Neo H POCT device to measure blood glucose and β-HBA. (With kind permission from Abbott GmbH, Wiesbaden, Germany)

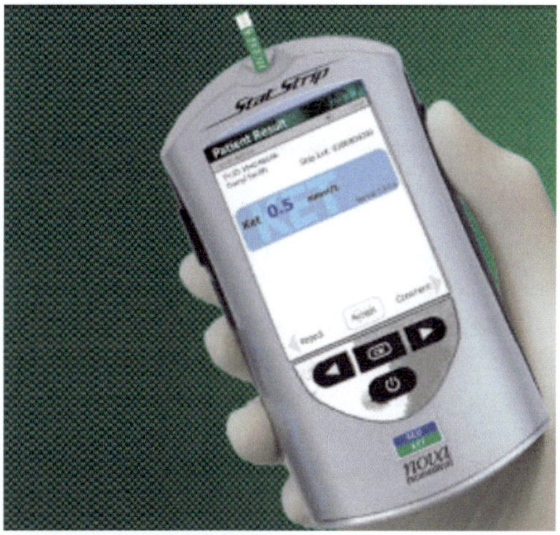

Fig. 19.3 StatStrip Glucose/Ketone POCT devices to measure blood glucose and ketones. (With kind permission from Nova, Mörfelden-Walldorf, Germany)

19.1.6 Bilirubin

Bilirubinuria can generally be detected if the plasma concentrations of conjugated (direct) bilirubin is >2 mg/dL. It occurs in patients with parenchymal liver damage (e.g. hepatitis, cirrhosis, intoxications), impairment of bilirubin excretion (e.g. Dubin-Johnson syndrome) or cholestasis. In healthy individuals, no bilirubin or only traces thereof are excreted into the urine. Therefore, a positive result should generally be followed up by further internal medicine and clinical chemical studies.

Direct bilirubin reacts with diazotized dichloroaniline in a coupling reaction at an acidic pH. This results in the formation of azobilirubin in different color gradations [11]. The sensitivity is 0.4 mg bilirubin/dL urine. Ascorbic acid (>250 mg/dL urine) and nitrite as well as prolonged storage of urine in direct sunlight can all cause false-negative results.

19.1.7 Urobilinogen

Urobilinogen is formed in the intestines by bacterial action on bilirubin and bile. Urobilinogen is re-absorbed into the blood circulation and is either metabolized in the liver or excreted in the urine. Urobilinogenuria can occur with or without concurrent bilirubinuria. Increased amounts of urobilinogen are eliminated in the urine when, in its enterohepatic circulation of bile pigments, the functional capacity of the liver is limited, overloaded or the liver is bypassed. This can be due to primary liver disease (e.g. hepatitis, cirrhosis, congested liver, intoxications), excessive breakdown of hemoglobin (e.g. in hemolytic and pernicious anemia, intravascular hemolysis, polycythemia and re-absorption of extravasated blood), as well as a result of cholangitis. A total absence of urobilinogen in the urine is caused by a lack of bile production in the liver and by impaired bile secretion into the small intestine as well as by a lack of bacterial bilirubin reduction in the bowel. A further cause can be a blocked bile duct (ductus choledochus), often due to calculi.

Urobilinogen reacts with para-dimethylaminobenzaldehyde in the presence of a color enhancer or with p-methoxybenzene diazonium tetrafluoroborate in a strong acidic milieu to form a pinkish red color complex [12]. The sensitivity is 0.2 mg urobilinogen/dL urine.

Urobilinogen is easily oxidized and can therefore lead to false-negative results if the urine is stored for too long.

19.1.8 Nitrite

Gram-negative pathogens (e.g. E. coli, Proteus mirabilis) are the most common and major causes of urinary tract infections (UTI). These bacteria reduce nitrate to nitrite. The detection of nitrite in urine is therefore a reliable sign of the presence of bacteria and other urinary pathogens. An average of 50–60 % of all UTI are diagnosed using the nitrite test – under favorable conditions this percentage can run as high as 90 % (first morning urine, high pathogen count). However, this is predicated on a sufficient amount of nitrate being consumed orally and that the urine stays in the bladder long enough for a nitrate reduction to occur (approx. 4–6 h). The detection of leukocyturia is an additional supporting result. The diagnosis of a UTI should be verified by a pathogen count. Nitrites are not present in the urine of a healthy individual [2].

Sulfanilamide reacts with nitrites in an acidic environment (diazotization). In a further reaction, the diazonium salts produced react with a chromogen (in this case benzoquinone derivative) to form a pink colorant. The detection of nitrite is specific for gram-negative bacteria; 80 % of all pathogens causing a UTI produce nitrite [3]. The sensitivity is 0.06–0.10 mg nitrite/dL urine. Negative results can occur when high concentrations of ascorbic acid (>25 mg/dL urine) are present, in a UTI caused by non-nitrite producing bacteria such as enterococci, staphylococci and pseudomonas spp. as well as in the case of frequent voiding where sufficient amounts of nitrite cannot be produced (<4 h).

19

19.1.9 pH

The pH of urine is influenced by diet and medications and is subject to diurnal variations. Fresh urine normally has a pH of 5.0–8.0. Physiologically low pH values are typical in meat-rich diets, whereas a vegetable-rich diet (e.g. in vegetarians) tends to cause a slightly alkaline pH. Persistently acidic samples occur when there is an increased breakdown of endogenous protein (e.g. due to fasting, diarrhea or high fever), but also in diabetic ketoacidosis. Alkaline urine points to a UTI (particularly by Proteus spp.) as a result of the urea-cleaving activity of certain bacteria. Storing urine for a longer time can cause alkalization of the sample. Urine pH monitoring is important in kidney stone prophylaxis to avoid recurrences.

Indicators such as methyl red and bromothymol blue are mostly used in combination. This allows pH values between 5.0 and 8.5 to be detected with a color change to indicate the result.

19.1.10 Erythrocytes/hemoglobin (Hb)

Hematuria is an accompanying symptom of many urological and internal diseases (◘ Tab. 19.1)

The blood test field can detect intact erythrocytes (hematuria) and free Hb as a result of lysis (hemoglobinuria), but also myoglobin. Due to the catalytic reaction of pseudoperoxidase of Hb and myoglobin, color indicators such as o-toluidine or 3,3',5,5'-tetramethylbenzidine are oxidized by cumene hydroperoxide and cause a color change. Intact erythrocytes appear as green dots on the reaction zone; the detection threshold is 5–20 erythrocytes/µL urine. The detection threshold for free Hb and myoglobin is 0.02–0.06 mg/dL urine. The test is equally sensitive for Hb and myoglobin [3]. False-negative results or too low readings are obtained at high ascorbic acid concentrations. False-positive results can be obtained when the samples are contaminated with oxidizing cleaning products containing hypochlorite or H_2O_2.

Tab. 19.1 Causes of hematuria

Renal causes	Post-renal causes	Extra-renal causes
Glomerulonephritis	Urinary tract infections	Hemorrhagic diathesis
Pyelonephritis	Urolithiasis	Toxic and pharmaceutical effects
Renal stones	Bladder tumors	Physical exertion
Renal tumors	Malformation of the urinary tract	March hematuria
Trauma	Others	Others
Others		

19.1.11 Leukocytes

Leukocyturia is an important symptom of inflammation of the kidneys and urinary tract. It is manifest in:

- Bacterial infections (e.g. acute and chronic pyelonephritis, cystitis, urethritis)
- Non-bacterial infections (e.g. fungi, yeast)
- Parasitic diseases
- Analgesic nephropathy
- Intoxication and
- Urinary flow disorders (obstructive uropathy)

The majority of positive leukocyte results are caused by bacterial urinary tract infections. In most cases, the leukocytes excreted in the urine are granulocytes. When leukocytes undergo physiological lysis, an esterase is released that catalyzes the hydrolysis of a pyrrole amino acid ester to generate 3-hydroxy-5-phenylpyrrole (indoxyl), which subsequently oxidizes to form a dimeric product, indigo – a dark blue pigment. The sensitivity of the test is 5–20 leukocytes/µL urine. The leukocyte test field cannot differentiate between lysed and intact cells.

19.1.12 Creatinine

Creatinine excretion is primarily dependent on individual muscle mass and does not change over the course of the day. It undergoes free glomerular filtration and not tubular re-absorption. Urinary creatinine therefore serves as a surrogate for albumin excretion (calculation of the albumin/creatinine ratio), whereby the effect of diuresis can be eliminated [10]. In normal diuresis, the creatinine concentration (as a function of the patient's size, age and weight) in spontaneous urine is 40–130 mg/dL. Creatinine catalyzes the reaction of diisopropylbenzene dihydroperoxide with 3,3',5,5'-tetramethylbenzidine in the presence of copper to form a dye. The sensitivity is 15 mg creatinine/dL urine.

19.1.13 Specific gravity

The specific gravity (density) is primarily dependent on the volume of excreted fluid and can therefore vary considerably (1.005–1.040). The test has lost its importance as a measure of renal function in terms of concentration capability. However, accounting for the specific gravity can aid the interpretation of other urine parameters. For example, a negative test strip result for protein should be interpreted differently in a strongly diluted urine sample compared to a concentrated one. It is important to know the specific gravity of in addictive substance analysis to detect any sample tampering (▶ Chapter 18).

A polyelectrolyte, e.g. polymethyl vinyl ether/maleic anhydride, takes up cations from the urine and releases the equimolar amounts of protons. These cause the color to change (bromothymol blue) on a pH indicator [5]. The test measures the ion concentration in the urine and correlates well with specific gravity between 1.000 and 1.030. Strongly alkaline urine (pH ≥8.0) leads to falsely low results, while strongly acidic urine (pH <5.0) produces overly high results.

19.1.14 Medications and narcotic drugs

Medications, narcotic drugs and doping substances can be detected in the urine near the patient by special diagnostic methods. Numerous qualitative or semi-quantitative test systems are available. However, test results used for addiction therapy can be falsified by adding diverse substances such as bleach, soap or salt. Most urinary toxicology screening tests will target a number of legal and illegal drugs simultaneously (e.g. opiates, methadone, propoxyphene, benzodiazepines, cocaine, amphetamines, cannabinoids). Positive results will indicate intake of drugs of abuse and cannot be produced by indirect contact with illegal drugs. Many prescription drugs and over-the-counter medications can interfere in these tests and produce a false-positive result (▶ Chapter 18).

19.1.15 Human chorionic gonadotropin (HCG)

The immunological detection of human chorionic gonadotropin (HCG, also abbreviated as β-HCG) forms the basis for all POC pregnancy tests. When sensitive tests are used, HCG can be successfully detected in concentrated morning urine around the time of the expected menstrual bleeding (▶ Chapter 24). The HCG production in the trophoblast is doubled every 24–36 hours; it is therefore useful to repeat the test after 48 hours if the initial test was negative. Although false-positive tests are rare, a quantitative confirmatory blood test is recommended, which needs to be carried out in the laboratory and is not available as POCT.

HCG can be detected in concentrated morning urine approx. 14 days after fertilization of the egg. The method involves an immunochromatographic assay (▶ Chapter 9). The sample is chromatographically separated on the test strip and converted to a visible color complex via an antigen-antibody reaction. The sensitivity of such rapid tests is 25 IU/L. If the test is negative it should be repeated after 48 hours. Heavily blood-stained urine samples can give false-positive results due to overload of the test strip field.

19.1.16 Other parameters

Over the past few years, POC tests have become available for the detection of bladder cancer. Utilizing immunological assay systems, these tests like the **UBC® Rapid Test** (urinary bladder cancer antigen rapid test) can detect nuclear matrix protein 22 (**NMP22**) or cytokeratin fragments 8 and 18 in urine already at the early stages of cancer [14, 13].

Numerous other urine rapid tests are offered on the market, e.g. a rapid HIV detection test. The sensitivity of such POCT methods is insufficient due to the low concentration of HIV antibodies in the urine. This is why such tests are not permitted in the EU. Another test on the market is designed to detect the stress of free radicals on the body tissue [4], although free radicals cannot be measured directly. The clinical benefit of such tests is, however, doubtful.

19.1.17 Pre-analytical phase

Fresh urine samples should be tested, which means max. 2 hours after sample collection. If the analysis cannot be carried out immediately, the urine should be stored at +4 C° to avoid bacterial growth that would consequently lead to false pH and nitrite concentration results. Urine should not be kept in direct sunlight due to the disintegration of bilirubin that can imitate falsely low concentrations. The addition of preservatives can affect individual test reactions; therefore the manufacturer's instructions for sample tubes should be followed strictly.

19.1.18 Quality control of urine test strips

The quality assurance of urinalysis is referred to in part B2 "Qualitative laboratory tests" of the RiliBÄK guideline [15] (▶ Chapter 38).

The manufacturer's specifications for internal quality assurance should be observed. Independently of this, quality assurance of the tests mentioned in RiliBÄK Tab. 2–1 should be carried out. An additional test to ensure the accuracy of the result is not needed if controls are already integrated into the analysis system (e.g. in many pregnancy tests). Participation in an interlaboratory testing is compulsory for all of the tests listed in RiliBÄK Tab. 2–2.

19.2 Fecal analyses

19.2.1 Tests for occult blood

Colorectal cancer (CRC) is the second leading cause of cancer mortality in Germany–after lung cancer. The lifetime risk is approx. 5 % and doubles after the age of 50, with an increase in incidence and mortality with every decade of life [12]. Colorectal cancer develops slowly; therefore, it is preferable to detect the tumor at an early stage through screening in order to increase the effectiveness of therapeutic intervention.

Colonoscopy is seen as the "gold standard" of all diagnostic procedures, however, it is only partially accepted by patients for various reasons [10].

As an alternative to colonoscopy, fecal occult blood tests (FOBTs) were introduced into Germany's early cancer detection program in 1977. Yearly screening is recommended for individuals from age 50 and 2-yearly from age 55. The costs are covered by health insurance companies. The basis for FOBTs is the observation that colorectal cancer has a tendency to bleed more heavily and frequently than healthy colonic mucosa. As the bleeding is often intermittent, the reliability of the test is increased when the testing is repeated a few times on different days. A positive result should be followed up with colonoscopy and not by just repeating the FOBT.

To date, CRC screening tests have been performed on a **guaiac basis**, as supplied by various manufacturers (e.g. Haemoccult, Hemocare, Hemofec) and approved by the German National Association of Statutory Health Insurance Physicians (KBV). The sensitivity threshold of the tests was set to indicate a positive result when two to three times the mean red blood cell excretion in a healthy digestive tract (0.5–1.0 mL blood) was exceeded.

These guaiac-based POCT methods were widely used in private medical practices; performing the tests was simple: Two small stool specimens taken from three consecutive defecations were spread onto a filter paper, impregnated with guaiac resin (test kit) and after a drying period of no less than 48 hours, a drop of hydrogen peroxide was applied. In the presence of blood, a blue coloration, caused by a pseudoperoxidase effect of heme, occurs on the test section. The test was read as positive when a blue coloration occurred in at least one test section, even when only faint. The diagnostic sensitivity for colon cancer was 20–40 %, the specificity about 95 % [7, 12].

The test validity was, however, compromised by various factors. Confounding factors such as meat and meat product, particular medications, ascorbic acid can affect the test [3, 5, 7, 8]. It has moreover been shown [11] that errors in the interpretation of the results play a significant part. The test should therefore be interpreted by experienced staff under suitable conditions such as bright light with no blue objects on the workbench. Furthermore, Germany had no external quality controls for the detection of fecal occult blood; neither does RiliBÄK Part B2 (qualitative tests) contain any specifications. Disappointingly, interlaboratory tests in Great Britain showed problems with the performance of several of the tests [2].

This situation has changed fundamentally since 21 April 2016 when the Federal Joint Committee (G-BA) resolved to approve only **quantitative fecal immunochemical tests**

(FITs or iFOBT) for CRC screening that are carried out in medical laboratories [4]. Therefore, fecal occult blood tests are no longer possible as near-patient testing.

Excerpts from the resolution dated 21 April 2016 are as follows:

> "The early cancer diagnosis guidelines issued on 18 June 2009 (German Federal Gazette (BAnz.) No. 148a dated 2 October 2009) are being changed:
> ... 3. Section 39 shall be amended as following:
> "Performance of fecal occult blood tests
> (1) The test for fecal occult blood test is performed on a stool specimen using a quantitative immunological test. Only tests that meet the following criteria are permitted:
> – A sensitivity of at least 25 % and a specificity of at least 90 % are achieved using only one stool specimen for the detection of colorectal cancer or advanced adenoma.
> – The stool specimen collection system should include clear instructions for use.
> – The stool specimen collection system needs to have a simple and hygienic stool collection receptacle as well as ensure the transfer of a defined stool amount onto a special specimen buffer system. It also needs to guarantee a stability of the specimen for at least 5 days within the given cut-off range at room temperature
> – Fulfillment of the aforementioned requirements must be proven by at least one conclusive study in which colonoscopy was used as a reference method."

Such immunological procedures (qualitative as well as quantitative) for the detection of fecal occult blood have been under development since the mid-1990s. Compared to the guaiac tests, they feature a significantly higher sensitivity (66–100 %) and specificity (overview in [12]). Dietary preparation of the patient is not required.

19

In Germany, qualitative iFOBTs for home testing are supplied by various manufacturers (e.g. AccuTell FOB test, FOB ideal, FOB cassette test, möLAB FOB test). The tests are not approved for the early cancer detection program and are not reimbursed by health insurance companies.

All tests rely on an immunochromatographic assay for human hemoglobin detection. In one meta-analysis of 12 studies, Lee et al. [6] showed that quantitative iFOBTs, referenced against colonoscopy, had an overall pooled sensitivity of 79 % and a specificity of 94 %.

To date, there are no larger studies that show a reduction in mortality when FITs are used. In Japan, however, a 60 % reduction in CRC mortality was demonstrated by more recent studies when quantitative FITs were performed yearly [9].

19.2.2 Leukocyte markers

The leukocyte proteins **calprotectin** and **lactoferrin** are stable proteins from neutrophilic granulocytes, which can be detected at higher amounts in stool when inflammatory or neoplastic processes are taking place. These markers therefore gain importance in the initial and follow-up diagnostics for inflammatory bowel disease (IBD) [13]. Specifically, evidence on calprotectin from studies is very positive. In the differential diagnosis of chronic IBD versus irritable bowel syndrome, the fecal concentration of calprotectin has adequate sensitivity and specificity. It is also a valuable marker for assessing the disease activity and the treatment response. Healing of mucosal lesions normalizes calprotectin concentrations [14]. Self-monitoring by affected patients is conceivable here.

Although calprotectin and lactoferrin show sensitivities for CRC that are comparable to iFOBT, at <70 % the specificities are unsatisfactory, making this parameter unsuitable for use in screening tests.

The same holds true for the dimeric form of pyruvate kinase isoenzyme (M2-PK), which is useful as a marker of IBD but cannot be used as a tumor marker [12]. The ScheBo M2-PK Quick Test (ScheBo Biotech, Giessen, Germany) has been evaluated in a study that demonstrated a good specificity of 87 % for tubular colonic adenomas. Its sensitivity (approx. 40 %) was, however, poor making it unsuitable as a screening test [1].

19.2.3 Molecular markers

Molecular markers for detecting neoplastic processes in the gastrointestinal tract are the subject of intensive research and regarded as "markers of the future". There is, however, as yet no consensus on the selection of promising genes (including APC, K-ras, P53) and structuring groups of markers (overview in [12]). Other limiting factors are at present the high equipment and personnel outlay. Although it is conceivable that POCT applications could be used in the future, they are not currently viable either from the clinician's or the patient's perspective.

References

Cited in 19.1

1. Berg B (1986) Ascorbate interference in the estimation of urinary glucose by test strips. J Clin Chem Clin Biochem 24: 89–96
2. Boege F, Schmidt-Rothe H, Scherberich JE (1993) Harnwegsdiagnostik in der ärztlichen Praxis. Dtsch Ärztebl 90: A1653–1667
3. Braun JS, Straube W (1975) Die Diagnostik der Mikrohämaturie mit einem neuen Teststreifen. Ein Vergleich mit mikroskopischen Untersuchungsmethoden. Dtsch Med Wschr 100: 87–89
4. Döll M (2003) Wie kann man eine Belastung des Körpers mit freien Radikalen feststellen? In: Döll M (Ed.) Das Antioxidantienwunder. Herbig, Munich, p. 262–266
5. Dorizzi RM, Caputo M (1998) Measurement of urine relative density using refractometer and reagent stripes. Clin Chem Lab Med 36: 925–928
6. European Confederation of Laboratory Medicine (2000) European urinalysis guidelines. Scand J Clin Lab Invest Suppl 231: 1–86
7. Haas M, Besenthal I, Renn W, Schmülling R, Eggstein M (1994) Mikroalbuminurie-Screening: Eine vergleichende Untersuchung von drei Schnelltests gegen ein nephelometrisches Verfahren. Diab Stoffw 3: 61–65
8. Heinemann L, Asche W, Withold W, Berger M (1994) Quantitative Beziehung zwischen der mit vier Schnelltests bestimmten Ketonurie und der gleichzeitig vorliegenden Ketonämie. Diab Stoffw 3: 339–342
9. Hofmann W, Edel, HH. Guder WG, Ivandic M, Scherberich, JE (2001) Harnuntersuchungen zur differenzierten Diagnostik einer Proteinurie: Bekanntes und Neues zu Teststreifen und Harnproteinen. Dtsch Ärztebl 98: A-756/B-618/C-578
10. Hoffmann W, Regenbogen C, Edel H, Guder WG (1994) Diagnostics strategies in urinalysis. Kidney Int 46 (Suppl 47): 111–114
11. Jendrassik L, Gróf P (1938) Vereinfachte photometrische Methoden zur Bestimmung des Bilirubins. Biochem Ztschr 297: 81–89
12. Kutter D, van Oudheusden APM, Eisenberg K, Hennecke A, Helbing R, Busch EW (1973) Die Brauchbarkeit eines neuen Teststreifens zum Nachweis von Urobilinogen im Harn. Dtsch Med Wschr 98: 112–118
13. Lüdecke G (2012) Comparative analysis of sensitivity to blood in the urine for urinebased point-of-care assays (UBC rapid, NMP 22 BladderCheck and BTA-stat.) in primary diagnosis of bladder carcinoma. Anticancer Research 2012(32): 2015–2018
14. Oehr P, Schroeder A (2006) Nutzen des qualitativen NMP22 BladderChek Tests für Diagnostik von Patienten mit Hämaturie und Verdacht auf Harnblasentumor: Ergebnisse einer praxisorientierten Ringstudie. Tumordiagn Ther 27: 205–210
15. Richtlinie der Bundesärztekammer zur Qualitätssicherung laboratoriumsmedizinischer Untersuchungen (2011) Teil B2 -Qualitative laboratoriumsmedizinische Untersuchungen Dtsch Ärztebl 108: A1647–A1651

Cited in 19.2

1. Cho CH, Kim J, Jang MA, Lee BJ, Park JJ, Lim CS (2016) Evaluation of the Performance of a Fecal Tumor M2-PK Rapid Kit Using Stool Specimens for Detection of Colorectal Tumors. Ann Clin Lab Sci 46(2):154–160
2. Duncan A, Hill PG (2007) A review of the quality of gastrointestinal investigations performed in UK laboratories. Ann Clin Biochem 44:145–158
3. Fraser CG (2008) Faecal occult blood tests – eliminate, enhance or update? Ann Clin Biochem 45:117–121
4. Hecken J (2016) Beschluss des Gemeinsamen Bundesausschusses über eine Änderung der Krebsfrüherkennungs-Richtlinie: Bewertung eines iFOBT-basierten Darmkrebsscreenings im Vergleich zu einem gFOBT-basierten Darmkrebsscreening. https://www.g-ba.de/downloads/39-261-2572/2016-04-21_KFE-RL_Bewertung-iFOBT.pdf (accessed on 6 July 2016)

5. Kahi CJ, Imperiale TF (2004) Do aspirin and non-steroidal anti-inflammatory drugs cause false-positive fecal occult blood test results? A prospective study in a cohort of veterans. Am J Med 117: 837–841
6. Lee JK, Liles EG, Bent S, Levin TR, Corley DA (2014) Accuracy of fecal immunochemical tests for colorectal cancer: systematic review and meta-analysis. Ann Intern Med 160(3):171
7. Ransohoff DF, Lang CA (1997) Screening for colorectal cancer with the fecal occult blood test: a background paper. American College of Physicians. Ann Int Med 126: 811–822
8. Rozen P, Knaani J, Samuel Z (1999) Eliminating the need for dietary restrictions when using a sensitive guajac fecal occult blood test. Dig Dis Sci 44:756–760
9. Saito H (2000) Screening for colorectal cancer: current status in Japan. Dis Colon Rectum 43: S78–84
10. Schneider AR, Caspary WF (2003) Diagnosis of colorectal carcinoma. An update. Radiologe 43: 105–112
11. Selinger RR, Norman S, Dominitz JA (2003) Failure of health care professionals to interprete fecal occult blood tests accurately. Am J Med 114: 64–67
12. Stein J, Loitsch SM, Shastry Y (2008) Nicht-invasive Diagnostik kolorektaler Tumore – hat der Guajac Test ausgedient? J Lab Med 32:158–167
13. Vermeire S, van Assche G, Rutgeerts P (2006) Laboratory markers in IBD: useful, magic, or unnecessary toys? Gut 55: 426–431
14. Walsham NE, Sherwood RA (2016) Fecal calprotectin in inflammatory bowel disease. Clin Exp Gastroenterol 9:21–29

Infectious diseases

Enno Stürenburg, Frank T. Hufert

P. B. Luppa, R. Junker (Eds.), *Point-of-Care Testing*
https://doi.org/10.1007/978-3-662-54497-6_20

20.1 Introduction

The measurement and monitoring of biochemical parameters such as blood glucose, blood gases, electrolytes and cardiac markers have particularly benefited from recent advances in POCT. More and more, this form of analysis is now positively impacting the field of infectious diseases as well. This trend has been driven, on the one hand, by methodical improvements that make it possible to miniaturize and simplify test systems (▶ Chapter 9, 10) and by the constant demand from many physicians for immediate test results, on the other. Meanwhile, the range of microbiological parameters available at the POC has been extended considerably (◘ Table 20.1). A POCT result, however, should always be regarded and interpreted in the context of the clinical symptoms and the current epidemiological situation. Proper handling of POCT systems and strict consideration of pre-analytical limitations are decisive in obtaining reliable test results.

Many microbiological POCT systems tend to be designed as rapid tests. In general, such tests can be carried out in a few simple steps by medical assistants or physicians without laboratory equipment or laboratory experience. An evaluable test result is delivered at the patient's bedside within not more than one hour. Under certain circumstances, patients can even carry out the test themselves, for example, as with some rapid HIV or malaria tests. Rapid microbiological tests usually verify microbial antigens. Less often, diagnostics is based on the detection of antibodies, e.g. in the rapid HIV test or rapid detection of heterophilic antibodies for diagnosing infectious mononucleosis (◘ Table 20.1).

Respiratory tract secretions are often used to detect microbial antigens in patients with respiratory tract infections (influenza virus, respiratory syncytial virus, Streptococcus pyogenes). Similarly, stool specimens of patients with gastrointestinal tract infections (Helicobacter pylori, Shiga toxin-producing E. coli, adenovirus, rotavirus) are examined. However, the corresponding urine tests are also available to detect antigens in legionella and pneumococcal infections [14, 19, 27, 40]. Blood samples

◘ Table 20.1 Rapid microbiological-virological tests in POCT format (as per 2016)

Type of infection	Virology	Microbiology
Sexually transmitted infections	HBV, HCV, HIV, HPV	GBS [21], CT [29, 35, 37, 49, 50], NG, MG, TV
Respiratory infections	Influenza A/B [7, 23, 31, 39], RSV	GAS [36], Legionella [14], Pneumococci [19, 27, 40], MTB/RIF
Gastrointestinal infections	Norovirus, Rotavirus [6, 8, 30, 48], Enterovirus	CDI [33, 45, 46], EHEC [25, 28, 44]
Nosocomial infections	Norovirus, Rotavirus [6, 8, 30, 48]	CDI [33, 45, 46], MRSA, VRE, CARBA-R
Tropical medicine and veterinary virology	Dengue Virus [2], avian Influenza [4], Yellow Fever Virus [12], MERS Corona Virus [3], Foot-and-Mouth Virus [1], Ebola Virus [16]	Malaria [11, 18, 22, 34, 38]
Antibody detection	HIV [9], EBV [10, 15, 20, 42]	

CARBA-R carbapenem-resistant Enterobacteriaceae; *CDI* Clostridium difficile; *CT* Chlamydia trachomatis; *EBV* Epstein-Barr Virus; *EHEC* enterohemorrhagic E. coli; *GAS* group A Streptococci; *GBS* group B Streptococci; *MRSA* methicillin-resistant Staphylococcus aureus; *MG* Mycoplasma genitalium; *MTB/RIF* Mycobacterium tuberculosis/rifampicin-resistance; *NG* Neisseria gonorrhoeae; *RSV* Respiratory Syncytial Virus; *TV* Trichomonas vaginalis are usually used to detect antibodies (HIV, infectious mononucleosis), in particular cases, saliva samples are also used, as with some rapid HIV self-tests [9, 10, 15, 20, 42].

20.2 POCT-guided therapy

From the perspective of an infectious disease specialist, POC testing is always indicated in the case of life-threatening infections that require immediate and targeted treatment.

This is where the time advantage associated with POCT over conventional diagnostics truly comes to bear: The total analysis time needed for rapid tests is usually only 15–30 minutes. Even under optimal conditions, analysis in a central laboratory cannot compete with this. Unless tests are carried out directly in the hospital's own lab, it often takes at least 1–2 hours to transport the specimens to the laboratory. Then, further time is needed for the actual analysis, including the necessary preparation (pipetting, incubation etc.), in addition to the time taken to report the results back to the requesting physician. Therefore, even for urgent requests, the total analysis time can be 1–2 days.

There are many well-documented cases in infectious disease medicine, and particularly in critical care medicine, where such time savings can confer a very beneficial effect on therapeutic outcomes. Kumar et al. [26] have shown that the survival rate of sepsis patients in intensive care correlated directly to early, clinically effective initial antibiotic treatment. A treatment delay of more than 2 hours can lower the survival rate to as much as <60 %.

Not only decisions about the treatment of life-threatening or highly acute infections such as sepsis, but also general decisions that need to be made within a short time frame are considerably facilitated by POCT. The immediate availability of rapid HIV test results, for example, is helpful for decision-making about prophylactic antiretroviral treatment during childbirth or for people with occupational exposure to HIV. Intrapartum detection of group B streptococci in women in labor or the detection of plasmodia spp., including self-testing by patients with suspected malaria infection, are similar situations where POCT proves its merits. Here, too, the presence of a pathogen can be detected immediately using an antigen test whereupon early, targeted antimicrobial therapy can lessen or even prevent infection. POCT is also useful for guiding treatment of viral infections. The effectiveness of zanamivir or oseltamivir in influenza treatment depends on them being given no later than 36–48 hours after the first symptoms develop [32]. This approach is similar with the respiratory syncytial virus (RSV). Studies have shown that treatment with ribavirin in RSV bronchopneumonia is only successful if initiated early enough [5].

20.3 Transmission prophylaxis through POCT

Besides playing a role as a decision-making aid for individual patients, POCT is also designed to prevent the spread of infection (transmission prophylaxis). Such targeted situations not only arise in hospitals where there is the risk of an undetected pathogen spreading from patient to patient. There are also constellations when a transmission risk is associated with outpatients. For example, it is well known that a high percentage of patients coming to an HIV or sexual health clinic will skip their follow-up appointment because they fear an unfavorable diagnosis [43]. Although this avoidance behavior is understandable, it is a problem insofar as test results usually confirm infection (HIV, gonorrhea or chlamydia infection). As such, this has far-reaching consequences for the patients themselves as well as for their sexual partner(s). Data from the USA spotlight the extent of this problem. In one study of an HIV clinic, more than a quarter of the 68,000 people seeking advice and undergoing a conventional HIV test did not attend their scheduled 2-week follow-up appointment to pick up the test result. This was different when the rapid HIV test was performed. Only 2.3 % of the 33,000 people that had the rapid test left the clinic before receiving their test result [47].

20.4 Pre-analytical confounders and influencing factors

Generally, rapid microbiological tests – like all testing methods – are subject to a variety of pre-analytical (and analytical) confounders and influencing factors that can negatively impact the diagnostic conclusiveness of the findings. This problem can be clearly illustrated using a rapid influenza diagnostic test as an example [39]. The following have a significant impact:

- Choice of specimen and where taken (nasal wash is better than throat swab)
- Specimen-taking instruments (swabs with gel are generally less useful than those without)
- The patient's activities immediately before the test. For example, the amount of detectable virus is reduced if the patient has eaten, drunk or gargled.

Other influencing factors include the time of sample collection – preferably the first 2–3 days after disease onset as virus shedding then declines rapidly – and the age of the person – children shed influenza viruses at a higher rate than adults. Other rapid tests are also affected by similar confounders and influencing factors.

20.5 POC test handling

The fact that many rapid microbiological tests are supplied in apparently easy-to-use formats, e.g. in the form of strips, cassettes or cartridges with pre-packaged diagnostic reagents for single-use on disposable devices, should not deceive. The handling of such tests is not trivial at all. Rather, it can be problematical, especially when it comes to sample collection. Indeed, evaluation studies on rapid detection tests for streptococci have clearly shown that the quality and reliability of test results is essentially dependent on the training and experience of the person taking the throat swab or performing the rapid test. Moreover, some rapid streptococcal tests have relatively subjective reading endpoints, often making interpretation prone to errors [36].

In particular, an increased infection risk for the examiner is one of the disadvantages specifically afflicting infectious disease medicine. This is not completely avoidable as, during testing, the examiner is exposed to the patient's specimen (respiratory secretion, stool, urine, blood), which potentially contain pathogens.

20.6 Performance capability of POCT diagnostics

POCT methods have been continuously developed and improved over the past 10–15 years. The majority of tests nowadays use immunochromatography (▶ Chapter 9) with moderate to high sensitivity (70–90 %) and relatively high specificity (>95 %). In exceptional cases, like with rapid HIV tests, today's POCT methods can achieve results that are indeed as reliable as those of conventional diagnostics. Some POCT systems with potentially limited sensitivity to a suspected pathogen (e.g. influenza viruses, RSV, rotaviruses, noroviruses or adenoviruses) may be compromised by patterns of seasonal variations in the prevalence of that organism. This has a considerable impact on the clinical reliability of the test, as exemplified by the immunochromatographic rapid influenza diagnostic test (Fig. 20.1). A low prevalence (2.5 % in the example given) is expected at the start of a seasonal flu outbreak. Despite high sensitivity (80 % in this example) and very high specificity (95 %), the negative predictive value (99 %) of the test is expected to be significantly higher than the positive predictive value (29 %). This is because the rapid test shows false-positive (49-fold) more often than true-positive (20-fold) results under these circumstances. Not until the prevalence rises to 10 %, which can certainly happen during a severe flu outbreak, does the positive predictive value increase

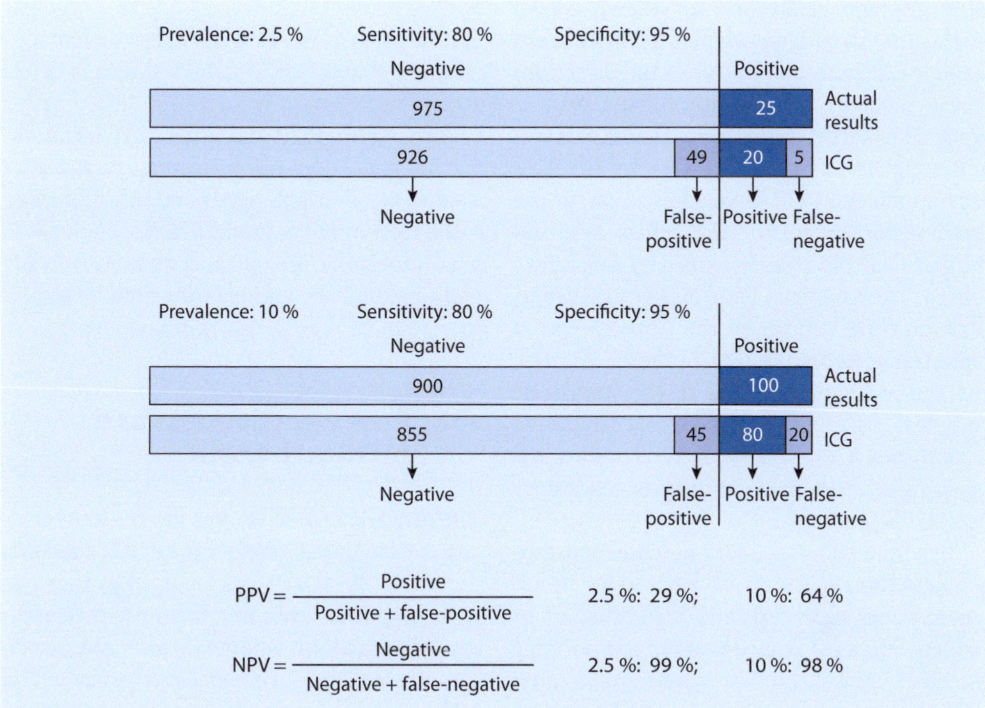

Fig. 20.1 Correlation between prevalence (2.5 % vs. 10 %), test sensitivity and specificity, negative predictive value (NPV) and positive predictive value (PPV) in a simulated rapid influenza diagnostic test based on immunochromatography (ICG)

(64 %). Of course, the biostatistical relationships shown here hold true for other infectious diseases that follow a seasonal pattern as well. In other words, the use of tests with limited sensitivity can yield more false results at the start of an outbreak when prevalence is still low. This, in turn, could certainly lead to an underappreciation of the extent of a disease [17] and must be taken into account when using POCT.

Conventional rapid tests are unable to detect any prevailing antibiotic resistances. Therefore, important information about therapeutic options may be lacking. Not only this, but epidemiological patterns regarding the development of resistant bacterial strains are no longer detected or else do not become apparent until later.

20.7 Molecular biological (PCR) tests

Immunological testing has certain limitations. Detection is difficult owing to the limited amount of microbial antigens released with particular regard to colonization of mucous membranes associated with low pathogen counts (e.g. group B streptococcus in the vagina) or intracellular pathogens (e.g. chlamydia in cervical smears). New testing concepts had to be developed in order to detect these pathogens. Given the degree of analytical sensitivity required, PCR methods and other nucleic acid amplification techniques (NAT) with greater sensitivity have been added to the diagnostic arsenal and further developed for faster and easier use.

The most important innovation in this context was the development of fully automated

PCR kits and ready-to-use, reagent-coated single-use cartridges where all PCR steps (sample exposure, amplification and detection) run in sequence without any further manual input. The market leader is the GeneXpert system, supplied by Cepheid (▶ Chapter 10). This fully automated PCR device is so easy to use that it is possible for trained staff to carry out the test near the patient, outside a laboratory setting. The American Food and Drug Administration (FDA) approved the GeneXpert as a rapid test according to CLIA criteria. At present, the technique is listed in the "moderate complex" category, reserved for accredited laboratories. The "CLIA-waived" classification, meaning without being subject to laboratory use, is still pending.

In addition to its primary function of pathogen detection, PCR can also be used for simultaneous analysis of resistance determinants or virulence factors. Linked assays incorporating multiplex technology allow more complex questions to be processed within one PCR run (e.g. detection of the following: S. aureus plus detection of methicillin resistance (mecA); two determinants (vanA and vanB) of vancomycin resistance in enterococci; clostridium difficile toxin B, also encompassing binary toxins and the tcdC deletion to detect highly virulent variants; mycobacterium tuberculosis also encompassing the rifampicin resistance (rpo) for detecting multi-drug resistant tuberculosis strains; simultaneous detection of influenza A, A/H1N1 and influenza B). The recently available i-system by Alere, which relies on isothermal amplification to detect the influenza A/B viruses and group A streptococci, (▶ Chapter 10) can boast significantly better sensitivity and specificity compared to the antigen detection method with the correspondingly more reliable results [7, 23, 31]. Besides its i-system, Alere also offers the q-Analyzer as a fully automated NAT platform.

Rapid molecular biological test methods on an **isothermal basis** are increasingly used for the diagnostics of tropical viral infections as well as in veterinary virology. Supplied in a portable case lab, they provide an advantage in areas with little infrastructure and can deliver reliable results [1, 2, 3, 4, 12, 13, 16]. An overview of the available assays for near-patient molecular biological test systems is shown in ◘ Tab. 20.2.

Progress in PCR and other NAT technologies has made near-patient testing accessible for a series of other pathogens and infectious diseases. Currently, the existing lack of clarity as to what molecular biological tests are medically and economically feasible for which patients is in need of further scientific evaluation.

20.8 Cost effectiveness and medical benefit

The disadvantage cited and the constant criticism raised in regards to most POCT methods pertain to the extra costs incurred by new systems [24]. Even assuming that near-patient diagnostics saves on laboratory tests and associated costs, POCT processes in general, and molecular assays in particular, are significantly more expensive than conventional (laboratory) tests. Moreover, POCT can create extra work for staff that were not previously involved in diagnostic tasks and possibly needs to be considered in job planning [24]. The question inevitably arises, as to how far near-patient microbiological analysis actually adds value to justify the added financial expenditure. Comprehensive reviews of this topic are rare and, if available, do not give the full picture either. The main difficulty is that the issues relating to the weighing of cost-effectiveness against the medical benefit of near-patient testing are multifactorial and complex, with all aspects above and beyond this being closely interlinked. It is therefore not possible to find a global solution. In fact, a well-justified estimation depends much more on the circumstances of the individual case. In addition, international study data cannot simply be applied to the situation in Germany because hospitals charge health insurance companies, using the German (diagnosis-related group) DRG system.

However, it is to be expected that POCT systems will soon play an increasing role in gen-

Tab. 20.2 Analytical spectrum of molecular biological near-patient diagnostics (as per 2016)

System	Multiplexity	Virology	Microbiology	Human genetics/ oncology
Alere i-System	Single test	FLU A/B	GAS	
bioMérieux FilmArray	Multiplex assay	Gastrointestinal panel: 22 commonly occurring gastrointestinal pathogens (viruses, bacteria, protozoans) Respiratory panel: 20 respiratory viruses and bacteria		
Roche Cobas LIAT	Single test	FLU A/B	GAS	
	Multiplex assay	FLU A/B + RSV		
Atlas io	Single test	NORO	CT, NG, TV, MG, CDI, MRSA	
	Multiplex assay		CT + NG + TV + MG	
Spartan RX	Microarray			CYP 2C20
Cepheid GeneXpert	Single test	HIV [a], HBV, HCV, HPV, FLU A/B, EBO, EV, NORO	GBS, CT, NG, TV, MTB/RIF, CDI, MRSA, VRE, CARBA-R	BCR-ABL
	Multiplex assay	FLU A/B + RSV		Factor II + V mutation

BCR-ABL transcription product BCR-ABL; *CARBA-R* carbapenem-resistant Enterobacteriaceae; *CDI* Clostridium difficile; *CT* Chlamydia trachomatis; *FLU A/B* Influenza A/B; *EBO* Ebola Virus; *EV* Enterovirus; GAS group A Streptococci; *GBS* group B Streptococci; *MRSA* methicillin-resistant Staphylococcus aureus; *MG* Mycoplasma genitalium; *MTB/RIF* Mycobacterium tuberculosis/ rifampicin resistance; *NG* Neisseria gonorrhoeae; *NORO* Norovirus; *RSV* Respiratory Syncytial Virus; *TV* Trichomonas vaginalis
[a] also quantitative, as viral load test.

eral medical care in the future as the density of physician coverage is projected to decrease. This will lead to a significant deterioration in patient care, particularly in the lowlands and rural areas if not counteracted. New ways must be found to maintain high quality care in non-metropolitan areas. POCT-based laboratory diagnostics will then gain in relevance.

20.9 Molecular MRSA screening

A closer look at molecular MRSA screening illustrates the conflict between cost-effectiveness on the one hand and medical benefits on the other [41]. With modern PCR tests, nasal MRSA colonization can be detected faster and more reliably than with many routine culture methods. It is indisputable that the sooner appropriate hygienic measures are put in place after a positive MRSA status identification, the lower is the risk that the infection will spread to other patients [41].

In terms of the cost of an MRSA culture (approx. € 3–5; approx. € 10–15 for a positive result) versus PCR (single-test cartridge approx. € 30–40), PCR is at least 2–3 times more expensive than the culture method. However, this additional cost for PCR testing may be justified when considering that every MRSA transmission prevented by early detection can save a hospital several thousand euros in added costs. Therefore, the cost-benefit ratio shifts in favor of PCR. Even when each case is balanced against a successful coding in the DRG flat rate payment system, the added costs incurred by MRSA transmission are only partially compensated. Furthermore, every new, preventable case of MRSA has the potential to impact negatively on public image and may result in lost revenue due to canceled admissions, which most hospitals should be eager to avoid [41].

Since the screening of all patients in a hospital is not financially viable, the cost-benefit consideration ultimately focuses on which patient group should be targeted using PCR analysis in order to establish the MRSA status more quickly and, then, which of them should be screened with a traditional culture method. A definitive answer (e.g. from larger meta-analyses) is not yet available. Nevertheless, the current data suggest that the benefit of molecular MRSA screening will only outweigh the costs for patients with a particularly high MRSA risk, e.g. in areas with a high MRSA prevalence [41]. At present, German hospitals remain hesitant. As long as there are no generally applicable recommendations issued (e.g. from the Robert-Koch-Institute, Berlin) on the use of molecular biological tests, the high costs and organizational constraints of MRSA-PCR – similar to a lack of effective hygiene management structures – will prevent screening from being carried out at all or, when, only to answer specific questions, mostly for quickly managing bed capacities.

References

1. Abd El Wahed A, El-Deeb A, El-Tholoth M, et al. (2013) A Portable Reverse Transcription Recombinase Polymerase Amplification Assay for Rapid Detection of Foot-and-Mouth Disease Virus. PLoS One 8:e71642
2. Abd El Wahed A, Patel P, Faye O, et al. (2015) Recombinase Polymerase Amplification Assay for Rapid Diagnostics of Dengue Infection. PLoS One:e0129682
3. Abd El Wahed A, Patel P, Heidenreich D, Hufert FT, Weidmann M (2013) Reverse transcription recombinase polymerase amplification assay for the detection of middle East respiratory syndrome coronavirus. PLoS Curr 5
4. Abd El Wahed A, Weidmann M, Hufert FT (2015) Diagnostics-in-a-Suitcase: Development of a portable and rapid assay for the detection of the emerging avian influenza A (H7N9) virus. J Clin Virol 69:16–21
5. Adcock PM, Stout GG, Hauck MA et al. (1997) Effect of rapid viral diagnosis on the management of children hospitalized with lower respiratory tract infection. Pediatr Infect Dis 16: 842–846
6. Bon F, Kaplon J, Metzger MH, Pothier P (2007) Evaluation of seven immunochromatographic assays for the rapid detection of human rotaviruses in fecal specimens. Pathol Biol (Paris) 55: 149–153
7. Bosevska G, Panovski N, Janceska E, Mikik V, Topuzovska IK, Milenkovik Z (2015) Comparison of Directigen Flu A+B with Real Time PCR in the Diagnosis of Influenza. Folia Med (Plovdiv) 57:104–110
8. Brandt CD, Arndt CW, Evans GL et al. (1987) Evaluation of a latex test for rotavirus detection. J Clin Microbiol 25: 8000–8002
9. Branson BM (2003) Point-of-care rapid tests for HIV antibody. J Lab Med 27: 288–295
10. Bruu AL, Hjetland R, Holter E et al. (2000) Evaluation of 12 commercial tests for detection of Epstein-Barr virus-specific and heterophile antibodies. Clin Diagn Lab Immunol 7: 451–456
11. Cruciani M, Nardi S, Malena M, Bosco O, Serpelloni G, Mengoli C (2004) Systematic review of the accuracy of the ParaSight-F test in the diagnosis of Plasmodium falciparum malaria. Med Sci Monit 10: MT81–MT88
12. Escadafal C, Faye O, Sall AA, et al. (2014) Rapid molecular assays for the detection of yellow fever virus in low-resource settings. PLoS Negl Trop Dis 8:e2730
13. Euler M, Wang Y, Heidenreich D, et al. (2013) Development of a panel of recombinase polymerase amplification assays for detection of biothreat agents. J Clin Microbiol 51:1110–1117
14. Ewig S, Tuschy P, Fätkenheuer G (2002) Diagnosis and treatment of Legionella pneumonia. Pneumologie 56: 695–703
15. Farhat SE, Finn S, Chua R et al. (1993) Rapid detection of infectious mononucleosis-associated heterophile antibodies by a novel immunochromatographic assay and a latex agglutination test. J Clin Microbiol 31: 1597–1600
16. Faye O, Faye O, Soropogui B, et al. (2015) Development and deployment of a rapid recombinase polymerase amplification Ebola virus detection assay in Guinea in 2015. Euro Surveill 20
17. Friedewald S, Finke EJ, Dobler G (2006) Near patient testing in exceptional situations. J Lab Med 30: 211–218
18. Gatti S, Gramegna M, Bisoffi Z et al. (2007) A comparison of three diagnostic techniques for malaria: a rapid diagnostic test (NOW Malaria), PCR and microscopy. Ann Trop Med Parasitol 101: 195–204
19. Gutiérrez F, Masiá M, Rodriguez JC et al. (2003) Evaluation of the immunochromatographic Binax NOW assay for detection of Streptococcus pneumoniae urinary antigen in a prospective study of

community-acquired pneumonia in Spain. Clin Infect Dis 36: 286–292
20. Gutierrez J, Rodriquez M, Maroto C, Piedrola G (1997) Reliability of four methods for the diagnosis of acute infection by Epstein-Barr virus. J Clin Lab Anal 11: 78–81
21. Honest H, Sharma S, Khan KS (2006) Rapid tests for group B streptococcus colonization in laboring women: a systematic review. Pediatrics 117: 1055–1066
22. Iqbal J, Khalid N, Hira PR (2002) Comparison of two commercial assays with expert microscopy for confirmation of symptomatically diagnosed malaria. J Clin Microbiol 40: 4675–4678
23. Jokela P, Vuorinen T, Waris M, Manninen R (2015) Performance of the Alere i influenza A&B assay and mariPOC test for the rapid detection of influenza A and B viruses. J Clin Virol 70:72–76
24. Junker R, Schlebusch H, Luppa PB (2010) Point-of-care testing in hospitals and primary care. Dtsch Ärztebl Int 107:561–567
25. Kehl KS, Havens P, Behnke CE, Acheson DW (1997) Evaluation of the premier EHEC assay for detection of Shiga toxin-producing Escherichia coli. J Clin Microbiol 35: 2051–2054
26. Kumar A, Roberts D, Wood KE et al. (2006) Duration of hypotension before initiation of effective antimicrobial therapy is the critical determinant of survival in human septic shock. Crit Care Med 34: 1589–1596
27. Lasocki S, Scanvic A, LeTurdu F et al. (2006) Evaluation of the Binax NOW Streptococcus pneumoniae urinary antigen assay in intensive care patients hospitalized for pneumonia. Intensive Care Med 32: 1766–1772
28. Mackenzie AM, Lebel P, Orrbine PC et al. (1998) Sensitivities and specificities of premier E. coli O157 and premier EHEC enzyme immunoassays for diagnosis of infection with verotoxin (Shiga-like toxin)-producing Escherichia coli. The SYNSORB Pk Study investigators. J Clin Microbiol 36: 1608–1611
29. Mahilum-Tapay L, Laitila V, Wawrzyniak JJ et al. (2007) New point of care chlamydia rapid test – bridging the gap between diagnosis and treatment: performance evaluation study. BMJ 335: 1190–1194
30. Nguyen TA, Khamrin P, Takanashi S et al. (2007) Evaluation of immunochromatography tests for detection of rotavirus and norovirus among Vietnamese children with acute gastroenteritis and the emergence of a novel norovirus GII.4 variant. J Trop Pediatr 53: 264–269
31. Nguyen Van JC, Camelena F, Dahoun M, et al. (2016) Prospective evaluation of the Alere i Influenza A&B nucleic acid amplification versus Xpert Flu/RSV. Diagn Microbiol Infect Dis 85(1):19–22
32. Nicholson KG, Aoki FY, Osterhaus AD et al. (2000) Efficacy and safety of oseltamivir in treatment of acute influenza: a rondomized controlled trail. Neuraminidase inhibitor flu treatment investigator group. Lancet 355: 1845–1850
33. O'Connor D, Hynes P, Cormican M, Collins E, Corbett-Feeney G, Cassidy M (2001) Evaluation of methods for detection of toxins in specimens of feces submitted for diagnosis of Clostridium difficile-associated diarrhea. J Clin Microbiol 39: 2846–2849
34. Palmer CJ, Lindo JF, Klaskala WI et al. (1998) Evaluation of the OptiMAL test for rapid diagnosis of Plasmodium vivax and Plasmodium falciparum malaria. J Clin Microbiol 36: 203–206
35. Rani R, Corbitt G, Killough R, Curless E (2002) Is there any role for rapid tests for Chlamydia trachomatis? Int J STD AIDS 13: 22–24
36. Reinert RR (2007) Rapid streptococcal antigen detection tests. J Lab Med 31: 280–293
37. Saison F, Mahilum-Tapay L, Michel CE et al. (2007) Prevalence of Chlamydia trachomatis infection among low- and high-risk Filipino women and performance of Chlamydia rapid tests in resource-limited settings. J Clin Microbiol 45: 4011–4017
38. Schmidt WP (2003) Malaria rapid tests – perspectives for malaria endemic and non-endemic regions. J Lab Med 296–301
39. Schweiger B (2006) Influenza rapid tests – advantages and limitations. J Lab Med 30: 219–225
40. Smith MD, Derrington P, Evans R et al. (2003) Rapid diagnosis of bacteremic pneumococcal infections in adults by using the Binax NOW Streptococcus pneumoniae urinary antigen test: a prospective, controlled clinical evaluation J Clin Microbiol 41: 2810–2813
41. Stürenburg E (2009) Rapid detection of methicillin-resistant Staphylococcus aureus directly from clinical samples: methods, effectiveness and cost considerations. Ger Med Sci 7:Doc06
42. Svahn A, Magnusson M, Jägdahl L, Schloss L, Kahlmeter G, Linde A (1997) Evaluation of three commercial enzyme-linked immunosorbent assays and two latex agglutination assays for diagnosis of primary Epstein-Barr virus infection. J Clin Microbiol 35: 2728–2732
43. Swain GR, McDonald RA, Pfister RJ, Gradus MS, Sedmak GV, Singh A (2004) Decision analysis: point-of-care chlamydia testing vs. laboratory-based methods. Clin Med Res 1: 29–35
44. Teel LD, Daly JA, Jerris RC et al. (2007) Rapid detection of Shiga toxin-producing Escherichia coli by optical immunoassay. J Clin Microbiol 45: 3377–3380
45. van den Berg RJ, Bruijnnesteijn van Coppenraet LS, Gerritsen HJ, Endtz HP, van der Vorm ER, Kuijper EJ (2005) Prospective multicenter evaluation of a new immunoassay and real-time PCR for rapid diagnosis of Clostridium difficile-associated diarrhea in hospitalized patients. J Clin Microbiol 43: 5338–5340

46. Vanpoucke H, De Baere T, Claeys G et al. (2001) Evaluation of six commercial assays for the rapid detection of Clostridium difficile toxin and/or antigen in stool specimens. Clin Microbiol Infect 7: 55–64
47. Warpakowski A (2006) Ärztezeitung online. Schnelltest – mehr Patienten erfahren HIV-Status. Edition of 21 August 2006; https://www.aerztezeitung.de/medizin/krankheiten/infektionskrankheiten/aids/article/415433/schnelltest-patienten-erfahren-hiv-status.html
48. Weitzel T, Reither K, Mockenhaupt FP et al. (2007) Field evaluation of a rota- and adenovirus immunochromatographic assay using stool samples from children with acute diarrhea in Ghana. J Clin Microbiol 45: 2695–2697
49. Widjaja S, Cohen S, Brady WE et al. (1999) Evaluation of a rapid assay for detection of Chlamydia trachomatis infections in outpatient clinics in South Kalimantan, Indonesia. J Clin Microbiol 37: 4183–4185
50. Yin YP, Peeling RW, Chen XS et al. (2006) Clinic-based evaluation of Clearview Chlamydia MF for detection of Chlamydia trachomatis in vaginal and cervical specimens from women at high risk in China. Sex Transm Infect 82 (Suppl 5): v33–v37

Emergency medicine

Walter Schaffartzik, Christian Müller, Tobias Lindner, Julia Searle, Martin Möckel

P. B. Luppa, R. Junker (Eds.), *Point-of-Care Testing*
https://doi.org/10.1007/978-3-662-54497-6_21

21.1 Preclinical emergency care

21.1.1 Emergency ambulance and transport systems

Preclinical rescue systems in medicine comprise emergency ambulances and vehicles staffed by clinicians. These include rapid-response vehicles and air rescue ambulances or intensive care helicopters. Helicopters are used to transport the emergency care team to the incident, but also to transport critically ill patients between hospitals under intensive care conditions.

In larger cities, the emergency clinician or paramedic usually arrives at the scene within 10 minutes of the alarm, whether by road or air. The time to reach the scene or transport times can take significantly longer in rural areas.

21.1.2 Responsibilities of an emergency clinician

The main task of an emergency clinician is to restore and stabilize the vital functions of the patient. This requires an initial diagnosis to be established, considering various differential diagnoses, alongside the quick and efficient administration of appropriate treatment. Finally, the clinician must identify and transport the patient to the nearest suitable hospital so that therapy can continue without delay. At the scene of the accident, the emergency clinician needs to make decisions concerning medical procedures like whether to intubate or ventilate – on-scene interventions viewed as lifesaving for patients with multiple trauma. The decision as to whether the patient requires endotracheal intubation and ventilation is based on physical examination of the patient and determination of their vital signs like blood pressure, heart rate and pulse-oximetrically measured arterial oxygen saturation (sO_2).

Monitoring encompasses the following parameters:

- Blood pressure (measured by oscillometry)
- Electrocardiographic parameters
- sO_2
- End tidal CO_2 capnography
- Blood glucose

Cardiac biomarkers are currently not widely checked by emergency clinicians. Further data from studies on the merits and usefulness of troponins in preclinical use are given in ▶ Chapter 17.

Endotracheal intubation and ventilation are the "gold standard" in the management of trauma patients. However, evidence has been mounting that the time taken for intubation on-scene would be better spent transporting the patient to the hospital more quickly [2, 11]. Pre-procedural preparation and subsequent intubation and ventilation undertaken by the emergency clinician called to the site can delay transport by about 25 minutes. The emergency clinician cannot presume that the interventions he renders will influence patient survival every time [43]. Failure to carry out endotracheal intubation and ventilation at the scene of the incident is seen as critical. The emergency patient's condition may deteriorate during transport. Therefore, many cases require intubation and ventilation in order to ensure adequate oxygenation of vital organs [40]. To render his time-pressured decisions like whether to intubate and ventilate, the emergency clinician is informed by objective criteria including cardiovascular status, oxygen saturation (sO_2) – but also guided by his own subjective assessment of the situation. That said, decisions are made intuitively as well [49]. Objective criteria such as Hb concentration, partial pressure of oxygen (pO_2) and carbon dioxide (pCO_2) along with acid-base status, particularly base excess (BE), help the emergency clinician make these important decisions on-site [54]. Key POCT analytes important to preclinical emergency care are summarized in ◻ Table 21.1, specifically taking into account on-scene times and transport times.

Every delay in treatment and in prognostic stratification of a patient should be seen as very critical for the quality of the clinical outcome [12]. In cases where the required treatment of critically ill patients on the ICU is delayed, the outcome is a prolonged hospital stay along with

Table 21.1 POCT analytes in preclinical emergency care

Areas of application	Parameters
Acid-base status	pH, HCO_3^-
Blood gases	pO_2, pCO_2, BE
CO oximetry	Hemoglobin (Hb), COHb, MetHb, sO_2
Electrolytes	Na^+, K^+, Cl^-, Ca^{2+}
Metabolites	Glucose, lactate
Cardiac markers	Cardiac troponins
Narcotics (qualitative detection)	Alcohol, cannabinoids, cocaine, amphetamines, opiates, barbiturates, benzodiazepines

higher mortality [6]. It has not yet been proven to date whether this observation has unlimited validity in preclinical emergency situations. However, it is clearly indicated that emergency patients should receive definitive hospitalized treatment as rapidly as possible. In this indication, preclinically obtained laboratory results, e.g. biochemical parameters, such as blood gases, acid-base status and cardiac biomarkers, can be extremely helpful.

21.1.3 Application of POCT in preclinical emergency care

Blood glucose measurement is routinely available in preclinical emergency care. However, apart from sO_2 monitors, the use of no other POCT devices has become established. There is one exception: some emergency care helicopters are equipped with POCT devices for measuring pH, pCO_2, pO_2, Na^+, K^+, glucose and hematocrit [10]. On-scene and transport times have to be specifically taken into account when POCT devices are considered for preclinical emergency care. In metropolitan areas, the on-scene and travel times are usually so short that even if the biochemical result were available, their therapeutic implications might be obviated by admission. By contrast, in rural areas it is critically important for an emergency clinician to decide whether the patient needs to be taken to a distant cardiology center or not.

Most on-scene calls for emergency clinicians involve diagnoses in conservative medical specialties. In this regard, a POCT device able to measure cardiac biomarkers could therefore be very helpful for emergency clinicians. Cardiac biomarkers are of particular importance in patients with symptoms suggestive of a myocardial infarction, but without ECG changes. It is well documented that only about half of patients suffering from coronary heart disease who develop symptoms of a heart attack, promptly call an emergency clinician. Therefore the emergency clinician must consider that the acute event may have happened a while in the past, making diagnosis and treatment of the patient even more urgent. The timely diagnosis of a myocardial infarction similarly opens up a wider window for timely treatment (► Chapter 17).

In **trauma patients,** on-scene assessment of Na^+, Cl^-, K^+ and Ca^{2+} does not have implications for initial trauma management. Conversely, knowledge of Hb and glucose levels as well as pO_2 and pCO_2 can sometimes be helpful to limit the effects of trauma on the body and improve the cost-effective use of resources [1]. BE, which is calculated from pCO_2, Hb and pH, is an important predictor of mortality. Using BE, early trauma-related oxygen deficits can be addressed by administering targeted treatment [40] (► Chapter 14).

The recognition of **flue gas inhalation victims** can cause problems in emergency care. Just a few years ago, it became possible to measure increased concentrations of COHb in blood at the POC (► Chapter 11 and 14). Knowledge of the COHb level can guide treatment. For example, decisions about the primary need for hyperbaric oxygen administration in a pressure chamber can already be made on-scene.

In attending to emergency patients, but especially traumatized ones, the attention of the emergency clinician should be directed as the possibility that the victim has used **narcotic drugs** prior to the accident. The number of people in Germany with alcohol dependency is estimated to be around 2.5 million [4, 9]. In the 15–25 age group, 20 % consume alcohol regularly. Nearly 5 million people show a risky abuse pattern [18], of whom 3 million also consume cannabis, cocaine or the illegal narcotic Ecstasy. A high degree of under-reporting can be presumed, particularly for easily obtainable drugs such as alcohol and cannabis [27]. Chronic drug users have complex physical and psychological co-morbidities and are therefore high-risk patients in trauma situations. Drug screening in preclinical emergency care can help identify existing risks, is as effective as screening in the central laboratory and its results are available more quickly [40]. Drug screening results can assist also the emergency clinician at the scene when attending to patients with impaired consciousness (▶ Chapter 18).

Note

POCT should be used in preclinical settings, as results can guide decisions for triaging, treatment, risk assessment and have implications for early interventions and procedures, improving both clinical outcome, survival quality and cost effectiveness.

21.2 POCT in the interdisciplinary rescue center and emergency department

The content of this chapter was partially taken from a publication in the Deutsche Medizinische Wochenschrift by Müller, Lindner, Searle and Möckel [29].

21.2.1 Introduction

Laboratory tests in emergency and peri-interventional intensive care

In hospital emergency care, laboratory tests are one of the most important aspects of diagnostics alongside clinical assessment, ECG and imaging (ultrasound, radiology). In this context, lab workups are primarily used to aid diagnosis when the clinical picture is unclear, to support initial therapeutic interventions as well as for risk stratification. Progress monitoring during therapy is only required for a small number of selected parameters such as electrolytes, lactate, troponin or Hb (e.g. post-transfusion). By contrast, it is rare in emergency medicine that more detailed differential diagnoses are rendered or microbiological cultures taken. Toxicological testing is a category in and of itself; with a few exceptions, such as alcohol and paracetamol, its importance in primary emergency care is undergoing intense debate.

In the area of peri-interventional intensive care, the focus – depending on the clinical picture and type of intervention – shifts to the ongoing evaluation of the patient's clinical progress, particularly during the intervention. In this setting, POCT is very frequently indispensable, as the turn-around time (TAT) in a central laboratory would be too long in many cases. At the forefront stands coagulation testing (▶ Chapter 6 and 15).

Choice of parameters and test methods

The selection of parameters and methods is dictated by the aim of the test. Is the aim only to avert an immediate threat to life? Or to render the appropriate primary diagnostics that streamline the duration of inpatient treatment within a cross-sectoral process? At what later point is an elective laboratory service available? In addition, the processes and timing of emergency and acute care need to be taken into account. The latter is determined by the duration of typical tasks, shift workers (necessary handover), but also by receipt procedures, acceptable waiting times for results and, at times, by political dictates. Generally, in emergency departments and ICUs, laboratory test results are expected within an hour. Tests are often not requested if time-to-result is longer than 4 hours. In planned interventions, such as acute operations or cardiac catheterization, the laboratory service must follow the intervention schedule. As with all diagnostic procedures, their applicability is dependent on their availability.

> **Overall, laboratory diagnostics must be aligned with the emergency care process, which may also entail laboratory testing or imaging procedures as the circumstances dictate. The general process of emergency care has been described by Möckel et al. [32].**

POCT is increasingly at the forefront of diagnostics, especially in emergency departments and ICUs. The reason for this is an increasing centralization of laboratory institutes with the resultant increase in TAT in addition to the growing use of time-critical laboratory parameters and algorithms. The POCT TAT – often only a few minutes – is not achievable in large hospital laboratories or when the tests are outsourced to central laboratories. This chapter therefore primarily focuses on current options offered by POCT diagnostics in the acute setting.

In the following, the most important laboratory parameters for emergency and peri-interventional intensive care are presented according to urgency or their function within the management process, and their provision as POCT explained.

21.2.2 Mandatory emergency parameters

Testing for mandatory emergency parameters should be provided in all emergency departments and ICUs at all times, as the results can have immediate therapeutic implications. A distinction is made here between emergency parameters that need to be available quickly (within an hour) and acute biomarkers that are not as time-critical, but should still be available within 4 hours.

Emergency parameters

Emergency parameters usually need to be provided as POCT, otherwise their timely availability cannot be guaranteed. The equipment to test for these laboratory values is regarded as the basic standard to any emergency department in Germany. Checklist 1 gives a selection of mandatory emergency parameters.

Checklist 1

Criteria governing mandatory emergency parameters with urgent analysis required within 1 hour and therefore usually as POCT:

- A test is absolutely imperative to diagnose acute clinical symptoms or an emergency situation.
- The findings of the test have immediate therapeutic implications.
- The parameter helps monitor the success or progress of treatment during an emergency or acute therapy.

Coma and impairment of consciousness Not only blood glucose, but also serum sodium concentrations are important parameters for assessing a patient with acute impairment of consciousness or in a coma. Poisoning, e.g. from carbon monoxide, should also be considered here and can be diagnosed by measuring COHb [55] (▶ Section 21.1.3 and ▶ Chapter 11).

Electrolyte imbalance Hyponatremia can cause nausea, vomiting, headaches, dizziness and even seizures and coma where severe cerebral edema develops. Depending on the speed at which the sodium serum level falls, a number of severe neurological symptoms can develop and fast, controlled correction can save lives [22]. Abnormal potassium levels can also lead to life-threatening symptoms, particularly those affecting the electrical conduction system of the heart, and require immediate treatment [44]. Although severe cardiac dysrhythmias are more often caused by hyperkalemia, hypokalemia can also trigger ECG changes and ventricular extrasystoles, even causing dangerous arrhythmias (ventricular tachycardia and ventricular fibrillation) [26].

Acute cardiac pathologies These particularly refer to symptoms of acute myocardial infarction, where a rise in troponin levels is part of the diagnostic definition [42, 50]. Important here is not only the early "rule-in" diagnosis of an acute myocardial infarction, indeed the early exclusion of that diagnosis in patients exhibiting symptoms of acute coronary syndrome is becoming increasingly decisive as well (▶ Chapter 17). In view of the increasing overcrowding in emergency departments, the large number of patients presenting with chest pain, of whom only relatively few have a myocardial infarction, new exclusion criteria with highly sensitive measurements of troponin or copeptin serum concentrations have been included in the ESC guidelines [31, 42]. A further important test in acute cardiological patients is that for thyroid-stimulating hormone (TSH).This marker not only supports a differential diagnosis of atrial fibrillation [46], but also helps exclude thyroid dysfunction prior to administering contrast media, e.g. for coronary angiography or computerized tomography. Procalcitonin (PCT) could play a future role in differentiating pulmonary congestion from infection in acute heart failure. Studies have shown that early initiation of an appropriate treatment strategy can improve patient survival rates as well as avoid unnecessary antibiotic use [25, 45].

Acute infections The benefit that knowledge of PCT confers on patients with lower respiratory tract infections is undisputed, achieving a safe and reduced duration of exposure to antibiotics, albeit with no effect on patient outcomes [7]. Moreover, PCT plays an important role in the early identification of patients with sepsis [47, 52]. This is of particular interest, as symptoms of impending sepsis are often non-specific in older patients and those with multiple comorbidities; although the clinical course can run dramatically and be fatal within a few hours. In addition to PCT, the early determination of lactate levels is also very useful [5]. Other parameters important for assessment of acute infections in emergency care include urine tests to rule out acute urinary tract infection as well as cell counts in aspirates such as cerebrospinal fluid and ascites.

Trauma and general surgery In the emergency department, POCT is also crucial for trauma and surgical patients e.g. to determine the need for a blood transfusion to treat acute bleeding, as recommended in the cross-sectoral guidelines on therapy with blood components and plasma derivatives issued by the German Medical Board (BÄK) [3].

Hemostasis The assessment of coagulation status (▶ Chapter 6 and 15) has important therapeutic implications, particularly for anticoagulated patients. For them, it is often essential to know the result before invasive procedures can be performed. For example, in patients who are suffering a stroke, an unclear coagulation status is often one reason why the door-to-needle time for thrombolysis is prolonged [51]. The new anticoagulants (NOACs) present a novel challenge in this context [37].

Pregnancy POCT methods also deserve mention in connection with the early confirmation or exclusion of pregnancy. This is not only important in diagnosing pregnancy complications in patients presenting with unclear abdominal symptoms [35], but also key to decisions on teratogenic drug administration or radiological procedures. ◘ Tab. 21.2 lists the relevant parameters in these categories.

Acute parameters

POCT methods to test for acute parameters must also be available to guarantee timely esults. They are essential for the diagnosis and indirect initiation of therapy for commonly presenting diseases. As these tests are not imminently necessary to avert life-threatening situations, a TAT of 4 hours has been established and can be regarded as an international consensus. In England, a maximal waiting time of 4 hours in emergency care for 95 % of patients has even been anchored in law [34]. Checklist 2 details the basis for the selection of parameters that must be available in

Tab. 21.2 Mandatory emergency parameters which should be available within ≤1 hour. (Mod. [29])

Parameters	Specimen material	Analysis	Typical relevance	References
Glucose	1; 3	POCT	Coma diagnostics	
Sodium	1	POCT	Impaired level of consciousness, heart failure (HF)	[22, 33, 44]
Potassium	1	POCT	Cardiac arrhythmias	[26, 44]
Lactate	1	POCT	Sepsis, shock	[5, 8]
BGA	1; 2; 3	POCT	Hypoxemia, hypercapnia/hypocapnia, pH, BE	
Hb	1	POCT	Bleeding	[3]
COHb	1	POCT	Poisoning	[55]
Troponin	1	POCT, CentLab*	Myocardial infarction (MI), pulmonary embolism	[42, 50]
β-HCG	1; 4	POCT	Pregnancy; before radiation exposure, ectopy	[21, 35]
Copeptin	1	POCT, CentLab**	Fast rule-out MI	[31, 42]
PCT	1	POCT, CentLab	Pneumonia, sepsis, MI	[7, 25, 45, 47, 52]
TSH	1	CentLab	Contrast media administration, atrial fibrillation	[46]
Urine dipstick	4	POCT	Urinary tract infection	
Cell count	5; 6; 7	CentLab***, POCT	Meningitis, SBP	
TCT, PTT, INR	1	POCT, CentLab****	Bleeding, stroke, anticoagulation	[13, 37, 51]

BE base excess; *BGA* blood gas analysis; *CentLab* central laboratory; *COHb* carboxy hemoglobin; *β-HCG* human chorionic gonadotropin; *SBP* spontaneous bacterial peritonitis; *TCT* thrombin clotting time; *TSH* thyroid-stimulating hormone
1 = venous blood; 2 = arterial blood; 3 = capillary blood; 4 = urine; 5 = cerebrospinal fluid; 6 = ascites; 7 = aspirate
* Guidelines recommend POCT explicitly, if CentLab has a TAT of >60 min
** POCT not available yet, however, reasonable for the „fast rule-out" process if myocardial infarction suspected
*** Generally not available as POCT
**** High-quality POCT methods are under development; for stroke; for multiple trauma and bleeding management, POCT shows absolute utility.

every emergency department and ICU in Germany.

> **Checklist 2**
> Mandatory acute parameters for urgent assessment within 4 hours; if one point applies, this parameter often requires POCT. The parameter …
> - … may be required to diagnose acute or emergency clinical symptoms.
> - … has therapeutic implications.
> - … has clinical utility for assessing the severity of clinical symptoms and to guide further appropriate diagnostics and/or disposition.

As with the emergency parameters, a number of acute biomarkers are recommended by national and international guidelines and firmly anchored in the algorithms for guideline-compliant management of specific diseases.

Natriuretic peptides In the European guidelines on the treatment of acute and chronic heart failure, natriuretic peptides (BNP, NT-pro-BNP, pro-ANP) are recommended as an integral part of the algorithm for assessing presumed acute heart failure, particularly when echocardiography is not available [28].

D-dimer testing The measurement of D-dimer is recommended explicitly in the European guidelines for the diagnosis and management of acute pulmonary embolism to avoid radiation exposure from unnecessary imaging procedures [20].

HIV testing HIV testing can also be classed with acute biomarkers in presumed exposures such as health profession-related vein punctures, needle-stick or cut injuries. The German-Austrian recommendations for post-exposure prophylaxis of HIV infection state that the "consent of the exposed person to an HIV test (in order to document the negative HIV status at the time of exposure) [is] a requirement for PEP [post-exposure prophylaxis]". The use of rapid testing is emphasized here to render the "PEP indication as definitively as possible" [15]. Timely testing by the central laboratory is often not possible, particularly if required outside of regular working hours.

Hemostasis parameters With regard to laboratory diagnostics for managing the severest of injuries, European consensus recommendations [48] include an early coagulation assessment (i.e. in the shock room) (▶ Chapter 6 and 15). This is because up to one-third of all patients admitted to the hospital have preexisting coagulation disorders [24]. These parameters include prothrombin time (PT), activated partial thromboplastin time (aPTT), platelet count and fibrinogen concentration [48]. As these tests only cover the initial coagulation phase, supplementary viscoelastic testing methods or platelet function analyses are recommended [23, 48] to classify the coagulation disorder and enable more targeted therapy (▶ Chapter 15). These complex POCT assays require whole blood samples. Two different devices are currently available for the clinical use [23]. However, despite good evidence, they are not widely used yet.

Blood gas analysis To assess the extent and impact of bleeding, measurements of lactate and base excess are recommended as further acute biochemical markers [48]. ◘ Tab. 21.3 lists the parameters derived from the application of Checklist 2. Here, it becomes clear that the required diagnostics are linked to the determinations carried out in the emergency department, ICU or in interventional/surgical unit. Measurement of fibrinogen, for example, is not required unless care for multiple trauma injuries is provided.

21.2.3 Process-streamlining laboratory parameters

Laboratory parameters are useful for streamlining processes or within a defined treatment pathway. However, they do not prevent immi-

Tab. 21.3 Mandatory acute parameters, required within ≤4 hours

Parameters	Standard material	Analysis	Typical relevance	References
CRP	1	CentLab, POCT*	Infection	
NP	1	CentLab, POCT**	Heart failure	[28]
D-dimer	1	CentLab, POCT***	Pulmonary embolism	[20]
HIV	1	CentLab, POCT****	AIDS, injury	[15]
Alcohol	1; 2	CentLab, POCT#	Intoxication	
Paracetamol	1	CentLab	Intoxication	
Fibrinogen	1	CentLab*****	Multiple trauma	[17]
Blood count	1	CentLab, POCT***	Anemia, infection	

CRP C-reactive protein; *NP* natriuretic peptide; *HIV* human immunodeficiency virus
1 = venous blood; 2 = exhaled air
* POCT depends on TAT of CentLab
** Use as POCT depending on the patient population, medical expertise and further diagnostic options (i.e. echocardiography) on site
*** POCT, if not available in CentLab within 4 hours
**** Following the guidelines of the professional associations, the result of an HIV test (needle-stick injury) must be available within 2 hours; POC tests for this are being developed further
***** In multiple trauma care
# Exhaled air testing sufficient for diagnostics; blood sample necessary for preservation of evidence.

nent threat to life. Process-streamlining parameters are required for certain steps of the process and it makes sense that these should be applied peri-interventionally or in emergency departments. It may not be possible to streamline the processes relating to the management of a certain clinical picture aimed at avoiding over-diagnostics, misdiagnoses or suboptimal hospital stays with the associated risks of nosocomial infections or complications if the test results are not available.

Checklist 3 summarizes the general criteria for the selection of process-streamlining parameters.

Checklist 3

Process parameters for urgent testing: if one point applies, the parameter is imperative and possibly required as a POCT method. The parameter …

- … impacts hygiene measures (example: isolation of patients infected with multiresistant bacteria).
- … has implications for therapy, the initiation of which can be streamlined when an early diagnosis is rendered.
- … supports the differential diagnosis and, under certain circumstances, dictates the treatment pathway and further diagnostics and/or disposition.

Process-streamlining parameters collectively lack urgent implications for therapy, but nonetheless influence further inpatient stay and management (Tab. 21.4). Of particular importance in this context are isolation measures to protect the patient and those around them. Indirectly, the results influence patient flow in the emergency department, as potentially infectious patients, often with resistant patho-

Tab. 21.4 Process-streamlining emergency parameters

Parameters	Standard material	Analysis	Relevance	References
MRSA	1	CentLab, POCT*	Hygiene	[19]
MSSA	1	CentLab, POCT*	Hygiene	[19]
Influenza viruses	1	CentLab, POCT*	Hygiene	[41]
Drug screening	2; 3	CentLab, POCT**	Intoxication	[53]

MRSA methicillin-resistant Staphylococcus aureus; *MSSA* methicillin-sensitive Staphylococcus aureus; *HBV* = hepatitis B virus; *HCV* hepatitis C virus
1 = swabs; 2 = venous blood; 3, urine
* POCT only valuable if a reliable PCR method is available and the screening process has been established.
** POCT qualitatively unsatisfactory to date, possible process-related utility, depending on the TAT of the central lab.

gens, cannot be transferred to a general ward or to a higher-level care institution. Then, isolation in single rooms is required, resulting in high staff and logistics demands and therefore the danger of infection spread [19].

On the other hand, in cases of suspected influenza, diagnostics has direct therapeutic implications. Here, rapid diagnostics can be helpful in at-risk patients or in the case of severe disease progression [41].

In addition to common toxic compounds such as alcohol and paracetamol (Tab. 21.3), the number of cases of intoxication involving illegal substances has increased exponentially [16]. The options for narcotics and medication screening have taken on a new dimension thanks to the introduction of high-performance liquid chromatography (HPLC) and mass spectrometry. The medical or legal implications of test results are crucial to emergency diagnostics [14]. Furthermore, it should be considered that non-targeted screening without a selective narrowing of possible substance use in the patient's medical history is very costly [36] and complex techniques are only available at a few centers.

21.2.4 Procedural aspects of POCT

POCT was initially introduced as a bedside test, whose prototype was blood glucose strip testing. In the meantime and partly as a result of high regulatory requirements (as set out in Germany in the guideline of the German Medical Association, RiliBÄK), advanced devices are used on site as complete IT integrated satellite laboratories. Nowadays classical bedside testing is used less frequently in emergency admissions than in outpatient settings [38].

The integration of fully automated, low-maintenance devices in the **laboratory information system** (LIS) and connection to the **hospital information system** (HIS) or its clinical front-end system (e.g. Ecare in emergency departments) has led to a significant improvement in analytical quality. Hence, POCT results do not usually need to be confirmed or re-tested by laboratory-based methods. Even in an optimal scenario where POCT and central laboratories are cooperatively integrated, increased efforts and increased costs are still present when compared with central sample processing using large-scale devices. Therefore, POCT should only be introduced if there is no viable alternative. General selection criteria can be found in Möckel et al. [30]. Fig. 21.1 illustrates the concept of streamlined POCT implementation.

In principle, the required parameter and its clinically acceptable TAT should be specified. The TAT is dictated by disease- as well as by process-related factors. If the TAT of a test cannot be attained or reliably guaranteed by a cen-

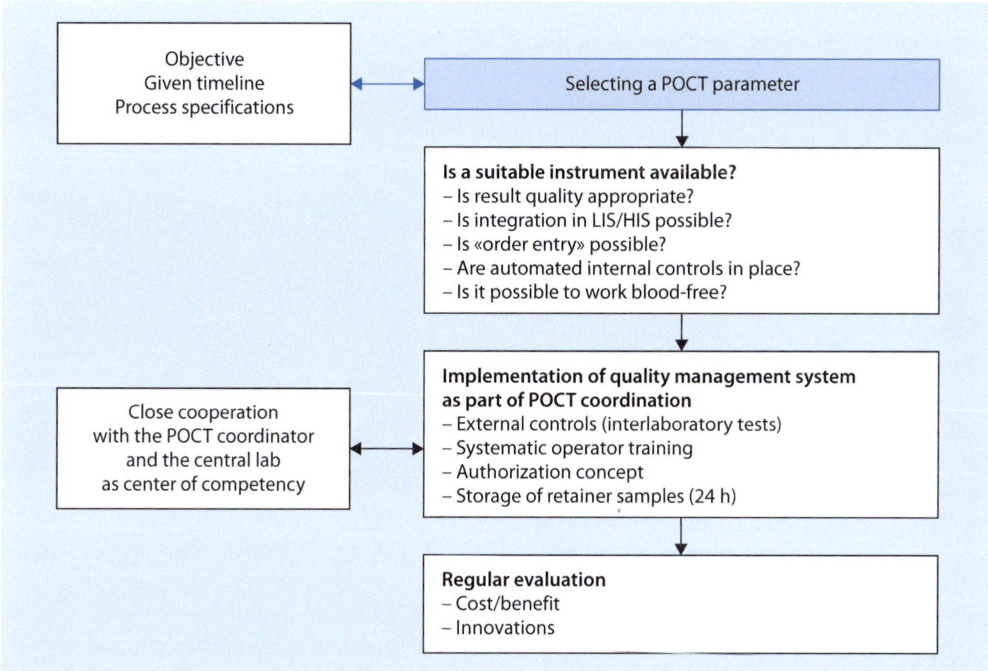

Fig. 21.1 Implementation of POCT methods in the emergency department and ICU

tral laboratory, an appropriate technical POC solution should be sought.

> **Essential requirements for using POCT methods are appropriate quality of results, the possibility of integrating requests ("order entry") and transmission of results into the LIS and/or HIS.**

Automated internal controls are indispensable for **quality assurance** without the need for extra staff. "Blood-free" working is desirable, i.e. the use of closed blood tubes, which reduces the risk of infections by minimizing the need for handling and pipetting. To maintain quality standards for POCT results, the device should undergo additional external inspections after implementation by the POCT coordinator (▶ Chapter 31). Systematic and regular staff training is needed and a clear policy for device use should be established in order to reduce the erroneous result rates due to the device operator error. This can be achieved, for example, by using personalized bar codes that allow only qualified staff to access the device. The storage of retainer samples is also very important to avoid measuring errors, but also to be able to identify any sample mix-ups.

POCT should be closely evaluated and its cost-benefit ratio critically appraised on a regular basis with regard to current innovations (▶ Chapter 31).

21.2.5 Discussion

It has been demonstrated that laboratory parameters for POCT in the emergency department and in peri-interventional intensive care should be selected on the basis of diagnostic necessity, therapeutic implications and general care processes. It is also of decisive importance to consider the aim and possible implications of a biochemical test so that the most appropriate methods can be laid down for the determination of that parameter (by POCT or at a central laboratory).

The patient population and the tasked responsibilities of the healthcare center play a decisive role. For instance, if an integrated chest pain clinic is in operation, there are already published recommendations on what laboratory services must be provided [38, 39]. Examples of general minimum requirements here include electrolytes, creatinine, blood count, CRP, coagulation status, D-dimer and cardiac troponins (ideally measured with highly sensitive assays), with additional recommendation for natriuretic peptide and copeptin. As regards timing, the following is clearly formulated: "For rapid laboratory diagnostics, a 24-hour emergency laboratory service is required. The time from blood draw to result availability should not exceed 45–60 min and must be reviewed regularly. If this is not possible, a POCT device must be used on site to determine cardiac markers. The result must be given quantitatively. The analysis of blood gases must be possible within 15 minutes" [38].

Besides the choice of laboratory parameters, their exact interpretation and streamlined assessment is required. It is vital that there is close cooperation between the emergency/intensive care team and the central laboratory. Here, the POCT methods employed must meet the same high quality standards as the central laboratory's and will usually be managed jointly.

Laboratory profiles

In an emergency, every diagnostic procedure is subject to strict criteria and subordinated to the general process. It is useful if standardized laboratory profiles are requested early, guided by the cardinal symptoms in order to streamline patient throughput times. For example, the emergency department at the Charité, a major hospital in Berlin, has had good experience over many years using this approach. A number of POCT methods are also included in the laboratory profile. With the request order, suitable instruments are entered into the device's work list. The blood is drawn into a small tube labeled with the corresponding bar code and scanned for identification by the device's bar code reader. The result appears with other data from the central laboratory on an (electronic) finding report.

Key messages

- Alongside the clinical examination, ECG and any available radiological imaging, laboratory tests represent a key pillar of the primary patient assessment.
- The choice of laboratory parameters for use in emergency care depends on their direct relevance to treatment, their diagnostic necessity and their utility in the process.
- At present in Germany, the use of POCT is coordinated and organized according to the RiliBÄK guideline. Strictly bedside methods are hardly relevant in emergency departments or in peri-interventional intensive care and are mainly used in outpatient settings.
- POCT is undergoing technological development and its application has to be constantly checked and adapted. With increasing aggregation in the central laboratory sector, POCT is currently growing in importance.

References

1. Asimos AW, Gibbs MA, Marx JA, Jacobs DG, Erwin RJ, Norton HJ, et al. (2000) Value of point-of-care blood testing in emergent trauma management. J Trauma [Internet] 48(6):1101–1108
2. Bochicchio G V, Ilahi O, Joshi M, Bochicchio K, Scalea TM (2003) Endotracheal intubation in the field does not improve outcome in trauma patients who present without an acutely lethal traumatic brain injury. J Trauma 54:307–311
3. Bundesärztekammer (2014) Querschnitts-Leitlinien zur Therapie mit Blutkomponenten und Plasmaderivaten
4. Bundesministerium für Gesundheit. Drogen- und Suchtbericht 2016. www.bmg.bund.de, www.drogenbeauftragte.de
5. Casserly B, Phillips GS, Schorr C, Dellinger RP, Townsend SR, Osborn TM, et al. (2015) Lactate measurements in sepsis-induced tissue hypoperfusion: results from the Surviving Sepsis Campaign database. Crit Care Med [Internet] 43(3):567–573
6. Chalfin DB, Trzeciak S, Likourezos A, Baumann BM, Dellinger RP, Group D-E study (2007) Impact of

delayed transfer of critically ill patients from the emergency department to the intensive care unit. Crit Care Med [Internet] 35(6):1477–1483
7. Christ-Crain M, Jaccard-Stolz D, Bingisser R, Gencay MM, Huber PR, Tamm M, et al. (2004) Effect of procalcitonin-guided treatment on antibiotic use and outcome in lower respiratory tract infections: cluster-randomised, single-blinded intervention trial. Lancet [Internet] 363(9409):600–607
8. Dettmer M, Holthaus C V, Fuller BM (2015) The impact of serial lactate monitoring on emergency department resuscitation interventions and clinical outcomes in severe sepsis and septic shock: an observational cohort study. Shock [Internet] 43(1):55–61
9. Deutsche Hauptstelle für Suchtfragen. Alkohol in Europa. 2006
10. DiSerio F, Petronelli MA, Sammartino E (2010) Laboratory testing during critical care transport: point-of-care testing in air ambulances. Clin Chem Lab Med [Internet] 48(7):955–961
11. Dretzke J, Sandercock Bayliss S Burls A (2004) Clinical effectiveness and cost-effectiveness of prehospital intravenous fluids in trauma patients. Heal Technol Assess [Internet] 8(23):1–118
12. Engoren M (2005) The effect of prompt physician visits on intensive care unit mortality and cost. Crit Care Med [Internet] 33(4):727–732
13. European Stroke Organisation Executive C, Committee ESOW (2008) Guidelines for management of ischaemic stroke and transient ischaemic attack 2008. Cerebrovasc Dis [Internet] 25(5):457–507
14. Fabbri A, Marchesini G, Morselli-Labate AM, Ruggeri S, Fallani M, Melandri R, et al. (2003) Comprehensive drug screening in decision making of patients attending the emergency department for suspected drug overdose. Emerg Med J [Internet] 20(1): 25–28
15. Gemeinsame Erklärung der Deutschen AG, Osterreichischen AG (2009) [Post-exposure prophylaxis of HIV infection. German-Austrian recommendations, update September 2007]. Dtsch Med Wochenschr [Internet] 134 Suppl:S16–33
16. Heyerdahl F, Hovda KE, Giraudon I, Yates C, Dines AM, Sedefov R, et al. (2014) Current European data collection on emergency department presentations with acute recreational drug toxicity: gaps and national variations. Clin Toxicol [Internet] 52(10):1005–1012
17. Hilbert-Carius P, Hofmann G, Stuttmann R (2015) [Hemoglobin-oriented and coagulation factor-based algorithm : Effect on transfusion needs and standardized mortality rate in massively transfused trauma patients]. Anaesthesist [Internet] 64(11):828–838
18. Jage J, Heid F (2006) Anästhesie und Analgesie bei Suchtpatients. Grundlagen zur Erstellung einer "standard operating procedure". Anaesthesist 55:611–628
19. Kommission für Krankenhaushygiene und Infektionsprävention (KRINKO) beim Robert Koch Institut (2014) Empfehlungen zur Prävention und Kontrolle von Methicillinresistenten Staphylococcus aureus-Stämmen in medizinischen und pflegerischen Einrichtungen Empfehlung der Kommission für Krankenhaushygiene and Infektionsprävention (KRINKO) beim Robert Koch-Institut
20. Konstantinides SV (2014) ESC Guidelines on the diagnosis and management of acute pulmonary embolism. Eur Hear J 35:3145–3146
21. Lazarenko GC, Dobson C, Enokson R, Brant R (2001) Accuracy and speed of urine pregnancy tests done in the emergency department: a prospective study. CJEM 3:292–295
22. Lindner G, Exadaktylos AK (2013) [Disorders of serum sodium in emergency patients : salt in the soup of emergency medicine]. Anaesthesist [Internet] 62(4):296–303
23. Maegele M, Inaba K, Rizoli S, Veigas P, Callum J, Davenport R, et al. (2015) [Early viscoelasticity-based coagulation therapy for severely injured bleeding patients: Report of the consensus group on the consensus conference 2014 for formulation of S2k guidelines]. Anaesthesist [Internet] 64(10): 778–794.
24. Maegele M, Lefering R, Yucel N, Tjardes T, Rixen D, Paffrath T, et al. (2007) Early coagulopathy in multiple injury: an analysis from the German Trauma Registry on 8724 patients. Injury [Internet] 38(3): 298–304.
25. Maisel A, Neath SX, Landsberg J, Mueller C, Nowak RM, Peacock WF, et al. (2012) Use of procalcitonin for the diagnosis of pneumonia in patients presenting with a chief complaint of dyspnoea: results from the BACH (Biomarkers in Acute Heart Failure) trial. Eur J Heart Fail [Internet] 14(3): 278–286
26. Marti G, Schwarz C, Leichtle AB, Fiedler GM, Arampatzis S, Exadaktylos AK, et al. (2014) Etiology and symptoms of severe hypokalemia in emergency department patients. Eur J Emerg Med [Internet] 21(1):46–51
27. Mastrovitch TA, Bithoney WG, DeBari VA, Gold NA (2002) Point-of-care testing for drugs of abuse in an urban emergency department. Ann Clin Lab Sci [Internet] 32(4):383–386
28. McMurray JJ, Adamopoulos S, Anker SD, Auricchio A, Bohm M, Dickstein K, et al. (2012) ESC guidelines for the diagnosis and treatment of acute and chronic heart failure 2012: The Task Force for the Diagnosis and Treatment of Acute and Chronic Heart Failure 2012 of the European Society of Cardiology. Developed in collaboration with the Heart. ESC Eur J Hear Fail 14:803–869

29. Möckel M, Müller C, Lindner T Searle J (2016) Laboruntersuchungen – was ist sinnvoll, was ist überflüssig? (Fast diagnostics in the emergency department: Laboratory testing – what we need and what we don't). Dtsch med Wochenschr 141(5):322–328
30. Möckel M, Searle J (2014) [Point-of-care testing in preclinical emergency medicine]. Med Klin Intensivmed Notfmed [Internet] 109(2):100–103
31. Möckel M, Searle J, Hamm C, Slagman A, Blankenberg S, Huber K, et al. (2015) Early discharge using single cardiac troponin and copeptin testing in patients with suspected acute coronary syndrome (ACS): a randomized, controlled clinical process study. Eur Heart J [Internet] 36(6):369
32. Möckel M, Searle J, Huttner I, Vollert JO (2015) Qualitative process analysis and modelling of emergency care workflow and interface management: identification of critical process steps. Eur J Emerg Med [Internet] 22(2):79–86
33. Mohammed AA, van Kimmenade RR, Richards M, Bayes-Genis A, Pinto Y, Moore SA, et al. (2010) Hyponatremia, natriuretic peptides, and outcomes in acutely decompensated heart failure: results from the International Collaborative of NT-proBNP Study. Circ Hear Fail [Internet] 3(3):354–361
34. National Health Service England (2015) A & E attendances and emergency admissions: June 2015 monthly report. Monthly report
35. Newbatt E, Beckles Z, Ullman R, Lumsden MA, Guideline Development G (2012) Ectopic pregnancy and miscarriage: summary of NICE guidance. BMJ [Internet] 345:e8136
36. Pemmerl S (2013) [Initial diagnosis and treatment for poisoning]. Med Klin Intensivmed Notfmed [Internet] 108(6):459–464
37. Pollack Jr. CV (2016) Coagulation assessment with the new generation of oral anticoagulants. Emerg Med J [Internet] 33(6):423–430
38. Post F, Giannitsis E, Darius H, Baldus S, Hamm CW, Hambrecht R, et al. (2015) Kriterien der Deutschen Gesellschaft für Kardiologie - Herz- and Kreislaufforschung für „Chest Pain Units" Update 2015. Kardiologe 9:171–181
39. Post F, Gori T, Giannitsis E, Darius H, Baldus S, Hamm C, et al. (2015) Criteria of the German Society of Cardiology for the establishment of chest pain units: update 2014. Clin Res Cardiol [Internet] 104(11): 918–928
40. Rixen D, Raum M, Bouillon B, Neugebauer E, der Deutschen Gesellschaft für Unfallchirurgie AGP (2002). Der Base-Excess als Prognose-Indikator bei Polytrauma-patients. Anästhesiol Intensivmed Notfallmed Schmerzther 37:347–349
41. Robert Koch Institut (2015) Influenza
42. Roffi M, Patrono C, Collet JP, Mueller C, Valgimigli M, Andreotti F, et al. (2015) 2015 ESC Guidelines for the management of acute coronary syndromes in patients presenting without persistent ST-segment elevation: Task Force for the Management of Acute Coronary Syndromes in Patients Presenting without Persistent ST-Segment. Elev Eur Soc Cardiol (ESC) Eur Hear J. DOI: http://dx.doi.org/10.1093/eurheartj/ehv320 ehv320
43. Ruchholtz S, Waydhas C, Ose C, Lewan U, Nast-Kolb D, German WGMT (2002) Prehospital intubation in severe thoracic trauma without respiratory insufficiency: A matched-pair analysis based on the trauma registry of the German Trauma Society. J Trauma-Injury Infect Crit Care [Internet] 52(5): 879–886
44. Schmidt BMW (2015) The most frequent electrolyte disorders in the emergency department. What must be done immediately? Internist [Internet] 56(7):753
45. Schuetz P, Kutz A, Grolimund E, Haubitz S, Demann D, Vogeli A, et al. (2014) Excluding infection through procalcitonin testing improves outcomes of congestive heart failure patients presenting with acute respiratory symptoms: Results from the randomized ProHOSP trial. Int J Cardiol [Internet] 175(3):464–472
46. Selmer C, Olesen JB, Hansen ML, Lindhardsen J, Olsen AM, Madsen JC, et al. (2012) The spectrum of thyroid disease and risk of new onset atrial fibrillation: a large population cohort study. BMJ [Internet] 345:e7895
47. Simon L, Gauvin F, Amre DK, Saint-Louis P, Lacroix J (2004) Serum procalcitonin and C-reactive protein levels as markers of bacterial infection: A systematic review and meta-analysis. Clin Infect Dis [Internet] 39(2):206–217
48. Spahn DR, Bouillon B, Cerny V, Coats TJ, Duranteau J, Fernandez-Mondejar E, et al. (2013) Management of bleeding and coagulopathy following major trauma: an updated European guideline. Crit Care [Internet] 17(2):R76
49. Stengel D, Porzsolt F (2003) Evidence-based medicine: Randomisierung nicht zwingend erforderlich. Dtsch Ärztebl 100:2145
50. Thygesen K, Alpert JS, Jaffe AS, Simoons ML, Chaitman BR, White HD, et al. (2012) Third universal definition of myocardial infarction. Eur Hear J 33: 2551–2567
51. Van Schaik SM, Scott S, de Lau LM, Van den Berg-Vos RM, Kruyt ND (2015) Short door-to-needle times in acute ischemic stroke and prospective identification of its delaying factors. Cerebrovasc Dis Extra [Internet] 5(2):75–83
52. Wacker C, Prkno A, Brunkhorst FM, Schlattmann P (2013) Procalcitonin as a diagnostic marker for sepsis: a systematic review and meta-analysis. Lancet Infect Dis [Internet] 13(5):426–435
53. Wu AHB, McKay C, Broussard LA, Hoffman RS, Kwong TC, Moyer TP, et al. (2003) National academy of clinical biochemistry laboratory medicine prac-

tice guidelines: Recommendations for the use of laboratory tests to support poisoned patients who present to the emergency department. Clin Chem [Internet] 49(3):357–379

54. Zander BE (2003) Diagnostische and therapeutische Bedeutung von Base Excess und Laktatkonzentration. Anästhesiol Intensivmed Notfallmed Schmerzther 37:343–346
55. Zorbalar N, Yesilaras M, Aksay E (2014) Carbon monoxide poisoning in patients presenting to the emergency department with a headache in winter months. Emerg Med J [Internet] 31(e1):e66–70

Neonatology

Norbert Gässler

P. B. Luppa, R. Junker (Eds.), *Point-of-Care Testing*
https://doi.org/10.1007/978-3-662-54497-6_22

22

22.1 Introduction

In 2014, 715,000 children were born in Germany – a small increase in the trend compared to the 1950s [13]. Of these, approximately 10 % were born prematurely, meaning before the 37th gestational week, often weighing less than 2,500 g. The incidence of premature births as well as that of premature infants with a very low birth weight (VLBW) of <1,500 g has been increasing significantly for years. In 2014, premature births amounted to approx. 8,000 children. VLBW infants, in particular, benefit significantly from recently improved diagnostic and treatment options. Hospital costs for inpatient treatment of infants with a diagnosis of "prematurity/low birth weight" account for over € 700 million per annum [4]. Infants with a birth weight of under 1,000 g, born in the 28th gestational week (extremely low birth weight, ELBW), spend 10–12 weeks on the intensive care unit (ICU) or on a general ward. The costs for an uncomplicated inpatient stay are approx. € 100,000. Morbidity, nursing care needs and mortality rise with increasing prematurity. Although less than 1 % of all infants are born with a birth weight of <1,500 g, diagnostics and laboratory test-guided therapeutic monitoring are particularly extensive in this very group.

Tab. 22.1 gives an overview of the most important laboratory tests and the associated clinical problems seen on a neonatal ICU [10]. Generally, the analysis spectrum is similar to that for adults. Despite this, a number of specific analytical problems exist as a result of the following situations:

- Very often, only extremely small sample volumes are available
- Capillary blood is mostly used
- During the first days of life, high hematocrit values are present in whole blood analyses
- Different measurement ranges are relevant or apply (e.g. for blood glucose, bilirubin and creatinine clearance)
- To some extent, higher concentrations of interfering substances can be present (e.g. bilirubin, Hb, glutathione)

Small sample volumes and the predominant use of capillary blood put neonatology into the POCT domain. Frequent blood draws for therapy monitoring of ventilated premature infants, in particular, are unavoidable. Obviously, bloodless continuous monitoring of metabolic parameters would be ideal. Until now, this was only routinely possible for bilirubin, pO_2, pCO_2 (within limits) and for sO_2 (▶ Chapter 14).

The blood volume of a neonate makes up 7–10 % of its body weight. This means that a premature infant weighing about 1500 g has a blood volume of max. 150 mL. In a very premature infant with a birth weight of <1000 g (ELBW), the blood volume is around 80 mL at the most. Unlike in mature newborns, iatrogenic anemias in premature infants increase erythropoiesis only slightly. Therefore, a larger blood loss may also have to be compensated for with blood transfusions. The hormone erythropoietin is additionally given as treatment in some cases.

Anemias in premature infants have to be cautiously avoided as they can lead to complications and low weight gain.

The demand for intensive metabolic control collides with the concept of minimal handling. Each intervention on a sick, premature infant can worsen its status. Indeed, simply touching the infant increases stress and risk of infection. This means that ideally only as few tests as possible should be carried out, but ones with the greatest conclusiveness: Whether at the POC or in the laboratory, sample taking for various tests should be carefully coordinated and at best timed together. The number of blood draws and the necessary blood volume can, without doubt, be reduced by streamlining the workflow.

The jury is still out on the limit values and therapeutic target concentrations to be set for a number of parameters, particularly in premature infants. An extensive collection of clinical chemistry reference ranges can be found in the book by Soldin [12]. Further important data are to be found in references [5] and [9].

Application of the proper technique to draw capillary blood from the heel in the least traumatic way is the ultimate key to obtaining ther-

Tab. 22.1 Laboratory tests in a neonatal ICU

Tests	Monitored physical status	Clinical problems
Blood gas analysis	Acid-base balance	Acidosis Hyperkalemia Hypoxia
	Breathing	Lung surfactant deficiency Respiratory distress syndrome Apnea attacks
	Circulation, blood pressure	Shock Hypoxia Cerebral bleeding Necrotizing enterocolitis
	Therapy monitoring of ventilated neonates	
Na^+, K^+, Ca^{2+}, Cl^-, Mg^{2+}, PO_4^{3-}, Zn^{2+}, trace elements (partially also in urine)	Renal excretion	Edema Electrolyte balance Acute renal failure
	Diet	Low weight increase Osteopenia of prematurity
Creatinine, urea, bilirubin, protein, blood glucose	Excretion	Acute renal failure
	Metabolism	Catabolism Jaundice Hypoglycemia
Cholesterol, bile acids, bilirubin, alkaline phosphatase	Function of gallbladder, liver and pancreas	Cholestasis
CRP, leukocytes, differential blood count, IL-6, IL-8, PCT	Immunity	Pneumonia Sepsis Meningitis
Therapeutic drug monitoring	Antibiotics	Infection
	Indomethacin	Ductus arteriosus occlusion
	Digoxin	Heart failure
	Phenobarbital	Seizures
	Caffeine	Respiratory insufficiency
Hb, hematocrit, transferrin, ferritin	Erythropoiesis	Anemia
Global coagulation tests, platelet count	Immature hemostasis system, therapy monitoring	
Urine test strips, urine status	Urinary tract infection	Sepsis
Swab with Gram staining	Bacterial infection	Sepsis

apeutically evaluable data. In particular, a sufficient blood volume (max. 600 µL) should be taken without squeezing the tissue to avoid hemolysis that interferes with the measurement of a number of parameters (e.g. K^+ or bilirubin). See also Fig. 4.1 in ▶ Chapter 4.

The following parameters can presently be measured by POCT on a neonatal general ward or ICU: pH, blood gases, Hb, Hb fractions, electrolytes, metabolites and global coagulation. Metabolites can be measured with dedicated devices designed for single-parameter assessments (e.g. blood glucose, bilirubin) or using device configurations that can measure various parameters simultaneously – partially with options for individual analysis – e.g. blood gas analyzers (▶ Chapter 6). The required sample volume is the most important selection criteria when choosing a blood gas analyzer. However, a flexible selection of parameter options and an integrated automated quality control should also be considered. Analytical criteria should be discussed with the POCT coordinator in the central laboratory.

The currently unmet needs for further development of POCT in the field of neonatology do not differ from other disciplines: smaller, simpler, faster, less sample material and ability to measure additional parameters, particularly some inflammatory markers such as cytokines. Jocelyn M. Hicks, one of the prominent representatives of POCT in pediatrics, predicts a complete transition from pediatric laboratory diagnostics to noninvasive test methods and analysis with biosensors [8]. However, if and when this vision will become reality is not yet clear.

22.2 Blood glucose monitoring

In the first hours of life, newborns have transient low blood glucose levels with values typically ranging from 40–60 mg/dL. Particularly in premature infants, values can fall as low as 10 mg/dL or lower without necessarily triggering significant clinical symptoms. Irrespective of controversial discussions about the definition of neonatal hypoglycemia, a concentration of 40 mg/dL is generally seen as a critical cut off for therapeutic action [7]. As severe hypoglycemia can cause neurological damage, methods used for blood glucose monitoring in neonatology have to be reliable, particularly in the range of 20–80 mg/dL and should be available within the shortest period of time. Interference by hematocrit is common in many POCT blood glucose analyzers. In the past, it caused great difficulties in measuring blood glucose correctly, particularly in the lower concentration range. Since then, devices have been developed that compensate for this electronically (▶ Chapter 5), making them suitable for use in neonatology.

22.3 Bilirubin determination

In approximately 60 % of all healthy newborns, a physiological jaundice occurs in the first few days after birth, with bilirubin rising up to a concentration of 12 mg/dL. It reaches its maximum around the 4th day of life only to drop continuously thereafter. For a variety of reasons, however, a small number of newborns develop a dangerous hyperbilirubinemia. Given the associated risk of bilirubin encephalopathy, also called kernicterus, such hyperbilirubinemia requires treatment. The Bhutani nomogram [1, 2, 3] has been largely established for deciding whether phototherapy or exchange transfusions are indicated. Other empirical nomograms are also used, e.g. for infants with additional risk factors such as premature birth, sepsis, hemolysis among others.

In general, the plasma of jaundiced newborns almost always contains unconjugated (or indirect) bilirubin. Depending on the measurement method, up to 1.5 mg/dL bilirubin is detectable as "direct" bilirubin in neonatal samples. Due to a lack of specificity of the measurement method, this fraction is not conjugated (glucuronidated) but, in most cases, unconjugated bilirubin. Higher amounts of conjugated bilirubin only occur in neonatal hepatitis, chronic hyperbilirubinemia due to marked erythroblastosis or in premature infants with toxic liver damage – mostly caused by medication or longer lasting parenteral feeding – but also in some other very rare diseases. Only total bilirubin can be measured at the POC. Methods to differentiate direct from indirect or conjugated from unconjugated bilirubin do not exist. Three different methods

are available to determine total bilirubin in neonates:
- Transcutaneous measurement
- Direct photometry in undiluted serum/plasma (bilirubinometer)
- Direct photometry in blood (CO oximetry module in blood gas analyzers)

22.3.1 Transcutaneous measurement

For transcutaneous bilirubinometry (▶ Chapter 11), two devices have been in use for some time now. Reflection densitometric measurements of the yellowish discolored skin at the forehead or over the sternum are used, with corrections being made for the intrinsic skin color. Transcutaneous bilirubin measurement is widely used for screening in both outpatient and hospital settings. In order to ensure reliable decision-making about the use of phototherapy, however, a "bloody" bilirubin determination is required if the transcutaneously measured value is 2–3 mg/dL below the respective cut off for phototherapy. After phototherapy, further measurements are carried out exclusively on the blood.

Many neonatal departments use a simple variant of direct spectrophotometry for determinations in the plasma (▶ Chapter 8). Bilirubinometers are filter photometers, where the absorption of plasma is measured at 455 nm, i.e. near the absorption maximum of bilirubin. The spectral interference of Hb is compensated for by performing a second measurement at 575 nm. Generally, a blood sample of 20–30 µL is sufficient. POCT of neonatal bilirubin is further simplified by the Bilimeter 3 (Pfaff Medical, Neuburg-Schrobenhausen, Germany).

Some blood gas analyzers allow direct spectrophotometric measurement of total bilirubin in whole blood (▶ Chapter 14). While the devices from Roche and Siemens are only designed for measuring bilirubin in newborns, the devices from Radiometer and IL can also measure adult samples. The measurement principle is identical for all devices [6]: The CO-oximetry module determines bilirubin in addition to Hb fractions by multi-wavelength measurements. Although the spectra of bilirubin and Hb differ greatly, high demands are placed on the measurement due to the large difference between bilirubin and the "interfering" Hb concentration. The bilirubin concentration is calculated from the absorption measurements, using multi-component analysis. Usually 35–100 µL of blood is sufficient for a measurement. The measuring range for most devices is between 3 and 30 mg/dL.

Note

The transcutaneous bilirubin assay is a mere screening method; its results need to be checked with a different method before any therapeutic intervention is undertaken [11]. This can be performed at the POC on undiluted plasma using photometry (bilirubinometers) or in blood using blood gas analyzers. Wet chemical checks are only necessary in exceptional cases. Recommendations about timing and frequency of bilirubin measurements – either transcutaneously or in blood – can be found in current guidelines [2].

References

1. American Academy of Pediatrics (2004) Management of hyperbilirubinemia of the newborn infant 35 or more weeks of gestation. Pediatr 114: 297–316
2. Berns M, Bührer C (2010) Leitlinie zur Hyperbilirubinämie von Neugeborenen. Monatsschrift Kinderheilkunde 158: 1117–1124
3. Bhutani VK, Vilms RJ, Hamerman-Johnson L (2010) Universal bilirubin screening for severe neonatal hyperbilirubinemia. J Perinatol 30(Suppl): S 6–15
4. Gässler N (2007) POCT in der Neonatologie, Seminar 412, "Point-of-Care-Testing (POCT) – Aktueller Stand" auf der 39. MEDICA, 17.11.2007, Düsseldorf
5. Green A, Morgan I, Gray J (eds) (2003) Neonatology & Laboratory Medicine. ACB Venture Publications, London
6. Hallemann H, Putz K, Schweiger G, Spitzer S, Strohmeier M (2005) Technical aspects of bilirubin determination in whole blood. Point of Care 4: 9–10

7. Haymond MW (1989) Hypoglycemia in infants and children. Endocrin Metab Clin North Am 18: 211–252
8. Hicks J (2005) Point-of-care Testing: Keynote lecture at the AACC Symposium Oak Ridge Conference, 20 April 2005, altimore, MD
9. Mäkelä E, Takala T, Suominen P et al. (2008) Hematological parameters in preterm infants from birth to 16 weeks of age with reference to iron balance. Clin Chem Lab Med 46: 551–557
10. Schlebusch H, Fahnenstich H, Niesen M (1993) Laboratoriumsuntersuchungen bei Frühgeborenen. DG Klinische Chemie Mitteilungen 24: 47–67
11. Schweizerische Gesellschaft for Neonatology (2006) Abklärung und Behandlung von ikterischen Neugeborenen ab 350/7 Schwangerschaftswochen. Paediatrica 17: 26–29
12. Soldin SJ, Brugnara C, Wong EC (eds) (2007) Pediatric Reference Intervals, 6th edn. AACC Press, Washington, DC
13. Statistisches Bundesamt, Wiesbaden. www.destatis.de

High-performance and elite sports

Silvia Achtzehn, Holger Broich, Joachim Mester

P. B. Luppa, R. Junker (Eds.), *Point-of-Care Testing*
https://doi.org/10.1007/978-3-662-54497-6_23

23.1 Introduction

Alongside complex laboratory methods, POCT devices have been used for many years to analyze biomarkers in the field of high-performance and elite sports. The aim of such testing is to optimize training routines and assess endurance- and regeneration-related processes in order to prevent overexertion reactions in athletes [56]. Because the spectrum of biomarkers measurable without specialized medical and laboratory personnel has been continually expanded, high-performance and elite sports have also benefitted from the continual development of POCT. The concept of POCT has a clinical origin and has hardly been used in sports settings, which is why there has been a paucity of publications to date on studies in high-performance and elite athletes where POCT is explicitly mentioned as measuring methodology. For this reason, the present chapter aims to provide an overview of the different areas that studies have focused on where POCT devices were used in high-performance and elite sports.

23.2 Areas of application for POCT

23.2.1 Capturing data on health and performance status

Depending on its extent and intensity, training can lead to acute physiological reactions and chronic adaptation processes which are intended to enhance athletic performance. The athlete's healthy constitution is a prerequisite for good performance potential and its enhancement by means of the different training models. Nevertheless, during both the training and competition phases, physical stress situations occur which can jeopardize an athlete's health.

To capture complete data on the health status of an athlete, the full range of routine diagnostics is certainly needed. POCT can likewise be used to measure several biomarkers for "near-athlete" evaluation of the health status in a time-critical fashion and independently of any central laboratory, meaning directly at the location where training is taking place. In this context, **inflammation** and **iron deficiency markers** are primarily in focus. Inflammatory processes need to be detected in an early state because they may prevent enhancement of physical performance or diminish performance status. Given that iron metabolism is particularly promoted by an elevated iron turnover and an elevated synthesis of iron-containing proteins, iron metabolism in high-performance and elite athletes should be checked regularly. An elevated iron requirement develops due to the buildup of additional hemoglobin mass and increased iron loss through sweat and urine (hematuria), by hemolysis (footstrike hemolysis) and micro-hemorrhages (in the gastrointestinal tract) [24, 44]. An elevated demand and/or loss of iron in conjunction with an inadequate supply of dietary iron can cause high-performance athletes to suffer from iron deficiency. In turn, hemoglobin synthesis is inhibited, thereby reducing the blood's ability to transport oxygen. Oxygen uptake plays a major role, especially in endurance sports, and represents a limiting factor in physical performance potential [6].

For over 60 years now, **lactate diagnostics** has served as an important parameter for capturing athletes' performance status. By means of various stress tests, an exercise-dependent lactate performance curve is generated and different biological models applied to assess an athlete's performance potential. Calculated thresholds are defined as the stress boundaries between purely aerobic and partially anaerobic lactic acid energy provision and thus serve as indicators of endurance performance capacity [23, 29]. Currently, the validity of the different threshold concepts is the subject of controversial debate due to the different influencing factors, with lactate being attributed with a positive signaling action in physiological adaptation processes [16, 18, 61]. That said, these concepts are employed regularly at performance sport centers and in sport science.

23.2.2 Training optimization, stress and regeneration

This refers to the long-term sports medical and sport scientific management of both individual high-performance and elite athletes and entire sport discipline-specific teams of players. This management can take place before, during and after competitive phases but also in studies on training interventions carried out under laboratory conditions.

When coaching and steering the training of individual athletes or entire teams, it is best to employ test methods and laboratory values that are specific to that sports discipline to obtain conclusive data on training effects, stress, fatigue levels and/or the degree of recovery and regeneration [35]. In **endurance athletes,** the emphasis of training is primarily placed on elevating aerobic energy provision. In this context, the training sessions can take place under both standardized conditions in the laboratory (e.g. on the treadmill or bicycle ergometer), out in the field or in one of the many domestic or international trainings centers. Particularly athletes in endurance sports will, moreover, benefit from an altitude exposure that helps them elevate their performance potential. By contrast, the training of **athletes in strength and speed-strength sports** is designed more to stimulate muscle growth and/or improve the body's ability to induce explosive muscle contractions under cyclic and acyclic conditions.

The level of physical exertion thus varies immensely depending on the type of training. Attempts to quantify these levels are undertaken by means of simple laboratory methods. For this, the frequency of POCT use keeps growing and offers multiple advantages during any long-term management of athletes. The relatively easy handling of POC tests, their low susceptibility to failure and their low-maintenance makes them readily accessible to both trainers and sport scientists. Preferably, devices are used that require low volumes of blood, i.e. allowing capillary blood draws. The testing of capillary blood samples, in turn, allows blood draws at close time intervals (ideally daily) throughout the entire period of training management, without having to excessively burden the athletes with venous blood draws. The transportable POCT devices can accompany the athletes in every setting before, during and after a training or competitive period and at the training camp, while allowing a time-critical assessment of the measured results.

POCT can also be used to regularly monitor the health status as well as long-term sport management measures. To obtain information relating to the effects of training on the aerobic performance capability in endurance athletes, lactate performance diagnostics are carried out together with the measurement of **total hemoglobin mass** (tHb). This analytical approach significantly correlates (r = 0.97) with performance capability and/or maximum oxygen uptake [52]. In endurance, strength and speed-strength sports, metabolic and skeletal muscle markers of POCT can be used to monitor both stress and exhaustion as well as the degree of recovery. Additional information relating to an athlete's exertion status can be supplied from noninvasive saliva samples by determining biomarkers of immunity and acute or chronic stress. This is likewise feasible with POCT devices. An overview of the parameters relevant during a training management session is provided by the diagnostic triangle (◘ Fig. 23.1).

Training intervention studies can be designed to investigate physical loading during and after various interventions as well as their acute and chronic effects on physical performance capability. Furthermore, variously designed recovery phases are the subject of studies aimed at verifying different regeneration strategies. The main areas of emphasis in studies from recent years have been high-intensity training (HIT) vs. high-volume training (HVT), altitude training, electromyostimulation (EMS), vibration and blood flow restriction (BFR) training, compression clothing as well as active and passive recovery phases.

Most interventional studies are conducted in the laboratory under standardized conditions and the effects measured on various physiological levels using complex laboratory

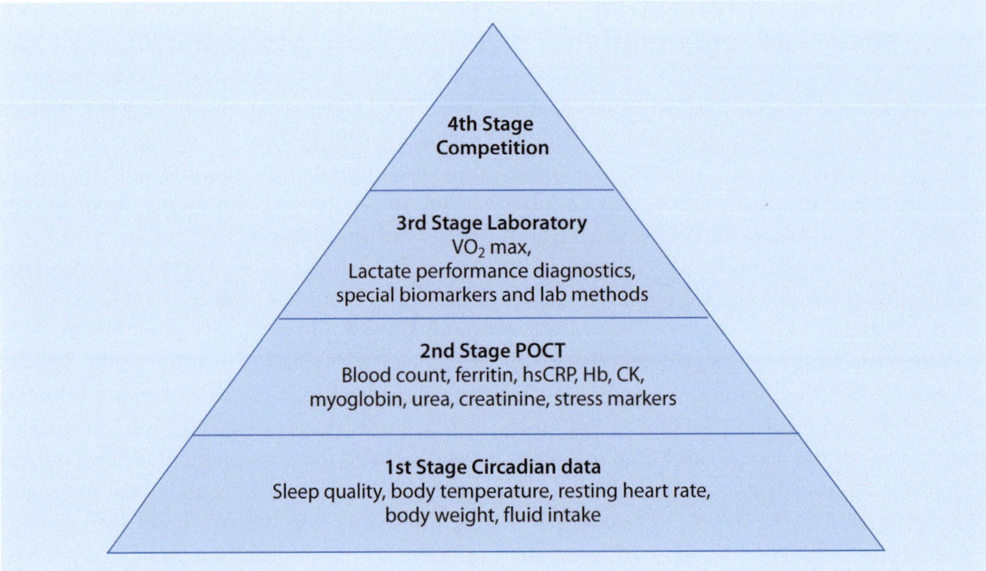

Fig. 23.1 Diagnostic triangle

methods. But also POCT methods are used and help to measure the previously mentioned biomarkers. In addition, myocardial markers are also the subject of interventional studies because they can provide data on myocardial and muscular loading. The analysis of saliva samples can provide conclusions about the states of stress or any adaptation to different training interventions.

In interventional studies, the use of POCT also offers the advantage that a minimally invasive or noninvasive sample collection can be conducted at very fine time intervals without any burden on the athletes. The sample collections are thereby carried out shortly before the intervention, several times within a few hours and up to several days after the exertion, while particularly allowing the different kinetics of the various biomarkers to be captured.

23.2.3 Injury prevention and individualized profiles

Depending on the extent and intensity of their training sessions and during competitive phases, high-performance and elite athletes can suffer significant biomass destruction. The stress and damage to skeletal muscle tissue is of particular interest here. To a certain degree, the stress triggers adaptation processes and produces higher functionality in muscles [10]. However, if this level is exceeded, the athlete runs the risk of injury. Assessment of the degree of muscular damage and the subsequently obtained knowledge about states of exhaustion, risk of injury or any targeted level of recovery has been the focus of scientists and trainers for several years now [35]. Metabolic and muscular stress markers are collected for this purpose and, ultimately, to prevent injury. The excursion of "**muscle damage markers**" has been investigated in many studies and their measurement by POCT is currently finding broad use in many disciplines. The level of the increase in these biomarkers can serve as an index of muscle damage resulting from chronic or acute mechanical loading [3, 8]. The interpretation of these data should provide time-critical decision-making aids for setting duration and intensity of training sessions as well as for assessing the performance capacity during a competitive event. However, this is only possible when monitoring is conducted at intervals of several days and laboratory results like those obtained by POCT are rapidly available.

In a further step, the results are assessed based on individual profiles given that individual athletes as well as athletes playing in team sports train according to individualized training regimens, with each being subjected to their own individual stresses in the competitive situation. These are mostly detached from the reference levels that were gathered from an "untrained" sampling of the normal population [20]. The more measurements the profiles include and the more samples can be collected weekly, the more powerful the conclusions. Multiple measurements are only feasible by use of POCT.

Advantages of POCT in high-performance and elite sports

- POCT devices can be operated and maintained without any medical technician's training.
- No medical personnel is needed to be available to take the minimally invasive capillary blood samples or non-invasive saliva samples.
- Sample collection is possible at close intervals thanks to the low sample volumes required in conjunction with minimally invasive or non-invasive sample collection.
- POCT is implementable in every setting and at the site of the athlete's performance (laboratory, training center, training camp, competition venue).
- "Athlete-focused" results allow both direct assessment of physical exertion as well as time-critical monitoring of training.
- Method-related differences in the measurement results can be minimized and training effects more effectively detected when transportable POCT devices (the same measurement method) accompany the athletes in every setting.

23.3 Areas of investigational focus

23.3.1 Inflammation

Examples of inflammatory markers that can be measured by POCT in the high-performance and elite sports include **leukocytes** (WBC) and **C-reactive protein** (CRP). The methods for CRP measurements must be highly sensitive (hsCRP) in order to detect even the minutest fluctuations. On the one hand, inflammatory processes can be detected early and at the site of training in case the athletes show symptoms or states of exhaustion. On the other hand, it is also known that intensive physical exertion can activate inflammatory processes [43, 44] that can be investigated and quantified with the help of POCT in every training setting and in training studies. The literature cites innumerable such studies on this subject: To date, however, none have reported on POCT as a measurement method. After an ultramarathon (130 km in two days), for example, the WBC (mean ± SD) of $4.9\pm0.2\times10^9$/L (baseline) were elevated to $11.8\pm0.8\times10^9$/L (after crossing the finish line) [2]. During a 100-km ultramarathon, the WBC increased after 75 km to $>14.0\times10^9$/L and 14 hours after the end of the race, elevated hsCRP levels of >5.00 mg/L were demonstrated [27]. After an (Ironman) triathlon, WBC levels were detected that were elevated by 237°% (directly thereafter) and 56°% (a day later) and hsCRP levels that were elevated by 543°% (directly thereafter) and 7,702°% (one day later) [40]. One study on footballers who played three games within one week was conducted to gather data on the required time to achieve recovery: elevated WBC were measured directly after the three games ($6.5\pm0.1\times10^9$/L as baseline vs. 12.9 ± 0.3; 13.7 ± 0.3; $13.2\pm0.3\times10^9$/L). One day after the games, elevated CRP levels were demonstrated (0.9 ± 0.0 mg/L at baseline vs. 3.5 ± 0.1; 4.1 ± 0.1; 3.7 ± 0.1 mg/L) [38].

During interventional studies aimed at investigating the effects of different training protocols on physical performance capability, likewise inflammatory markers are collected in order to be able to estimate the hormonal and metabolic stress of different stress protocols. For several years, these tests have centered on the different effects of HVT vs. HIT (sometimes also referred to as high-intensity interval training or HIIT), with the latter being of much shorter duration, but very high intensity of the intervals. In one study, for example, elevated WBC were measured three hours after performing an HVT protocol on the treadmill (two hours at 55°% of VO_2max) and after a HIT protocol (four times 30-second intervals until total exhaustion ("all out") with 7.5 minutes active pause) (5.6±1.7 vs. 8.0±1.9 and 5.0±0.6 vs. 8.4±1.7×10^9/L) [32].

23.3.2 Iron deficiency

Noteworthy markers of an iron deficiency that can be measured by POCT include **hemoglobin (Hb) concentrations**, **hematocrit** (Hct) and the iron storage protein **ferritin** (FER).

In iron deficiency, FER is lowered initially, whereas lowered (anemic) Hb and Hct levels are not detectable until the iron deficiency is already clinically detectable. Therefore, FER is the parameter most commonly used to assess iron metabolism in athletes. Even when Hb and Hct are normal, lowered FER levels can already point to a pre-latent iron deficiency. In studies on high-performance and elite athletes, FER is not only measured for its mere comparison with reference levels. In longitudinal studies, FER was also monitored regularly to determine if an elevated need for iron is given during intense physical exertion and to ensure a balanced diet. Multiple studies have demonstrated that lowered FER levels can diminish performance potential [11]. In one of the author's own studies on athletes from different sport disciplines, 13.9°% of the male and 40.2°% of the female athletes also had lowered FER levels (≤ 30 mg/dL) irrespective of their Hb levels [1]. In the literature, differences relating to the type of sport practiced are also described. Runners exhibit significantly lower FER levels than cyclists, female rowers lower levels than female basketball players, long-distance runners lower levels than short- and medium-distance runners and professional athletes lower levels than recreational athletes. A connection to the extent or intensity of training can therefore be inferred. It was also demonstrated that FER decreases over the course of a training session, suggesting an elevated iron requirement (summarized in [1]).

Notably, CRP should be monitored in athletes with normal FER and lowered Hb levels because FER increases in inflammatory processes and can thus mimic sufficiently repleted iron stores. Acute extreme physical exertion can also activate inflammatory processes and lead to short-term elevations in FER. After a two-day ultramarathon (130 km distance), the FER levels increase, for example, from 47±7.7 µg/L (pre-marathon) to 78.5±1.8 µg/L (day 3) intercurrent with a minor increase in CRP levels and markedly elevated interleukin-6 levels [2]. For this reason, POCT devices per se offer the combined measurement of FER and CRP.

In addition, CRP can be measured in several POCT devices concurrently with hematological parameters. Here, it must be pointed out that a very broad spectrum of POCT devices is available for measuring the hematological and biochemical markers presented here and in the following. The range extends from small devices for measuring individual parameters to benchtop devices that can be used to conduct variously targeted single or panel tests. It is left to the discretion of the user to decide which device is the most suitable to meet the test requirements. In a sport science context, the greatest role is played by the requirements for necessary sample material (venous or capillary) and the required sample volumes, transportability and parametric spectrum. Here, consideration must be given to the purchase costs and price per measurement, which also dictates sample frequency and can vary greatly in the described area of application (◘ Fig. 23.2).

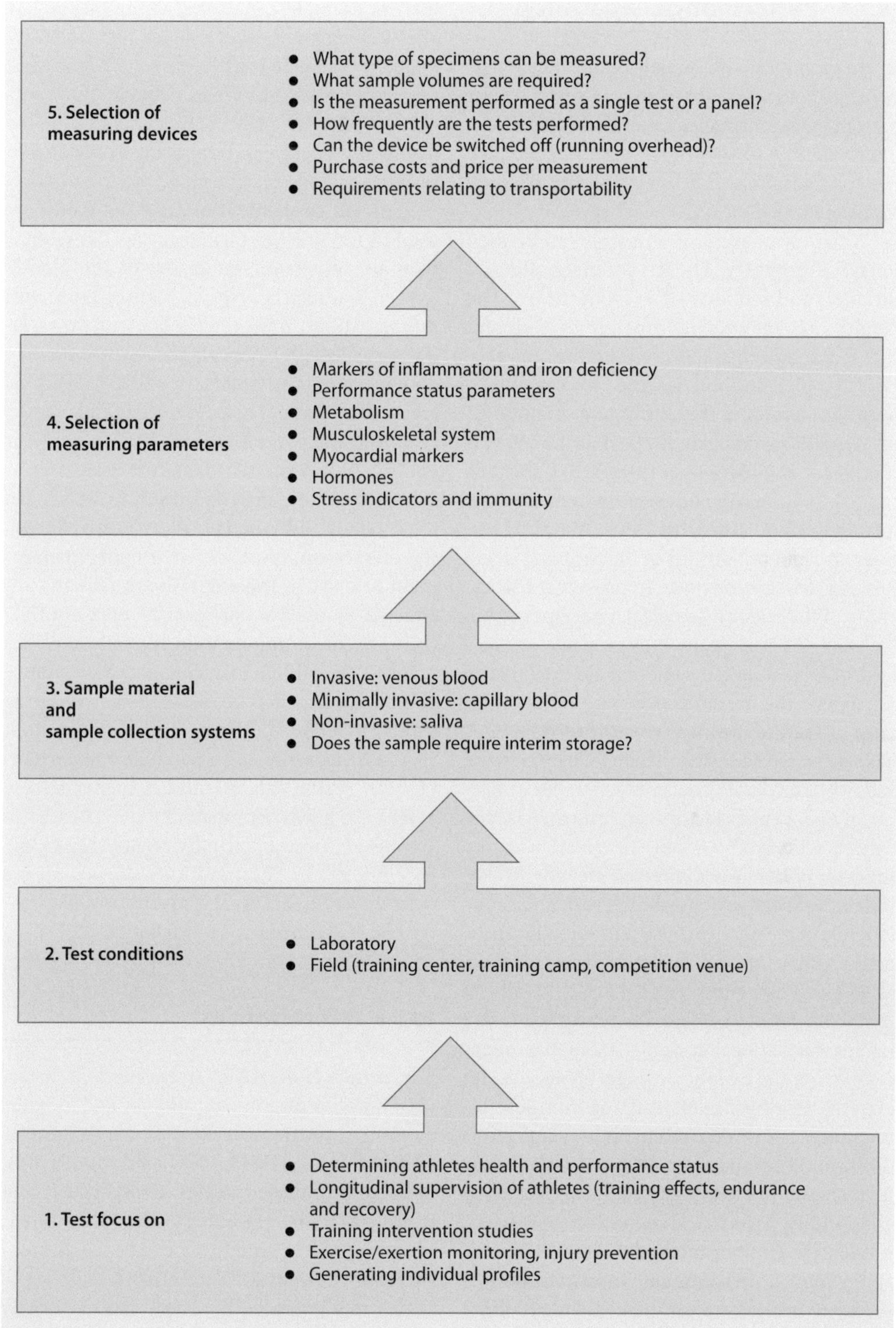

Fig. 23.2 Decision-making aids for POCT devices

23.3.3 Performance capability

Spiroergometry with lactate performance diagnostics belongs to the standard program for measuring performance potential. In addition, biomarkers of oxygen transport capacity, like Hb concentration, Hct levels and the tHb, are monitored.

A variety of exercise protocols can be used in **spiroergometry**. The test protocols differ in duration and intensity of exertion levels. The results are analyzed according to different threshold concepts and computation models. For example, these are designed to compute an **aerobic/anaerobic threshold** (e.g. 4 mmol/L) of energetic metabolism [31]. After the calculation, recommendations with regard to the output (e.g. watt or m/s) or exertion intensity (e.g. heart rate) are issued for future training sessions. A right or left shift of the thresholds will provide information on an improvement/worsening of the endurance performance potential.

Over the past years, however, new models have also been investigated that are designed to document the accumulation and elimination rates of **lactate** during physical exertion. This has been spurred by the debate focusing on the validity of threshold concepts given that a number of endogenous factors can influence lactate concentrations; also, a new perspective on the function of lactate has emerged recently. Today, lactate is viewed as a metabolite with an important role in metabolism, now additionally attributed a controlling and regulating signal function in adaptive tissue responses [18, 61]. In all test protocols, lactate performance diagnostics are mainly carried out on capillary blood samples taken from the earlobe or fingertip and, under extreme time constraints, ranging from several-minute intervals (depending on the staged test) down to 30 seconds (e.g. Wingate test). These tests should possess high measuring accuracy and validity, particularly in low lactate concentrations [17]. Minor methodology-related variability in measurements can impact the performance or intensity recommended for a training session. Depending on the test protocol, however, high lactate concentrations are evaluated that, for example, can exceed 20 mmol/L lactate after implementation of a short, high-intensity regimen (Wingate test). In these test protocols, the numerous blood draws taken in quick succession likewise place high demands on the measuring devices to be quickly available and ready to take measurements.

The **Hb concentration** and **Hct value** are used to monitor performance capability, since they are important indicators of the blood's oxygen binding capacity. As is known from numerous studies, however, Hb does not necessarily correlate with performance potential and/or maximum oxygen uptake ($r = 0.25$ [51]). This is based on changes in the overall blood volume (BV) and the plasma volume (PV) as acute or chronic physiological adaptation processes to intensive or long-duration training, which can affect relative Hb concentration (g/dL) [47, 49]. For this reason, sport science training management primarily measures tHb but remains unaffected by the PV changes. To measure tHb, the CO rebreathing method is employed: This involves the athlete inspiring a low amount of CO, which can then be measured as CO-Hb in capillary blood, allowing conclusions about tHb to be drawn using a computational model [12, 50]. Using this method, a higher tHb of 1,048±126 g was demonstrated in triathletes as compared to untrained individuals (913±133 g). By comparison, the triathletes showed an Hb concentration of 14.6±0.9 g/dL and the untrained individuals of 15.3±0.8 g/dL [23].

23.3.4 Metabolism

The primary metabolic stress markers measured both for monitoring athletes and in interventional studies on POCT are creatinine (CREA), urea (UREA/BUN), bilirubin (BIL), blood gases and electrolytes. **CREA** is measured in athletes both to monitor the renal status regularly as well as acutely after extreme physical exertion subjecting the electrolyte balance to high stress levels. CREA levels can rise briefly after intense exertion, only to drop again over the course of an entire competitive season [3].

After a cycling race across the Alps that covered a distance of 525 km at an altitude of 300–2750 m above sea level on a circuit including 11 mountain passes, CREA levels were elevated, e.g. from 0.95±0.17 mg/dL to 1.26±0.21 mg/dL. Twenty-four hours later, they had returned to baseline (0.94±0.17 mg/dL) [41].

UREA or **BUN** are similarly measured in combination with CREA for renal monitoring [3]. Unlike CREA, UREA does not return to baseline until later after strenuous exercises. In the above-mentioned cycling race, the pre-exertion UREA levels were 35±8 mg/dL and 68±24 mg/dL directly thereafter. Twenty-four hours later they were still at 48±17 mg/dL. By contrast, after 24 hours, the CREA levels had returned to baseline (0.95±0.17 vs. 1.26±0.21 and 0.94±0.17 mg/dL), respectively) [41]. After a 130-km ultramarathon lasting two days, the BUN levels rose from 16.7±1.0 to 23.4±1.2 mg/dL and were still elevated one day later (21.96±1.25 mg/dL) [2].

BIL levels are closely connected to the physiological turnover or destruction (hemolysis) of RBC. In high-performance athletes, hemolysis of RBC can be triggered for different reasons (footstrike hemolysis, mechanical destruction from impact with the ground and muscle contractions, oxidative membrane damage) [3]. After the previously mentioned 130-km ultramarathon, the total bilirubin (T-BIL) concentrations rose from 0.51±0.03 mg/dL to 1.12±0.08 mg/dL after crossing the finish line again up to 1.51±0.12 mg/dL by the next morning [2]. An elevation in T-BIL from 1.0±0.1 to 3.1±0.4 mg/dL was measured in ultramarathon athletes taking part in a "Spartathlon" (which lasts between 32 and 36 hours) [53]. BIL measurements can thus be used as an indicator of an elevated RBC conversion rate.

In high-performance and elite sports, CREA, UREA/BUN and BIL are currently measured, e.g., by applying the Reflotron (Roche Diagnostics, Mannheim, Germany) or the Spotchem (Arkray Inc., Kyoto, Japan) clinical chemistry analyzers. The advantage of this lies in the targeted measurement of one individual parameter in a capillary blood sample. More modern devices offer an array of panels that contain additional (albeit non-relevant) parameters for high-performance and elite sports, but they are accordingly more expensive.

Another investigative field constitutes the measurement of parameters such as blood gases, acid-base balance, buffer systems, pH and electrolytes. These parameters help assess the metabolic regulatory systems under high physical loading and record athletes' overall stress tolerance. The analyses include both interventional studies (e.g. SO_2 in altitude training) as well as complex performance diagnostics (e.g. spiroergometry, repeated sprints, Wingate test). For example, investigations aim to find out whether high physical exertion causes metabolic acidosis (measurable by a drop in pH, base excess and bicarbonate concentration), which in turn, can lead to a reduction in physical performance capability [21]. This can be conducted both in true training and competitive situations as well as in interventional studies with different training protocols.

In one study that compared HIIT with HVT, higher metabolic stress was demonstrated by the Wingate test (HIIT) as compared to a 130-minute protocol (HVT, at 55°% of peak power output). The mean minimal pH of all subjects across several measuring time points was 7.19±0.06 in the Wingate test and 7.39±0.01 in the 130-minute protocol. The base excess values were −15.4±3.5 vs. −1.6±0.0 mmol/L [63]. A positive role in adaption processes has been attributed to the acute metabolic changes occurring during and after high-intensity interval training. During extreme physical exertion, loss of fluids due to sweating can lead to dehydration. Additionally, electrolytes are lost in the sweat. Electrolyte measurement has likewise been the subject of many exercise science studies and is possible in real settings thanks to POCT methods. In one study on ultra-endurance swimmers, for example, the i-STAT (Abbott) demonstrated a loss of sodium (hyponatremia). The pre-race and post-race values were 140.0±1.6 mmol/L 137.8±2.5 mmol/L for male and 139.3±1.7 and 135.2 mmol/L for female swimmers, respectively [60].

23.3.5 Musculoskeletal system

Examples of musculoskeletal stress markers measured in high-performance and elite sports include aspartate aminotransferase (AST), alanine aminotransferase (ALT), lactate dehydrogenase (LDH), creatine kinase (CK) and myoglobin (Mb). These biomarkers can be measured with POCT devices because they also occur in hepatic tissue and cardiac muscle and thus belong to the spectrum of parameters used in clinical emergency diagnostics.

Depending on its scope and intensity, continual training – as well as the extreme exertion of competition phases – can lead to muscle injuries. The muscle damage is caused both by mechanical and metabolic influences that, in turn, damage the muscle membrane triggered by direct or indirect factors. As a result, intracellular muscle components can migrate into the extracellular space [8]. During this process, the levels of enzymes and proteins from skeletal muscle tissues physiologically measurable in the serum can increase to varying extents, depending on the degree of injury to the muscle membrane. The measurement of musculoskeletal stress markers is designed for the assessment of the extent of muscular injury as well as the assessment of a recovery phase and regeneration.

AST, ALT, LDH, CK and Mb represent the classic measurands of muscle stress in high-performance and elite sports. Numerable studies have been conducted on the increase in these parameters in the blood of high-performance athletes as a function of the nature and duration of each training session or competition-level exertion [3, 8, 9]. In marathon runners, for example, an increase in AST from 29.3±12.8 U/L to 51.6±18.0 U/L (four hours after the race) and to 106.9±55.7 U/L (24 hours after crossing the finish line) was measured [30]. After a 130-km ultramarathon, the values increase from 24.5±2.6 U/L to 94.7±14.0 U/L one day after the marathon. The LDH levels increased from 217.1±12.6 U/L to 381.1±20.9 U/L (directly after crossing the finish line) [2]. After an ultramarathon at a distance of 308 km, the LDH levels were increased from 385.6±57.6 U/L to 1,002.3±224.6 U/L [54]. LDH can additionally be used as an indicator of intravascular hemolysis, which can be induced by very strenuous exercises.

An elevation in CK and Mb is especially measurable after long competitions in endurance sports (ultra-distance marathons or triathlons) and eccentric training (e.g. downhill running in ultramarathons); in untrained subjects, the values may show a greater increase than in trained athletes [3, 59]. After finishing the above-mentioned 130-km ultramarathon, for example, markedly elevated CK levels of 3,716±493 U/L were measured, as compared to a baseline of 214±47 U/L [2]. In another study, CK levels elevated to above 20,000 U/L were even detected after ultra-high performance in ultra-bikers and ultra-runners [13]. Very high CK levels with a peak of 28,545±33,611 U/L were also demonstrated in one interventional study on whole-body EMS [28]. CK measurements are primarily taken as part of longer-term training management to monitor muscular stress. In one of the author's own studies, CK was measured daily by POCT in 9 (middle-distance) track and field athletes at a three-week altitude training camp. This made it possible to monitor and steer training. The lowest values were determined to be 182±76 U/L on the first day and the highest were 452±184 U/L on the ninth day [56].

One special feature characterizing the different biomarkers is their unique kinetics, as evidenced by a varying rate of increase and decrease in their serum levels or by an accumulation of the biomarkers in the serum as a function of the breadth, intensity and frequency of muscular strain. For this reason, instead of measuring one single biomarker, a panel of several is preferred. For laboratory studies, blood is drawn at close intervals over several hours to identify the various types of kinetics, whereas in high-performance athletes, the blood draws are carried out on a daily basis.

In practice, measurements of **CK** and **Mb** are regularly used for monitoring muscular stress. The measurement of both parameters not only allows the evaluation of muscle damage that has taken place recently, but also that having occurred further in the past. Mb increases relatively quickly after strenuous exercise and can quickly return back to baseline as well. Depending on the type of exercise, its peak can be measured between 45 minutes and up to three hours afterwards, usually declining back to baseline within eight hours [3, 26]. A measurable elevation in CK occurs later. When and at what level the peak CK value is reached, however, depends on the type of muscular stress. In untrained males, doubly elevated CK levels are demonstrated as early as eight hours after a strength training session (133±88 U/L vs. 278±175 U/L) [25]. After endurance training, a peak is detectable after 24 hours with values frequently being elevated up to 48 hours thereafter. After intensive eccentric training, the values may even continue to rise up to four days later [3, 62]. In one of the author's own tests, for example, the CK levels after exercise on the bicycle ergometer with EMS increased from 462±384 U/L to 786±505 U/L after four hours and up to 2,196±2168 U/L after 24 hours. After four hours, the Mb levels exhibited a peak of 150.9±121.5 ng/mL (baseline: 27.5±21.0 ng/mL), after which they dropped slightly, but were still elevated 24 hours later (97.8±84.4 ng/mL) [62]. It should be noted that, depending on the repetition of the training units, these curves cannot necessarily be reproduced – a phenomenon various authors have dubbed a physiological adaptation to training ("repeated bout effect") [9, 10, 28, 43].

The combined measurement of CK and Mb using POCT is receiving ever-greater attention in high-performance and elite sports. There are, however, only a few POCT devices that allow individual measurements of CK using capillary blood (e.g. the Roche Reflotron and the Arkray Spotchem EZ SP-4430). Mb can be measured with the Cobas h232 (Roche) and the i-Chroma Reader (BodiTech Med Inc., Chuncheon, South Korea, ▣ Fig. 23.3).

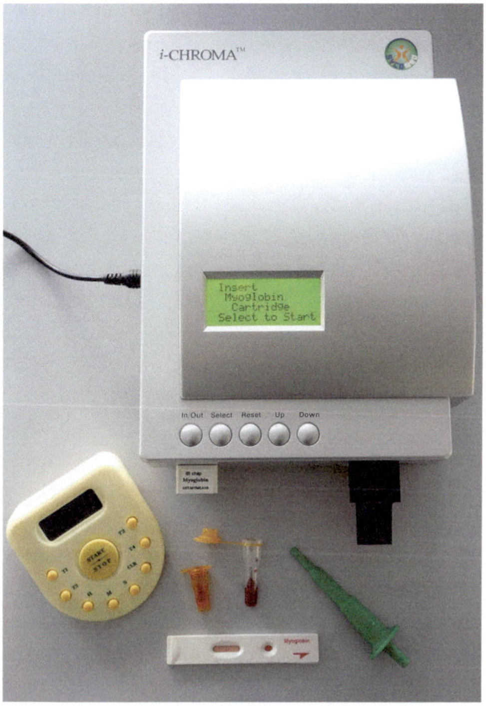

▣ **Fig. 23.3** i-Chroma Reader

23.3.6 Myocardium

The high physical loading that high-performance athletes are exposed to during training or competition can induce myocardial stress that, in turn, can lead to elevation of myocardial markers in the blood. Since POCT tests are carried out for diagnostics in clinical emergencies to detect heart disease, some of them can also be used in sport science studies. The myocardial markers that have already been investigated in conjunction with intensive or prolonged training include the **cardiac troponins** (cTnT and cTnI) and **NT-proBNP** (N-terminal pro brain-type natriuretic peptide).

Troponins can occur in skeletal and cardiac muscle (sTn and cTn). Theoretically, sTn could also be used to monitor muscle injuries. However, since sTn is not an emergency parameter in clinical diagnostics, there are no POCT tests available today. The opposite applies to the two cTn species (cTnT and cTnI). Elevated cTn concentrations could not be detected post-intensive physical exertion until highly sensitive immunoassays became available. Many POCT devices are ascribed this feature.

BNP is a peptide hormone produced by cardiac muscle cells that exert a protective function during myocardial stress. Its cleavage product, NT-proBNP, is detectable in the blood for a longer period of time and can be measured with numerous POCT devices. The literature has described an increase in cTnT/cTnI and NT-proBNP after intensive and enduring physical exertion in athletes, indicating a high cardiovascular requirement [5, 37, 45, 48, 65]. After an ultramarathon, for example, an elevation in NT-proBNP from 38.1±4.8 pg/mL to 1,280.6±259 pg/mL was demonstrated [58]. The elevation in biomarkers, however, is not automatically a signal of a myocardial hazard, but rather may reflect a physiological reaction to intense cardiac activity and the regulatory process of myocardial adaptation [3].

23.3.7 Saliva analysis (stress indicators, immune defense)

Many sport science studies have focused on physiological stress reactions and the immune status of athletes in different stress situations. In this context, the analysis of biomarkers obtained from the saliva has been gaining importance over the past years. The foremost biomarkers of immunity measurable in the saliva are the stress hormones **cortisol** (sC), **alpha-amylase** (sAA) and **immunoglobulin A** (sIgA). Their measurement permits the potential consequences for the immune status – e.g. as a reflection of psychological stress before competitions, on long trips to a training camp or under certain training conditions – to be estimated. Stress markers can also be collected to investigate whether a training intervention has triggered any adaptations. Depending on the intensity and duration of exertion, sC and sAA may show either acutely or chronically levated levels. In one study on tennis players, cortisol concentrations that increased concurrent with increasing training intensity were measured during a five-week training phase [19]. Significant correlations (r=0.71) between the levels and the stress perceived by the subjects. In professional skiers, who performed six high-intensity training sessions under hypoxic conditions on a ski ergometer over a period of two weeks, an increase in sAA from 962±508 U/mL to 1,462±792 U/mL was observed after the last intervention [7].

The initially increasing and then declining values over the course of recurrent training exertion were suggestive of an adaptive process to stress. Permanently lowered levels can be indicators of chronic stress (states of exhaustion, overtraining). A combined measurement of both parameters makes sense because they exhibit different secretion rates (sAA rises faster than sC). Both the sympathetic adrenal medulla axis (sAA) and the hypothalamic pituitary adrenal (HPA) axis (sC) react to stress. The variable sIgA is measured an indicator of the mucosal immune systems and is investigated to assess training stress and the burden of different interventions (altitude exposure, HIT, eccentric training) have on the immune system or to monitor health status [7, 19, 33]. The primary aim is to identify negative effects like acute or chronic mental stress that can lower sIgA and increase susceptibility to infections of the upper respiratory tract [7, 42, 64].

The non-invasive determination of these biomarkers from saliva samples features the advantage that the samples can be collected in every type of training situation or investigative setting. The athletes themselves can even perform sampling on their own. Moreover, unlike venous blood draws, saliva sampling poses much less psychological stress to the athletes (particularly relevant when measuring stress markers). To date, more complex laboratory methods (e.g. ELISA) have tended to be used for analysis in studies with saliva samples as well. For uncomplicated and fast measurement of sC, sAA and sIgA that can be equated with POCT, rapid lateral flow tests manufactured by iPRO Interactive (England) have been available for several years now. More and more, these methods are being employed in the monitoring of individual athletes or entire teams. However, it should be noted that only one study can currently be mentioned that investigated the IgA test with regard to its validity and reliability and compared it with an established method (ELISA) [14].

23.4 Pre-analytic phase

In the pre-analytical phase, attention must be paid to the fact that the high level of physical activity in high-performance athletes can have a particular influence on the measured results. One aspect relevant for measuring Hb/Hct is that changes in overall blood volume can be triggered by the acute or chronic physiological adaptation processes during intensive training loads or long training sessions. This particularly applies to plasma volume, which causes a change in the relative Hb concentration (g/dL) [49]. Because acute physiological adaptation processes take place, reductions in plasma volumes and briefly elevated Hb/Hct levels are measurable after training sessions. In comparison, the chronic adaptations to prolonged training periods lead to an elevation in total blood volume. During these processes, the plasma can increase more than the red compartment and lowered Hb/Hct levels are measurable over the long term. These lowered values led to the term "pseudoanemia" or "athlete's anemia" to indicate that no anemia is present in the classical sense [54].

A change in body position or increase in water uptake can cause short-term changes in Hb/Hct. In the author's own trials, it was demonstrated that acute adaptations can cause a short-term change in the Hb value up to 1 g/dL [1]. For this reason, blood must be drawn in order to monitor athletes under highly standardized conditions.

Moreover, from a purely physical standpoint, reductions in plasma volumes after strenuous exercises lead to increased concentrations of biomarkers in the serum. For this reason, the concentrations measured after a training session are corrected according to the changes in plasma volumes [15].

When drawing capillary blood to measure lactate during strenuous physical training sessions, consideration should be given to the fact that sweat contains very high lactate concentrations [46]. Therefore, care must also be taken to thoroughly clean the puncture site when taking samples under great time pressure. Moreover, sport science studies have compared the results measured in blood samples taken from different collection sites. Higher values were found in sample collected from finger tips as compared to earlobes [17].

Furthermore, consideration must be given to the fact that physical stress, e.g. as occurs during marathons or endurance training, can activate the coagulation systems and fibrinolysis [34, 57, 66]. In the authors' own studies on endurance athletes, clot formation in the capillary was frequently observed during capillary blood draw. For this reason, the need for very thorough mixing of the sample in heparinized capillary tubes was emphasized. Capillary tubes containing an elevated portion of Li-heparin are commercially available as well.

In longitudinal studies and when defining individual profiles, the pre-analytical influence of variations has moreover been observed in the circadian rhythm of several parameters. Therefore, the time of day at which sampling was performed should be documented and then accounted for when interpreting the results. When conducting training intervention studies, the time of day of sample collection should be well planned and standardized. It is conceivable that intervention-related changes in biomarkers are masked by the biomarkers' natural, circadian variations.

23.5 Analytic/post-analytic factors

In high-performance and elite sports, as in clinical diagnostics, high demands are placed on the measurement precision and validity of POCT devices. Indeed, even minor changes in longitudinally measured values should be attributable to biological variability. Different measurement methods on the different POCT devices can cause methodological variability. Therefore, a switching of devices within long-term studies should be avoided for generating individual profiles and assessing performance development. This allows differences in measurands relating to loading or training stimuli to be assessed reliably. Measurement variability, e.g. in lactate performance diagnostics, can have direct implications for the recommended heart rate or running speed of a training regimen.

At the high concentrations to be expected in the biomarkers presented here, it must be noted that POCT measuring devices, like large autoanalyzers in a laboratory, have an upper measurement range limit. In this case, for measuring highly concentrated biomarkers in plasma or serum, there is the option to centrifuge whole capillary blood samples (e.g. using the suitable GK 150 collection system manufactured by Kabe) and analyze the diluted supernatant.

For lactate performance diagnostics, small hand-held devices that rely on dry chemical test strips are used on the one hand, while benchtop devices measuring by wet chemical methods are used on the other. The difference between the two variants is that the hand-held devices measure the lactate plasma concentration, whereas the benchtop devices measure both the plasma and the erythrocyte concentration of lactate after hemolysis of the blood sample. The latter is systematically lower. For this reason, longitudinal athlete studies and management measures must apply one measurement method in order to minimize variability in measurement methodology and to enable a reliable evaluation of lactate concentrations [17].

When planning to use POCT in high-performance and elite sports, it should be noted that some devices can only measure the aforementioned parameters as a panel, i.e. in combination with other parameters. These are frequently irrelevant to the sport context and result in higher costs. On the other, there are devices with which permit an individual measurement of CK and Mb, for example, although no device features the two single measurements simultaneously. From the perspective of users in high-performance and elite sports, a combination would be much more desirable.

Users of POCT in high-performance and elite sports also need to be aware that several POCT devices block the reagent once it passes its expiration date. RiliBÄK has laid down this programmed locking for laboratories in Germany that analyze blood samples for pathological diagnostics.

It has been sufficiently investigated that biomarkers of individuals who are a continually exposed to high levels of physical stress are not comparable with the reference ranges registered in the normal population due to both their physiological as well as pathophysiological differences. For this reason, they should be compared with control ranges taken from a similar population. For example, it is known that, because they train regularly, athletes competing in different disciplines have consistently higher CK and CREA levels than in the untrained population [3, 4, 9, 36, 39]. Biomarkers in high-performance and elite athletes are additionally subject to high inter- and intra-individual variances, depending on the breadth and intensity of a training session. On the one hand, high inter-individual variability can be explained by varying levels of stress across the different disciplines and training programs and, on the other, by the genetic predispositions. The high intra-individual variability is seen as a function of training versus competition. Ideally, measured values of athletes should be assessed on an individual profile and determined at close intervals. They can only be interpreted in conjunction with individual levels of loading. Genetic predisposition also explains the variability of biomarkers in patients; given comparable stress levels, these individuals turn out to be so-called "responders" or "non-responders", thus leading to different profiles.

References

1. Achtzehn S (2013) Hämoglobinkonzentrationen, Hämatokrit-Werte und Retikulozyten in der Sportwissenschaft. Der Andere Verlag, Uelvesbüll
2. Arakawa K, Hosono A, Shibata K et al. (2016) Changes in blood biochemical markers before, during, and after a 2-day ultramarathon. Open Access J Sports Med 7: 43–50
3. Banfi G, Colombini A, Lombardi G et al. (2012) Metabolic markers in sports medicine. Adv Clin Chem 56: 1–54
4. Banfi G, Del F M, Lippi G (2009) Serum creatinine concentration and creatinine-based estimation of glomerular filtration rate in athletes. Sports Med 39(4): 331–337
5. Banfi G, Lippi G, Susta D et al. (2010) NT-proBNP concentrations in mountain marathoners. J Strength Cond Res 24(5): 1369–1372
6. Bassett J R, Howley E T (2000) Limiting factors for maximum oxygen uptake and determinants of endurance performance. Med Sci Sports Exerc 32(1): 70–84
7. Born D P, Faiss R, Willis S J et al. (2016) Circadian variation of salivary immunoglobin A, alpha-amylase activity and mood in response to repeated double-poling sprints in hypoxia. Eur J Appl Physiol 116(1): 1–10
8. Brancaccio P, Lippi G, Maffulli N (2010) Biochemical markers of muscular damage. Clin Chem Lab Med 48(6): 757–767
9. Brancaccio P, Maffulli N, Buonauro R et al. (2008) Serum enzyme monitoring in sports medicine. Clin Sports Med 27(1): 1–18, vii
10. Brentano M A, Martins K L F (2011) A review on strength exercise-induced muscle damage: applications, adaptation mechanisms and limitations. J Sports Med Phys Fitness 51(1): 1–10
11. Burden R J, Morton K, Richards T et al. (2015) Is iron treatment beneficial in, iron-deficient but non-anaemic (IDNA) endurance athletes? A systematic review and meta-analysis. Br J Sports Med 49(21): 1389–1397
12. Burge C M, Skinner S L (1995) Determination of hemoglobin mass and blood volume with CO: evaluation and application of a method. J Appl Physiol (1985) 79(2): 623–631
13. Chlíbková D, Knechtle B, Rosemann T et al. (2015) Rhabdomyolysis and exercise-associated hyponatremia in ultra-bikers and ultra-runners. J Int Soc Sports Nutr 12: 29
14. Coad S, Mclellan C, Whitehouse T et al. (2015) Validity and reliability of a novel salivary immunoassay for individual profiling in applied sports science. Res Sports Med 23(2): 140–150
15. Dill D B, Costill D L (1974) Calculation of percentage changes in volumes of blood, plasma, and red cells in dehydration. J Appl Physiol 37(2): 247–248
16. Faude O, Kindermann W, Meyer T (2009) Lactate threshold concepts: how valid are they? Sports Med 39(6): 469–490
17. Faude O, Meyer T (2008) Methodische Aspekte der Laktatbestimmung. [Methodological Aspects of Lactate Determination]. Dtsch Z Sportmed 59(12)
18. Gladden L B (2004) Lactate metabolism: a new paradigm for the third millennium. J Physiol 558 (Pt 1): 5–30
19. Gomes R V, Moreira A, Lodo L et al. (2013) Monitoring training loads, stress, immune-endocrine responses and performance in tennis players. Biol Sport 30(3): 173–180
20. Halson SL (2014) Monitoring Training Load to Understand Fatigue in Athletes. Sports Med 44(S2): 139–147
21. Harper LD, Clifford T, Briggs MA et al. (2015) The effects of 120 minutes of simulated match-play on indices of acid-base balance in professional academy soccer players. J Strength Cond Res
22. Heck H, Beneke R (2008) 30 Jahre Laktatschwellen – was bleibt zu tun? Dtsch Z Sportmed 59(12): 297–302
23. Heinicke K, Wolfarth B, Winchenbach P et al. (2001) Blood volume and hemoglobin mass in elite athletes of different disciplines. Int J Sports Med 22(7): 504–512
24. Hinton P S (2014) Iron and the endurance athlete. Appl Physiol Nutr Metab 39(9): 1012–1018
25. Hurley BF, Redmond R A, Pratley R E et al. (1995) Effects of strength training on muscle hypertrophy and muscle cell disruption in older men. Int J Sports Med 16(6): 378–384
26. Kahanov L, Eberman LE, Townsend C M et al. (2012) Creatine Kinase and Myoglobin as biochemical markers of muscle damage in division-1 collegiate football players. Online Journal of Medicine and Medical Science Research 1(5)
27. Kasprowicz K, Ziemann E, Ratkowski W et al. (2013) Running a 100-km ultra-marathon induces an inflammatory response but does not raise the level of the plasma iron-regulatory protein hepcidin. J Sports Med Phys Fitness 53(5): 533–537
28. Kemmler W, Teschler M, Bebenek M et al. (2015) (Very) high Creatinkinase concentration after exertional whole-body electromyostimulation application: health risks and longitudinal adaptations. Wien Med Wochenschr 165(21–22): 427–435
29. Kindermann W (2004) Anaerobe Schwelle. Dtsch Z Sportmed 55(6): 161–162
30. Kratz A, Lewandrowski KB, Siegel A J et al. (2002) Effect of marathon running on hematologic and biochemical laboratory parameters, including cardiac markers. Am J Clin Pathol 118(6): 856–863

23

31. Mader A, Liesen H, Heck H et al. (1976) Zur Beurteilung der sportartspezifischen Ausdauerleistungsfähigkeit im Labor. Sportarzt und Sportmedizin (27)
32. Mathes S, Mester J, Bloch W et al. (2017) Impact of high-intensity and high-volume exercise on short-term perturbations in the circulating fraction of different cell types. J Sports Med Phys Fitness 57: 130–137
33. McKune A, Starzak D, Semple S (2015) Repeated bouts of eccentrically biased endurance exercise stimulate salivary IgA secretion rate. Biol Sport 32(1): 21–25
34. Menzel K, Hilberg T (2011) Blood coagulation and fibrinolysis in healthy, untrained subjects: effects of different exercise intensities controlled by individual anaerobic threshold. Eur J Appl Physiol 111(2): 253–260
35. Meyer T, Kellmann M, Ferrauti A et al. (2013) Die Messung von Erholtheit und Regenerationsbedarf im Fußball. [The Measurement of Recovery and Regeneration Requirements in Football.] Dtsch Z Sportmed 2013(01): 28–34
36. Meyer T, Meister S (2011) Routine blood parameters in elite soccer players. Int J Sports Med 32(11): 875–881
37. Michielsen, Etienne CHJ, Wodzig, Will KWH, van Dieijen-Visser M P (2008) Cardiac troponin T release after prolonged strenuous exercise. Sports Med 38(5): 425–435
38. Mohr M, Draganidis D, Chatzinikolaou A et al. (2016) Muscle damage, inflammatory, immune and performance responses to three football games in 1 week in competitive male players. Eur J Appl Physiol 116(1): 179–193
39. Mougios V (2007) Reference intervals for serum creatine kinase in athletes. Br J Sports Med 41(10): 674–678
40. Neubauer O, Konig D, Wagner K H (2008) Recovery after an Ironman triathlon: sustained inflammatory responses and muscular stress. Eur J Appl Physiol 104(3): 417–426
41. Neumayr G, Pfister R, Hoertnagl H et al. (2005) Renal function and plasma volume following ultramarathon cycling. Int J Sports Med 26(1): 2–8
42. Owen A L, Wong D P, Dunlop G et al. (2014) High intensity training and salivary immunoglobulin-a responses in professional top-level soccer players: effect of training intensity. J Strength Cond Res 30(9): 2460–2469
43. Papacosta E, Nassis GP (2011) Saliva as a tool for monitoring steroid, peptide and immune markers in sport and exercise science. J Sci Med Sport 14(5): 424–434
44. Peake J, Nosaka K, Suzuki K (2005) Characterization of inflammatory responses to eccentric exercise in humans. Exerc Immunol Rev 11: 64–85
45. Peeling P, Dawson B, Goodman C et al. (2008) Athletic induced iron deficiency: new insights into the role of inflammation, cytokines and hormones. Eur J Appl Physiol 103(4): 381–391
46. Regwan S, Hulten E A, Martinho S et al. (2010) Marathon running as a cause of troponin elevation: a systematic review and meta-analysis. J Interv Cardiol 23(5): 443–450
47. Röcker K, Dickhuth H-H (2001) Praxis der Laktatmessung. Dtsch Z Sportmed 52(1): 33–34
48. Sawka MN, Convertino VA, Eichner ER et al. (2000) Blood volume: importance and adaptations to exercise training, environmental stresses, and trauma/sickness. Med Sci Sports Exerc 32(2): 332–348
49. Scharhag J, Urhausen A, Schneider G et al. (2006) Reproducibility and clinical significance of exercise-induced increases in cardiac troponins and N-terminal pro brain natriuretic peptide in endurance athletes. Eur J Cardiovasc Prev Rehabil 13(3): 388–397
50. Schmidt W (1999) Die Bedeutung des Blutvolumens für den Ausdauersportler. Dtsch Z Sportmed 11+12: 341–349
51. Schmidt W, Prommer N (2005) The optimised CO-rebreathing method: a new tool to determine total haemoglobin mass routinely. Eur J Appl Physiol 95(5–6): 486–495
52. Schmidt W, Prommer N (2010) Impact of alterations in total hemoglobin mass on VO2max. Exerc Sport Sci Rev 38(2): 68–75
53. Schulpis K H, Tsironi M, Skenderi K et al. (2008) Dramatic reduction of erythrocyte glucose-6-phosphate dehydrogenase activity in athletes participating in the ultradistance foot race "Spartathlon". Scand J Clin Lab Invest 68(3): 228–232
54. Shaskey D J, Green G A (2000) Sports haematology. Sports Med 29(1): 27–38
55. Shin K-A, Park KD, Ahn J et al. (2016) Comparison of changes in biochemical markers for skeletal muscles, hepatic metabolism, and renal function after three types of long-distance running: Observational study. Medicine (Baltimore) 95(20): e3657
56. Sperlich B (2016) Exercise physiology – integrative & experimental. Dtsch Z Sportmed 2016(02): 25–26
57. Sperlich B, Achtzehn S, Marées M de et al. (2016) Load management in elite German distance runners during 3-weeks of high altitude training. Physiol Rep 4(12): e12845
58. Sumann G, Fries D, Griesmacher A et al. (2007) Blood coagulation activation and fibrinolysis during a downhill marathon run. Blood Coagul Fibrinolysis 18(5): 435–440
59. Tchou I, Margeli A, Tsironi M et al. (2009) Growth-differentiation factor-15, endoglin and N-terminal pro-brain natriuretic peptide induction in athletes participating in an ultramarathon foot race. Biomarkers 14(6): 418–422

60. Vincent HK, Vincent K R (1997) The effect of training status on the serum creatine kinase response, soreness and muscle function following resistance exercise. Int J Sports Med 18(6): 431–437
61. Wagner S, Knechtle B, Knechtle P et al. (2012) Higher prevalence of exercise-associated hyponatremia in female than in male open-water ultra-endurance swimmers: the 'Marathon-Swim' in Lake Zurich. Eur J Appl Physiol 112(3): 1095–1106
62. Wahl P, Bloch W, Mester J (2009) Moderne Betrachtungsweisen des Laktats: Laktat ein überschätztes und zugleich unterschätztes Molekül. Schweizerische Zeitschrift für Sportmedizin und Sporttraumatologie 57(3)
63. Wahl P, Hein M, Achtzehn S et al. (2015) Acute effects of superimposed electromyostimulation during cycling on myokines and markers of muscle damage. J Musculoskelet Neuronal Interact 15(1): 53–59
64. Wahl P, Mathes S, Köhler K et al. (2013) Acute metabolic, hormonal, and psychological responses to different endurance training protocols. Horm Metab Res 45(11): 827–833
65. Walsh NP, Gleeson M, Pyne DB et al. (2011) Position statement. Part two: Maintaining immune health. Exerc Immunol Rev 17: 64–103
66. Weippert M, Divchev D, Schmidt P et al. (2016) Cardiac troponin T and echocardiographic dimensions after repeated sprint vs. moderate intensity continuous exercise in healthy young males. Sci Rep 6: 24614
67. Weiss C, Bärtsch P (2003) Aktivierung der Blutgerinnung und Fibrinolyse durch körperliche Belastung. Dtsch Z Sportmed 54(5): 130–135

POCT in obstetrics and gynecology

Vanadin Seifert-Klauss

P. B. Luppa, R. Junker (Eds.), *Point-of-Care Testing*
https://doi.org/10.1007/978-3-662-54497-6_24

24.1 Introduction

One of the first POCT systems in gynecology was the pregnancy test that measures the human chorionic gonadotropin (hCG) excreted in the urine. For decades, this test has rapidly and consistently influenced diagnostic and therapeutic algorithms in general practice, outpatient centers and hospitals. In the last decades, dozens of these tests that have a sensitivity for hCG of between 5 and 50 IU/L have become available in pharmacies and, in the meantime, also from drugstores. Numerous clinical and innumerable personal decisions made by female patients are based on these tests – be it the cancellation of X-rays for non-urgent radiological indications or in the differential diagnosis of ectopic pregnancy as the cause of abdominal pain. In addition, POCT plays an increasing role in the care of pregnant patients, those giving birth and those who want to become pregnant or have menstrual cycle complaints, as well as in other exceptional situations.

Studies on quality indicators comparing POCT and tests carried out in a central laboratory showed major differences, particularly in pre-analytical areas: Patient identification, measurements without evaluable results and insufficient sample volumes. By contrast, the analytical quality indicators in the POCT systems fared better than in the samples analyzed in the laboratory [2]. This reflects the fact that although laboratory staff and human error in sample assessments are reduced, POCT technology has simultaneously shifted the verification work previously performed by the laboratories over to clinical caregivers whose responsibilities and tasks are thereby multiplied.

24.2 Rapid pregnancy tests

A recently published comparative study of 8 commercially available hCG tests showed that many of the tests were not as accurate as hoped for: Agreement between expected and actual results ranged between 95 % and <75 %, and in some cases, was a mere 33 % and 39 % [8]. Even at institutions where such tests create a relative cost factor due to high-volume turnover, compromises sometimes have to be made between sensitivity and quality of the measurement. In several cases, the use of tests with a sensitivity of 50 IU/L hCG led to false-negative pregnancy results, which were noticed when simultaneously reported hCG serum tests were carried out to diagnose acute clinical symptoms. In a few cases, a pregnancy was diagnosed after elective surgery that must have already existed prior to the operation. The ensuring uncertainty over the validity of test systems used in the hospital led to duplicated tests being performed over a longer period and thereby to higher costs. Ultimately, a particularly low-priced test was rejected in favor of a more costly but reliable system. A publication by Johnson has shown that cost pressure in the retail sector (over-the-counter, OTC sales) has also compromised the quality of tests [8]. The consequences of false-negative tests in wider practice are not known.

24.3 POCT in low-resource environments

An argument is made for POCT use in third world counties, where access to medical care is very limited. There are very good examples of this. A more recent study investigated the validity of a urine test used as a pointer to asymptomatic bacteriuria in pregnant women. Screening would be useful given that such infections, if left untreated, can cause premature births and low birth weight in infants. In parallel with the microbiological test, screening for nitrite and leukocyte esterase was performed. Under ideal conditions, these surrogate parameters show a strong agreement or high correlation with the infection. The authors concluded that diagnostic bacterial cultures should nevertheless be continued, as many false-positive and false-negative results occur, suggesting a low sensitivity and a low predictive value of the urine POC test [3].

24.4 POCT in rupture of membranes

When dealing with borderline cases, where a rupture of membranes cannot be ruled out unequivocally, POCT is of pivotal importance in the care of pregnant women worldwide. In confirmed amniorrhexis where labor has not yet started, infection parameters are closely monitored and, if necessary, timely antibiotic treatment is started. This is required due to a higher risk of ascending infection and the risk of a dangerous amniotic infection syndrome in mother and child that can develop without the protective barrier of the amniotic sack. Various commercially provided tests offer assessment of biomarkers that are specific for amniotic fluid, e.g. insulin-like growth factor-binding protein (IGFBP-1) or placental alpha microglobulin (PAMG-1). Published in 2014, one meta-analysis of 17 studies encompassing 2,147 pregnant women found PAMG-1-based tests to be more accurate. The authors, however, stress that no women with bleeding were included in any of the studies. Blood is known to interfere with such amniotic fluid tests and can affect their results considerably. The authors of the meta-analysis therefore warned that a substantial number of women presenting to the gynecologist with suspected ruptured membranes and simultaneous vaginal bleeding in real clinical practice were not represented in those studies [12].

24.5 Tests for estimating the risks for preeclampsia

Preeclampsia is another dangerous syndrome in pregnancy that can now be diagnosed early by testing for serum biomarkers. Its detection is more difficult in regions with a suboptimal medical infrastructure and low resources. Therefore, affordable, sensitive and specific POCT procedures are required that can be performed by medical support staff or by the pregnant women themselves [1]. Initial approaches entailed urine-based diagnostic algorithms that were suitable to differentiate uncomplicated pregnancies from those in pregnant women requiring special treatment. Whilst the discrimination of podocytes in urine sediment samples was very good [6], the method was not suitable for POCT due to a lack of sufficiently automated microscopic detection of podocytes. The protein-to-creatinine ratio, although specific, was not sufficiently sensitive either [13]. Newer developmental approaches combine a mixture of POCT and mobile health: a smartphone app that can detect a color reaction in urine in seven steps at the POC (Congo Red Dot) and one that quantifies the color intensity of a smartphone-acquired image to determine particular misfolded protein concentrations [9].

24.6 POCT in the delivery room – pH meters

A delivery room without POCT is unimaginable nowadays: prenatal and immediate postnatal determinations of umbilical cord pH and base excess (as well as secondary pCO_2, pO_2) have been key to clinical procedure for decades. In the event of pathological cardiotocography (CTG) during labor, compensatory capabilities and reserves of the unborn infant are monitored by fetal blood sampling from the scalp. This is possible if the cervix is dilated sufficiently (from approx. 4 cm). Testing is predicated on the membranes rupturing; otherwise, the amniotic sack is opened and the fetal blood sampled. If physiological pH levels are present but the ongoing CTG readings pathological, fetal blood sampling is repeated every 30–60 min, depending on the levels and the clinical status. A fall in the pH, e.g. below 7.25, may have direct clinical implications that, depending on the progression and stage of the delivery, can culminate in initiation of an accelerated or vaginal operative delivery (based on uterus height, involving vacuum or forceps extraction). Where the cervix fails to dilate sufficiently, an emergency cesarean section may be mandated. Short distances, reliably serviced pH meters and rapid analysis of capillary blood samples are essential for optimal management of delivery room POCT given that sudden changes and often multiple deliveries can occur at the same time. Requirements for quality management (e.g. the introduction of individual codes for each device user) sometimes clash with unexpected situations to be mastered under huge time pressure, for example, when non-permanent staff, insufficiently trained for a certain device, are tasked to carry out the analyses.

A recent Cochrane meta-analysis assessed the merits of lactate instead of conventional pH measurements in the delivery suite [4]. Various studies had postulated an advantage of lactate measurements due to lower blood volume requirements. To obtain a whole capillary filled with fetal blood is challenging: the sample site on the fetal scalp has to be cleaned with a paraffin-soaked swab and prepared for a small, controlled cut while the laboring mother experiences contractions. In some cases, fetal blood sampling is unsuccessful. The attempt to sample intrapartum fetal scalp blood was successful in 98.7 % (lactate) versus 79.4 % (pH). This result applied to one of the two evaluated studies reporting this information. The meta-analysis covered 3,348 mother-baby pairs in two randomized controlled studies. The results favoring lactate measurement were significantly biased, as three neonatal deaths that were not linked to the birth pH occurred in the group that was allocated to pH measurement. Two infants had prenatally unknown diaphragmatic hernias and one had congenital cardiac fibrosis [4]. The replacement of pH by lactate assessments is not to be expected soon, given that birth pH and base excess have been essential parts of neonatal statistical surveys for years and are themselves seen as quality performance indicators of entire obstetric departments.

24.7 Issues associated with unregulated test kit availability

In community women's health, POCT products sold in pharmacies, drugstores and supermarkets are becoming increasingly relevant. A publication that critically looked at this development warned that the pre-analytical and post-analytical problems already associated with conventional laboratory requests, which will be greatly intensified by this trend [10]. An example to be cited are hormone tests offered by some pharmacies that do not meet test requirement standards, e.g. progesterone determination in the menstrual cycle phases where no analyte concentration is to be expected (pre-analytical problem), whereby a "deficiency" is defined that can potentially be physiologically normal (post-analytical misinterpretation). Concerns with sampling of unsuitable biological materials (e.g. spontaneous urine instead of morning urine samples) add to the list of problems with which some over-the-counter POCT systems are fraught. Other issues include false test execution, erroneous result readings, missing warnings in the event of critical values and the poor ability to verify results [11] immediately after testing (e.g. via chip technology on serial test devices).

24.8 POCT for contraception or the desire to become pregnant

A new – numerically – successful area of application for POCT technology features ovulation monitors: Some use urine dipsticks to calculate the ratio of estriol-3-glucuronide (E3G) and luteinizing hormone (LH) or others work with indicators like core body temperature, cervical mucus or saliva to determine the ovulation probability. Various models used for both natural family planning and fertility monitoring are marketed according to the desired sensitivity and software programming. These POCT systems have been used successfully in observational studies on ovulation and anovulation with scientific objectives to determine the appropriate time for blood sampling after presumed ovulation or to determine the frequency of anovulatory vs. ovulatory cycles over a longer period in a decentralized way [11]. In the meantime, urine dipsticks are also being supplied without mini-computers and can be combined with smartphone apps.

A recently published review on menstrual cycle computers and apps included eight temperature computers (among them a vaginal ring, a vaginal tampon and a wireless biosensor), three hormone computers, six mini-microscopes and two resistance measuring devices that allowed self-measurements in saliva or vaginal secretion [5]. According to the authors, the latter cannot be recommended for use in contraception, but improvements in the area of hormone computers are to be expected. Amongst fertility apps, pure menstruation calendar apps are distinguished from apps for natural family planning, which also record symptoms and/or body signs such as basal temperature, cervical secretions and cervix data. Sometimes courses and advice/counseling sessions are offered along with the tests. Several apps can be synchronized with an online cycle management setup. In one study, the Natural Family Planning (NFP) working group calculated a PEARL index of 0.4 for "perfect use" of their course and advice/counseling method. The menstrual cycle databank of myNFP consists of 200,000 cycles (https://mynfp.de). Critics have voiced concerns about the legal aspects of data security.

24.9 POCT in medical abortion aftercare

Studies have also looked at POCT-supported aftercare in medical abortions where self-performed testing of hCG levels was carried out by the patients themselves. Their experiences were not completely positive: Some women became more worried from having to perform the tests themselves. Indeed, the burdensome psychosocial dimension of this aftercare as well as by the impact of POCT results in other situations should not be underestimated [7].

References

1. Acestor N, Goett J, Lee A, Herrick TM, Engelbrecht SM, Harner-Jay CM, Howell BJ, Weigl BH (2016) Towards biomarker-based tests that can facilitate decisions about prevention and management of preeclampsia in low-resource settings. Clin Chem Lab Med 54(1):17–27
2. Cantero M, Redondo M, Martín E, Callejón G, Hortas ML (2015) Use of quality indicators to compare point-of-care testing errors in a neonatal unit and errors in a STAT central laboratory. Clin Chem Lab Med 53(2):239–247
3. Demilie T, Beyene G, Melaku S, Tsegaye W (2014) Diagnostic accuracy of rapid urine dipstick test to predict urinary tract infection among pregnant women in Felege Hiwot Referral Hospital, Bahir Dar, North West Ethiopia. BMC Res Notes 7:481
4. East CE, Leader LR, Sheehan P, Henshall NE, Colditz PB, Lau R (2015) Intrapartum fetal scalp lactate sampling for fetal assessment in the presence of a non-reassuring fetal heart rate trace. Cochrane Database Syst Rev 5:CD006174
5. Freundl G, Gnoth C, Krahlisch M (2016) Zykluscomputer und -Apps. Gynäkologische Endokrinologie 14:93–104
6. Garovic VD, Wagner SJ, Turner ST, Rosenthal DW, Watson WJ, Brost BC, Rose CH, Gavrilova L, Craigo P, Bailey KR, Achenbach J, Schiffer M, Grande JP (2007) Urinary podocyte excretion as a marker for preeclampsia. Am J Obstet Gynecol 196(4):320.e1–7
7. Hassoun D, Périn I, Hiên H, Demars HH (2016) Feasibility of self-performed urine pregnancy testing for follow-up after medical abortion. Eur J Obstet Gynecol Reprod Biol 197:174–178
8. Johnson S, Cushion M, Bond S, Godbert S, Pike J (2015) Comparison of analytical sensitivity and women's interpretation of home pregnancy tests. Clin Chem Lab Med 53(3):391–402
9. Jonas SM, Deserno TM, Buhimschi CS, Makin J, Choma MA, Buhimschi IA (2016) Smartphone-based diagnostic for preeclampsia: an mHealth solution for administering the Congo Red Dot (CRD) test in settings with limited resources. J Am Med Inform Assoc 23(1):166–173
10. Lippi G, Plebani M, Favaloro EJ, Trenti T (2010) Laboratory testing in pharmacies. Clin Chem Lab Med 48(7):943–953
11. Niethammer B, Körner C, Schmidmayr M, Luppa PB, Seifert-Klauss VR (2015) Non-reproductive effects of anovulation: bone metabolism in the luteal phase of premenopausal women differs between ovulatory and anovulatory cycles. Geburtshilfe Frauenheilkd 75(12):1250–1257
12. Palacio M, Kühnert M, Berger R, Larios CL, Marcellin L (2014) Meta-analysis of studies on biochemical marker tests for the diagnosis of premature rupture

of membranes: comparison of performance indexes. BMC Pregnancy Childbirth 14:183
13. Sanchez-Ramos L, Gillen G, Zamora J, Stenyakina A, Kaunitz AM (2013) The protein-to-creatinine ratio for the prediction of significant proteinuria in patients at risk for preeclampsia: a meta-analysis. Ann Clin Lab Sci 43(2):211–212

Legal and organizational framework

Content

Medical device legislation and POCT

Folker Spitzenberger, Claus Langer, Ullrich M. Gassner

P. B. Luppa, R. Junker (Eds.), *Point-of-Care Testing*
https://doi.org/10.1007/978-3-662-54497-6_25

25.1 Introduction

Legally regulated quality assurance in POCT can basically be differentiated into two areas which derive from the regulations in European law that govern the placing of in vitro diagnostic medical devices (IVD) on the market. POC diagnostic devices fall under the purview of such legislation. These regulations are part of a new European approach to technical and legal harmonization aimed at realization of the single European market for a number of products including IVD. ▶ Section 25.2 of this chapter explains this approach, its revision in 2008 and its application within the medical device sector.

On the one hand, extensive requirements governing quality assurance during the production of IVD can be directly derived from European IVD legislation. In conjunction with the conformity assessment of IVD, inter alia, manufacturers are obliged to apply and document a quality assurance system. ▶ Section 25.3 of this chapter describes the requirements relating to the conformity assessment of IVD, how these requirements are laid down in the European IVD Directive [13] and transposed into German national law by the second law amending the German Medical Devices Act (MPG) [30].

On the other hand, the quality assurance measures meanwhile legally governing the operation and use of diagnostic devices indirectly derive from the regulations laid down in the IVD Directive. These include the regulations subordinate to MPG [20], i.e. the Medical Devices Operator Ordinance (MPBetreibV) [29] and, anchored therein, the Guideline of the German Medical Association on Quality Assurance of Medical Laboratory Examinations (RiliBÄK) [5]. Furthermore, accreditation of medical laboratories plays a role. Although accreditation in Germany takes place on a voluntary basis, it is legally binding in some subspecialties of medical laboratory diagnostics or related areas such as neonatal screening and paternity testing. In connection with the conformity assessment of IVD and other medical devices, MPG also legally governs the approval of testing laboratories, with said approval generally being based on an accreditation. In ▶ Section 25.4 of this chapter, IVD-relevant principles of MPG and its subordinate regulations are described.

▶ Section 25.5 concludes with an outlook on the pending revision of European legislation on medical devices and the most important regulatory consequences for POCT resulting therefrom.

25.2 European legislative framework for medical devices

25.2.1 "New approach" and "global approach" to legislative harmonization within the European market

Free movement of goods, persons, services and capital are the four fundamental freedoms of the single European market in accordance with the provisions of the Treaty on the Functioning of the European Union (TFEU) [19]. The European Union (EU) takes measures for the approximation of laws, regulations and administrative provisions of the Member States in order to abolish restrictions on trade and thereby realize the single market. The most important measures relate to directives which EU Member States have to transpose into national law.

EU directives can embody detailed regulations of facts and harmonization strategies associated therewith. Such directives exhibit high regulatory density and are historically based on an older approach to European legislation. Some examples of products affected by this harmonization of legislation are the drugs currently regulated by Directive 2001/83/EC on the Community code relating to medicinal products for human use [24]. The essential characteristic involves the official testing and granting of marketing authorization of the affected products before they are placed on the market. In this way, product quality, safety and efficacy can best be ensured.

Furthermore, there are extant harmonization directives in accordance with the new approach to the harmonization of legislation – a regulatory strategy based on a Council decision taken in 1985 [11] incorporating the following principles:

- Legislative harmonization is limited to the adoption of the essential requirements regarding legal health or environment protection issues with which products put on the market must conform in order to enjoy free movement within the European market.
- Technical specifications and detailed operating procedures conforming to the essential requirements established by the directives are drawn up into harmonized standards by European or international standardization committees rather than by European legislators. Although the application of these standards is not legally mandated, products manufactured in conformity with harmonized standards are presumed to conform to the essential requirements. This is referred to as presumption of conformity.
- Free movement of products within Europe is achieved by the mutual recognition of conformity assessments which are generally carried out on the product manufacturer's sole responsibility. In cases where serious health or safety risks might arise due to product- and production-related defects, independent testing institutions, called conformity assessment bodies, must be additionally involved in the conformity assessment. Conformity assessment procedures result in declarations of conformity which all EU Member States recognize. The affixing of the CE marking on the products by the manufacturer is the tangible sign of their conformity to European legal requirements and is a prerequisite for free movement of the product (CE = French abbreviation for "conformité européenne").

The "Global approach to the assessment of conformity" amended the new approach [2, 7, 10] and lets manufacturers choose between different conformity assessment procedures. These are structured in a modular fashion so that the appropriate modules can be used for the design and production stages of the goods. The relevant harmonization directives contain information about the modules and conformity assessment procedures basically applicable to the products.

Moreover, the global approach not only envisages conformity assessments for the products in question, but also for the conformity assessment bodies themselves. These include certification bodies and stake-holding subcontractors, e.g., test laboratories. Since these conformity assessment bodies are primarily involved in the assessment of products with respect to quality and safety requirements, a high level of quality and competence is demanded. The competence criteria, which conformity assessment bodies must fulfill, are documented in harmonization directives. Conformity with these criteria is usually demonstrated by designations as a conformity assessment body issued by the national authority and is valid Europewide. As appropriate, the designation procedure can take results from parallel accreditation procedures undertaken by national accreditation bodies into account. This is how the Member States notify the European Commission of certification bodies designated by this procedure. It is also why they are referred to as "notified bodies" [10, 22].

Aimed at further harmonization and heightened consistency of regulations regarding free movement of goods, the new and global approach was extensively modernized in 2008. Compiled from three European Legal Acts, the "New Legislative Framework" was designed to create a European accreditation concept for evaluation of the competence of conformity assessment bodies, to harmonize market surveillance strategies in the Member States while paying attention to current product safety requirements and strengthening the significance and reliability of the CE marking for products traded on the European market [3, 27, 28] (▶ Chapter 39).

25.2.2 Harmonization of medical devices and its significance for POCT

Medical devices are also classified into the over 20 different product groups regulated by the harmonization directives within the new approach. They are the most sensitive products regulated by this approach in terms of requirements for quality, safety and effectiveness and/or health protection performance.

Three European directives form the basis of the regulatory framework for medical devices. Directive 90/385/EEC encompasses regulations for active implantable medical devices [8]; Directive 93/42/EEC regulates the manifold product range of medical devices [9] which can constitute a broad array of products from gauze bandages, condoms to contact lenses. The third is Directive 98/79/EC which ultimately regulates IVD. The directives were amended and changed several times over the past years, lastly in 2007 [25] and 2011 [26] (◘ Fig. 25.1).

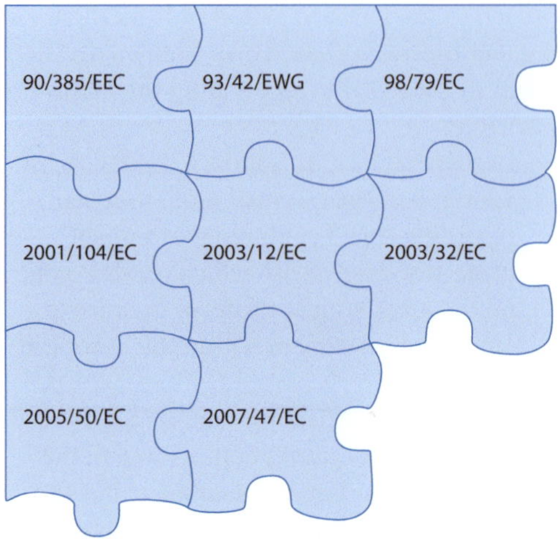

◘ **Fig. 25.1** The European medical device legislation is based on the three fundamental Directives 90/385/EEC, 93/42/EEC and 98/79/EC. In the last years, they were amended by the directives 2001/104/EC on substances derived from human blood or human plasma, 2003/12/EC on the reclassification of breast implants, 2003/32/EC on medical devices manufactured utilizing tissues of animal origin, 2005/50/EC on the reclassification of hip, knee and shoulder joint replacements and 2007/47/EC amending directives 90/385/EEC and 93/42/EEC. In 2011, Directive 2011/100/EU replaced Directive 98/79/EC of the European Parliamen.

In Germany, as in other European countries, European harmonization of legislation in the medical device sector had wide-ranging consequences. Formerly, for example, the German requirements governing IVD were laid down heterogeneously in different regulations such as the Medicinal Products Act, the Weights and Measures Act, the Equipment Safety Act and the Medical Devices Ordinance. By contrast, the national transposition of the European directives was unified by the MPG which contains regulations for all product groups in the medical device sector. In some EU Member States, such regulations had not existed at all, meaning that harmonization of legislation in this field accomplished a unified standard across Europe for all product manufacturers, on the one hand, and for patient and user groups on the other.

According to the legal definition in Section 3 (4), an IVD is "a medical device which is intended to be used, alone or in combination with others, as a reagent, reagent device, calibrator material, control material, kit, instrument, apparatus, equipment or system, according to the intended purpose specified by the manufacturer, for the in vitro examination of specimens derived from the human body, including blood and tissue donations, purely or mainly with a view to providing information

- on physiological or pathological conditions
- congenital abnormalities or
- to investigate the safety of or tolerance by potential recipients or
- to monitor therapeutic measures."

Consequently, all systems or devices used in the field of POCT have to be classified as IVD. For this classification according to the IVD Directive, it is irrelevant whether the devices used for diagnosis are applied in medical laboratories by trained personnel (laboratory diagnostics devices), on the ward (POC diagnostic devices) or for self-testing (home diagnostic devices, "home-use"). Regarding the CE marking and the associated free movement on the market,

POC diagnostic devices must fulfill the requirements for quality, safety and performance as laid down in the IVD Directive.

25.3 Requirements of the European directive on in vitro diagnostic medical devices (IVD Directive)

25.3.1 Product categories and conformity assessment procedures

IVD are not divided into classes just as with active implantable medical devices (placed on the market according to Directive 90/385/EEC), but unlike all other medical devices (placed on the market according to Directive 93/42/EEC). However, the IVD Directive does distinguish between four groups of IVD which have to follow different conformity assessment procedures depending on their hazard potential. The following groups are distinguished:

- Devices according to Annex II, List A of Directive 98/79/EC,
- Devices according to Annex II, List B of Directive 98/79/EC,
- Devices for self-testing (except self-testing devices for blood-glucose monitoring) and
- other IVD.

Devices according to Annex II List A are referred to as **high-risk devices**. Their malfunctioning can have serious consequences for the health of users, patients and/or the general public. Devices according to Annex II List B are referred to as **risk devices**. They have an enhanced risk potential, but it is much lower than the List A devices (▶ Section 25.1 and ◘ Tab. 25.1). Since promulgation of the IVD Directive, there has been criticism that no truly discernible systematic classification of risk potential exists for List B devices like for the high-risk devices in List A.

The third product group sub-classifies devices for self-testing, called **home-use devices**. These are devices used by laypersons instead of trained laboratory personnel to determine specific diagnostic markers. However, devices for self-testing listed in Annex II, List B of the IVD Directive are excluded from this group, namely devices for blood sugar monitoring.

All other diagnostics are classified as devices with the lowest hazard potential. Following the principle of the IVD Directive, all devices neither listed in Annex II of the Directive nor used as self-testing devices are automatically classed as "other" IVD medical devices, thereby obviating any obligation on the part of the manufacturer to categorize them in further detail.

The result is that POC diagnostics are essentially not allocated to a specific product group according to the regulations of the IVD

◘ **Tab. 25.1** High-risk and risk devices according to Annex II of Directive 98/79/EC (IVD Directive). Reagents and reagent products, including related calibrators and control materials

List A	List B
Determination of the following blood groups: ABO system, rhesus (C, c, D, E, e), Kell system Detection, confirmation and quantification in human specimens of markers of HIV infection (HIV 1 and 2), HTLV I and II, and hepatitis B, C and D.	Determination of Duffy and Kidd blood group systems Determination of irregular anti-erythrocyte antibodies Detection and quantification of congenital infections by rubella and toxoplasmosis Determination of phenylketonuria Determination of infections by cytomegalovirus and Chlamydia Determination of HLA tissue groups: DR, A, B Determination of PSA Software for evaluating the risk of trisomy 21 Devices for the self-diagnosis of blood sugar

Directive. Rather, the intended purpose of a POC system in terms of the diagnostic variable to be determined dictates its allocation to a product group. Several of the devices listed in Annex II of the IVD Directive are also used in POCT: for example, "bedside" tests available on the market for years to confirm blood type or more modern rapid HIV tests (waived tests). However, most diagnostic products used at the POC are categorized as "other" IVD.

The regulatory characteristics for the four product groups concern the selection of conformity assessment procedures as stipulated by the Directive. Depending on the procedure, these in turn dictate the different requirements governing design, manufacturing and verification of the devices manufactured. Additionally, the need to involve notified bodies differs. Notified bodies must only be involved in conformity assessments for devices listed in Annex II and for devices for self-testing, but not in the case of other IVD.

For the manufacturer, the safety concept of the IVD Directive also specifies different scopes for getting notified bodies involved in certification. For high-risk devices, a notified body has to evaluate the quality assurance system (or an equivalent alternative) encompassing the design and production phase in addition to a design examination and a verification of the manufactured devices ("batch release"). Manufacturing of risk devices needs certification of a comprehensive quality assurance system (or an equivalent alternative), however, without design examination and verification of the manufactured devices by a notified body. Ultimately, only an examination of the design of self-testing devices by a notified body is mandatory.

Compared to other IVD, higher safety requirements are placed on self-testing devices due to the fact that laypersons are performing the diagnostic tests outside of medical laboratories. A comparable safety requirement for POC diagnostic devices routinely used by personnel, who are not trained laboratory staff, has not yet been demanded in the IVD Directive.

25.3.2 Essential requirements

Irrespective of the choice of conformity assessment procedure, the essential requirements laid down in Annex I of the IVD Directive must be fulfilled by all IVD, including POC diagnostic devices for them to enjoy free movement on the market. These essential requirements stipulate, among others, that IVD design and manufacturing must integrate safety principles, taking into account the generally accepted state-of-the-art. The device must perform according to the manufacturer's intended use, specifically in terms of the manufacturer's stated sensitivity, specificity, accuracy, repeatability, reproducibility including interference control and detection limits and conform to the generally accepted state-of-the-art. Further requirements concern chemical and physical characteristics of IVD, requirements on mitigation of infections and contamination risks, construction safety, precision of measuring devices, protection against radiation, safety of devices equipped with energy sources, labeling and packaging, including the device's instructions for use.

Conformity with the essential requirements is not only documented by the manufacturer's certificate of conformity ultimately issued, but by a comprehensive technical documentation of the device, equally mandatory for all IVD. According to Annex III of the IVD Directive, the technical documentation must include the results of the risk analysis on the product as well as data from the performance evaluation studies conducted (evaluation of the diagnostic trials). Although, the conduct of a formal certification procedure is not necessary according to Annex III, every IVD manufacturer is still required to put a quality assurance system in place and demonstrate its existence with technical documentation.

Accordingly, POCT diagnostic devices are subject to the same essential requirements that apply to all IVD. However, special requirements for the conformity assessment, especially in the fields of risk management and performance evaluation tests, result from the characteristics relating to intended use and application of these devices (by non-laboratory personnel at the bedside).

25.3.3 Harmonized standards and "common technical specifications"

Also in the case of IVD and according to the new approach, standards are used to technically specify the requirements laid out in the IVD Directive. These standards are drafted by the European standards organizations CEN (CEN = French abbreviation of Comité Européen de Normalisation) and CENELEC (Comité Européen de Normalisation Electrotechnique) under contract from the European Commission and subsequently recognized by all Member States. Therefore, they are called mandated harmonized standards. Based on the Vienna Agreement of 1991, the work results of the European standards organizations and the international standards organization ISO [23] are adopted in parallel to this.

Periodically, the Commission publishes the titles and references of the harmonized standards. To date, more than 35 standards have been published within the framework of the implementation of the IVD Directive [6]. Examples of particular importance are the standards EN ISO 13485 [16], EN 13612 [14], EN ISO 14971 [17] and EN 14136 [15]. EN ISO 13485 documents requirements for quality management systems which manufacturers are required to install. EN 13612 describes criteria for performance evaluation and EN ISO 14971 includes requirements for risk management during production. EN 14136 does not primarily address product manufacturers, but interlaboratory test providers. This standard describes requirements governing the design, performance and evaluation of interlaboratory tests for in vitro diagnostics and therefore considers the significance of external quality assessment schemes within the IVD vigilance system (reporting system).

A great number of harmonized standards are also applicable on POC diagnostic devices, even though none of these standards have yet to be aligned with the special aspects of these devices. By contrast, several standards exist that include the requirements for self-testing devices, e.g. EN ISO 15197 which describes the requirements for blood glucose monitoring systems as self-tests [18]. Due to the increasing importance of POCT, an intensification of standardization projects in this area can be expected in the future.

The "Common Technical Specifications (CTS)" represent a special case where a European expert group draws up specifications to define quality criteria and the minimum scope of the performance evaluation for devices of Annex II, List A of the IVD Directive [1]. Similar to standards, the application of CTS creates a presumption of conformity. Nevertheless, CTS hold a superior rank over standards. As conventional IVD, POC diagnostic devices used for measurement of List A diagnostic markers must meet the CTS.

Although no special product standards have been laid down for POCT diagnostic devices so far – neither on the European nor the international level – there is an international standard for the clinical application of these devices: EN ISO 22870 "Point of Care Testing (POCT) – requirements for quality and competence" [12] (► Chapter 39).

25.4 Act on Medical Devices (Medical Devices Act – MPG) and subordinate regulations

The MPG of 1994 transposes EU law governing the medical device sector into German law [20] mostly by direct reference to the relevant provisions and annexes of the relevant EU directives. IVD also became affected by the harmonization of legislation upon enactment of the second law amending the Act on Medical Devices of December 2001. For that reason, it has been required since 7 December 2003 and 7 December 2005, respectively, that the relevant conformity assessment procedures be conducted and the CE marking correctly affixed to the product prior to the first placing on the market or first putting into service of an IVD.

The MPG contains the regulations on the free movement of medical devices, while laying down further requirements for the tasks and responsibilities of the competent authorities, the operation and use of the devices along with how medical device vigilance is to be conducted.

The competent authority for the designation of notified bodies in the entire medical device sector is the Central Authority of German Federal States for Health Protection with regard to Medicinal Products and Medical Devices (Zentralstelle der Länder für Gesundheitsschutz bei Arzneimitteln und Medizinprodukten, ZLG). Besides the authorities of the German Federal States responsible for supervision, the Federal Institute for Drugs and Medical Devices (Bundesinstitut für Arzneimittel und Medizinprodukte, BfArM) and the Paul Ehrlich Institute (PEI) have tasks within the vigilance system for medical devices which includes risk assessments of medical devices. The PEI is responsible for all IVD according to Annex II, List A of the IVD Directive as well as many devices listed in Annex II, List B, whereas the BfArM is responsible for all other IVD, a smaller number of devices listed in List B and all other medical devices [4, 21].

The MPG delegates a number of powers to issue ordinances that are transposed into national regulations. For example, Section 37 (5) sub-section 2 MPG empowers the German Federal Ministry for Health "to issue ordinances... stipulating requirements regarding the quality system in operating and using in vitro diagnostic medical devices,...".

Lastly, these requirements result from the safety approach of the IVD Directive based on the consideration that "the large majority of such devices do not constitute a direct risk to patients and are used by competently trained professionals". Thereunder, the involvement of notified bodies regarding the assessment of IVD is only required for relatively few devices mainly used in transfusion or infectious disease medicine (Annex II, List A and B of the IVD Directive) or for self-testing. Concurrently, however, quality assurance during operation and use of the devices is an implicit part of the European regulation.

The requirements governing quality assurance are currently laid down in the MPBetreibV [29], which mandates measures for internal and external quality control with reference to the RiliBÄK. On 1 April 2010, it also became mandatory to implement a comprehensive quality management system (QMS) for IVD. The entire POCT sector is subject to the measures for implementation of a QMS.

25.5 Reform of the European legislation on medical devices

A fundamental revision of the European legislation on medical devices has been recently enacted. Since May 2017, IVD have become subject to a separate, directly applicable European regulation, Regulation (EU) 2017/746.

There were several reasons behind the planning of the revision: Over the years, the directives had come in need of consolidation because the legal wording was fragmented by continual amendments. Concurrently, the technological and scientific developments of the past years and decades had exposed deficiencies in current legislation governing medical devices on the European level. Moreover, the establishment of high standards for the quality and safety of IVD aims to eliminate the general safety concerns sometimes expressed by the public. Finally, the international market demands progressive harmonization between the EU and non-member countries. Given the above, the guidelines developed on an international level, especially those drafted by the Global Harmonization Task Force (GHTF) and its follow-up initiative, the International Medical Devices Regulators Forum, are also applied to IVD. This is particularly true for regulations governing the unique device identification (UDI), general safety and performance requirements, technical documentation, conformity assessment procedures, the clinical evaluation and criteria for classification. In the future, classes A, B, C or D will apply instead of the previous listing system [21]).

The IVD Regulation ("IVDR") contains stricter regulatory requirements, also for manufacturers of POC diagnostic devices (Regulation (EU) 2017/746 of The European Parliament and of the Council of 5 April 2017 on in vitro diagnostic medical devices and repealing Directive 98/79/EC and Commission Decision 2010/227/EU). In future, a new product category "**devices for near-patient testing**" will be established, certainly to account for the increasing importance of POC diagnostics. This category contains devices which are not intended for self-testing, but for the application outside the (conventional) laboratory environment, generally near or at the patient. For such class B, C and D devices, it is stipulated that a notified body will be required to audit the **design dossier**. Devices for near-patient tests are not assigned a priori to a certain class, but put in a class of their own according to the intended purpose of use as claimed by the manufacturer. Product information (labeling, instructions for use, identification) is provided in the language(s) of the Member State(s) in which the intended users will be supplied with the device. Moreover, devices for near-patient tests are additionally tested on their performance in different medical environments (e.g. patients' homes, emergency departments, outpatient centers). Generally, they must be designed and manufactured in such a way that they perform appropriately for their intended purpose. The skills of and the means available to the intended users must also be considered alongside the impacts of fluctuations to be reasonably anticipated in the techniques and environments of the intended users. Therefore, the information and instructions provided by the manufacturer should be easy to understand and apply by the intended users. This includes information about the level of the users' education, qualification and/or experience.

References

1. Commission Decision of 27 November 2009 amending Decision 2002/364/EC on common technical specifications for in vitro diagnostic medical devices (2009/886/EC) (OJ L 318, 4.12.2009, p. 25)
2. Council Decision of 22 July 1993 concerning the modules for the various phases of the conformity assessment methods and the rules for the affixing and use of the CE conformity marking, which are intended to be used in the technical harmonization directives (93/465/EEC) (OJ L 220, 30.8.1993, p. 23)
3. Decision No 768/2008/EC of the European Parliament and of the Council of 9 July 2008 on a common framework for the marketing of products, and repealing Council Decision 93/465/EEC (OJ L 218, 13.8.2008, p. 82)
4. Bornhak H, Dörr V, Halbauer J, Meyer-Lüerßen D, Siekmeier R, Will HG (2002) Die Anforderungen der Medizinprodukte-Sicherheitsplan-Verordnung für In-vitro-Diagnostika im Rahmen des Medizinproduktegesetzes. MedizinProdukteRecht 2: 1–14
5. German Medical Association (2014) Guideline of the German Medical Association on Quality Assurance of Medical Laboratory Examinations (RiliBÄK 2014). Dtsch Ärztebl 111: A1583–A1618
6. Commission communication in the framework of the implementation of the Directive 98/79/EC of the European Parliament and of the Council of 27 October 1998 on in vitro diagnostic medical devices (OJ C 226, 10.7.2015, p. 43)
7. Commission communication to the Council of 15 June 1989: A global approach to certification and testing – Quality measures for industrial products (OJ No C 267, 19.10.1989, p. 3)
8. Council Directive of 20 June 1990 on the approximation of the laws of the Member States relating to active implantable medical devices (90/385/EEC) (OJ L 189, 20.7.1990, p. 17), last amended by Directive 2007/47/EC of the European Parliament and of the Council of 5 September 2007 (OJ L 247, 21.9.2007, p. 21)
9. Council Directive 93/42/EEC of 14 June 1993 concerning medical devices (OJ L 169, 12.7.1993, p. 1), last amended by Directive 2007/47/EC of the European Parliament and of the Council of 5 September 2007 (OJ L 247, 21.9.2007, p. 21)
10. Council Resolution of 21 December 1989 on a global approach to conformity assessment (OJ C 10, 16.1.1990, p. 1)
11. Council resolution of 7 May 1985 on a new approach to technical harmonization and standards (OJ C 136, 4.6.1985, p. 1)
12. DIN EN ISO 22870:2006: Point-of-care testing (POCT) – Requirements for quality and competence

13. Directive 98/79/EC of the European Parliament and of the Council of 27 October 1998 on *in vitro* diagnostic medical devices (OJ L 331, 7.12.1998, p. 1), last amended by Commission Directive 2011/100/EU of 20 December 2011 (OJ L 341, 22.12.2011, p. 50)
14. EN 13612:2002: Performance evaluation of in vitro diagnostic medical devices
15. EN 14136:2004: Use of external quality assessment schemes in the assessment of the performance of in vitro diagnostic examination procedures
16. EN ISO 13485:2012): Medical devices – Quality management systems – Requirements for regulatory purposes
17. EN ISO 14971:2012: Medical devices – Application of risk management to medical devices
18. EN ISO 15197:2013: In vitro diagnostic test systems – Requirements for blood-glucose monitoring systems for self-testing in managing diabetes mellitus
19. Consolidated versions of the Treaty on European Union and the Treaty on the Functioning of the European Union (OJ C 326, 26.10.2012, p. 1)
20. The Act on Medical Devices (Medical Devices Act – MPG) in the version of 7th August 2002 (Federal Law Gazette I, p. 3146), last amended by Article 278 by the ordinance of 31st August 2015 (Federal Law Gazette I p. 1474)
21. GHTF/SG1/N045:2008: Principles of In Vitro Diagnostic (IVD) Medical Devices Classification
22. The 'Blue Guide' on the implementation of EU product rules 2016. http://ec.europa.eu/DocsRoom/documents/16210
23. ISO/CEN: Agreement on technical cooperation between ISO and CEN, (23.12.2002), Chapter 4. http://www.cenorm.be/boss/supmat/refdoc/archive/ms/ms002.htm
24. Directive 2001/83/EC of the European Parliament and of the Council on the Community code relating to medicinal products for human use (OJ L 311, 28.11.2001, p. 67), last amended by Directive 2012/26/EU of the European Parliament and of the Council of 25 October 2012 (OJ L 299, 27.10.2012, p. 1)
25. Directive 2007/47/EC of the European Parliament and of the Council of 5 September 2007 amending Council Directive 90/385/EEC on the approximation of the laws of the Member States relating to active implantable medical devices, Council Directive 93/42/EEC concerning medical devices and Directive 98/8/EC concerning the placing of biocidal products on the market (OJ L 247, 21.9.2007, p. 21)
26. Commission Directive 2011/100/EU of 20 December 2011 amending Directive 98/79/EC of the European Parliament and of the Council on in vitro diagnostic medical devices (OJ L 341, 22.12.2011, p. 50)
27. Regulation (EC) No 764/2008 of the European Parliament and of the Council of 9 July 2008 laying down methods relating to the application of certain national technical rules to products lawfully marketed in another Member State and repealing Decision No 3052/95/EC (OJ L 218, 13.8.2008, p. 21)
28. Regulation (EC) No 765/2008 of the European Parliament and of the Council of 9 July 2008 setting out the requirements for accreditation and market surveillance relating to the marketing of products and repealing Regulation (EEC) No 339/93 (OJ L 218, 13.8.2008, p. 30)
29. Ordinance concerning the setting up, operation and application of medical devices (Ordinance on Operators of Medical Devices – MPBetreibV) of 21 August 2002 (Federal Law Gazette I p. 3397), last amended by Article 3 der Ordinance of 11 December 2014 (Federal Law Gazette I p. 2010)
30. Second law amending the German Act on Medical Devices (2nd MPG-ÄndG) of 13 December 2001 (Federal Law Gazette I p. 3586), amended on 23 May 2002 (Federal Law Gazette I p. 167)

Liability issues relating to POCT

Ulrich M. Gassner

P. B. Luppa, R. Junker (Eds.), *Point-of-Care Testing*
https://doi.org/10.1007/978-3-662-54497-6_26

26.1 Introduction

Should personal injury arise from the use of POCT, liability is determined by the general regulations governing medical devices given that POC devices qualify as in-vitro diagnostic medical devices (IVD) and therefore fall under this legal classification (▶ Chapter 25). Most claims for damage compensation are filed against:

- Manufacturers (producers) by patients, operators and injured third parties
- Users and operators by patients and injured third parties
- Notified bodies by patients, producers, users and injured third parties
- Competent authorities by patients and manufacturers

The first two sections of this chapter will only address the first two groups of respondents in line with their relevance to legal reality. Essentially, the respondent is sued by the injured party to provide restitutio in integrum. This means restoring the position of a contractual partner's legal interest that exists outside of the boundaries of the contractual relationship – in this case his physical integrity – to the original state it would have been in had that particular damage not occurred. This tier of liability is subject to statutory regulation. The injured party has a right to contractual and non-contractual (tortious) claims for damage compensation. Unlike with pharmaceuticals, there is no separate liability regime governing the product liability to which medical device manufacturers (producers) are subject. Similarly, user liability and operator liability are also regulated in the general laws.

Additionally, two further tiers of liability can be distinguished. One tier of liability is of an administrative or regulatory nature: It concerns the liability of the responsible parties for the obligations incumbent upon them under medical device laws. Pursuant to Section 5 (1) of the German MPG [8], the manufacturer or its authorized representative, but also under certain circumstances, the importer (Section 5 (2) MPG) are the responsible parties. Should a responsible party breach his obligations, he becomes subject to actions by the competent authority. The other tier of liability concerns administrative and criminal sanctions (Sections 40 et seq. MPG). The competent authority can impose fines whenever the provisions of MPG are violated. An administrative offense shall be deemed to be committed by any person who violates the provisions of Section 30 (1) MPG by willfully or negligently failing to appoint a safety officer or by failing to do so in time (Section 42 (2), No. 13 MPG). Persons committing serious violations may even receive a prison sentence. According to Section 40 (1) No. 4 MPG, any person who endangers patients, employees or third parties by operating or using defective medical devices shall be punished by the imposition of a monetary fine or by imprisonment for up to three years. Notwithstanding, in practice, these two tiers of liability are much less relevant than the liability of manufacturers, users and operators under civil law. For this reason, these tiers will not be ddressed further in the following.

26.2 Superimposition of liability spheres

During application of medical devices in hospitals, especially complex liability constellations occur since the responsibilities of the physician, hospital and manufacturer superimpose. On the one hand, all participants must respect specific protection obligations to patients, but are dependent on cooperation. The physician must rely on the manufacturer of the medical device concerning information about its quality, application and related risks in order to use the device properly and be able to inform patients correctly. Consequently, the manufacturer is liable for clear malfunctioning of medical devices and not the physician. At least, if the fault would have been detectable by the physician, he is also partly liable [14, 16].

Even minor breaches of duty of care can constitute grounds for liability. Therefore, strict compliance with RiliBäK and other legal provisions is mandated. Avoidance of liability is a result of cooperation of all participants starting with the clinic management over chief physician, POCT coordinator and POCT supervisor up to the attending physician and nursing staff.

Manufacturers, users and operators of medical devices mutually form a community of responsibility together with patients that works proactively to mitigate risk.

26.3 Manufacturer's liability

26.3.1 Liability regimes

Civil liability of the manufacturer is regulated, on the one hand, in the general regulations of the German Civil Code (BGB) [2] and, on the other hand, in the German Product Liability Act (ProdHaftG) [7]. The main constructional difference between the two liability regimes is that liability under the ProdHaftG applies irrespective of the actual fault (absolute or strict liability) versus the tortious liability of the BGB. When it comes to liability in legal practice, differences between absolute and tortious liability have mostly been equalized due to the presumptions of fault arising from the many legal precedents.

26.3.2 Liability under the German Product Liability Act (ProdHaftG)

Under the ProdHaftG, the manufacturer bears liability for any personal injuries caused by his defective device. In the narrower sense, the ProdHaftG defines a manufacturer as "a person who has produced the final product, a raw material or a component part" (Section 4 (1) sentence 1 ProdHaftG). The proprietary production of IVD medical devices at healthcare facilities as defined under Section 3 (22) MPG can justify the definition that said facility counts as a manufacturer, if the patient "must bring himself to within that sphere of control" for diagnostics [5].

According to Section 3 (1) ProdHaftG, a product is defective when it does not provide the safety, taking all circumstances into account, in particular

- its presentation,
- the use to which it could reasonably be expected that it would be put and
- the time when it was put into circulation,

whereas said safety can be reasonably ascribed to the product. In this context, the product's safety is hereby not defined as a subjective expectation of safety on the part of individual consumers, rather as an expectation that is objectively justified. The safety expectations of that group of persons targeted by the product's manufacturer are decisive. The general public and especially patients and physicians expect that medical devices will not pose risks beyond a reasonable extent as set forth in Section 4 (1) No. 1 MPG. This similarly above implies that the expectation of complete and absolute safety is not justified in the sense of Section 3 (1) ProdHaftG.

According to Section 3 (1) ProdHaftG, evaluation of a defect must also consider the use to which the product "could reasonably be expected to be put" alongside its presentation as reflected in instruction manuals, advertising claims and so on. This wording clearly illustrates that, besides proper use as intended, objective misuse should also be accounted for, say e.g., the predictable false handling of plug connections [1]. Moreover, the time when the product was put into circulation is important for fault evaluation. Accordingly, any increased safety expectations that arise later – for example those resulting from improvements to the product – shall not be considered. A product is consequently not defective for the sole reason that a better product has subsequently been put into circulation.

Further aspects to be considered in line with Section 3 (1) ProdHaftG are, as an example, the familiarity of users with a product or scientific technical standards as reflected in harmonized standards as set forth in Section 8 (1) MPG [4].

In practice, three defect categories have emerged for the evaluation of a defect as defined under Section 3 (1) ProdHaftG: defects in construction, production and instructions [11]. In accordance with a recent judgement by the Court of Justice of the European Union (CJEU), a potential defect of implantable medical devices can justify liability [4]. However, this legislation cannot be applied to IVD medical devices, inter alia, since, according to their intended purpose, such devices do not have any contact with patients at all, but only with the specimens obtained from them.

In accordance with Section 1 (1) ProdHaftG, a claim for damage compensation is only legitimate when the damage is caused by a defect in the product and covered by the protective purpose of this provision. That means, inter alia, that other causes must be excluded [21].

Finally, it should be mentioned that liability can also be ruled out under Section 1 (2) ProdHaftG. As defined therein, the producer is not liable if "the state of scientific and technical knowledge at the time the producer put the product into circulation was not such as to enable the defect to be discovered" (Section 1 (2) No. 5 ProdHaftG). Knowledge gained later does not mean that the product was originally defective. Consequently, the manufacturer does not bear the development risk [4].

26.3.3 Liability under tort law

Section 823 "Liability in damages" of the German Civil Code (BGB) is the pivotal standard governing liability. Liability as defined by Section 823 (1) BGB requires:

- Violation of a certain legal interest
- Intent or negligence (culpability)
- Unlawfulness of the act of injury
- Damage (injury)
- Causality of the act of injury in relation to the violation of a certain legal interest (causal relation justifying liability)

The decisive aspect is whether the manufacturer has infringed upon any specific legal duties to maintain safety incumbent on him. Said duties are reflected in the typical sources of error formulated in the legislation: Defects in design, production and instructions. In addition to the product liability arising from these defect categories under the ProdHaftG, the manufacturer is liable tortiously for errors in product vigilance and organization [11, 17].

According to Section 823 (2) BGB, someone "who culpably commits a breach of a statute that is intended to protect another person is liable to make compensation." The "protection law" concept describes a certain provision of the affected law characterized by protection not only relating to the general public, but specifically to the individual. In the MPG, the following statutes meet this requirement: Section 4 (1) No. 1 and 2, (1) No. 1 to 3); Section 6 (1) sentence 1, (2) in conjunction with Section 7 (1); Sections 11 (2), 12, 20, 21, 30 (1) and 4 MPG as well as nearly all criminal statutes and provisions on administrative fines (Sections 40, 41, 42 (1) and (2) No. 1). Further protection standards are codified in the Medical Devices Operator Ordinance (MPBetreibV) (Sections 4 (1) and (4), 8 (2), 10 (1) and (2), 15 (1) [22]. Individual protection standards can also be found in the German Medical Devices Regulation (MPV) [17] (Sections 3, 13 (1)) [8, 11]. Finally, most provisions of the RiliBÄK [3] in conjunction with Section 9 (1) MPBetreibV qualify as protection laws as defined by Section 823 (2) BGB.

26.4 User and operator liability

26.4.1 Liability regimes

Legislation on medical devices does not distinguish special provisions with regard to the civil liability of operators and users [14, 15]. Therefore, the general liability standards apply. Applicable are ProdHaftG, tort law (Sections 823 et seq. BGB) as well as statutory claims arising from the existing treatment contract between the involved parties (Sections 280 (1) in conjunction with Sections 630a et seq. BGB). The two latter liability standards have in common that they presume (negligent or intentional) fault (Section 276 (1) sentence 1 BGB). According to Section 276 (2) BGB, someone acts negligently if they fail to exercise reasonable care. Any type of negligence satisfies Section 823 (1) BGB so that the slightest breach of duty gives rise to liability.

Even the most minor breaches of duty give rise liability.

26.4.2 Absolute liability

Operators and users become self-producers if they violate the provision of Section 4 (4) MPBetreibV by disregarding the intended purpose of the original manufacturer. At the same time, they assume the manufacturer's responsibility for the product (see above). In practice, this is particularly important in relation to the application of stand-alone software [12].

26.4.3 Fault-based liability

Treatment contract

Any physician who treats a patient has de facto entered into a treatment contract (Section 630a (1) BGB). From this contract arises an obligation on the part of the physician vis-à-vis the patient that treatment must be rendered according to the medical standards that are generally recognized at the time of the treatment (cf. Section 630a (2) BGB). Contrary to the common vernacular, the treating party is not the person performing the medical intervention but the party promising to provide the treatment. This must not necessarily be one and the same person. Rather, a legal entity not capable of performing the action can be the contractual party, e.g. the body responsible for the hospital.

Legal duties to ensure safety

Under on the treatment contract entered into, both hospital operator and user are obliged to deploy the medical device with due diligence and to not use defective devices. This is an ancillary contractual obligation that is reflected in the form of a legal obligation to ensure safety. The principle is based on the concept that anyone creating sources of danger has to take the necessary precautions to safeguard third parties. In general, they are reflected in the protection statutes as set forth in Section 823 (2) BGB (see above). An operator or user who culpably violates such obligations is contractually obliged to compensate for the damage and pay for pain and suffering (Sections 630a, 280 BGB).

Attribution

The culpability of the acting nursing staff or treating physician, respectively, shall be attributed to the hospital operator provided said persons are acting as the operator's vicarious agents (Section 278 sentence 1 alternately 2 BGB). This liability for the fault of others presumes that the "principal" freely decided to delegate responsibility, chose his vicarious agents of his own volition and has authority to issue instructions to said agents. However, the vertical authority (privity) to issue instructions is initially only given in his own department. According to the application of POCT in inpatient departments, RiliBÄK offers a centralized and a decentralized model for the allocation of responsibilities for quality assurance. The centralized organization concept is characterized by the laboratory management, or on behalf of it, the POCT coordinator, being responsible, by contrast, the transfer of responsibility and organization on the POCT coordinator by the clinic management is not possible in the decentralized concept, where responsibility is on the departmental level (► Chapter 31, [20]).

Non-contractually, hospital operators, physicians as well as nursing staff may also be personally liable by virtue of legislation governing torts (Sections 823 ff. BGB). In spite of their responsibility to implement the tasks they are charged with, employees are hereby entitled to be generally exempted from liability by their employer. As principal, the hospital operator is liable for the action of his vicarious agents (Section 831 (1) sentence 1 BGB). In derogation from Section 278 BGB, however, the principal's liability does not extend to faults by others, but only to his own fault, namely faults in the selection and guidance of the vicarious agent as well as in the procurement of materials for said agent. However, this does not apply, if he can prove that he exercised reasonable care (Section 831 (1) sentence 2 BGB). As long as the principle has done this, he has no fault of his own. Decentralized exculpatory evidence is possible in larger hospitals, if the hospital clinic management exculpate itself in its relationship to the persons responsible for selection and supervision. Notwithstanding, organizational culpability under Section 823 BGB may in turn apply (see below).

Malpractice

Medical malpractice can be classified as follows: errors in diagnosis (diagnostic error); errors in the selection of therapy and diagnostics; errors in collecting findings required for therapy and diagnosis (medical examination fault), incorrect performance of the correctly selected therapy or diagnostics, errors in follow-up care, in ensuring the success of treatment (obligation to provide information) and in errors in the termination of treatment [10, 14].

Assessment standard

An objective standard of care has to be applied to assess errors. Insofar, special circumstances relating to the treatment situation or a particular individual's level of proficiency, like inadequate training or experience, are not usually exculpable for the physician [14]. The fact that the manufacturer has complied with the pertinent safety requirements by affixing the CE mark to the POCT device is equally as irrelevant [11, 16].

According to a more recent, albeit hitherto isolated judgment, a treating physician faced with a conflict with requirements in medical device legislation may act primarily in accordance with the medical standard of care if necessary to reinstate the health of the patient [13].

POCT-relevant liability constellations

In terms of medical malpractice, the right selection of tests is of paramount important in the use of POCT. The physician is obliged to check that the test procedure used for diagnosis or therapy monitoring corresponds to the state of the art in science and technology. In spite of lower quality compared to laboratory tests, POC tests must satisfy an adequate medical standard for the concrete medical problem. This standard may not be sacrificed to expediency [19].

> **The selection of the suitable test for a medical problem must satisfy an adequate medical standard.**

Users and operators of POC devices must primarily follow the requirements of MPBetreibV. In this context, Sections 3 and 4 MPBetreibV contain the most important general requirements in terms of person-related and substantive issues.

Medical devices may only be operated, used and maintained by persons who have the required training or knowledge and experience (Section 4 (2) MPBetreibV in conjunction with Section 14 MPG). For example, an operator can also be a physician with hospital rights who brings and operates his own POCT devices there (Section 4 (5) MPBetreibV).

Not only operators, but users are also responsible for compliance with the substantive requirements. In particular, medical devices may only be operated and used for their intended purpose (Section 4 (2) MPBetreibV). Furthermore, the generally accepted standards of technology (good engineering practice) must be observed (Section 4 (1) MPBetreibV). Beside every operator (cf. Section 3 (1) MPBetreibV) every user must additionally check the functionality and proper condition of the device and follow the instruction manual as well as all other safety-related information and maintenance instructions (Section 4 (6) sentence 1 MP-BetreibV). The function test must not only cover the medical device but also include all medical products, accessories including software or other parts connected to or with the medical device and its application alongside any combination thereof (Section 4 (6) sentence 2 MPBetreibV).

Private and statutory health insurance companies and nursing care insurers are not operators of medical devices, but – in the interests of their insured persons – must assume the obligations of an operator with respect to medical device safety (Section 3 (1) sentence 1 MPBetreibV). Since the entitled medical care is generally provided by third parties such as medical supply houses, the tasks arising from the operators' obligations can be transferred to these third parties (Section 3 (1) sentence 2 MPBetreibV). In such cases, the provider initiating the provision must take the necessary precautions to ensure that these tasks can be accomplished properly (Section 3 (1) sentence 3 MPBetreibV).

No POCT device should be operated without a prior function test.

Moreover, MPBetreibV contains further provisions relevant to liability, such as its Section 7 which regulates the requirements on maintenance measures in more detail. According thereto, the operator is liable for the proper conduct of maintenance when employing personnel not authorized by the manufacturer [23].

The RiliBÄK provisions are also particularly relevant to the application and operation of POC devices. As set forth in Part B, observance of the minimum requirements governing internal and external quality assurance of the results of quantitative and qualitative POC tests deserves mention here.

Only strict compliance with the RiliBÄK and the provisions of the legislation on medical devices protects against liability.

Organizational culpability

Organizational culpability exists when the responsible party has not organized the delegation of obligations in a way that ensures proper selection and continuous supervision of operational processes and activities of the personnel and prevents damage to third parties within the realms of possibility [14].

RiliBÄK contains the relevant minimum standards. In particular, responsibilities for the conduct of POC tests and individual tasks like those charged to the POCT coordinator or POCT supervisor must be clearly defined. Training, continuing education and instruction of the personnel must be in guideline-compliant. Section 12 (1) German Occupational Protection Law (ArbSchG) [6] stipulates that the employer shall give any users who are employees sufficient and appropriate training in device handling [23]. Finally, the provisions of German act governing medical technical assistants (MTAG) [9] have to be observed when skilled and competent persons are employed. Essentially, only the following specially qualified persons may conduct POC tests:

- Persons who, by virtue of their university diplomas, have the required expertise, capabilities and skills (physicians, natural scientists) as well as alternative medical practitioners (Section 10 (1) No. 1 MTAG).
- Medical technical laboratory assistants (Section 9 (1) No. 1 MTAG),
- Persons undergoing an education in the above-mentioned profession (e.g. medical students) (Section 10 (1) No. 2, 5 MTAG),
- Persons with diplomas for other medical training, such as nursing staff, who without being authorized under Section 10 (1) No. 1 to 5 MTAG, work under the supervision and responsibility of a person defined under Section 10 (1) sub-section 1 MTAG (Section 10 (1) sub-section 6 MTAG) [19].

References

1. German Federal Court of Justice (BGH) (1994) Neue Juristische Wochenschrift 47: 3349–3351
2. German Civil Code (BGB) in the version promulgated on 2 January 2002 (Federal Law Gazette [BGBl.] I p. 42, 2909; 2003 I p. 738), last amended by Article 2 of the Act dated 21 February 2017 (BGBl. I p. 258)
3. German Medical Association (2014) Guideline of the German Medical Association on Quality Assur-

ance of Medical Laboratory Examinations (RiliBÄK 2014). Dtsch Ärztebl 111: A1583–A1618
4. CJEU, Judgment Boston Scientific Medizintechnik, C-503/13 and C-504/13, EU:C:2015:148
5. CJEU, Judgment Veedfald, C-203/99, EU:C:2001:258 [Judgement in English: http://eur-lex.europa.eu/legal-content/EN/TXT/PDF/?uri=CELEX:61999CJ0203]
6. Act on the Implementation of Measures of Occupational Safety and Health to Encourage Improvements in the Safety and Health Protection of Workers at Work (ArbSchG) of 7 August 1996 (Federal Law Gazette (BGBl.) I p. 1246), last amended by Article 427 of the regulation dated 31 August 2015 (BGBl. I p. 1474) http://www.gesetze-im-internet.de/englisch_arbschg/englisch_arbschg.html#p0080; last accessed on 21 Sept. 2017
7. Act on Liability for Defective Products (Product Liability Act – ProdHaftG) of 15 December 1989 (Federal Law Gazette [BGBl.] I p. 2198), last amended by Article 180 (5) of the regulation of 31 August 2015 (BGBl. I p. 1474) http://www.gesetze-im-internet.de/englisch_prodhaftg/index.html; last accessed on 22 Sept. 2017
8. Act on Medical Devices (Medical Devices Act – MPG) in the version of 7 August 2002 (Federal Law Gazette [BGBl.] I p. 3146), last amended by Article 16 of the Act of 23 December 2016 (BGBl. I p. 3191)
9. German Act Governing Medical Technical Assistants (MTAG) of 2 August 1993 (Federal Law Gazette [BGBl.] I p. 1402), last amended by Article 21 of the Act on 18 April 2016 (BGBl. I p. 886)
10. Greiner HP (2014) § 839 Haftung bei Amtspflichtverletzung. In Spickhoff, A (Ed.) Medizinrecht. 2nd Edition, CH Beck, Munich, Germany
11. Heil M (2003) § 22 Haftung der Hersteller, Betreiber und Anwender für Medizinprodukte. In: Dieners, P, Anhalt, E (ed) Handbuch des Medizinprodukterechts. CH Beck, Munich, Germany
12. Kaiser J, Gassner UM, Reng M, Prokosch HU, Bürkle T (2012) Risiken und Nebenwirkungen der Integration medizinischer Software in klinische IT-Strukturen – Erlanger Memorandum. GMS Med Inform Biom Epidemiol 8: 1–9
13. Regional Court of Berlin, Kammergericht (KG) (2014) MedizinProdukteRecht 14: 120–126
14. Koyuncu, A, Dahm-Loraing, R (2009) Die Haftung des Arztes und Krankenhauses für Medizinprodukte – Teil 1: Grundlagen und Behandlungsfehlerhaftung. PHI 16: 172–179
15. Koyuncu A, Dahm-Loraing R (2009) Die Haftung des Arztes und Krankenhauses für Medizinprodukte – Teil 2: Aufklärungsfehler, Prozessrecht und Ausblick. PHI 16: 218–224
16. Koyuncu A, Dahm-Loraing R (2010) Haftung für Medizinprodukte – Teil 1: Haftung des Herstellers nach dem ProdHaftG. PHI 17: 108–116
17. Koyuncu A, Dahm-Loraing R (2010) Haftung für Medizinprodukte – Teil 2: Die deliktische Produzentenhaftung gem. § 823 BGB. PHI 17: 142–153
18. German Medical Devices Ordinance (Medizinprodukte-Verordnung) of 20 December 2001 (Federal Law Gazette [BGBl.] I p. 3854), last amended by Article 3 (5) of the Ordinance of 27 September 2016 (BGBl. I p. 2203)
19. Meyer-Lüerßen D, Meyer-Lüerßen I (2006) Rechtssicherheit von Point-of-Care-Tests. J Lab Med 30: 230–233
20. Müller-Plathe O, Briedigkeit L, Schlebusch H, Ziems J (1999) Patientennahe Laboratoriumsdiagnostik (Point-of-Care Testing). II. Rechtliche Aspekte. J Lab Med 23: 600–603
21. Regional Court (OLG) of Cologne, Judgment of 23 September 2009 – 5 U 220/08 –, juris
22. Ordinance Concerning the Setting Up, Operation and Application of Medical Devices (Ordinance on Operators of Medical Devices – MPBetreibV) of 21 August 2002 (Federal Law Gazette [BGBl.] I p. 3397), last amended by Art. 1 and 2 of the regulation of 27 September 2016 (BGBl. I p. 2203)
23. Weimer T (2007) Medizinproduktehaftung – Straf- und zivilrechtliche Haftung der Anwender und Betreiber von Medizinprodukten – Teil 2: Medizin-ProdukteRecht 7: 1–14

POCT and data management

Peter B. Luppa, Christoph Braun, Andreas Bietenbeck

P. B. Luppa, R. Junker (Eds.), *Point-of-Care Testing*
https://doi.org/10.1007/978-3-662-54497-6_27

27.1 Introduction

This chapter aims to discuss the options available through the electronic networking of POCT devices in hospitals and inpatient settings. Here, the term connectivity is explicitly not to be understood as an electronic data connection of decentralized analytical systems in satellite laboratories to a central laboratory, frequently also described as "POCT connection to a central laboratory" [3]. Rather, the concept is understood to mean a complex architecture of online devices distributed peripherally across departments linked to a central information system. Connections to external stakeholders are also optional solutions. This entails the use of a server from an external supplier (e.g. private-practice laboratory). Advantages of networking include

- Fast, simple and error-free data transfer
- Electronic transmission of diagnostic data
- Obviation of manual tasks
- Central control/configuration of devices
- Central operator management
- Central equipment and batch management
- Invoicing of services
- IT-supported implementation of quality management schemes (e.g. RiliBÄK requirements in Germany)
- Generation of performance statistics

27.2 The POCT data manager

Characteristically, the IT infrastructure of a hospital consists of multiple components. The hospital information system (HIS) is the overarching system for the collection, processing and transfer of medical and administrative data. A laboratory information system (LIS) is used to manage the workflow of a clinical laboratory test from the receipt of the order to release of the test result. The POCT data manager forms the interface between POCT devices and the hospital's IT systems. The data manager has three main basic functions:

- Connecting POCT devices to an IT network (across manufacturers or manufacturer-bound)
- The POCT findings are integrated in the laboratory report and/or the HIS-based electronic medical record
- POCT coordination and organization (devices, operator, patients, quality controls, reagents, device control, device reports etc.)

Accordingly, the POCT data manager should meet the following requirements [11]:

- Bidirectional connection to the LIS/HIS to capture patient master data and transmit to the individual devices
- Transfer of the POCT patient data from different devices, possible further validation and release
- The patient's measured values are transmitted as POCT reports to LIS and HIS for further processing in the patient's medical record.
- Automatic monitoring of quality controls
- Report on quality control data as table or chart (the documented reports must be archived on the server for years)
- Graphical presentation of the values
- Understandable color-coding to convey incomplete or deficient quality controls to the POCT coordinator
- Locking of systems/individual analytes in the event of deficient quality control
- Remote-access to POCT devices (control and remote maintenance of individual devices, e.g. sensor calibration of blood gas systems)
- Central master data administration of POCT locations, devices, reagents, control materials etc.
- Operator management, including operator authorization and transferring this information to the individual devices
- Independent monitoring of all process sequences and report on status and error messages generated by the networked POCT devices

It is complicated and time-consuming to assign authorizations to every POCT operator. Therefore, modern systems enable the allocation of operators to groups. These groups can in turn be assigned roles with the required authorizations. Operator accounts that are not assigned to any one specific person, but to an entire ward, for example, pose impediments to quality assurance and should therefore be precluded. Neither is it generally sufficient to work with numbers as identifiers and not release the real names of the operators. Exactly as with every other medical report on findings, a medical laboratory test must be clearly attributable to the person responsible. An institutional agreement is a suitable recommendation to preclude the unauthorized use of operator data, e.g. for performance evaluation of staff. Operator data can be manually entered into the POCT data manager. Alternatively, the appropriate interfaces can be used to connect the devices to a hospital's centralized HR department. This helps avoid redundant data entries.

Tab. 27.1 contains a current market overview of the POCT data management systems available in Germany and Europa [11]. Larger manufacturers of POCT systems frequently supply a matching POCT data manager with their devices. If several devices from various manufacturers are used together, it may be necessary to run several POCT data managers in parallel and not to compile the data until they are aggregated in the LIS. By supporting a broad portfolio of POCT devices, manufacturer-independent POCT data managers claim to avoid such complicated and time-consuming duplicate structures. In Europe, device-independent software products such as POCTopus by OSM (Essen, Germany) and POCcelerator by Conworx (Berlin, Germany) have a strong share of the market, whereas the likes of Quick-Linc/QML, a comprehensive POCT data management and connectivity system by TELCOR Inc. (Lincoln, NE, USA) [2] and RALS-Plus, a remote automated laboratory system, from Alere, formerly Medical Automation Systems (Charlottesville, VA, USA) [12] are more frequently encountered in the USA. Further service providers are listed in the references [5].

27.3 Connecting POCT devices to a network

POCT devices designed especially for professional use are interoperable to enable networked operations. The following routes of

Tab. 27.1 Market overview of data management systems in Germany (status: April 2018)

Manufacturer	Proprietary, one method/device	Proprietary, several methods/devices	Independent of device manufacturer
IMP Computersysteme AG (OSM Group)	–	–	POCTopus
Conworx (now Siemens)	–	–	POCcelerator
TELCOR	–	–	QML
Abbott	QC Manager Info HQ Manager	–	–
Siemens	–	RapidComm	
Instrumentation Laboratory (Werfen Group)	GEMweb Plus	–	–
Radiometer	–	AQURE POC	–
Roche Diagnostics	–	Cobas IT 1000	
Alere Informatics	–	Alere RALS	

transmission can be used to integrate the POCT systems into the in-hospital network:
- Serial (RS-232) to ward PC
- Local area network (LAN) box; converts serial interfaces to a network protocol
- Cable-bound via router
- Wirelessly via WLAN (a proper encoding, e.g. WPA2 protocol, must be used to protect sensitive patient data. Furthermore, sufficient net coverage must be ensured so that the data can be transmitted without any time lags).

Frequently, docking stations are also used for charging the batteries of a mobile POCT device and exchanging its data at the same time. In these devices, the data are not transmitted continuously, rather transmission takes place once the device is placed in the docking station. Many POCT devices additionally have interim storage for data in case the network connection is interrupted. The data exchange then takes place once the connection is reinstated. Other wireless standards (Bluetooth, mobile communications) are not widely used for the transfer of POCT data in the hospital or the physician's practice.

In an environment with several operators and patients, the device operator and patient whose values are measured can be entered. Usually, this is done by scanning barcodes.

Paired with the matching software, a suitable POCT device can then support the following functions:
- Uploading operator data from POCT software
- Uploading patient data from POCT software
- Patient query (e.g. devices query the POCT software for a patient name belonging to a case number)
- Lock or release function on device according to operators or operator groups
- Uploading configuration data from POCT software
- Locking of device or measuring channel upon QC excursion
- Individual entry of operator ID when performing QC as well as patient tests
- Commenting any QC violation on device by operator
- Discarding/deleting measurements
- Remote access (e.g. for control, remote maintenance, recalibrations)
- Warning displays on limit excursion or after the difference between two values exceeds defined delta check criteria

Generally, this type of connectivity is not possible with simple tests such as lateral flow assays. Increasingly, however, automated read-out units that enable networking, are sold [1].

27.4 Networking strategies in inpatient settings

27.4.1 Networking strategies with internal POCT data management

The following system structure is usually applied for **networking** decentralized in-hospital POCT devices [8]:

> POCT devices ↔ Data management system ↔ LIS ↔ HIS.

That way, decentralized POCT analyses can be carried out in a hospital and then centrally monitored and documented by the central laboratory [9]. Nevertheless, this results in an additional outlay for logistics, materials and human resources for the laboratory. The described network options represent the most feasible process for complete documentation of POCT quality control data. Any other non-information technology-driven solutions without transfer of quality control data to the central laboratory, for example with manually generated lists, are not only time-consuming for the operators, but also prone to errors and more likely to discourage operating staff.

Exceptions include operation in a central laboratory of POCT devices like blood gas analyzers, viscoelastic coagulation analyzers or read-out devices for urine dipsticks. These tests

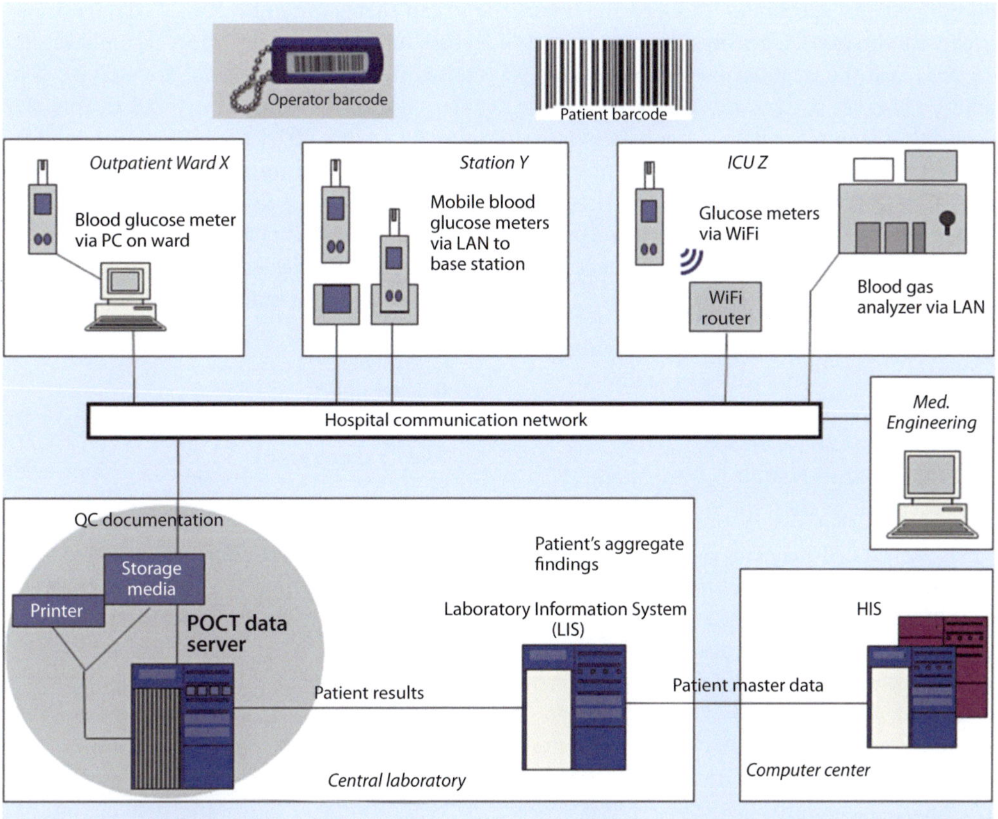

Fig. 27.1 Example of a POCT network at a medical center. *HIS* Hospital information system; *HCS* Hospital communication system; *QC* Quality control

meet all the criteria for a POCT system and could also be conducted near the patient. If, nevertheless, these tests are run in a central laboratory – usually for organizational reasons –, the devices can naturally be ported directly with the LIS just like any other normal laboratory devices.

Depending on the hospital's setup, barcode numbers can be used to identify the patients and the POCT operators. All patient-related POCT measured values are designed to be stored long-term on the POCT server and made available as secured patient identity after transmission to the clinical wards as separate POCT report for the electronic medical record [13].

A statement issued by the Working Group for POCT of the German Society for Clinical Chemistry and Laboratory Medicine (DGKL) from 2005 [11] emphasized the advantages of such an IT network, namely the significant time-savings on the part of both the ward staff operating the devices and the POCT coordinator in the central laboratory, while stressing the advantages in terms of completeness of data collection and documentation [10]. In Fig. 27.1, a schematic is presented depicting an example of intranet connectivity for POCT with blood glucose and blood gas analyzers operated on different wards and managed by a centralized POCT server.

After linking up with the matching patient master data, coupling the POCT server to the LIS gives the central laboratory the option to generate a cumulative POCT findings report of the results from the measurements performed on the patients' specimens and feed the report back to the hospital. Occasionally, these reports do not reach the treating physician in a timely manner because of the sometimes considerable delays that take place before the data is transferred to the POCT server. In some hospitals, POCT results are therefore transferred in parallel a priori to other subsystems such as a patient monitoring system on an ICU (Patient Data Management System, PDMS).

> **In the case of networked POCT devices, hazards associated with any failure of the IT infrastructure should be taken into account. It should be avoided that a POCT measurement is thereby made impossible to perform.**

Many devices display the measured data and have internal memory that stores the results until to data transfer functions work again. The effectiveness of many process sequences at a medical center may nevertheless be significantly limited. Annotations on the "Application of risk management for IT-networks incorporating medical devices" can be found in the DIN EN 80001-1 standard.

27.4.2 Strategy for networking with an external POCT data management system

If the objective is to find an external solution for POCT data management, it entails using the server of a service provider situated outside the hospital. The service provider (e.g. a laboratory physician in private practice) is thus responsible for the proper functionality of both software and hardware. The hospital is spared the costs for both hardware and software maintenance and care. This solution involves the installation of a decentralized data management system where the patient's findings and quality control data are stored and secured in compliance with the regulatory rules. This system transmits the patient data via an HL7 interface to the HIS/LIS. The POCT coordinator in the hospital is granted access to the data management system and can thereby monitor the QC data. It is obvious that data security must be upheld and maintained. For that reason, any service provider connecting several hospitals in this way must be ensure strict separation of hospitals' measurement and operator data (multi-client capability).

Prior to the technical implementation, several issues must be cleared up. For this purpose, the following systematics have proven their merits:
- Checklists
- Project plan
- Data flow chart
- Master data record

Checklists

Questions relating to IT structure
- How does the flow of patient data work?
- How does the data flow of requested orders work?
- Is the existing IT infrastructure adequate?
- Are enough electrical connections available?
- Is additional IT hardware (switches or hubs) required?
- Which IP addresses have been assigned for the devices in the hospital?
- Will it be necessary to convert the network data, e.g. Network Address Translation (NAT) and who will complete this?
- Which ports will facilitate the access of the devices to the data management system? Which ports must be activated?
- How many digits and which syntax is needed for operators' barcode numbers?
- How many digits and which syntax is needed for the patient number (case number)?
- Should pseudo case numbers be stored for emergencies?
- Are the names of staff documented by the data management system in addition to the numbers? If not, who manages the assignment?
- Should data on findings be captured and transmitted to a LIS or HIS?

- Should patient data from the LIS or HIS be sent to the POCT data manager software?
- Are the parameters stored in the LIS or HIS?

Aspects relating to IT organization

- Who is the hospital's POCT coordinator?
- Has a POCT committee been constituted?
- How does the operator log himself on? Barcode + keyboard/barcode only – device-independent?
- How is the case number captured? Barcode + keyboard/barcode only – device-independent?
- Who provides training on which devices?
- Who regulates the implementation of Part A of the RiliBÄK?
- Which devices should be networked with which parameters?
- Should devices be monitored offline?
- Should only QC data be captured and evaluated?
- Definition of responsibilities for:
  - Quality control
  - Devices (maintenance, exchanges, repairs)
  - IT (hardware, software, network)
  - Organization/master data entry/ maintenance

Project plan

To ensure streamlined implementation, the hospital should prepare a project plan outlining the necessary steps and the way this will be processed in a time-limited manner. In this case, the contact person for all IT issues should be a staff member of the in-house IT departments. The POCT coordinator can assume responsibility for timely implementation of the project plan. In this context, there are also aspects that need to be clarified individually in checklists.

Data flow chart

With a view to the proper flow of measured patient and quality control data, it is important that the hospital be presented with a graphical chart of the data flow. This graphic should be structured as simply and understandably as possible. Next, it is then the responsibility of each IT department to set up their own network according to this simple, easy-to-understand chart. It is recommended to document the corresponding network configuration on this chart.

Master data record

A master data record should be designed to present all data keys to the installation and configuration of devices in a manageable form. Likewise, the contents here can best be reviewed in advance by means of checklists. When scrupulously maintained, this master data record can also be used as a statistical instrument.

It is the electronic data connection between hospital and supplier that is of paramount importance for the technical implementation of the decentralized solution. Generally, an adequately sized data line is required between the hospital and the service provider. The communication between POCT devices and the hospital's laboratory and the data management system should be realized via this data transmission line.

In addition to the IT network, questions of logistics must be addressed and their implementation defined in detail.

It takes complex systematics to network a large number of POCT devices. Not only is it imperative to ensure that POCT devices are networked in order to comply with all relevant directives, but also that the service be setup satisfactorily. Service plays a substantial role in the system being accepted by the nursing staff.

27.4.3 POCT networking in private practitioners' offices

The obligation to control quality also applies to the general practitioner's setting. Documentation in the practices is primarily written by hand. Depending on type and number of the measuring devices, this handling method is unproblematic. In practices that run more complex POCT devices (e.g. blood gas or hematology devices), manual documentation and interpretation is very time-consuming. In this context, it would be logical to network the devices with a data management system; however, this is usually difficult to implement in the physician practice setting. Alternately, Internet platforms are offered (e.g. Conworx, Berlin, Germany) that perform the analysis after entry of the QC data. In addition to professional and accurate analysis, the IT service provider also guarantees legally compliant data security. These costs for the physician are calculable and manageable.

27.5 POCT1-A standard

Generally, the POCT server will be connected to the HIS via a Health Level 7 (HL7) protocol, whereas an American Society for Testing and Materials (ASTM) protocol or the POCT1-A standard will be used for connection to the LIS. The POCT1-A medical communication protocol was developed to standardize the communication pathways between POCT devices and HIS, while ensuring quality assurance that complies with the statutory requirements [15]. POCT1-A derived from draft standard developed, prototyped and piloted by the Connectivity Industry Consortium (CIC) membered with medical devices manufacturers, in-vitro diagnostics companies and vendors in the healthcare system. This CIC was founded in 1999 after market research initiated by the Industry Liaison Division of the American Association for Clinical Chemistry (AACC) had shown that at any point in time in the USA only around 17 of all POCT devices in a hospital had connectivity with a POCT server [8]. As early as 2001, an international specification emerged from collaboration with the Clinical Laboratory Standards Institute (CLSI), HL7 and Institute of Electrical and Electronics Engineers (IEEE), that was recognized by the CLSI [6] and was congruent with the statutory provisions of the participating countries [4]. Integrating the Healthcare Enterprise (IHE) – an international initiative of users and manufacturers with the objective of standardizing the data sharing between IT systems in the healthcare system – similarly designed their Laboratory Point-of-Care Testing (LPOCT) profile according to this specification.

The standard consists of two communication interfaces: a device interface (DI) and an observation reporting interface (ORI). The DI links the POCT device with the POCT server and describes the transfer of measured data via an existing infrastructure (hospital network). The ORI is concerned with the transfer of these data to the HIS. POCT connectivity according to the CIC standard is presented in ◘ Fig. 27.2.

The POCT1-A standard envisages the use of messaging in the XML format for sharing data between a POCT device and the server. POCT1-A is aligned along the structure of HL7 messages. The second part of the standard describes the further processing of the measured values in HIS and LIS as well as how POCT1-A data is converted into HL7-compliant syntax.

The POCT1-A standard simplified the connection of individual devices, but did not create any plug-and-play functionalities [14]. This standard allows the individual manufacturer room for interpretation in POCT implementation. Therefore, extensive adjustments can be required, particularly when POCT systems of several manufacturers are operated simultaneously. However, the introduction of a new POCT device is not a commonplace event. Indeed, it should be a well thought-through step that is preceded by a long decision-making phase. Against this backdrop, any true plug-and-play functionality is of secondary importance.

27.6 eLearning and POCT

The use of POCT extends the circle of persons performing medical laboratory testing considerably. Moreover, a high-frequency fluctuation of staff can take place. Therefore, regular training of all operators is associated with high expenditure. As a consequence, many POCT software packages feature integrated eLearning solutions to help structure training courses

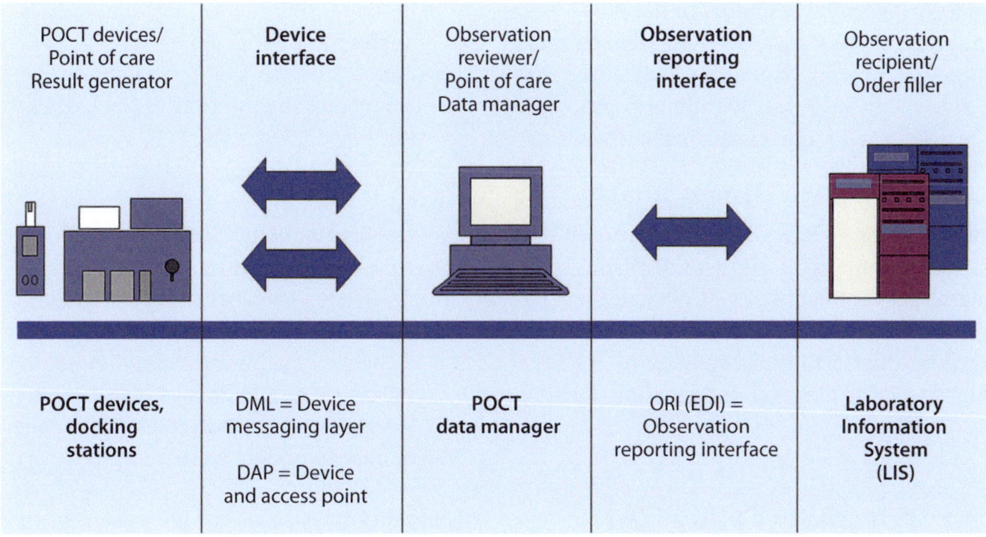

Fig. 27.2 POCT connectivity according to the CIC standard/IHE Laboratory Point of Care Testing (LPOCT) profile

more efficiently. Alternatively, eLearning can be run independently of the POCT software and thereby be available for other training courses at a hospital (Fig. 27.3).

If possible, the contents of a training course should be supplied by the POCT system manufacturer and then adapted by the operators to the local circumstances. The Sharable Content Object Reference Model (SCORM) offers such a data format. A successful training course is distinguished not only by a sensible selection of contents, but by the didactic teaching thereof. After completion of an eLearning course, individual authorizations can be adapted in the POCT software accordingly. For this purpose, the users of the eLearning training course must

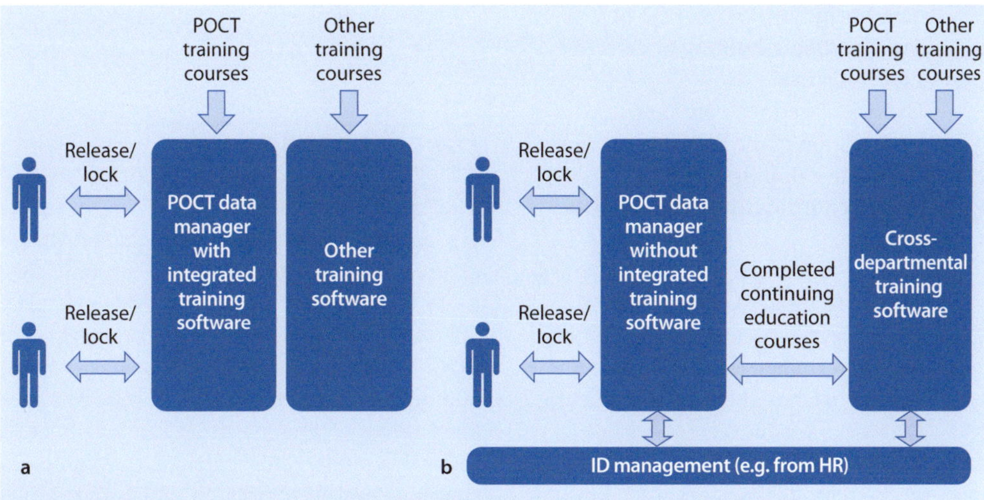

Fig. 27.3a,b eLearning and POCT data managers in the IT system landscape of a hospital. **a** POCT data manager with integrated eLearning. **b** Modular setup with cross-hospital educational software that is not only used for POCT training courses exclusively

be assigned to the operators in the POCT data manager. In principle, it is possible to make these assignments manually. A centralized electronic ID management with interfaces to both the POCT and the eLearning software can nevertheless facilitate this task immensely. Valuable information on the implementation of an eLearning POCT educational concept can be found in the application guide titled "Training of professional users of devices for near-patient testing" (VDE-AR-E 2411-2-101) issued by the German Association of Electrical Engineering, Electronic Information Technology (VDE).

27.7 Advantages and disadvantages of a POCT network

The **advantages** of an IT solution for POCT are as follows:

- **For POCT operators:**
  - Time savings for ward and nursing staff
  - Information about the quality control status per device (time at which the next quality control measurement is scheduled, any upper and lower excursions from the permissible measuring range)
  - Electronic, storage of quality control data (including operator ID codes)
  - Timely troubleshooting of defective POCT devices
  - Use of pre-defined comments
  - Aggregation of the POCT results into a cumulative finding report
- **For POCT coordinators in the central laboratory:**
  - Remote function for the POCT coordinator – facilitates implementation of higher-level organizational administrative tasks (management of operator IDs, control batches, the measuring device ID, working groups etc.)
  - Time savings for the POCT coordinator in the central laboratory
  - Option to adapt control- and batch-specific data
  - Option to adapt quality control rules
  - Online recalibrations, e.g. of blood gas analyzers
  - Quick problem-solving in the case of erroneous quality controls (including remote locking of the POCT device when appropriate)
- **For the hospital:**
  - Generation of performance statistics, costs-benefit profiles etc.
  - As-needed guidance for reinvestment into new POCT equipment
  - Enhanced quality awareness among the nursing staff and thereby the associated intensification of outwardly presentable quality assurance measures

Disadvantageous aspects of networking all POCT devices, however, include:

- The added financial outlay for purchasing of network-compatible blood glucose and blood gas analyzers as well as for retrofitting existing systems for online operations,
- Possible difficulties with POCT data management when networking certain models of manufacturers with each other and
- Cementing the spectrum of device – switching a POCT system to another incurs financial and organizational outlay (manufacturer-independent POCT server software however will facilitate such a change in terms of connectivity).

27.8 The road ahead

The legislative requirements in terms of ensuring high analytical reliability in qualitative and quantitative laboratory tests will certainly continue to mount in the future. This primarily affects POCT. Currently, the documentation of POCT results (complete patient master data) and quality assurance are both areas that have been recognized as needy of improvement in many hospitals. This particularly applies to simple lateral flow tests, in which the results frequently must be manually documented. **Automated read-out systems** offer substantially more safety and convenience [7].

The turnaround time (TAT) of a POCT measurement often only allows retrospective quality assurance. That means that an isolated erroneous measurement might not be identified until a time when the result has already had implications on the patient's further medical treatment. This risk can be reduced by **electronic warnings**. When a POCT measurement e.g. deviates too strongly from a previous value, the proper software will directly trigger a warning on the POCT device. In a hospital, the same analytes are frequently measured both by POCT and the central laboratory. By applying suitable mathematical methods such as "Data mining EMRs to Evaluate Coincident Testing" (DETECT), the two measurements can be compared to identify errors and inconsistencies [16].

The newly created models of **integrated care**, e.g. in the form of medical care centers, are especially interesting for hospitals which treat the chronically ill. Participating hospitals can offer new services with regard to the quality assurance of patient self-measurements. All necessary data can be exchanged from and to a POCT server over the Internet very simply from individual hospitals, patients and physician's practices. With suitable IT connectivity, preclinically obtained POCT results [18] can also be captured directly in the HIS. After highly specialized treatments, POCT offers the possibility for the hospital to monitor patients at home, i.e. home monitoring [17]. In addition to the IT issues to be solved, a number of organizational questions still need to be answered for these scenarios.

References

1. Chien T-I, Lu J-Y, Kao J-T et al. (2007) Comparison of three automated urinalysis systems – Bayer Clinitek Atlas, Roche Urisys 2400 and Arkray Aution Max for testing urine chemistry and detection of bacteriuria. Clin Chim Acta 377:98–102
2. Clarke B (2007) Point-of-Care Testing connectivity-The new generation. Point of Care 6:148–153
3. DiSerio F, Amodio G, Varraso L et al. (2005) Integration between point-of-care cardiac markers in an emergency/cardiology department and the central laboratory: methodological and preliminary clinical evaluation. Clin Chem Lab Med 43:202–209
4. Dyer K, Nichols JH, Taylor M et al. (2001) Development of a universal connectivity and data management system. Crit Care Nurs Q 24:25–38; quiz 22 p following 75
5. Gj K (2002) Principles and practice of point-of-care testing. Lippincott Williams & Wilkins, Philadelphia, PA, USA
6. Institute Clinical and Laboratory Standards (2001) CaLS Point-of-Care connectivity (POCT1-A). Wayne, PA, USA
7. Jang D, Shin SY, Seo DW et al. (2015) A smartphone-based system for the automated management of point-of-care test results in hospitals. Telemed J E Health 21:301–305
8. Jones RG, Johnson O, Hawker M (2010) Informatics in point-of-care testing. In: Price C, St. John A, LJ K (eds) Point-of-Care Testing. AACC Press, Washington DC, USA, p 287–309
9. Lewandrowski K, Gregory K, Macmillan D (2011) Assuring quality in point-of-care testing: evolution of technologies, informatics, and program management. Arch Pathol Lab Med 135:1405–1414
10. Luppa PB, Blobner M (2004) Online-Qualitätssicherung in der Sofortdiagnostik. Dtsch Ärztebl A399–A400
11. Luppa PB, Gässler N, Haeckel R et al. (2005) Comments by the "Point-of-Care Testing" Working Group of the DGKL on the subject of linking POCT instruments in hospitals with a central laboratory. J Lab Med 29:241–245
12. Menke G (2007) Medical Automation Systems and a Brief History of Point-of-Care Informatics. Point of Care 6:154–159
13. Messner B, Frick G, Blobner M et al. (2004) Online quality assurance of point-of-care testing by using the hospital information system. J Lab Med 28: 264–272
14. Nichols JH (2015) Point-of-Care Testing: Past experience and future challenges. Point of Care 14: 103–105
15. Nichols JH (2003) Quality in point-of-care testing. Expert Rev Mol Diagn 3:563–572
16. Schroeder LF, Giacherio D, Gianchandani R et al. (2016) Postmarket surveillance of Point-of-Care glucose meters through analysis of electronic medical records. Clin Chem 62:716–724
17. Song B, Schulze M, Goldschmidt I et al. (2013) Home monitoring and decision support for international liver transplant children. Stud Health Technol Inform 192:268–272
18. Venturini JM, Stake CE, Cichon ME (2013) Prehospital point-of-care testing for troponin: are the results reliable? Prehosp Emerg Care 17:88–91

Patient safety and POCT

Mario Plebani

P. B. Luppa, R. Junker (Eds.), *Point-of-Care Testing*
https://doi.org/10.1007/978-3-662-54497-6_28

28.1 Introduction

Economic pressures, patient empowerment and the increasing recognition of the need to reorganize the delivery of care based on patient-centered models are transforming healthcare. Although the dominant model of laboratory testing throughout the world remains the centralized laboratory, alternative models are being increasingly considered [26]. Recent advances in miniaturization, sensor technologies, microfluidics and wireless communications have allowed diagnostic testing to be moved closer to the patient and be used outside the traditional central laboratory setting.

Referring "to any diagnostic test performed at or near the patient" to help determine which medical care is needed [13], the term "point-of-care testing (POCT)" as such does not formally show up in the literature until 1994, even though blood glucose determination, urinalysis and pregnancy testing for human chorionic gonadotropin had all been available on POCT devices prior to 1994 [11]. The rapid and widespread implementation of POCT systems has led some to question whether we are using this technology appropriately.

As shown in ◘ Fig. 28.1, POCT impacts a wide variety of healthcare providers, patients, and settings: It affords access to laboratory services in low- as well as high-income countries [9], in primary care, home care and during disasters and pandemic outbreaks. POCT may provide an excellent solution to certain critical testing needs in rural areas and hard-to-reach places. Prime examples are the dispersed populations in Australia and on the African continent. Given this background, the term "point-of-need" might be more aptly than "point-of-care" to define this type of laboratory testing.

Notwithstanding the above, particularly in developed countries, the use of POCT has been

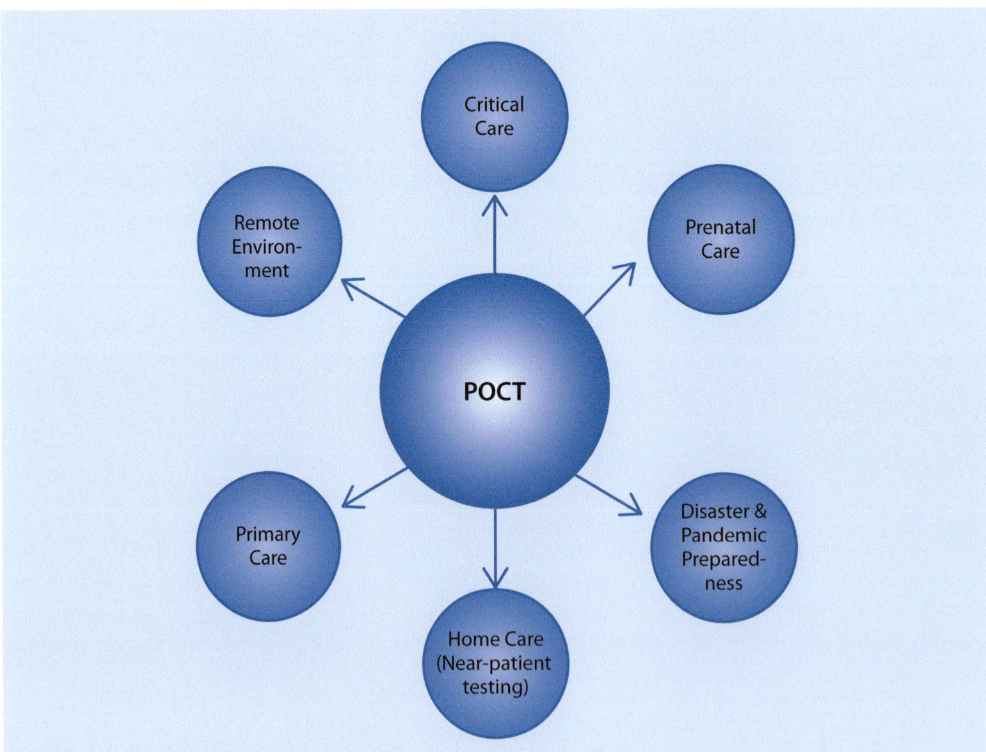

◘ **Fig. 28.1** Clinical settings of POCT deployment

widespread across hospital settings, namely in emergency departments, intensive care units and operating rooms for answering the clinical needs of critical care.

The use of POCT is particularly attractive as, at least from a theoretical viewpoint, it permits immediate access to test results for better and more effective management of patients. Yet, there is the need to minimize the time to treatment initiation by adopting POCT without compromising the quality of the laboratory information obtained.

28.2 POCT and performance criteria

"The core principle underlying POC measurement has been described as reducing turnaround time without compromising the quality of information on which clinical decisions for patients are based" [3]. This quote should guide our debate on the quality required to enable the right decision-making.

In turn, performance criteria or quality specifications are defined as "the level of performance required to facilitate clinical decision making" [4]. Performance characteristics fall into two categories:

- Practicability, i.e. the methodological details on how to execute the test, including speed of analysis, volume and type of sample required
- Reliability, i.e. the scientific facets of the methodology such as precision, bias, limit of detection and measuring range

Quality specifications governing the reliability of a laboratory test, particularly its precision and bias, are absolutely necessary for analytical quality management, for obtaining an appropriate answer to clinical questions asked and for meeting patient needs. In POCT, the considerations that speed and shorter turnaround times (TAT) surpass all others does not work. Indeed, the rapid availability of POCT results and the immediate diagnostic and therapeutic implications of many POCT tests amplify the likelihood that erroneous results from these tests will cause preventable adverse events [16]. This fact clearly represents a patient safety issue.

As mentioned above, theoretically, there is no reason why different quality specifications should apply to the same test when measured by POCT or in the central laboratory. To enable proper decision-making and monitoring, it is obvious that a test result should be comparable over time within the same laboratory and across different clinical laboratories. Therefore, results of clinical laboratory measurement procedures should be equivalent within clinically meaningful limits – a dictum that cannot be upheld without everybody adopting the same quality specifications.

As we move towards full electronic reporting of laboratory results, we appreciate more fully that differences in analytical and extra-analytical quality specifications may affect the interchangeability of laboratory information [22, 23]. This, in turn, requires a careful examination of the analytical characteristics of any method used in centralized laboratories or with POCT before its introduction into clinical practice. Evidence accumulated on this basis shows that POCT assures valuable quality specifications for some measurands (e.g. electrolytes, blood gas analysis, INR), but does not yet do so for others (e.g. high-sensitivity cardiac troponin). A recently published paper allows a better understanding of the implications of the previously described requirements: Mirzazadeh and colleagues demonstrated that the POCT of electrolytes and calcium is reliable and that there is a sufficient agreement between the results obtained by POCT and those produced by laboratory analyzers [17]. This conclusion is even more important given that previous studies had shown that only about one-third of clinicians relied on POCT to guide their clinical decisions, with most preferring to wait for laboratory confirmation before making important clinical decisions [6, 10].

One exception to the previously described rule regarding the need for equal analytical performance characteristics between POCT and central laboratory, however, deserves mention:

when the same measurand is used for different clinical needs and goals. **Glucose monitoring** embodies the paradigm of this situation. In fact, when used for monitoring, POCT may provide reliable analytical results [12], while the same technology cannot achieve the more stringent goals required for diagnosis [8]. It is not only the analytical quality of glucose meters that fails to meet the requirements for a diabetes diagnosis, but it is doubtful whether they can be safely used for insulin dose adjustment either [1].

28.3 POCT and patient safety

28

The link between POCT and patient safety appears at the intersection of technical limitations, rapid availability and the therapeutic implications of this seemingly new genre. POCT grouped into pre-, intra- and post-analytical phases according to the traditional framework adopted by central laboratories for evaluating the quality of the total testing process would help ensure patient safety (▶ Chapter 4). In theory, POCT does eliminate some of the more problematic steps in the testing process, including specimen collection, transport and result distribution. In practice, however, POCT creates other challenges from a risk management perspective, namely analyses performed by non-laboratory personnel, poor quality control, non-conforming quality assessment workflows and great vulnerability in the pre-analytical phase [20].

28.3.1 Pre-analytical sources of errors

Pre-analytical factors, such as altered sample integrity (e.g. hemolysis, lipemia, icterus, and so on) that would be identified and considered as causes of preanalytical errors in the laboratory setting where serum or plasma are tested, often go unrecognized in POCT systems that use whole blood samples or do not have a defined mechanism for assessing sample integrity. In fact, visible **hemolysis**, as a hallmark of a more generalized process of blood cell damage, is usually not apparent until the serum or plasma has been separated off. According to the literature:

- There is no quick and reliable way to determine whether hemolysis is present in a whole blood specimen.
- Hemolysis is more than 5 times higher in whole blood gas analysis than in serum.
- 42 % of potassium results are upgraded from normo- to hypokalemic or downgraded from hyper- to normokalemic [7, 15].

Tab. 28.1 POCT and potential errors in the pre-analytical phase (modified from [20])

Latent condition for error	Potential reduction with POCT
Inappropriate/excessive ordering	No
Mistimed testing	No
Patient identification	No/yes
Specimen identification	Yes
Specimen collection	No
Specimen attributes	No

Another issue affecting the pre-analytical phase is the parameter selection on some POCT devices that require a pre-defined number of tests regardless of the clinical question. A case in point is the simultaneous measurement of cardiac biomarkers with cardiac troponin, such as myoglobin and CK-MB, despite evidence that they do not add valuable information to the results. Similarly, such examples are available for other common clinical chemistry tests. The potential sources of errors in the pre-analytical phase are shown in Tab. 28.1.

28.3.2 Analytical sources of errors

Evidence is available to show that POCT presents a higher risk of errors in the intra-analytical phase than central laboratory testing [18].

As previously reported for some measurands (e.g. blood gas, electrolytes), analytical performance characteristics are valuable whereas they do not comply with current recommendations for other measurands (e.g. cardiac troponin) [19]. Additional evidence indicates that quality control procedures are poorly followed and documented [14].

28.3.3 Post-analytical sources of errors

In several studies, the post-analytical phase was an important source of errors and risk to patient management. In particular, despite progress in POCT device connectivity, **manual transcription of data** still remains a common procedure for POCT blood glucose and other tests, thus leading to a high risk of errors. In one study, 3.2 % of glucose concentrations were incorrectly reported, the time of blood sampling was recorded in an imprecise manner and 12.1 % of results were missed [2]. The survey from a more recent publication shows that the majority of results (77 % of respondents) were hand-transcribed into patient records even though most of the meters had connectivity capacity and that this clearly represented a risk of error [25]. Nevertheless, the most relevant source of errors derives from the rapid availability of results and the immediate therapeutic implications engendered by many POCT tests. This fact, in turn, amplifies the likelihood that erroneous results from these tests will cause preventable adverse events [16]. In the final phase of the testing process, all potential defects that can occur at previous steps should be recognized and fixed by repeating the analysis, removing the interference and by other appropriate corrective actions. As the rapid result availability has been found to be an amplifier of POCT error, appropriate strategies should be implemented to minimize the risk related to the notification of wrong results which may cause adverse events and patient harm.

28.4 Conclusions

Quality in laboratory medicine should be defined as the guarantee that each and every step along the total testing process is performed correctly and accurately to ensure valuable decision making along with safe and effective patient care [21]. This definition applies to tests performed in central laboratories as equally as to decentralized testing, including POCT. While it is often suggested that for POCT considerations of speed of analysis surpass all others, evidence shows that quality in the pre-, intra-, and post-analytical phases is required to assure clinicians receive valuable information and to avoid patient harm. Even if most laboratory errors do not impact patient outcomes, the potential to injure patients is inherent and, for that reason, every potential risk of error deserves commensurate investigation, response and prevention (▶ Chapter 40). POCT's advantages and disadvantages should be weighed carefully when implementing it in clinical practice; appropriate actions for assuring adequate training to operators, effective procedures for quality control and assessment should be put in place.

Finally, the outcomes of introducing POCT can be strongly influenced by local factors. The same testing approach may produce benefits in one setting but not in another [24]. This reinforces the view that clinical governance [5] as a systematic approach to quality assurance should be applied to all aspects of healthcare and laboratory testing, not only POCT.

References

1. Boyd JC, Bruns DE (2001) Quality specifications for glucose meters: assessment by simulation modeling of errors in insulin dose. Clin Chem 47:209–14
2. Carraro P, Plebani M (2009) Post-analytical errors with portable glucose meters in the hospital setting. Clin Chim Acta 404:65–7
3. Collinson PO (2006) Cardiac biomarkers. Pathophysiology and disease management. IN: Totowa NJ (ed): Humana Humana Press, pp 559–74
4. Fraser CG (2001) Biological variation: from principles to practice. AACC Press, Washington

5. Freedman DB (2002) Clinical governance: implications for point-of-care testing. Ann Clin Biochem 39:421–3
6. Gray TA, Freedman DB, Burnett D, Szczepura A, Price CP (1996) Evidence based practice: clinicians' use and attitudes to near patient testing in hospital. J Clin Pathol 49:903–8
7. Hawkins RC (2006) Correction and reporting of potassium results in haemolysed samples. Ann Clin Biochem 43:88–9
8. Hyltoft Petersen P, Brandslund I, Jørgensen L, Stahl M, de Fine Olivarius N, Borch-Johnsen K (2001) Evaluation of systematic and random factors in measurements of fasting plasma glucose as the basis for analytical quality specifications in the diagnosis of diabetes. 3. Impact of the new WHO and ADA recommendations on diagnosis of diabetes mellitus. Scand J Clin Lab Invest 61(3):191–204
9. Jani IV, Peter TF (2013) How Point-of-Care testing could drive innovation in global health. N Engl J Med 368: 2319–24
10. José RJ, Preller J (2008) Near-patient testing of potassium levels using arterial blood gas analysers: can we trust these results? Emerg Med J 25:510–3
11. Kazmierczak S (2008) Improving healthcare through advances in point-of-care technologies. Clin Chem Lab Med 46:1–2
12. Kristensen GB, Sandberg S (2010) Self-monitoring of blood glucose with a focus on analytical quality: an overview. Clin Chem Lab Med 48:963–72
13. Kost GJ (1995) Guidelines for point-of-care testing. Improving patient outcomes. Am J Clin Pathol 104:S111–S127
14. Larsson A, Greig-Pylypczuk R, Huisman A (2015) The state of point-of-care testing: a European perspective. Ups J Med Sci 120:1–10
15. Lippi G, Blanckaert N, Bonini P, Green S, Kitchen S, Palicka V, et al. (2008) Haemolysis: an overview of the leading cause of unsuitable specimens in clinical laboratories. Clin Chem Lab Med 46:764–72
16. Meier FA, Jones BA (2005) Point-of-care testing error: sources and amplifiers, taxonomy, prevention strategies, and detection monitors. Arch Pathol Lab Med 129:1262–7
17. Mirzazadeh M, Morovat A, James T, Smith I, Kirby J, Shine B (2016) Point-of-care testing of electrolytes and calcium using blood gas analysers: it is time we trusted the results. Emerg Med J 33:181–6
18. O'Kane MJ, McManus P, McGowan N, Lynch PL (2011) Quality error rates in point-of-care testing. Clin Chem 57:1267–71
19. Palamalai V, Murakami MM, Apple FS (2013) Diagnostic performance of four point ofcare cardiac troponin I assays to rule in and rule out acute myocardial infarction. Clin Biochem 46:1631–5
20. Plebani M (2009) Does POCT reduce the risk of error in laboratory testing? Clin Chim Acta 404:59–64
21. Plebani M (2012) Quality indicators to detect pre-analytical errors in laboratory test. Clin Biochem Rev 33:85–8
22. Plebani M (2013) Harmonization in laboratory medicine: the complete picture. Clin Chem Lab Med 51:741–51
23. Plebani M (2016) Harmonization in laboratory medicine: Requests, samples, measurements and reports. Crit Rev Clin Lab Sci 53:184–96
24. Rainey PM, Ulibarri M (2014) Point-of-Care Testing. Is faster better? Am J Clin Pathol 142:582–3
25. Sharp L, Farrance I, Greaves RF (2016The application of glucose point of care testing in three metropolitan hospitals. Pathology 48:51–9
26. St John A, Price CP (2014) Existing and emerging technologies for Point-of-Care Testing. Clin Biochem Rev 35:155–67

The importance of infection control in POCT

Axel Kramer, Eva Gruner

P. B. Luppa, R. Junker (Eds.), *Point-of-Care Testing*
https://doi.org/10.1007/978-3-662-54497-6_29

29.1 Tasks and objectives

One important feature of POCT is that findings obtained from near-patient tests translate into immediate implications for diagnosis and/or therapy.

Independent of this, basic hygiene measures for infection protection must be applied to near-patient laboratory diagnostics. Consequently, in the context of POCT, two important areas for infection control arise in settings where maintaining basic hygiene measures are paramount.

- Compliance with hygiene measures when using POCT on the patient
- Compliance with hygiene at the POCT workplace

It should be pointed out that POCT procedures are also developed as an option to assess infection risks. This is discussed in more detail in ▶ Section 29.4 (PCR-based risk-adapted screening).

Applying basic hygiene measures can prevent the transmission of most pathogens. Basic hygiene measures include hand disinfection, the use of personal protection equipment (PPE), environmental surface disinfection of near-patient areas as well as, potentially, those further away from patients, the reprocessing and handling of medical products (MP), alongside proper laundering, safe injection techniques and cough etiquette. Although not included in the CDC Directive [6], vaccination of staff and antiseptic measures are part of basic hygiene measures.

29.2 Hygiene compliance in POCT

The focus of hygiene and infection control in POCT diagnostics is to protect patients from infections and to protect microbiological samples from environmental contamination. Staff protection must also be ensured.

Note

Hygiene requirements are specified in the institutional hygiene plan. In consultation with the relevant staff, the plan must be updated annually as part of the enforcement of German Infection Protection Act (IfSG) [22].

Three obstacles that are the most frequent causes of neglected hygiene rules need to be overcome: Lack of time, shortages of personnel and lack of trained staff.

29.2.1 Personal hygiene

Personal hygiene measures, including occupational, specialist and protective clothing for medical staff, are intended to prevent nosocomial infections and their spread, as well as to protect staff from infections.

The principles of employee hygiene in Germany are specified in the following statutory provisions: IfSG [22], Federal Hygiene Directive, German Biological Agents Ordinance (BioStoffV) [30] and TRBA 250 [28], accident prevention guidelines for the health service (BGV C8) [11]. These include the KRINKO guideline [24], particularly the recommendations for hygiene requirements for the cleaning and disinfection of surfaces (2004) [15], for hospital hygiene, health and safety requirements, for hygienic work clothing and personal protective equipment (2007) [25], for hygiene measures in infections or colonization with multiresistant gram-negative rods (2012) [17], recommendations for the prevention and control of "methicillin-resistant Staphylococcus aureus (MRSA) in medical and care institutions (2014) [18], for the prevention of infection in the care and treatment of patients with infectious diseases (2015) [19], for hand hygiene in health care institutions (2016) [20]. Further details can be found in recommendations by specialist associations (e.g. AWMF, DGKH; selected recommendations are listed separately in the references) [8–10, 12–14].

Occupational clothing The legislators mandate that professionals in this field wear occupational clothing. From a hygiene perspective, short-sleeved tunic and trousers are most appropriate.

Protective clothing Protective clothing is to be worn in addition to occupational clothing for tasks that carry a higher risk of contamination (e.g. when attending to patients in isolation, intensive care patients, immunocompromised patients). Depending on the task, this includes non-sterile or sterile protective gowns, waterproof aprons, non-sterile or sterile gloves [21], face shields or masks, goggles and hair protection. Long hair should be tied back. In cases where close patient contact is not anticipated, non-sterile protective fleece gowns are sufficient for contact with patients in isolation; they should be disposed of when leaving the isolation or intensive care unit.

Non-sterile gloves must be donned before any planned or potential contact with blood, secretions, excretions, body fluids or pathogens which often occurs during POCT procedures. Gloves should only be kept on for dealing with one and the same patient. After the task has been carried out, the gloves must be disposed of.

A face shield or mask must be worn during tasks where aerosol generating or spray from infectious materials can be anticipated. The same applies when caring for patients in reverse isolation. Staff members who suffer from a cold should wear a face mask to prevent transmission to patients or to other staff members (e.g. influenza). Face shields or masks must fully cover the mouth and nose. They should be worn for a maximum of 2 hours and must be changed when over-moistened. It is not acceptable to re-doff and re-don (put on and off) masks or letting them hang around the neck like a bib.

Goggles must be worn where a possible spray event of infectious material is anticipated.

Hand hygiene [8, 20] In general, hand hygiene is one of the most important measures to prevent the spread of nosocomial infections. It is estimated that up to 90°% of all exogenously transmitted nosocomial infections are caused by transfer of pathogens from the hands.

Measures comprise hand disinfection and hand washing alongside skin protection measures and skin care.

29.2.2 Reprocessing of accompanying medical products for POCT

Basic hygiene rules apply when handling all kinds of POCT systems. When operating a device, gloves should only be worn when there is possible contact with sample material. It should be noted, however, that without interim disinfection, contamination with pathogens can occur via the device work surface.

Unit-use POCT devices mostly use individually packaged reagents and strips with analytical sensors incorporated into strips or cartridges that are discarded after analysis. When using strips, e.g. in blood glucose meters, gloves can be worn from the time of taking blood to the end of the analysis process [3].

Benchtop devices are usually located in **separate areas in** hospital departments or practices. When built-in centrifugation analyzers or micro ventilation and micro pumps are used, it is important to ventilate the system and avoid aerosol generation when processing potentially infectious material [3].

> **It is important to carry out a risk assessment of the POCT device to ensure hygiene safety and infection control by putting standard operating procedures (SOP) in place and offering regular infection prevention training [16].**

Work surfaces that have had contact with potentially infectious material should be routinely wipe-disinfected at least once daily as well as after visible contamination.

Blood glucose meters should, if possible, be personalized to each patient. If used for more than one patient, the device must be disinfected prior to use. Wipe systems are well-established, practical and therefore particularly suitable. It is essential they are stored in closed containers, their use-by dates checked and that they are cleaned with disinfectant after use; the compatibility of the disinfectant and wipes with the POCT devices should be checked as well (according to the manufacturer's instructions).

Adherence to the manufacturer's instructions is important as not all POCT devices are compatible with all disinfectants.

If such information is lacking, the manufacturer must be consulted before operating the device and its release documented. Any reprocessing must be documented in the hygiene plan. The required effectiveness against viruses and spores should also be considered in the case of reprocessing, as the spectrum is only guaranteed if stated by the manufacturer.

Failure to implement basic hygiene measures can, in individual cases, have legal consequences, e.g. in the event of non-compliant reprocessing or neglected hand disinfection. If an infection occurs from a preventable situation, exonerating evidence has to be provided by the user or the institution [26].

29.2.3 Disposal

Containers for the disposal of test strips or cartridges should be kept in close proximity to the device.

29.2.4 Vaccination protection

Vaccination is an important individual and collective preventive measure, also to protect staff working in POCT settings [7].

Vaccination is "publicly recommended" by the state health authorities pursuant to Section 20 (3) IfSG and must be offered to employees. Proof of effective vaccination protection is required for affected employees who have contact with immunodeficient patients.

Vaccination protection is recommended for the following diseases that carry a higher occupational risk: Hepatitis A and B, measles, meningococcal disease, mumps, pertussis, pneumococcal disease, rubella, varicella and viral influenza (for vaccination instructions, see STIKO 2015) [27].

29.2.5 Immediate action after accidental contamination

Note

Employees exposed to biologically hazardous materials must be educated about risks from accidental contamination as well as about the necessary prevention measures by means of operating instructions and hygiene plans. Post-exposure procedures should be practiced [5].

The focus is on the following protective actions: Structured, well-thought through working procedures, availability of break-proof and puncture-resistant containers for the disposal of used cannulas etc., the use of safety devices as well as the provision and use of PPE.

In the event of injury, the following immediate measures are recommended:

- After a needlestick or sharps injury, forced bleeding of the puncture site should be introduced, followed by rinsing with water or a skin antiseptic.
- Afterwards, a wound antiseptic should be applied.

After oral or eye exposure, rinsing with tap water followed by an antiseptic rinse containing a mucosal or eye antiseptic is recommended. Every exposure to blood or similar infectious material, even when the infection status of the index person is unknown, must be reported internally as a work-related accident. The **accident insurance physician** decides about further procedures regarding vaccination, post-HIV exposure prophylaxis and serological testing [5].

29.3 Compliance with hygiene measures when using/employing POCT devices and procedures

At POCT workstations on clinical wards and at outpatient centers, infection risks for the general public must be prevented due to the potential release of pathogens during analysis or from improper disposal due to the failure to use suitable protective measures [1].

The building-related requirements for working with microorganisms are specified in the German Biological Agents Ordinance (BioStoffV) Sections 10 and 11, in its Annex II and III as well as in Section 5.2. to 5.5. of the technical regulations for biological materials (TRBA 100). TRBA 100 and/or 250 substantiate the requirements of BioStoffV, particularly of Appendix II. All surfaces need to be waterproof, easy to clean and resistant to the cleaning products and disinfectants used.

Note

As part of an exposure assessment, prior to working with biological material as well as for major changes, the employer must specify appropriate protective measures in the hygiene plan prioritizing hand hygiene, PPE, disinfectant surface cleaning and waste disposal [1].

Surfaces where POCT procedures are carried out can be contaminated with pathogens. Therefore, the following basic rules need to be followed strictly:

- Spatial separation of workplaces where there is a burden from infectious versus non-infectious material.
- Regular cleaning and/or disinfection of work surfaces (after each procedure, at the end of the working day and immediately after any visible contamination).
- Prohibition of eating, drinking, smoking and storage of food items in the laboratory.
- While working with infectious material, neither the face nor technical devices such as telephone or computer keypads should be touched.

Aerosols can be generated when carrying out the following procedures: Opening sample containers that are closed with a lid, mixing of samples, pipetting and emptying of syringes and sample containers [1]. A microbiological safety bench may be required.

The principles of the German federal working group for waste (LAGA, 2015) must be followed for waste disposal [23] and specified in the hygiene plan.

POCT devices must be cleaned daily after use, when necessary by the operating personnel. See ▶ Section 29.2 for the relevant procedures. The disinfectant contact time should be 5 minutes for a bactericidal and fungicidal effect and 30 minutes to destroy HBV and HIV.

29.4 PCR-based, risk-adapted screening

The rapid detection of carriers such as MRSA, Enterococci or Gram-negative bacteria with particular resistance characteristics and preventive isolation is only possible if screening results are available quickly. This is done by using bedside tests that do not require sample preparation. Diseased carriers who suffer from toxin-producing strains of Clostridium difficile infection can be identified as quickly as patients infected with Noro or Influenza viruses. The list of rapidly detectable molecular biological pathogens is continually expanding (▶ Chapter 10 and 20). The simultaneous detection of the entire pathogenic species spectrum (multiplex analyses) for defined severe infections, e.g. of the lung or of the CNS, is important in this context. Ready-to-use reagents for diagnostics deliver results in approximately 1–2 hours. Major cost savings can be made if screening is restricted to patients with defined risk factors and not extended to the general patient population. At the same time, this reduces the nosocomial MRSA infection rate significantly [2, 4].

Alternatively, immunochromatographic tests (lateral flow assays) can be used as qualitative strip tests to identify single pathogen strains such as Clostridium difficile. Here, the results are available within 10–15 mins.

References

1. Caspari G (2016) Laboratorien mit Infektionsrisiken. In: Kramer A, Assadian O, Exner M, Hübner NO, Simon A (Ed.) Hospital - und Praxishygiene, 3rd Edition Elsevier, Munich, p. 615
2. Fenger S, Bartels C (2017) Unternehmen Krankenhaus unter besonderer Berücksichtigung der Hygiene im Krankenhaus. In: Eikmann T, Christiansen B, Exner M, Herr C, Kramer A (Ed.) Hygiene in Krankenhaus und Praxis.·Ecomed-Verlag, Landsberg am Lech.
3. Gruner E (2015) Hygieneregeln im Umgang with Point-of-Care testing (POCT). Pipette – Swiss laboratory medicine No 4: 10–1
4. Kramer A, Fleßa S, Hübner C, Hübner NO, Leesch L, Ryll S, Wegner C (2013) Die Greifswalder MRSA-Präventionsstrategie und die Rolle von POCT. In: Eiff W v. (Ed.) POCT Management. Heidelberg: Medhochzwei 20: 239–62
5. Kramer A, Wicker S (2016) Sofortmaßnahmen nach akzidenteller Kontamination: In: Kramer A, Assadian O, Exner M, Hübner NO, Simon A (Ed.) Hospital- und Praxishygiene, 3rd Edition Elsevier, Munich, p. 511–5
6. Siegel JD, Rhinehart E, Jackson M, Chiarello L (2007) Guideline for isolation precautions: Preventing transmission of infectious agents in healthcare settings. www.cdc.gov
7. Wicker S, Kramer A, Pitten FA (2016) Impfprophylaxe and Personalschutz. In: Kramer A, Assadian O, Exner M, Hübner NO, Simon A (Ed.) Hospital- und Praxishygiene, 3rd Edition Elsevier, Munich, p. 132–5

Laws, regulations, ordinances, guidelines, guidance statements, recommendations, standards

8. Working Group 'Hospital & Practice Hygiene' of AWMF (2015) Händedesinfektion und Händehygiene. AWMF Guidelines Register No. 029/027. 8/2015. HygMed 40(9) 369–85. http://www.awmf.org/uploads/tx_szleitlinien/029-032l_S1_Hygiene_respiratorisch_uebertragbare_Infektionen_2016-01.pdf
9. Working Group 'Hospital & Practice Hygiene' of AWMF (2015) Hausreinigung und Flächendesinfektion: Hygienische Anforderungen. AWMF Guidelines Register No. 029/006. 9/2015. HygMed 40(10): 418–24
10. Working Group 'Hospital & Practice Hygiene' of AWMF (2016) Hygieneanforderungen bei ausgewählten respiratorisch übertragbaren Infektions-Erkrankungen (aerogen und Tröpfchen). AWMF Guideline Register No. 029–032.1/2016
11. Berufsgenossenschaftliche Vorschrift für Sicherheit und Gesundheit bei der Arbeit (1997) Unfallverhütungsvorschrift Gesundheitsdienst vom Oktober 1982, in der Fassung vom 1. Januar 1997 (BGV C8)
12. Deutsche Gesellschaft für Krankenhaushygiene (DGKH) (2013) Konsensuspapier Blutzuckermessung. 2013. http://www.krankenhaushygiene.de/pdfdata/hm/HM_06_2013_blutzuckermessung.pdf
13. Deutsche Gesellschaft für Krankenhaushygiene (DGKH) (2015) Versorgung mit Arbeitskleidung (synonym Dienst- oder Berufskleidung) von Leasingpersonal und Hospitanten, Praktikanten, Aushilfen, Grünen Damen/Herren in Krankenhäusern/Pflegeeinrichtungen. http://www.krankenhaushygiene.de/ccUpload/upload/files/sektionen/2015_12_DGKH_kleidung_leasingpersonal.pdf
14. Deutsche Gesellschaft für Krankenhaushygiene (DGKH) (2015) Schutzkittel bei medizinischen und pflegerischen Tätigkeiten sowie bei Barrieremaßnahmen und Isolierungen. http://www.krankenhaushygiene.de/pdfdata/hm/2015_HM_01-2_schutzkittel.pdf
15. Recommendation issued by the Commission for Hospital Hygiene and the Prevention of Infection (KRINKO) at the Robert Koch Institute (RKI) (2004) on hygiene requirements for cleaning and disinfecting surfaces. BGBL Gesundheitsforsch Gesundheitsschutz 47:51–61
16. Recommendation issued by the Commission for Hospital Hygiene and the Prevention of Infection (KRINKO) at the Robert Koch Institute (RKI) and by the Federal Institute for Drugs and Medical Devices (BfArM) (2012) on requirements governing hygiene during the reprocessing of medical devices. BGBL 55:1244–310
17. Recommendation issued by the Commission for Hospital Hygiene and the Prevention of Infection (KRINKO) at the Robert Koch Institute (RKI) (2012) on hygiene measures infections or colonization with multi resistant gram-negative rods. BGBL 55:1311–54
18. Recommendation issued by the Commission for Hospital Hygiene and the Prevention of Infection (KRINKO) at the Robert Koch Institute (RKI) (2015) recommendations for the prevention and control of "methicillin-resistant Staphylococcus aureus (MRSA) in medical and care institutions. BGBL 2014; 57:696–732

19. Recommendation issued by the Commission for Hospital Hygiene and the Prevention of Infection (KRINKO) at the Robert Koch Institute (RKI) (2015) Prevention of infection within the scope of care and treatment of patients with transmissible diseases. BGBL 58:1151–70
20. Recommendation issued by the Commission for Hospital Hygiene and the Prevention of Infection (KRINKO) at the Robert Koch Institute (RKI) (2016) Hand hygiene at healthcare facilities. BGBL 59:1189–220
21. EN 374 (2003) Protective gloves against chemicals and microorganisms. Beuth, Berlin
22. Gesetz zur Änderung des Infektionsschutzgesetzes und weiterer Gesetze (IfSGuaÄndG) of 4 August 2011. BGBl I 2011; (41):1622 (2016) Verordnung zur Anpassung der Meldepflichten nach dem Infektionsschutzgesetz an die epidemische Lage (IfSG-Meldepflicht-Anpassungsverordnung – IfSG-MeldAnpV) of 18 March 2016. BGBL Part 13: 515
23. German Federal Working Group for Waste (LAGA) (2015) Vollzugshilfe zur Entsorgung von Abfällen aus Einrichtungen des Gesundheitsdienstes
24. Robert Koch Institute (2016) Richtlinie für Krankenhaushygiene und Infektionsprävention. Grundwerk inkl. Lieferung 18 zur Fortsetzung. Munich: Urban Fischer. http://www.rki.de/DE/Content/Infekt/Krankenhaushygiene/Kommission/kommission_node.html
25. Robert Koch Institute (2007) Anforderungen der Krankenhaushygiene und des Arbeitsschutzes an die Hygienebekleidung und persönliche Schutzausrüstung. Epid Bull subsec.1:3–4
26. Schneider A, Bierling G (2016) HuR Hygiene und Recht. Entscheidungssammlung – Richtlinien. Loose-leaf collection. mhp, Mainz
27. STIKO (2015) The Standing Committee on Immunisation (STIKO) at the Robert Koch Institute/Status: August 2015. Epid Bull subsec. 24: 327–60
28. Technical Regulations for Biological Materials 250 (TRBA 250) (2014) Biologische Arbeitsstoffe im Gesundheitswesen und in der Wohlfahrtspflege. GMBl No. 10/11 of 27 March 2014: 1–78; 1st amendment GMBl No. 25 of 22 May 2014: 535; 2nd amendment GMBl No. 29 of 21 July 2015: 577
29. Technical Regulations for Biological Materials 100 (TRBA 100) (2014) Schutzmaßnahmen für Tätigkeiten mit biologischen Arbeitsstoffen in Laboratorien. GMBl 2014, No. 51/52 of 17 October 2013: 1–50; amendment of 30 June 2014, GMBl No. 38: 814
30. Ordinance on Safety and Health Protection at Workplaces Involving Biological Agents (Biological Agents Ordinance - BioStoffV) of 15 July 2013 (Federal Law Gazette [BGBl.] Part I p. 2514)

Economic aspects of POCT

Norbert Gässler, Ralf Junker, Claus Langer, Birgit Schäfer

P. B. Luppa, R. Junker (Eds.), *Point-of-Care Testing*
https://doi.org/10.1007/978-3-662-54497-6_30

30.1 POCT cost analysis

Frequently, the higher price for POCT analysis is cited when comparing the costs of POCT versus laboratory diagnostics [4]. However, this perspective needs to be subjected to closer scrutiny because it generally only accounts for the proportion of costs spent on materials. When a hospital laboratory pays for an analyte, it is less expensive because it is usually purchased in large volumes and the measurements are performed with a high degree of automation [5]. The contrary applies to POCT where irregularly occurring single analyses are carried out [6]. Indeed, the cost calculation for the laboratory not only include purely material costs for reagents, control and calibration materials but also the device-specific costs like investments, maintenance, running overheads for electricity, water, consumables etc. On top of this come the not insignificant costs for laboratory staff. All of these factors must be seen in aggregate for each individual analysis. The administrative and management costs (overhead) should not be neglected either (Tab. 30.1).

Tab. 30.1 Cost types

Personnel	Salaries Allowances Continuing education Other personnel costs
Materials	Reagents (lab supplies) Consumables Office supplies Other supplies
Devices	Device maintenance Repairs Leasing agreements (for devices and reagents or price per result) Investment costs (amortized over the planned term, depreciation)
Other costs	Occupancy costs (costs directly attributable to maintenance, energy, telephone, cleaning etc.) Administration/management

30.2 Costs for the pre- and post-analytical phases

Besides the actual analytical process itself, the POCT cost analysis must also examine costs for the pre- and post-analytical phase. That means accounting for sample-taking and the subsequent transport of the sample to the laboratory, quality controls and transmission of the analytical findings to the sender, usually supported by IT systems. For this purpose, the corresponding costs for the infrastructure similarly should be included in the calculation [1, 3] (see below).

Costs for the pre- and post-analytical phases
- Blood draw systems
- Person-hours for blood draws
- Request forms or hospital information system HIS (order entry)
- Person-hours for laboratory request
- Person-hours for sample transport
- Laboratory information system (LIS)
- LIS / HIS server
- Printer, fax etc.
- Person-hours for documentation, filing of finalized results
- Person-hours for quality control measures

By definition, POCT analyzers are near or at bedside or the relevant analyzers are brought to the patient where the analysis is carried out. Marked time advantage can be achieved because capillary blood samples are usually taken [2].

In many hospitals in Germany, venous blood is generally drawn by the physician, whereas capillary blood draws are reserved for the nursing staff. Clear financial differences in sample collection result from this. As illustrated, sample transport is obviated and does not need to be accounted for financially. The subsequent processes performed in the laboratory such as technical and medical validation, printing out the findings etc. are frequently not considered in the POCT analysis on the wards and at outpatient centers. The personnel costs that are directly attributable to POCT are markedly lower than the comparable personnel costs incurred by the laboratory. Yet, the post-analytical phase, e.g. result transmission, technical and medical validation, quality control evaluation etc. must be taken into account for POCT analysis. According to RiliBÄK, these measures are mandatory and generally conducted by the POCT coordination office.

Not only the time advantage and the quality differences should be mentioned here, but also questions relating to the medical and economic utility. "An important criterion for point-of-care laboratory testing is the immediate communication of therapeutic recommendations based on the laboratory analysis." This is the literal definition found in the English version of RiliBÄK. This underscores the benefit conferred by POCT methods, provided the rapid analysis permits immediate medical action.

Using the example of capillary blood glucose testing, the POCT method is contrasted with conventional laboratory practice, while ◘ Tab. 30.2 compares the necessary person-hours and costs incurred. The costs for nursing and laboratory staff are calculated to average € 40,000 and € 50,000, respectively, per job and year. Since POC glucose testing is performed by the nursing staff exclusively as single measurements, the necessary person-hours can be multiplied directly by the gross hourly wages previously mentioned. The work expenditure for quality control and validation is calculated based on approx. 10 single measurements per day. In the central laboratory, the glucose measurements are generally processed in series; the average series length is assumed to be approx. 30 single samples per run. That is why the person-hours necessary for the individual glucose measurement are calculated proportionately (as based on the authors' time and cost calculations). Usually, several persons in the laboratory are responsible for capillary blood draws; it is estimated that an average of 10 patients are on their get lists. Transport times and blood draw per staff member and patient are set at 3 min, equaling € 1.56 personnel costs.

It can be stated that the costs for capillary blood glucose measurement at the POC are comparably as high as glucose testing in the central laboratory. This is due to the more than twice as high material costs for the test strips and the lower person-hours or personnel costs. The comparison however depends on each hospital's situation (e.g. on the question of whether a pneumatic delivery system is available).

When considering each single performance of a capillary blood glucose measurement, the fact that the fast availability of the analytical results not only enables direct therapeutic measures to be implemented, but is also important for the patient´s convenience should be additionally thrown into the balance. These are significant criteria for the entire treatment process, but are difficult to reflect on a simple spreadsheet.

30.3 Cost coverage for POCT services within the German healthcare system

Within the German healthcare system, reimbursement for (POC) diagnostics is predicated upon whether the test is performed on an outpatient or inpatient basis within the scope of statutory or private health insurance. Nevertheless, the (POC) test is always an integral part of the physician's services. In their procedures for approving new services for inclusion in the statutory health insurers' reimbursement catalog, the German Uniform Assessment Standard (EBM), the responsible parties within the physicians' self-governance do not fundamentally differentiate whether the laboratory service in question involves a method performed as POCT or in the central laboratory. It is necessary for diagnostic (POC) tests to be given an EBM remuneration code because otherwise they are subject to the "right to reserve permission" in the outpatient care setting. That means that reimbursement is only possible for services approved by the responsible bodies.

Which body is responsible for the procedure for approving a new diagnostic (POC) test

Tab. 30.2 Capillary blood glucose testing costs comparing POCT versus laboratory at a hospital site

Laboratory			POCT		
	Average times [s]	Average costs [€]		Average times [s]	Average costs [€]
Analysis order (order entry)	30	0.190[1]	Analysis order	20	0.138[1]
Preparation of a work-station list*	120	0.034[2]			
Travel time to blood draw (laboratory – ward)	60	0.521[2]	Transport time to the patient (ward – hospital room)	30	0.208[1]
Blood draw	60	0.521[2]			
Transport time after blood draw (ward – laboratory)	60	0.521[2]			
Preparation of analyzers*	360	0.104[2]	Quality control – device (technical status)**	120	0.083[1]
Analysis*	300	Personnel costs 0.086[2] Material costs 0.126	Blood draw and measurement	180	Personnel costs 0.832[1] Material costs 0.262
Quality control (daily)*	30	Personnel costs 0.008[2] Material costs 0.001	Quality control (1× weekly)**	180	Personnel costs 0.017[1] Material costs 0.008
Quality control (monthly)*	60	0.001[2]	Quality control (monthly)**	60	0.002
Technical validation*	120	0.034[2]	Technical validation**	30	0.019[1]
Medical validation*	20	0.011[3]	Medical validation**	20	0.034[3]
Result printout (on the ward)*	30	0.008[2]	Result documentation	60	0.417[1]
Presentation to physician	-		Presentation to physician	30	0.208[1]
Total personnel costs		2.039	Total personnel costs		1.955
Total costs***		0.127	Total material costs***		0.270
Aggregate total		2.166	Aggregate total		2.225

* Average series length: 30 single samples per run
** Average series length: 10 single samples per day
*** Exclusive repair/maintenance costs
[1] Personnel costs nurses, [2] Personnel costs laboratory, [3] Personnel costs academics

30

depends on the test's intended purpose. The Federal Joint Committee (G-BA) is the responsible authority when the service involves a new examination and treatment method, including diagnostic agent, or a laboratory diagnostic service as part of early detection (e.g. laboratory services within the scope of the maternity protection directive). A competence of the assessment subcommittee to approve new catalog entries is also given for all other laboratory test services rendered within the scope of care provided by physicians under contract with the National Association of Statutory Health Insurance Funds (GKV). This separation grew historically. The present trends, e.g. in personalized medicine, currently show a tendency in favor of the G-BA if it deems a diagnostic test method to be socially relevant, e.g. the laboratory test helps in the decision for or against adjuvant chemotherapy in breast cancer. In July 2015, Section 87 (3e) of the 5th Book of the German Code of Social Law (SGB V) adopted into law that associations, professional and medical societies as well as individual companies are entitled to information regarding the competence of the body approving a new product.

One of the main differences in responsibilities lies in the groups of persons who are permitted to apply for approval to perform a laboratory test: In the G-BA, this is reserved for the public benches, i.e. the GKV, National Associations of Statutory Health Insurance Physicians (KBV) and Dentists (KZBV), German Hospital Federation (DKG) as well as the patient advocacy group. That means that third parties only have the possibility to submit an application by proxy via one of the stakeholders.

If the procedure is held by the assessment subcommittee, the right to submit an application is transferred in the code of procedure to the stakeholder institutions, GKV and KBV for approving innovative laboratory parameters for the Chapter 32 of EBM. By contrast, medical device associations, medical and professional societies are granted the right to submit proposals; in other words, it is not mandatory that their proposals for accepting new laboratory services are to be considered.

Even the procedure differs substantially across institutions. The scientific expertise at the G-BA is fundamentally obtained by its commissioning of the Institute for Quality and Efficiency in Health Care (IQWiG); the assessment subcommittee has established a procedure that includes the laboratory center of competence.

The laboratory working group of the assessment subcommittee developed a questionnaire for approval of new laboratory services. The questions cover a new (POC) test's diagnostic performance, its costs-benefit ratio and any substitution potential.

Whereas the G-BA designs its method transparently, the decisions made by the assessment subcommittee are closed to the public. The reasons for an innovative (POC) laboratory service being approved for inclusion in or rejected from the catalog of services covered by the statutory health insurers therefore often remains unclear. Neither do relevant third parties have any legal remedy to object to assessment subcommittee decisions, whereas legal recourse is available to all stakeholders in many proceedings before the G-BA.

One fundamental rule for accepting a new (POC) laboratory test states that the name of the test manufacturer is not included in the invoicing code. Rather, the laboratory test method must be described under the performance category in the code item.

Mobile laboratory diagnostic systems fundamentally beg the question as to whether the associated material costs have already been included as a flat-rate in the remuneration code. This could theoretically lead to a disadvantage given that the materials for POCT methods mostly cost more than for conventional laboratory diagnostics. This is because the EBM standard does not generally differentiate its performance descriptions into "classic" and POC methods; there are only very few cases where the rendering of laboratory services on prefabricated reagent slides is expressly limited or excluded. One example, however, is the quantitative measurement of D-dimer which is explicitly prohibited by the EBM to be carried out as a POCT. Sometimes, such restrictions are also contained in the guidelines of the medical societies, even if they are not binding for the reimbursement. EBM (to date) has envisaged the levying of a possible material surcharge for the use of a POC system.

If a (POCT) method is not listed in the catalog of services of the Statutory Health Insurance Funds, it can alternatively be offered to the patients as a self-payer service (IGeL) to be invoiced by the fee schedule for privately rendered medical services.

Laboratory test services rendered within the scope of private health insurance are invoiced according to the German physicians' fee schedule (GOÄ). Chapter M lists laboratory diagnostics although invoicing using what is called an analogous code in Article 8 of the general provisions is taking on ever greater dimensions.

An established system for the approval of new services, as envisaged by the GKV does not exist for the GOÄ. The analogous invoicing set forth in Article 8 thereby allows a new service to be invoiced using a pre-existing code as long as it is equivalent "in type, costs and time expenditure". This invoicing option thus already allows today's innovative (POC) services to find relatively uncomplicated entry into the practices of physicians who treat the privately insured.

In hospitals, services are invoiced according to a rate per case or Diagnosis-Related Group (DRG) system, and nationwide defined additional charges. DRGs cover all services rendered in hospitals. Moreover, the main advantage over the outpatient care setting is that the principle of the mistake of law applies, i.e. only services explicitly excluded by the responsible committees may not be rendered. Diagnostic devices that bear the CE mark can therefore be invoiced in the inpatient setting.

New diagnostic POCT services can therefore also be introduced into the provision of care without an application procedure or official approval as a DRG being as the laboratory service is remunerated within the rate per case for the underlying disease the patient is receiving treatment for. The advantage of this aspect is that a hospital – in simple terms – more or less has the freedom to decide which methods it chooses to render the necessary laboratory diagnostics. In addition to a test's medical benefit or utility, the key factors for the decision by the hospital operator to use it are therefore fiscal ones. When a (POC) laboratory service burdens the DRG to the extent that the affected rate per case is no longer cost-effective, a medical benefit or a substitutions potential ought to be apparent elsewhere in order for the payor to decide to use that innovative (POC) method.

If the use of a new laboratory method causes the currently existing DRG to be exceeded, the hospital using it can nevertheless submit a DRG application to the Institute for the Hospital Remuneration System (InEK) for remuneration of a new examination and treatment method. The InEK is then tasked to further develop the DRG system to keep step with medical progress. If approved, then an annual additional charge must be negotiated with the statutory health insurers.

References

1. Anonymous (2007) MAGD-Modell – Vorschläge zur Vereinheitlichung der Kennzahlenermittlung in Krankenhauslabors. www.krankenhauslabor.de
2. Felder RA (2002) The distributed laboratory: Point-of-care services with core laboratory management. In: Price CP, Hicks JM (eds) Point-of-care Testing AACC Press, Washington, pp 99–118
3. Gässler N (2001) Personalkosten im diagnostischen Labor – Nicht-Vergleichbares vergleichen. Klinik Management Aktuell 55: 70–71
4. Junker R, Schlebusch H, Luppa PB (2010) Point-of-care testing in hospitals and primary care. Dtsch Ärztebl Int 107: 561–567
5. Plecko I, Pfeiffer A, Wieland E (2007) Laborautomation im Krankenhaus: Systeme, IT, Potentiale and Perspektiven. J Lab Med 31: 206–235
6. Schimke I, Griesmacher A, Schimke E, Müller MM (2006) Patientennahe Sofortdiagnostik (Point-of-Care Testing; POCT) im Krankenhaus – Ja oder Nein. Intensivmed 43: 143–155

Areas of application

Content

Implementation of POCT

Norbert Gässler, Peter B. Luppa, Andreas Bietenbeck, Astrid Petersmann, Alexander Pröbstl, Daniel Romann, Ralf Junker

P. B. Luppa, R. Junker (Eds.), *Point-of-Care Testing*
https://doi.org/10.1007/978-3-662-54497-6_31

31.1 Introduction

POCT has been employed in hospital settings for a long time now. In the times before clearly defined quality assurance standards and the DRG-based fee per case system were in place, the implementation of POCT methods into the clinical workflow was the almost exclusive domain of the clinical operators. Even when a central laboratory existed at the same time in the hospital, the devices used were neither evaluated nor were any costs analyzed. Often, quality controls were lacking to a significant extent, and the nursing personnel was rarely instructed or supervised accordingly. Meanwhile, clear rules have been established in Germany (RiliBÄK) and internationally that govern the quality management and quality assurance of laboratory testing, including within the scope of near-patient laboratory diagnostics. Guidelines and process descriptions on how to implement POCT in the hospital setting are now widely available. One example worthy of citing is a review by the Australasian Association of Clinical Biochemists (AACB) [5] that made a comparison of prescribed implementation procedures and quality specifications in different countries.

In terms of clinical utility and laboratory medicine, the undeniably relevant advantages of POCT diagnostics are the fact that no specimen transport is required and that there are fewer pre-analytical problems with unstable analytes. The rapid availability of results should also be mentioned here. Applying POCT can specifically lead to a reduction in the length of stay in emergency departments [13]. As repeatedly described in the literature, the time saved with POCT in the hospital, however, is not always associated with medical and economic advantages [8, 14, 15, 18, 19]. Moreover, a premature introduction of POC test methods not coordinated with all the organizational structures of the hospital will often lead to structural weaknesses [16].

31

Note

For the implementation of POCT in a hospital with the central laboratory's participation, it is therefore necessary to develop an integrative concept that coordinates all major stakeholders and demands associated with this diagnostic test. At the same time, quality management and quality assurance measures must be implemented to assure the quality of the clinical laboratory.

In each case, this concept must be adapted to space and resources available and account for the medical alignment of each hospital. At hospitals with modern pneumatic tubing and/or cassette conveyer systems and a well-organized central laboratory, sufficiently short laboratory test turnaround times are also generally guaranteed in emergencies and urgent cases [21]. Therefore, an introduction or expansion of POCT diagnostics at these hospitals needs critical scrutiny, especially in view of the costs [1, 12].

31.2 Stakeholders and responsibilities

In Germany, there are a large number of successful programs on the quality management of POCT in hospitals. Many were started years ago and already boast a track record of continuing further development. The pertinent progress reports and suggestions as to how these successes can be replicated elsewhere have been published [6, 10, 11, 17, 20].

The aggregation of all these reports is the unanimous consensus on how to form a suitable management structure [7]. This empirical evidence has been incorporated into the numerous international recommendations (▶ Chapter 38). The DIN EN ISO 22870 standard [3] clearly focuses on the responsibility of management for POCT in terms of organizational requirements. Central responsibility is transferred to the management of both the healthcare facility and of the laboratory. The hospital's management is expected to define the medical goals, provide the facilities, personnel and the necessary resources for POCT implementation. The laboratory's management is responsible for developing a concept for the selection and performance evaluation of POCT devices as well as for formulating quality requirements and objectives.

In order to organize POCT in a clinically and economically sensible way while complying with the statutory regulations for quality assurance, DIN EN ISO 22870 recommends that the hospital management appoints a **POCT committee**, headed by a POCT coordinator. Members of this committee are the:

- POCT coordinator
- Head of the central laboratory
- IT officer at the hospital
- Head of the hospital's quality and risk management
- Representatives of the departments implementing POCT (POCT officers)
- Representatives of the nursing directorate, pharmacy, and medical engineering department
- Representatives of the Commercial Director (responsible for purchasing)

Depending on the organizational structure and situation of the hospitals and/or the hospital group, additional persons can be appointed to the committee.

As defined by DIN EN ISO 22870, a qualified staff member is named **POCT coordinator** by the central laboratory manager and confirmed by the hospital's management. The POCT coordinator acts according to a written description of his responsibilities and competence, which should also be confirmed by the hospitals management. Since the introduction of the POCT concept has far-reaching impacts, the participation of the responsible quality and risk management representative on the committee is recommended [1].

The area of responsibility of the coordinator and the committee extends to the selection of equipment, the selection of the analytical procedures used as POCT as well as to the responsibilities and powers of the staff within the departments, including RiliBÄK-compliant quality control and the required process sequences. Furthermore, the POCT coordinator must also organize the complete documentation of certified end-users, along with their training and continuing further education. The POCT coordinator and the head of the central laboratory can also be one and the same person.

The **POCT officers** of the individual POCT-implementing clinical areas are crucial for the functioning of the system, as they are an intermediate link between POCT coordination and the end-users of the POCT devices. According to DIN EN ISO 22870, the multidisciplinary POCT committee must make and implement all decisions for the use of POCT procedures. The clinical requirements (indications) for POC tests, their financial impacts (costs/benefits), technical feasibility (resources) and integration into the functional processes of the individual departments must all be taken into account. Additional central tasks of the committee include the evaluation and selection of devices and systems for POCT. In addition to costs, the key criteria for the procurement of POCT devices are analytical performance, practicability and the possibility of integration into existing data processing systems. Frequent system changes should be avoided due to the associated (re-)training expense.

Note

In accordance with the administrative structures of the hospital institution, strategic orientation and related financial decisions should therefore be made at an early stage, in order to give the POCT committee a planning and financial scope for decision-making.

The formation of a POCT committee requires all parties involved to understand the necessity, commitment and patience involved to get such a body to become – and remain – operational.

Tab. 31.1 Different tasks and perspectives of the participants/stakeholders in the POCT process

Participants	Primary focus
Laboratory	Analytical process and quality assurance, patient safety, communication of rules in the event of system disturbances outside of regular working hours or worst-case failures (IT down) and (where applicable) training responsibility
Hospital	Fast results and uncomplicated organization
Medical Engineering Department	Resources (devices, consumables, and auxiliary materials) and their maintenance
Medical Director	Medical outcome, patient safety, and satisfaction, resources deployed
Director of Nursing	Integration of POCT analytics into existing workflows, reduction to the bare essentials to streamline nursing workflows, implementation on the operative level, the operator's viewpoint and (where applicable) training responsibilities
IT (Computer Center)	Integration of data into existing laboratory and hospital information systems (LIS & HIS)
Pharmacy, Laboratory and Economic Department (purchasing)	Cost-effective and efficient use of reagents and auxiliary materials along with the associated costs, negotiation of the best purchasing conditions, compliance with contract award regulations with the involvement of operators.
Patients	Benefits for diagnostics and therapy

A general policy decision made by the hospitals management will simplify the implementation of POCT.

It is important to note that the implementation of POCT procedures should not be discussed solely between interested clinical departments and the central laboratory [1]. In the long run, satisfactory solutions can only be found if all participants/stakeholders in the POCT process, with their different interests and views, are involved in the decision-making process and in solving fundamental problems and conflicts that arise (Tab. 31.1). The committee should not only meet in response to current incidents but also regularly (at least once a year).

31.3 Quality management and tasks of the POCT coordinators

In a hospital with a central laboratory, the individual clinics or clinical departments should be given the opportunity to either implement POCT on their own under full compliance with the RiliBÄK or – more expediently – to accept quality management by the POCT coordinators. In this case, the quality assurance measures – especially internal quality controls – demanded by the current RiliBÄK [2] are organized, monitored and documented by the POCT coordinator. Under certain conditions, the participation of the individual departments in external quality assessments is therefore not required (▶ Chapter 38).

The currently valid version of the RiliBÄK results in extensive tasks for the POCT coordinator:

- Internal quality control must be carried out twice daily (with the exception of most unit-use devices). The POCT coordinator should organize, monitor and document this. The POCT coordinator should be entrusted with the decision-making power and responsibility for the procurement and use of various materials for quality control.
- The results of the internal quality control must be evaluated retrospectively by calculating the root-mean-square of the

measurement deviation. The POCT coordinator must organize this calculation and its evaluation.
- Each POCT-operating organizational unit must participate in external quality assessments (exception: This requirement can be waived if different organizational units – together with the central laboratory – are combined to form one single hospital-wide organizational unit for clinical laboratory analyses. The organizational help of the POCT coordinator is also required for this.
- Each POCT-operating facility must have a **Quality Management Manual** describing its own organizational and analytical framework. Instructions and assistance from the POCT coordinator are required for the RiliBÄK-compliant creation of this manual.

Further responsibilities of the POCT coordinator are listed below [2]. Some of the aforementioned points are supported by software packages on POCT servers (▶ Chapter 27):
- **Spatial requirements.** POCT examinations must be performed in rooms and under ambient conditions conducive to work being carried out without impairing the quality of the analysis or the health and safety of staff and patients. The POCT coordinator has the right to monitor and, if necessary, issue specific instructions. After an initial audit, s/he can delegate the responsibility for ensuring that local regulations are complied with to the POCT representative of the respective hospital, who is then responsible and must notify the POCT coordinator of any changes.
- **Equipment.** All POCT devices must be recorded in an inventory list specifying the manufacturer, model and serial number or another unique identifier of the device, as well as the installation site and date of commissioning. In addition, records of periods for maintenance/servicing, malfunctions, repairs and the like must be kept in a device logbook. An operating manual must be accessible at all times.
- **Analytics.** Reagents, test equipment and performance of the device must be checked prior to routine use and all results documented. Records must be kept of the materials and reagents procured so that they can be traced back to each individual test during an audit. Up-to-date and detailed standard operating procedures (SOPs) must always be available for each test procedure.
- **Data processing of results.** The POCT coordinator organizes – as far as possible – the transfer of all patient results to the hospital information system so that the POCT results received can be called up at the wards at any time and documented in the medical record. In this context, the values must be labelled specifically as POCT results. Also, it is important that the examiner be identifiable.
- **Training and qualification.** The POCT coordinator is responsible for the critically important qualification of POCT operators. In consultation with the nursing service, s/he organizes regular training courses and competency tests for the users, which is generally associated with a high expenditure of time and effort due to the high staff turnover common in many hospitals. Training courses on handling the devices can be organized in cooperation with the device manufacturers, while important topics such as pre- and post-analytics must be dealt with by the POCT coordinator and/or his/her staff (▶ Chapter 4).
- In addition, efforts to ensure the quality-compliant POC testing can only be successful over the long term when informational and training events are repeated on a regular basis [14].

◘ Fig. 31.1 summarizes the tasks involved with central POCT coordination.

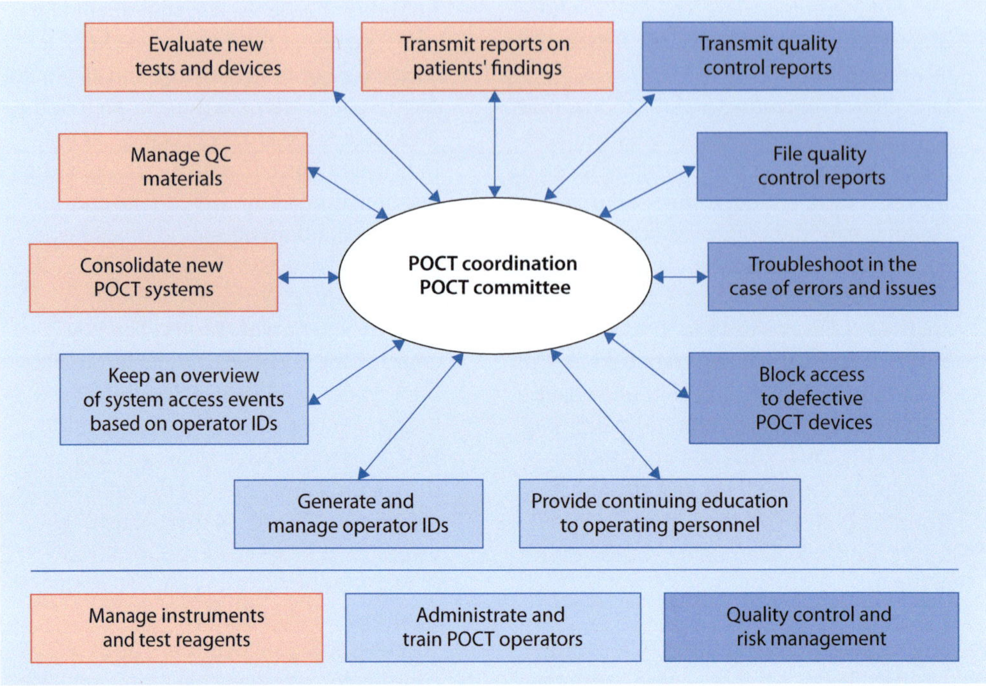

Fig. 31.1 Tasks of centralized POCT coordination

31

31.4 Quality assurance of POCT results, assessment criteria for comparison measurement and implementation

When a POCT system is introduced in a hospital, the same analyte is often analyzed both in the central laboratory and with the new POCT system. Besides the general requirements for clinical laboratory tests, this results in some additional special features relating to the evaluation of the new device.

Conformity with European directives for medical devices must be fundamentally ensured for any new POCT system. This is confirmed by the CE marking. The manufacturer also determines the medical intended use of the device for this conformity assessment.

Before purchasing a POCT system, the analytical requirements for that type of testing must be determined. The RiliBÄK already prescribes minimum requirements for important analytes. In some cases, however, even-higher analytical requirements may be necessary due to specific medical needs.

If the same analyte is measured both in the central laboratory and in a POCT system, both types of analysis should be entered into a central patient file (e.g. in the LIS), but they should still be clearly identified, thereby remaining distinguishable from one another. This enables the treating physicians to assess the parameters over time and to take into account the different metrological characteristics of both tests.

Deviations between the central laboratory method and the POCT method must be minimized in order to prevent incorrect medical evaluation. For this reason, the Working Committee on Point of Care Testing (POCT) (NA 063-03-11 AA) within the DIN Standards Committee Medicine (NAMed, NA 063) 2015 [4] crafted a document which establishes evaluation criteria for comparative measurements when implementing a POCT method. First, a sufficient number of specimens must be collected within the reference range, but also in the gray and the pathological ranges of the parameter to be examined. If possible, routinely collected specimens should be analyzed. However, POCT devices often measure in a different matrix than tests in the central laboratory. For example, capillary blood is used for POC testing of glucose, whereas traditional tests measure glucose in blood plasma. In order to compare methods, additional samples may have to be taken under certain circumstances; the ethics committee responsible may have to be involved for this purpose.

In any event, the comparative measurements must be carried out by those persons who also work with the devices during daily operations. If a new device has been introduced, it must be ensured that the operator has a good command of the new test before carrying out a method comparison.

When evaluating the results, it must then be examined whether both analytical methods lead to the same result interpretation and to the same therapeutic consequence. The comparison of methods takes place at the level of a medical **concordance analysis** and thus takes into account the different error components of POCT and laboratory methods (e.g. different pre-analytics, different competence of the operators).

If the medical question is aimed at a classification of patients (infected/non-infected; pregnant/non-pregnant), **4-field panels** are suitable for a concordance analysis. If measurement inaccuracies lead to a gray area, the 4-field panel can be extended accordingly. The quality of the agreement of the two tests can be appraised; this is done by using the kappa statistic to test the interrater reliability and then expressed as a **kappa coefficient** [9]. 100°% agreement produces a kappa coefficient of 1. Compared to other measurands, the kappa coefficient has the advantage that it delivers meaningful results even with an unbalanced number of samples in the individual ranges.

In order to visualize the conformity of the medical interpretation over the entire concentration range, the use of **error grids** is recommended. Error grids are complicated and take time to create while demanding vast clinical experience as well. Such error grids have already been published for some areas, e.g. the determination of blood glucose or INR. If these are adopted for in-house evaluation, it must be ensured that the medical procedure is comparable among that institution's treating physicians.

It is often difficult to define general acceptance criteria for the simultaneous operation of a POCT system and a laboratory method. For reliable operation of both tests, the concrete clinical situation is always important; the evaluation therefore requires close cooperation between the laboratory and treating physicians.

31.5 Nursing staff and POCT

The nursing staff is one of the key stakeholders in POCT. As a rule, their tasks include:

- Identifying the patient
- Preparing the patient
- Collecting the sample
- Carrying out the measurement
- Reporting the result to the physician and, as appropriate, documenting it in the patient's file

Additionally, further tasks can accrue. These are dictated by the organizational framework and human resources in the department where POCT is conducted. In particular, however, the ability or lack thereof to cooperative with other departments (e.g. laboratory, medical engineering, pharmacy) is a critical factor. Examples of such additional tasks include:

- Organization and execution of quality control according to regulatory demands if the hospital's own laboratory was closed or outsourced;
- Maintenance and ordering repairs on measuring instruments if a medical engineering department is not available, cannot or will not do maintenance or repairs.

- Monitoring of the ward's stocks of reagents and necessary auxiliary materials.

Performing both standard as well as additional tasks can place a significant time burden on nursing staff. In hospitals where a POCT committee is in place, one nurse should therefore be a member on the Committee in order to articulate the specific concerns of the nursing staff with regard to POCT.

The redistribution of tasks between the different professional groups can also make sense from an economic point of view. More and more responsibility is being placed on the medical and nursing staff – especially with respect to on-site quality assurance, handling of medical devices, the task of the Medical Device Act officer, as well as the acceptance of specific documentation and POCT device training for employees across all occupational groups.

31.5.1 Device selection

The analyses most frequently carried out by nursing staff on a routine basis are blood glucose and blood gas analyses, including electrolyte measurements. For the selection of devices, it is necessary to have a sustainable overall POCT concept in place. Automatic data processing of the patient- and user-related data, as well as software-supported quality control documentation and evaluation are becoming more and more important for various reasons (▶ Chapter 27).

The following must be included in any equipment planning decisions:

- Where should the device be used, e.g. peripheral ward, intensive care unit, outpatient clinics, surgical area?
- Which parameters should be measured?
- What are the analytical requirements?
- Which sample volume is required or only available in limited quantities?
- What is the training level of the operators?
- How much maintenance is required?
- How intuitive is the usability and, therefore, how much training is required?
- How can quality management be carried out in accordance with regulatory demands (RiliBÄK)?
- Is a IT network connection possible?
- Is a remote control connection available (▶ Chapter 27)?
- How are the issues of infection prevention associated with operation, cleaning and maintenance solved (▶ Chapter 29)?
- How is the economic advantage of the standardization of the devices evaluated?
- What are the follow-up costs, especially for accessories, data transfer and interfaces?

As an example, blood gas analysis systems will be presented in more detail in the following. The future of blood gas analysis lies in systems that require as little personal maintenance as possible, i.e. self-calibrating and self-monitoring devices with cartridge technology and integrated quality control systems that can be combined and assembled on-site by the operator with regard to parameters, number of measurements and the longest possible service lifetime. The less time spent on device operation and maintenance, the more time is freed up for patient-centered basic care and treatment.

The technical setup of complex blood gas analyzers can vary greatly in terms of maintenance requirements: **Labor-intensive servicing** is particularly necessary for multi-component systems. Frequently, regular checks of the filling levels of operating solutions and electrodes as well as visual inspection of sample inlets, hoses, connections and pumps for damage and wear are prerequisites for trouble-free operation. The presence of many individual components makes troubleshooting time-consuming and requires some experience – even after the necessary training – in order to quickly eliminate a problem and restore the device back to a ready-to-measure state.

Less time-consuming and complicated maintenance work is required for devices that consist of combinations of cartridges and individual components. The amount of time required depends largely on which parts of the device are equipped with automated components and to what extent errors can be remedied manually or by simply exchanging component cartridges.

A minimum of maintenance work is required for devices that contain all the reagents required for measurement in one cartridge. With this type of device, after a cartridge is installed, no further maintenance work is generally required during the entire lifespan of the cartridge.

31.5.2 Operator training

For efficient workflows and accurate results, a team must be put together that is responsible for operating the devices as well as for training operators and that "owns" these tasks.

Initially, the central training topics to be covered include:
- Software training for the responsible persons in the central laboratory
- Initial training (blood glucose devices, blood gas devices, ...) of the device representatives of the respective departments by the industry partners/suppliers
- Authorization and device training of the POCT operators by the POCT coordinator or their officers

Besides contact persons knowledgeable on the subject of POCT in the laboratory, there should be at least one person for each area of application who is highly familiar with POCT, its importance and all facets of its daily application. This will depend on the organization, allocation of responsibilities and the size of the department, but applies to conveying this POCT knowledge within one's own team. This function can very well be assumed by a member of the nursing staff. Those persons are to be thoroughly trained in the equipment and familiarized with the data processing of measured values, quality controls and the pitfalls of the systems so that they can act as multipliers. If the hospital decides to standardize with a single line of devices, it is advisable to involve competitors in the training of operators as early as the tendering process. On-site training should be a matter of course. The contractual conditions should also include the swift availability of the service technicians in the event of a malfunction. Frequently, the adoption of legal obligations is not sufficiently taken into account when new medical devices are introduced as part of larger-scale equipment standardization projects.

If there are suitable IT infrastructures in the hospital (e.g. as part of the hospital's own continuing education institute), **E-Learning** programs can help consolidate acquired knowledge in a simple way (recertification of staff members) and introduce new employees to the topic in the work area. More details are provided in ▶ Chapter 27.

31.5.3 Storage and care

Storage of reagents, control materials and accessories should be as centralized as possible, so that only daily-use items and consumables need to be stocked in individual areas to ensure smooth operation.

Possible central storage locations, depending on the structure of a hospital's purchasing and warehousing operations include:
- Central medical inventory
- Pharmacy
- Central laboratory

31.5.4 Quality assurance

Even years after its entry into force, many operators are still not truly familiar with the regulatory affairs, such as the RiliBÄK, which also govern POCT. A shift in thinking must take place in this regard, as only then can training measures really be successful. The RiliBÄK 2014 [2] places higher demands on the nursing staff involved in POCT quality assurance than before. In particular, a quality management manual summarizing all documents relevant to POCT is prescribed for each institution. Software products in which the QM documents are centrally managed are recommended. In this way, all documents for controlling POCT-relevant documents and their validity can be monitored at all times and made available within the scope of the certification aspired to by the organization. All quality management measures should be structured and implemented together with the POCT coordinator of the hospital (► Chapter 38).

References

1. Bietenbeck A, Junker R, Luppa PB (2015) Central laboratory service and Point-of-Care Testing in Germany – From conflicting notions to complementary understandings. Point of Care 14:1–11
2. Bundesärztekammer (2014) Richtlinie der Bundesärztekammer zur Qualitätssicherung laboratoriumsmedizinischer Untersuchungen. Dtsch Ärztebl 111: A1583–A1618
3. German Institute for Standardization (2006) DIN EN ISO 22870 "Point-of-care testing (POCT) – Requirements for quality and competence"
4. German Institute for Standardization (2014) DIN 58964:2015-09. Quality assurance of POCT results – Assessment criteria for comparison measurement and implementation Beuth-Verlag, Berlin
5. Farrance I (2012), on behalf of the Australasian Association of Clinical Biochemists (AACB). Policies, procedures and guidelines for Point-of-Care Testing. http://www.aacb.asn.au/documents/item/635
6. Gässler N, Luppa PB, Hafner G et al. (2006) Information der Arbeitsgemeinschaft "Point-of-Care Testing" der DGKL: Einführung of POCT innerhalb einer Klinik. Klinische Chemie Mitteilungen 37: 161–163
7. Gramz J, Koerte P, Stein D (2013) Managing the challenges in Point-of-Care Testing: An ecosystem approach. Point of care 12(2):76–79
8. Grodzinsky E, Wirehn AB, Fremner E et al. (2004) Point-of-care testing has a limited effect on time to clinical decision in primary health care. Scand J Clin Lab Invest 64: 547–551
9. Grouven U, Bender R, Ziegler A, Lange S (2007) Der Kappa-Koeffizient. Dtsch Med Wochenschr 132: e65–e68.
10. Gruszecki AC, Hortin G, Lam J et al. (2003) Utilization, reliability, and clinical impact of point-of-care testing during critical care transport: Six years of experience. Clin Chem 49: 1017–1019
11. Hänecke P, Haeckel R, Koschinsky T, Luppa P, Schlebusch H, Wahl HG (2004) Qualitätssicherung der patientennahen Sofortdiagnostik (Point-of-Care Testing) im Krankenhaus: Muster für eine hausinterne Richtlinie. J Lab Med 28: 256–263
12. Howanitz PJ, Jones BA (2004) Comparative analytical costs of central laboratory glucose and bedside glucose testing: A College of American Pathologists Q-Probes Study. Arch Pathol Lab Med 128(7): 739–745
13. Jang JY, Shin SD, Lee EJ, Park CB, Song KJ, Singer AJ (2013) Use of a comprehensive metabolic panel point-of-care test to reduce length of stay in the emergency department: a randomized controlled trial. Ann Emerg Med 61(2):145–151
14. Junker R, Schlebusch H, Luppa PB (2010) Point-of-care testing in hospitals and primary care. Dtsch Ärztebl Int 107:561–567
15. Lee-Lewandrowski E, Lewandrowski K (2009) Perspectives on cost and outcomes for point-of-care testing. Clin Lab Med 29:479–489
16. Luppa PB, Blobner M (2004) Online-Qualitätssicherung in der Sofortdiagnostik. Dtsch Ärztebl 101: A399–A400
17. Schimke I, Griesmacher A, Schimke E, Müller MM (2006) Patientennahe Sofortdiagnostik (Point-of-Care Testing; POCT) im Krankenhaus – Ja oder Nein. Intensivmed 43: 143–155
18. Schlüter B, Junker R (2003) Labordiagnostik – schneller ist nicht immer besser. Dtsch Ärztebl 100: A87–A89
19. St John A, Price CP (2013) Economic evidence and Point-of-Care Testing. Clin Biochem Rev 34(2):61
20. St-Louis P (2000) Status of Point-of-Care Testing: promise, realities, and possibilities. Clin Biochem 33: 427–440
21. Stotler BA, Kratz A (2012) Determination of turnaround time in the clinical laboratory: "Accessioning-to-Result" time does not always accurately reflect laboratory performance. A J Clin Pathol 138(5):724–729

POCT in the physician practice setting

Ralf Junker, Hans Günter Wahl

P. B. Luppa, R. Junker (Eds.), *Point-of-Care Testing*
https://doi.org/10.1007/978-3-662-54497-6_32

32.1 Introduction

POCT is already established in many healthcare sectors, for example in the private practitioner setting (▶ Chapter 1 and ▶ Chapter 2). As in other sectors, the following aspects should be considered whenever POCT is introduced:

- Diagnostic accuracy
- Organizational advantages
- Cost effectiveness
- Practical performance
- Quality assurance

32.2 Diagnostic implications

The most important advantage of POCT diagnostics in the hospital – faster delivery of diagnostic results near the patient – only plays a subordinate role in the private practitioner setting. Unlike the priorities of the inpatient environment, those of the private practitioner setting mainly focus on organizational aspects and the issue of patient satisfaction and/or compliance. There are isolated cases in which one could accordingly speak of a necessity for "emergency testing", e.g. aimed at detecting specific pathogens prior to antibiotic therapy or to determine the blood count before initiating chemotherapy.

As in hospital settings, the first step is to ask questions as to what are the possible indications for the corresponding test. Afterwards, one should look at whether the deployment of POCT will produce medical or organizational advantages.

The spectrum of possible analyses is huge. The overview below presents a selection of possible POCT applications in different medical disciplines.

Selected examples of POCT in the physician practice setting

- Urologist: Prostate-specific antigen (PSA) for cancer screening
- Gynecologist: Human chorionic gonadotropin (β-HCG) for pregnancy testing
- Cardiologist: Brain natriuretic peptide (BNP) for monitoring the clinical course in patients with heart failure
- Oncologist: Blood count prior to chemotherapy
- Sports physician: Lactate as performance check
- General practitioner: INR for monitoring anticoagulation
- Diabetologist: Blood sugar and HbA_{1c} for monitoring the clinical course of blood glucose control
- Pediatrician: Rapid streptococcal detection test prior to antibiotic therapy
- Nephrologist: Albumin (urine) as microalbuminuria screening in diabetic nephropathy
- Psychiatrist: Drug screening during therapy for withdrawal

The respective chapters of this book contain detailed descriptions of the individual diagnostic options. Here, the aim is merely to mention some examples of a few applications so that the reader gets a general impression of the multitudinous facets of the medical considerations relating to the use of POCT in the physician practice setting [6].

32.2.1 Cardiovascular markers

The test of cardiovascular markers in a physician's practice comprises BNP (brain natriuretic peptide), cardiac troponins and D dimer.

BNP is measured for the differential diagnosis as well as the long-term monitoring of heart failure. Nevertheless, it is questionable as to whether performing POC laboratory tests confers benefit in the private practitioner setting. Generally, the clinical care of the patients in emergency situations is required, regardless of any measured results, whereas the measurements in a central laboratory alone should then be sufficient when it comes to long-term monitoring. The same deliberations apply to D-dimer as a rule-out test for deep vein thrombosis (▶ Chapter 17).

The measurement of **cardiac troponins** represents the main diagnostic criterion in acute myocardial infarction. Measurements taken during acute emergency situations are not uncritical, given that the results produced may still be negative up to 2 h after the acute event, even if ultrasensitive tests are used. In other words, a negative result thus cannot be relied on to rule out an acute myocardial infarction. In the event of a myocardial infarction, after a repeat measurement in the practice 2 h later, valuable time would be lost up to hospital admission and initiation of therapy.

32.2.2 Infectious diseases

Alongside CRP assays, the direct or indirect detection of pathogens is an integral part of infectious diseases diagnostics. In particular, this involves the ability to make early decisions about the necessity for antibiotic therapy. Whether the use of POCT for infection markers actually lowers the amount of antibiotics prescribed or makes their administration more targeted is the subject of controversial debate.

In American studies, POCT is generally ascribed with cost effectiveness because it reduces analysis in the laboratory. However, this only applies when the costs for POCT are of a similar magnitude as the equivalent laboratory test [3, 5, 13].

By contrast, when assessing the concrete case of the frequently used rapid streptococcal antigen detection tests, the authors of different studies concluded that the result of POCT only insignificantly impacts clinical decisions for therapy, albeit with the same applying to tests done in the laboratory [9]. Moreover, if the POCT result is negative and complications arise, an additional laboratory test is recommended – meaning that duplicated analyses are sometimes performed [10].

In terms of qualitative aspects, numerous pathogen detection tests have been demonstrated to show sufficient analytical sensitivity and specificity [10, 11, 12]. Conversely, the German Federal Joint Committee (G-BA) of the statutory health insurances and physicians has issued a recommendation on the annual screening of genitalia for infections caused by Chlamydia trachomatis. This recommendation expressly points out that POCT systems should not be used. Although not explicitly mentioned, it can be assumed that the lack of sensitivity of the immunological methods compared to molecular biological tests is the reason for this (▶ Chapter 9 and ▶ Chapter 20; [4]). Ultimately, it is thus a matter of the underlying quality standard.

32.2.3 Diabetes monitoring

The regular self-monitoring of blood glucose by the patients has long become established practice. Physician's practices are nevertheless well advised to reserve the option for blood glucose measurements in emergency situations.

One important task incumbent upon practices specializing in diabetes is the inspection of the blood glucose meters used by their patients in the homecare setting. However, only analyzers that determine glucose by means of the reference method may be considered for use as comparator devices. There is, at present, no POCT device available that can do this (▶ Chapter 12).

Currently, POCT methods for measuring HbA_{1c} are primarily employed in diabetes outpatient centers. This fast feedback saves patients from having to wait unnecessarily for their laboratory result. In future, HbA_{1c} monitoring by POCT might gain greater importance now that it has been proven that POCT devices can be used for the diagnosis of diabetes mellitus (► Chapter 12; [8]).

32.3 Economic aspects

32.3.1 General considerations

In principle, the use of POCT has economic impacts across all levels of the healthcare systems. This can produce considerable conflicts of interests.

One key aspect that is currently preventing a further dissemination of POCT analyses in the private practitioner setting – at least within the German System – is seen in the limited options for invoicing the analytical tests.

Generally, POCT in the physician practice setting was initially characterized by markedly higher costs compared to conventional laboratory analyses, without being commensurately accounted for in the physician's invoicing system. For example, costs for the additional workload affecting the physician practice staff are added to the expenditure for the actual analysis (measuring instrument, reagents, test strips, control material, paper, electric current, disposal, documentation, room etc.).

32

Hence, on the level of physician's practice, any savings or at least a cost-effective performance of the tests are only achievable to a very limited extent, if at all.

Savings potentials can also be realized by reducing the number of second contacts with the patient, including telephone calls for discussing the results. This could likely trigger better compliance on the part of the patients, which in turn benefits the overall system. Greater patient satisfaction might also be another positive impact.

32.3.2 Remuneration

According to the German Uniform Assessment Standard (EBM) for patients covered by the statutory health insurers' schemes, laboratory tests as described above should either be carried out by each private practitioner themselves or can be delegated. The ways and means are not stipulated: It is left to each physician's discretion to render the services themselves, e.g. using in-house laboratory analysis systems or POCT, or alternatively, to have them carried out by laboratory physicians. One special group includes the specific codes for invoicing analyses performed as single measurements with unit-use reagents in the practice's own in-house laboratory. This mainly involves analyses classifiable as basic diagnostics (clinical chemistry, hematology etc.). In such cases, however, the remuneration is only marginally higher to account for the additional workload borne by the physician's practice. In most cases, it is therefore hardly possible for physicians to render these POCT services cost-effectively in their own practice. Unless a service ideology is given precedence, the services are therefore not rendered in the physician's practice.

However, for some time now, parameters like cardiac troponins or – just recently added – BNP tested outside the laboratory can be invoiced by the physician in private practice in addition to analyses classifiable as basic diagnostics. Whereas the in-practice testing of cardiac troponins is borderline cost-effective, the situation looks entirely different for BNP: The reimbursement allows the service to be provided in a cost-covering manner; macroeconomically, a positive development might result from savings on echocardiographic examinations [7]. On the other side of the balance hangs the fact that the physician's practice would have to forego income from echocardiographic examinations.

The German physicians fee schedule (GOÄ for privately insured patients) similarly incorporates the option to invoice laboratory analyses as optional-reserve services to be kept available in the practice. These are mainly basic diagnostic services. In general, however, there is the option that the patient can be invoiced for all payer-related analyses without consideration of the tested parameters. Given the low amount, it is less likely that the services can be performed cost-effectively [1]. As with the patients covered by the statutory health insurance, the service ideology is paramount.

32.4 Performance and organization

The practical performance of POCT in the physician's practice is usually tasked to the non-medical professionals in the physician's practice.

One major point for the successful application of POCT methods is the safe handling of devices, test strips and reagents – proper training is therefore a prerequisite. Whereas simple strip tests are essentially self-explanatory, more complex devices such as automated hematology devices require individual instructions. Special focus should be placed on quality assessment. Not least, the use of POCT depends on the infrastructural human resources and facilities. Therefore, a number of aspects need to be observed with regard to performance, among others:

- Are suitable workplaces available for carrying out the analysis?
- Who will perform the measurements (human resources)?
- Who is responsible for quality control?
- How are the measured and control values documented?
- What storage capacity is required at which temperature for reagents and accessories?
- How is the waste disposed of?

These questions should be answered parallel to the medical and economic issues because sensible and successful POCT integration is fundamentally only possible when suitable preconditions are installed.

32.5 Quality management

Within the German system, RiliBÄK [2] also governs the use of POCT in the physician practice setting. Accordingly, simplified rules are only applicable if unit-use reagents and the matching analytical systems are used. With regard to all other laboratory tests performed in the physician's practice (e.g. measurements using small hematology analyzers), the full scope of the provisions of RiliBÄK apply to the quality management and internal quality assurance, e.g. two controls within 24 h on each day of use, at the latest after 16 h. Further details can be found in ▶ Chapter 37.

References

1. Bundesärztekammer (2001) Gebührenordnung für Ärzte (GOÄ). Deutscher Ärzte-Verlag, Köln
2. Bundesärztekammer (2008) Richtlinie der Bundesärztekammer zur Qualitätssicherung laboratoriumsmedizinischer Untersuchungen. Dtsch Ärztebl 105: A341–A355. Errata: Dtsch Ärztebl 105; A 650 and Dtsch Ärztebl 2010;107, A 51–52
3. Dahler-Eriksen BS, Lauritzen T, Lassen JF, Lund ED, Brandslund I (1999) Near-patient test for C-reactive protein in general practice: assessment of clinical, organizational, and economic outcomes. Clin Chem 45: 478–485
4. Gemeinsamer Bundesausschuss der Ärzte und Krankenkassen (2007) Beschluss des gemeinsamen Bundesausschusses über eine Änderung der Richtlinien zur Empfängnisregelung und zum Schwangerschaftsabbruch sowie der Mutterschaftsrichtlinien: Screening auf genitale Chlamydia trachomatis-Infektionen bei Frauen. www.g-ba.de
5. Gulich MS, Matschiner A, Glück R, Zeitler HP (1999) Improving diagnostic accuracy of bacterial pharyngitis by near patient measurement of C-reactive protein (CRP). Br J Gen Pract 49: 119–121
6. Junker R, Schlebusch H, Luppa PB (2010) Point-of-care testing in hospitals and primary care. Dtsch Ärztebl Int 107: 561–567
7. Kassenärztliche Bundesvereinigung (2008) Einheitlicher Bewertungsmaßstab (EBM). Deutscher Ärzte-Verlag, Köln
8. Kerner W, Brückel J (2010) Definition, Klassifikation und Diagnostik des Diabetes mellitus. Diabetologie und Stoffwechsel 5 (Suppl 2) S109–S112; (updated version available on DDG website: www.deutsche-diabetes-gesellschaft.de/Evidenzbasierte Leitlinien/Definition

9. Nerbrand C, Jasir A, Schalén C (2002) Are current rapid detection tests for Group A Streptococci sensitive enough? Evaluation of 2 commercial kits. Scand J Infect Dis 34: 797–799
10. Reinert RR (2007) Streptokokken-Schnelltests. J Lab Med 31: 280–293
11. Schweiger B (2006) Influenza rapid tests – advantages and limitations. J Lab Med 30: 219–215
12. Seamark DA, Backhouse SN, Powell R (2003) Field-testing and validation in a primary care setting of a point-of-care test for C-reactive protein. Ann Clin Biochem 40: 178–180
13. Takemura Y, Kakoi H, Ishida H et al. (2004) Immediate availability of C-reactive protein and leukocyte count data influenced physicians' decisions to prescribe antimicrobial drugs for new outpatients with acute infections. Clin Chem 50: 241–244

Patient self-monitoring

Hannelore Rott, Theodor Koschinsky

P. B. Luppa, R. Junker (Eds.), *Point-of-Care Testing*
https://doi.org/10.1007/978-3-662-54497-6_33

33.1 Self-monitoring of glucose metabolism in diabetes mellitus

33.1.1 Self-monitoring requirements

At the beginning of near-patient diagnostics, the clinical course of diabetes was only monitored by the attending physicians or their medically qualified staff. As technology advanced – with easier- and faster-to-operate glucose monitoring systems – additional applications in the daily routine of people with diabetes were made possible.

However, as these devices are now mostly used by laypeople, additional criteria must be met in order to enable patients to self-monitor while preventing the risk of errors arising from the use and interpretation of glucose measurements. These are:

- Physicians should identify people with diabetes who are willing and able to perform glucose self-monitoring, as non-compliance, i.e. improper handling, may be associated with additional therapeutic risks and is therefore not cost-effective and ultimately futile.
- There must be a medical indication for glucose self-monitoring. Indeed, blood glucose does not always need to be monitored in the same way or with the same frequency, i.e. it is not necessarily indicated for all forms of metabolic control [5, 6].
- People with diabetes capable of taking an active and responsible role in their own treatment under the routine conditions of everyday should be offered motivation. Through intensive training courses, the patients can be educated in the technical basics of glucose self-monitoring, test results documentation and taught to immediately adjust their treatment individually [18].
- Regular medical follow-up is required, involving discussions about self-monitoring data and their use in treatment, e.g. diet, variable physical activity or decisions about calculating the right insulin dose. This also includes spot-checks for quality assurance of self-monitoring during medical consultations, during which parallel measurements of blood glucose are taken and the patients' technical performance during self-testing is observed.
- Rules should be applied that meet the standards of the Federal Joint Committee (G-BA) in Berlin and the respective insurance providers in Germany governing coverage of costs for glucose self-monitoring equipment, i.e. glucose meters, test strips and/or sensors, finger prickers, lancets for capillary blood sampling etc.

Two main test system principles are available to patients with diabetes mellitus to help them control their glucose metabolism: Self-monitoring of urine and of blood-glucose levels.

33.1.2 Urine glucose self-monitoring

The first glucose meters available for everyday use in Germany were mere semi-quantitative test strips to measure urine glucose. However, diagnosing hypoglycemia in a timely fashion or achieving targeted insulin dose adjustment was not possible because the urine glucose measurement only reflects a mean blood glucose level over a particular urine sampling period. Also, urine glucose is only detectable when the individually variable renal excretion threshold is exceeded. As a result, self-monitoring of urine glucose is no longer applied in Germany, although the costs are considerably lower than for the alternative, blood glucose self-monitoring. Nevertheless, this form of glucose self-monitoring is still used to a large extent internationally due to cost considerations and the lacking availability of blood glucose self-monitoring meters.

33.1.3 Blood glucose self-monitoring

Over the past 50 years, technological developments were driven not least by requirements at hospitals and in physicians' practices for rapid POC monitoring of real-time blood-glucose levels where immediate treatment adjustments can be made. Now, after a ten-year delay in development, test strips viable for everyday use can also be supplied to people with diabetes. At the beginning, it was test strips alone, but now these are always supplied in combination with the matching device.

The standard method usually involves single measurements taken on freshly sampled capillary blood, typically taken from the fingertip or earlobe. However, alternative sites can be used under specific circumstances (▶ Chapter 12). In addition, continuous blood glucose monitoring devices have been available for some years now, where the needle sensors measure glucose in the subcutaneous interstitial fluid with subsequent conversion to the equivalent blood glucose values (▶ Chapter 13).

A summary of the measurement technology principles deployed in these blood glucose meters was provided in ▶ Chapters 12 and 13 as these devices are also used for clinical course monitoring by healthcare staff. Blood glucose self-monitoring meters are medical devices governed by Section 3 (4) of the German Medical Devices Act (MPG). Quality criteria that need to be considered for CE mark approval are regulated according to DIN EN ISO standard 15197:2015. This is essentially identical to the international ISO standard 15197:2013 [8, 11]. It should be noted that such blood glucose meters are not licensed for the primary diagnosis of diabetes and are therefore not allowed for use in this indication.

Unlike laboratory devices, quality assurance measuring of market-approved meters used by patients for self-monitoring has not been legislated yet and therefore the related costs are not covered by the statutory health insurers. This is serious given that published data has highlighted considerable differences in the measurement accuracy (in some cases even exceeding the permitted error margins) between different measuring systems and across individual test strip batches of the same system [3, 9, 10]. To date, manufacturers only recommend regular inspections using their own device-specific control solutions at various concentrations to verify that devices are functioning correctly. However, these are not sufficient to detect the aforementioned variations in measurement accuracy.

As a result, there is a serious need – at least in Germany – to close this legal loop-hole as quickly as possible. In order for costs to be reimbursed by statutory or private healthcare payers, each new test strip batch should be subjected to measurement quality control by an independent institution and their permissibility published in a timely fashion, similar to the Scandinavian model, for example (SKUP) [15].

Errors in blood glucose self-monitoring are not only due to analytical errors made by the glucose measurement system but also result from pre-analytical errors caused by ineffective skin cleaning (or even skin disinfection) or by mistakes in handling the self-monitoring equipment [11]. Frequent sources of error (and their consequences) include, for example:

- Lack of hand washing after contact with fruit (glucose residuals on the puncture site)
- Hands not dried (dilution of the blood sample)
- Residuals of alcohol, lotions and soaps (reaction on the test strip, sample contamination)
- Test strip containers not closed (causing damage to the test enzymes, mainly due to moisture)
- Test strip storage at ↑ or ↓ temperature (interference with the enzyme activity)
- Test strip use after expiry date (↓ enzyme activity)
- Measurement by approx. >3000 m (↓ pO_2, ↓ temperature, ↓ humidity)

The following should also be considered: In general, localized or systemic infection risks of capillary blood sampling for everyday blood glucose self-monitoring can normally be reduced through compliance with the following measures regarding relevant skin site and sampling environment:

- Washing off contaminants such as soil, natural fertilizer, foreign blood or other dirt
- Thorough surface disinfection and drying of the skin
- Use of sterile lancets/needles
- Verified hygienic disposal of all lancets/needles, test strips/sensors and swabs contaminated with blood

The relevance of these measures – applied in isolation or together – in reducing the risk of localized or systemic infections in this context, however, has not yet been systematically evaluated. According to the current German Diabetes Association, further general skin disinfection is not necessary for everyday blood sampling other than in the specific examples given. This opinion is based on experience accumulated over decades [7].

Finally, blood sampling aids such as finger prickers and lancets can, when used incorrectly, lead to incomplete blood sampling. If unnecessary pain is caused, poor motivation can detract from compliance with the prescribed self-monitoring – all preventable by targeted patient education, including practical exercises.

To enable patients with diabetes to better operate such meters, it is also important to save blood glucose measurements with the date and time of day and, depending on the device, mark them as pre- or postprandial so that the mean values over 7, 14, 30 and 90 days can be calculated. Various analysis programs allow numerical and graphical presentation of blood glucose along with telemetric data transfer to the attending physician. Various manufacturers offer blood glucose self-monitoring systems with integrated calculators that indicate the appropriate insulin dose. However, such a variety of data analysis systems exist that there are no uniform standards. Alongside blood glucose measurements, some devices feature extra functions like voice output of test results or the measurement of additional blood parameters (e.g. ketones, hematocrit) [18].

33.2 POCT of INR during treatment with vitamin K antagonists

33.2.1 Importance of INR levels

Given the significant inter-individual variations in pharmacokinetics, patient self-management in vitamin K antagonist treatment requires regular INR monitoring, usually once per week. Although the conversion of prothrombin time into Quick values (in %) is still common in German-speaking countries, this measurement unit has globally been replaced by INR for monitoring treatment with vitamin K antagonists.

As early as 1983, the WHO introduced the International Normalized Ratio (INR) to improve the reliability of anticoagulation treatment with vitamin K antagonists and, in particular, to standardize the comparability of values determined with different thromboplastins at different laboratories using various methods [2, 19]. The prothrombin time, measured in seconds, is converted to INR based on the sensitivity of thromboplastin used in the test and calibrated against an international WHO reference thromboplastin. INR is therefore independent of method and laboratory, and hence also comparable in international studies.

By definition, INR is 1.0 in individuals with normal coagulation and not on medication. Treatment with vitamin K antagonists causes this value to rise. In most patients treated for almost all indications such as thromboembolism, atrial fibrillation and status post heart valve replacement, the target INR nowadays is usually 2.5 with a target range of 2.0–3.0. The target INR may be higher in isolated cases of severe thrombophilia (e.g. antiphospholipid syndrome with high titers of antiphospholipid antibodies) or particular cardiac and valvular diseases or cardiac procedures.

The prothrombin time measured by POCT is converted into the INR and thus allows self-monitoring and management at home.

33.2.2 Clinical importance of INR POCT

During vitamin K antagonist therapy, INR testing by the patients themselves at the point of care leads to significant improvements in the clinical course and quality of life. This has been proven unequivocally in numerous meta-analyses. A meta-analysis by Heneghan et al. [12] included 14 clinical studies with more than 1400 patients using POCT and a similarly large control group, whose INR was tested in a coagulation laboratory by conventional methods. The result showed unequivocally that the rate of thromboembolic events was halved and the rate of severe bleeding complications reduced by a third in the self-monitoring group. A significant reduction in the number of deaths on vitamin K antagonist treatment has also been shown [16].

This was associated with a significant improvement in the INR "time in therapeutic range" (TTR in %). In Germany, when INR is measured by general physicians and laboratories, the TTR is usually about 50–65 %, in INR POCT (depending on patients) about 70–90 %. Depending on the study, self-monitoring improved the TTR by 3–20 % [4].

33.2.3 INR POCT in Germany

In Germany, certain pre-conditions must be met before health insurance companies will cover the costs for educational programs, provision of devices and equipment such as test strips and lancets.

Note

The German Federal Gazette (BAnz.) dated 9 August 2002 (Volume 147, p. 18805) defines the indications:

- "Patients with mechanical heart valves directly post-operatively,
- or requiring other long-term vitamin K antagonist therapy. There must be a demonstrable therapeutic benefit and essential need for self-monitoring, e.g. status post complications such as thrombosis or bleeding on conventional treatment, other situations such as difficulties attending the physician's practice for regular INR checks (shift work, unfavorable local conditions, immobility, nursing care etc.), poor venous status, long-term anticoagulation in children".

Education centers have to check that patients are able to carry out the measurements, document the results in the patient's INR log book and subsequently dose the medication themselves. A distinction is made between INR self-testing (patients themselves measure the values, but the vitamin K antagonist dose is determined by the physician) and INR self-management (patients are responsible for both the INR measurement and dosing). In Germany, self-management is the preferred choice; however, health insurance companies only cover the costs after patients have undergone comprehensive standardized training courses.

It can sometimes be useful (for example, in elderly patients or those with poor venous access) to allow the patient to perform the measurement themselves or with help from healthcare staff in the physician's practice. The dosing of medication is determined by the physician. A special invoicing number was introduced for this reason (EBM 33026), so that now various physicians' practices and some hospitals can perform INR POCT themselves in difficult patients, urgent cases or acute emergencies such as bleeding, suspected overdoses etc.

In many typical vitamin K antagonist indications, the relevance of INR POCT has declined since the market launch of direct oral anticoagulants (NOAC) like rivaroxaban, apixaban, edoxaban, dabigatran. This does not include mechanical heart valves! At present, many patients who have problems taking vitamin K antagonists are treated with these newer drugs, which do not require treatment monitoring.

The most frequently used INR POCT devices in Germany are the CoaguChek XS Plus and the XS Pro from Roche Diagnostics. The INRatio from Alere was withdrawn from the market in 2016.

33.2.4 Measurement quality of INR POCT

The correlation between INR POCT and prothrombin-time measured in the laboratory using venous citrated blood samples is excellent (correlation coefficient 0.8–0.95 [1, 4, 17]. In 84–87 %, parallel measurements did not differ by more than 0.5 INR units [13, 14]. Therefore, INR measured with POCT can be deemed equal to that measured by the laboratory.

The test strip principles of INR analysis for POCT devices are explained in ▶ Chapter 6.

33

According to RiliBÄK 2014, regular quality controls must be carried out in near-patient INR measurement. In the physician's practice, weekly analyses with a manufacturer's control solution with known target values must be carried out when INR is measured with the CoaguChek system. A control analysis must also be carried out with each batch change. Participation in ring trials is not compulsory for physicians in private practices. This also does not apply to hospitals if the central laboratory manages the POCT coordination.

33.2.5 Costs for INR POCT

Costs for the CoaguChek (over 5 years) amount to approx. € 450–500. These include device, reagents, test strips (one measurement per week) and finger prickers. After the 5th year of depreciation of the device, the costs are approx. € 210.

When measured by the physician, the costs are currently estimated at € 4.70 per analysis. Therefore a two-weekly measurement per patient and per year would be € 122 (calculation by the Working Group coagulation and heart valve patients).

References

1. Azarnoush K, Camilleri L, Aublet-Cuvelier B, et al. (2011) Results of the first randomized French study evaluating self testing of the International Normalized Ratio. J Heart Valve Dis 20(5):518–525
2. Barthels M, Bergmann F (2016) Global- und Gruppentests. In: Barthels M (ed) Das Gerinnungskompendium. 2nd ed. Thieme, Stuttgart New York, p 359–370
3. Baumstark A, Pleus S, Schmid C et al. (2012) Lot-to-lot variability of test strips and accuracy assessment of systems or self-monitoring of blood glucose according to ISO 15197. J Diabetes Sci Technol 6:1076–1086
4. Dauphin C, Legault B, Jaffeux P, et al. (2008) Comparison of INR stability between self-monitoring and standard laboratory method: preliminary results of a prospective study in 67 mechanical heart valve patients. Arch Cardiovasc Dis 101 (11–12):753–761
5. Deutsche Diabetes Gesellschaft (2015) Praxisempfehlungen: Therapie des Typ-1-Diabetes. Diabetologie 10 (Suppl 2) S129–S139
6. Deutsche Diabetes Gesellschaft (2015) Praxisempfehlungen: Therapie des Typ-2-Diabetes. Diabetologie 10 (Suppl 2) S140-S151
7. Deutschen Diabetes Gesellschaft (2016) Stellungnahme 11 February 2016: Notwendigkeit der Desinfektion der Haut vor der Kapillar-Blutgewinnung für die Glukosemessung im Krankenhaus. http://www.deutsche-diabetes-gesellschaft.de/stellungnahmen/stellungnahmen.html
8. DIN German Institute for Standardization (2015) In vitro diagnostic test systems – Requirements for blood-glucose monitoring systems for self-testing in managing diabetes mellitus (ISO 15197:2013); English Version DIN EN ISO 15197:2015
9. Freckmann G, Schmid C, Baumstark A et al. (2012) System accuracy evaluation of 43 blood glucose monitoring systems for self-monitoring of blood glucose according to DIN EN ISO 15197. J Diabetes Sci Technol 6:1060–1075
10. Freckmann G, Link M, Schmid C et al. (2015) System accuracy evaluation of different blood glucose

monitoring systems following ISO 15197:2013 by using two different comparison methods. Diabetes Technol Ther 17:635–648
11. Freckmann G (2016) Patientensicherheit bei der Blutzuckermessung. Thieme Praxis Report 8 (3): 1–20
12. Heneghan C, Alonso-Coello P, Garcia-Alamino JM, Perera R, Meats E, Glasziou P (2006) Self-monitoring of oral anticoagulation: a systematic review and meta-analysis. Lancet 367(9508):404–411
13. Hentrich DP, Fritschi J, Muller PR, Wuillemin WA (2007) INR comparison between the CoaguChek S and a standard laboratory method among patients with self-management of oral anticoagulation. Thrombosis Research 119(4):489–495
14. Ryan F, O'Shea S, Byrne S (2010) The reliability of point-of-care prothrombin time testing. A comparison of CoaguChek S and XS INR measurements with hospital laboratory monitoring. Int J Lab Hematol 32 (1 Pt 1):e26–e33
15. Scandinavian evaluation of laboratory equipment for primary health care (SKUP): http://www.skup.nu
16. Siebenhofer A, Berghold A, Sawicki PT (2004) Systematic review of studies of self-management of oral anticoagulation. Thromb Haemost 91(2): 225–232
17. Torreiro EG, Fernandez EG, Rodriguez RM, Lopez CV, Nunez JB (2009) Comparative study of accuracy and clinical agreement of the CoaguChek XS portable device versus standard laboratory practice in unexperienced patients. Thromb Haemost 101(5):969–974
18. Verband der Diabetes-Beratungs- und Schulungsberufe in Deutschland e.V.(VDBD) (2014) Leitfaden zur Blutzucker-Selbstkontrolle in Beratung und Therapie 2–39; Eigenverlag
19. WHO Expert Committee on Biological Standardization (1983) Thirty-third report. World Health Organ Tech Rep Ser 687:1–184

POCT in non-medical settings

Norbert Gässler, Andreas Bietenbeck, Gerhard Eiselen

P. B. Luppa, R. Junker (Eds.), *Point-of-Care Testing*
https://doi.org/10.1007/978-3-662-54497-6_34

34.1 Introduction

Frequently, POCT is also carried out in non-medical settings. This chapter deals with POCT applications that mainly apply to pharmacies and nursing homes. Thematically, the patient self-tests discussed in ▶ Chapter 33 will be addressed separately.

Clearly defined rules for the application and quality assurance of POCT at old-age homes and special nursing care facilities (e.g. nursing homes for dementia patients) are lacking. The German non-profit Federal Association of Geriatric Care (DBVA) sees a need for action in this sector. There are hardly any other service providers in the healthcare system where the assured quality of a product is so pivotal to their interests as with pharmacies; indeed, the concepts of efficacy, safety and harmlessness are intricately linked with the concept of medicinal products like no other. Pharmacists are subject to perform quality assurance. How this is done is left to his own discretion.

34.2 POCT in the pharmacy setting

34.2.1 POCT framework

34

Blood and urine tests are offered as rapid, near-patient tests in pharmacies. Because individuals oftentimes would like to find out for themselves first whether they are suffering from a metabolic disorder or disease, they will frequently turn to the pharmacist first. The pharmacist will then carry out the rapid test in the pharmacy or sell the corresponding test kit to the patient for self-diagnosis at home. The classic among these diagnostic kits is the pregnancy test. The POC tests offered by pharmacies are frequently carried out there. A variety of specimens like blood, urine, stool, saliva, sweat or secretion are applied to the respective test procedure.

Prerequisites for the reliability of screening tests include the proper blood drawing technique, an error-free analytical system and its correct operation [9]. Quality assurance measures are also required. Enacted in June 2012, the new German Ordinance on the Operation of Pharmacies (ApBetrO) stipulates that self-inspection and regular participation in measures for external quality auditing should be conducted within the scope of any quality management system (ApBetrO Section 2a) [7].

34.2.2 Quality Control

The Central Laboratory of German Pharmacists (ZL) sends out interlaboratory test samples to enable the quality of the results to be verified. After the correct analysis result is submitted, it will then be certified by the ZL [4].

All pharmacies that routinely carry out blood tests and have in place or would like to introduce quality assurance measures for this sector are eligible to participate. Every registered pharmacy receives two interlaboratory test samples from a company commissioned by the ZL. The two interlaboratory test samples contain varying concentrations of glucose, HbA_{1c}, total and HDL cholesterol, triglycerides, liver enzymes ALT, AST and γ-GT, uric acid, creatinine, hemoglobin, CRP and microalbumin. The pharmacies are then supposed to measure these samples under routine conditions. These interlaboratory test samples are sent out once every quarter. The results obtained are entered into the results sheets provided and forwarded to the ZL.

The assessment is device-specific, i.e. a target value is determined for every model of POCT analyzer and every characteristic clinical chemistry and hematology variable. Despite identical samples, there are different target values for every measuring device. This is due to the strong influence of the sample matrix on the measured values; each matrix differs depending on the measuring device. The respective median of the results obtained by all interlaboratory test participants is taken as the target value. The acceptance limits also differ for each characteristic variable and are listed in the annexes to the guidelines of the Federal Chamber of

Pharmacists (BAK) [4]. The assessment limits for the relevant characteristic variables of the interlaboratory test are based on the requirements set forth in RiliBÄK, Table B1 [8].

In its guidelines on quality assurance, the BAK has set forth exact requirements governing the performance of blood tests. This describes a system check of the measuring devices, similar to the blood testing procedure. Beyond this, extensive commentaries and recommendations are made on the individual topics [4].

Internal quality assessment covers the inspection of the devices at regular intervals, including optical systems and system controls; here, the manufacturer's instructions should always be observed. The system controls should be conducted on each day of use at a minimum. Here as well, the requirements set forth in RiliBÄK [1] are binding. The percentage that the single values obtained from control samples deviate from the target value must not exceed the maximum permissible deviation listed for each analyte in Table B1 Column 3 of RiliBÄK [8], e.g. glucose 11 % and total cholesterol 7 %.

Beyond this, the "Model Statutes for the quality management systems of German pharmacies", enacted by ABDA and adopted by the Pharmacy Chamber of the German Federal States, can be considered as an informative quality assurance tool.

34.2.3 Test procedures/methods

The past few years have seen a fast growing trend that pharmacies are more and more becoming healthcare and communication centers. This also extends to medical-technical services. Section 1a of the German Pharmacies Act (Apothekerordnung) is the legal foundation governing the performance of services commonly rendered by pharmacies. Oftentimes, POCT tests represent a simple and fast way to measure such health variables [5].

The breadth of the spectrum of laboratory diagnostic parameters and the increase in specialized and cost-intensive tests also harbors the danger that laboratory tests will be ordered indiscriminately and in excessively high numbers. This applies to a high degree to the easy-to-run, rapid tests, i.e., that can be conducted by essentially anyone at any time.

34.3 POCT in nursing care facilities

Many elderly, multimorbid persons reside in nursing homes where they are cared for and supported in activities of daily life by professional nursing staff. The prevalence of diabetes mellitus among nursing home residents is particularly high, estimated at approx. 25 % [2]. In nursing homes, therefore, **blood glucose** is commonly measured by POCT methods (▶ Chapter 12). By contrast, other POCT tests, such as the urine stick test or INR measurements for monitoring oral anticoagulant therapy, hardly play any notable role in the nursing home setting.

In formal legal terms, the use of such quick tests in nursing homes continues to be regarded as patient self-testing, although there are comparably few systematic studies on the subject [3]. De facto, quality assurance in German nursing homes only takes place sporadically and can be especially challenging for the nursing home's management.

The mental, physical and manual abilities of nursing home residents can vary greatly. For example, an individual with impaired mobility might still be able to measure his blood glucose levels independently. One study conducted in the German county of Borken showed that 7.8 % of diabetics in inpatient nursing facilities self-tested their own blood sugar [6]. However, most residents are dependent on the support of nursing staff for taking measurements. In the nursing home setting, not rarely, the extent of support will increase with time. This is how helping a patient measure their blood glucose can turn into a measurement completely carried out by one of the caregivers. The legal requirements differ greatly when it comes to self-measurements and POCT tests performed on third parties. Even when self-measurements are ordered by a physician and instruction courses are recommended, there are no comprehensive quality assurance measures that are mandatory. However, measurements taken by caregivers are subject to the current version of the German Medical Devices Operator Ordinance (MPBetreibV).

Insofar as they were ordered by a physician, the requirements of RiliBÄK apply to all medical laboratory tests on patients, even when conducted as POCT (► Chapter 38). To navigate this minefield responsibly while ensuring optimal care of their residents, nursing homes should account for blood glucose tests in their quality management systems. Unlike in a hospital, residents will usually have their own measuring devices. If agreement can be reached for standardization of glucometers, it is easier to put uniform in-house rules in place. The devices in nursing homes are designed for private use; accordingly, they lack important functions like a barcode scanner for operator identification that would otherwise greatly facilitate quality management within the nursing home (► Chapter 31). RiliBÄK requires regular quality control measurements and participation in external quality assessments. In a nursing home, however, the test strips are prescribed to be patient-specific. Given the number of test strips, quality control measurements are not considered as part of the total testing volume.

Another neuralgic point is the interface between the physician attending to the patient and the caregiver who frequently carries out the actual POCT measurement. The blood glucose target values set by the physician are critical for interpreting the measured values. Oftentimes, however, this value is not known to the nurses. The LIVE-GERI study conducted in 2010 demonstrated that the targets were not communicated in 84 % of the patient cases [11]. Conversely, the values measured in the nursing homes were scrupulously taken down in the nurse's record, but in most of the instances, no standardized data exchange took place with the treating physicians. Communications are further complicated by the high number of physicians – sometimes averaging up to 20 – who work with the nursing home [11]. That brings with it the danger that deviations from target blood sugar levels are detected too late.

34.4 Conclusion

Decisions regarding the implementation of POCT must consider the fact that well-functioning internal and external quality assurance systems have been established for more than two decades in laboratory medicine, handled as by professionally run laboratories. Pharmacists, however, are not fundamentally bound to the requirements of RiliBÄK. The reason is that RiliBÄK is only valid for the practice of medical science being reserved for licensed physicians. Nevertheless, pharmacists are aware of their responsibility in handling patient data and, with it, the importance of applying internal and external quality controls [10].

In nursing care facilities, the blood glucose level is a decisive reference value guiding the treatment of diabetes mellitus. Despite the pitfalls described, it is therefore to be demanded for the future that quality assurance in blood glucose monitoring at nursing homes be improved by unequivocal rules issued by national medical associations. The communication problems between attending physicians and nursing care facilities described are not limited to POCT settings alone. For that reason, solutions should be sought in the telemedicine sector to at least reduce the interfacing problems in this area (► Chapter 35).

References

1. Anonymous (2014) Qualität durch Kompetenz. Pharm Ztg 49:52
2. Bahrmann A, Wörz E, Specht-Leible N, Oster P, Bahrmann P (2015) [Diabetes care and incidence of severe hypoglycemia in nursing home facilities and nursing services: The Heidelberg diabetes study]. Z Gerontol Geriatr 48:246–254
3. Balzer K, Butz S, Bentzel J, Boulkhemair D, Lühmann D (2013) Beschreibung und Bewertung der fachärztlichen Versorgung von Pflegeheimbewoh-

nern in Deutschland. Volume 125, 1st Edition, p. 1–348. Deutsches Institut für Medizinische Dokumentation und Information, Köln
4. Kommentar zur Leitlinie der Bundesapothekerkammer zur Qualitätssicherung – Physiologisch-chemische Untersuchungen – Durchführung der Blutuntersuchungen (Revision vom 15.05.2014) der BAK. www.abda.de/themen/apotheke/qualitaetssicherung0/leitlinien
5. Latsch H (2008) Screeningmethoden zur Untersuchung von Körperflüssigkeiten – Blutuntersuchungen in der Apotheke. PZ Prisma 201–211
6. Lederle M, Ettlinger G, Schäfer T (2008) Situation von pflegebedürftigen Diabetikern in Pflegeeinrichtungen im Kreis Borken. Diabetes Stoffwechsel Herz 3:169–176
7. Pfeil D, Pieck J, Gebler H, Schulte-Löbbert FJ, Tisch L, Brüggmann J (2013) Apothekenbetriebsordnung. Kommentar (Loseblattwerk). Govi-Verlag Eschborn, Pharmazeutischer Verlag. Current version dated 13 November 2013
8. Richtlinie der Bundesärztekammer zur Qualitätssicherung laboratoriumsmedizinischer Untersuchungen. Dtsch Ärztebl 105, Vol. 7 (2008) A341–355; last amended, promulgated: Dtsch Ärztebl 110, Vol. 39 (2013) A1822
9. Müller U (2001) Wie genau sind Blutuntersuchungen in der Apotheke? Pharm Ztg 23:39–40
10. Ruland U, Noltze I (2001) Blutuntersuchungen in der Apotheke sind zuverlässig. Pharm Ztg 51:40–42
11. Zeyfang A, Dippel FW, Bahrmann A, Bahr R, Feucht I, Hamann O, Hodeck K (2010) Aktuelle Versorgungssituation und Ressourcenbedarf bei insulinpflichtigen Typ-2-Diabetikern in ambulanter und stationärer Pflege: Ergebnisse der LIVE-GERI Studie. Diabet Stoffw 5:293–300

POCT in telemedicine

Andreas Bietenbeck, Siegfried Jedamzik

P. B. Luppa, R. Junker (Eds.), *Point-of-Care Testing*
https://doi.org/10.1007/978-3-662-54497-6_35

35.1 Introduction

Telemedicine makes it possible to overcome spatial separation between healthcare providers and patients by offering remote clinical diagnostics, consultations or medical emergency care via data and communication technologies. Particularly in rural areas, telemedicine has the potential to become an important component of medical care.

In laboratory medicine, the majority of samples are analyzed at specialized central laboratories, often far away from the patient. The results are then transmitted electronically to the attending physician. Support in interpreting the findings can be provided by the laboratory physician either personally or on the telephone as a teleconsultation. In that regard, telemedicine has long figured prominently in laboratory medicine, without being named as such. In the POCT sector, telemedicine fosters the development of new treatment and healthcare models [4].

Many countries politically encourage the increasing use of telemedicine to close gaps in service provision and design more cost-efficient medical treatments. In the USA, the HITECH Act (Health Information Technology for Economic and Clinical Health) was passed back in 2009, which was drafted to promote the increased use of data and communication technology in healthcare. In Germany, the Act on secure digital communication and applications in the health care system (E-Health Act) aims to promote networking between the various stakeholders. Depending on the type of application, the concept is either referred to as telediagnostics or telemonitoring.

35

35.2 Requirements governing POCT in telemedicine

With or without POCT, telemedicine requires a technical infrastructure and organizational setup. Data transfer in telemedical projects relies on multifaceted and extremely complex technologies. Before online transmission, however, data awaiting transmission are usually aggregated and encoded near the place where they are collected. Data should be protected from unauthorized access but still needs to be quickly and reliably accessible, if required. A measurement must always be reliably linked to the correct patient. A telematic infrastructure with this capacity is currently being set up across Germany.

In order for laboratory results to be exchanged, a universal interoperability standard must be defined. **LOINC** (Logical Observation Identifiers Names and Codes) has established itself worldwide as common terminology for laboratory test orders and results. This common language relies on six dimensions, called Parts: 1) Component (Analyte), 2) Property, 3) Time, 4) System (Specimen), 5) Scale and 6) Method.

As a set, they allow unique identification of each analysis. All of these dimensions can be represented in POC tests corresponding to their analogous analyses in the central laboratory. Despite this, the strength of a POCT finding is often lower, for example because of the compromises made in terms data accuracy as a result of the highly integrated design of a POCT device. POCT measurements also need to be identifiable as such after data transfer to ensure accurate interpretation of the findings.

35.3 Telediagnostics using POCT

POCT is primarily used for telediagnostics in emergencies or other unusual circumstances. Although a diagnosis is seldom based on POCT alone, its findings are a valuable component of any telemedically supported treatment regimen.

Telediagnostics have been successfully deployed in specialized emergency ambulances, for example, in mobile stroke units equipped with CT (computed tomography) scanner to diagnose strokes. The images taken of the patient on-site can be transmitted to a specialist who may be able to diagnose a stroke even before the patient is admitted to a hospital. Stroke-specific laboratory tests (e.g. small blood count, INR, aPTT, γGT, glucose), carried out as POCT procedures, complete the diagnosis and thereby make early thrombolytic therapy possible [3, 11].

POCT and telediagnostics also play pivotal roles in medical disaster assistance and humanitarian relief efforts [12]. Quick, essential laboratory tests can be carried out using POCT, even lacking any central laboratory infrastructure basics. Data exchange can also considerably facilitate coordination, logistics and treatment provision. Experts can also be called upon to render teletherapy, while utilizing limited resources efficiently [6]. In addition, telemedicine does not just facilitate the treatment of individual patients. The joint evaluation of all tests also helps to solve public health issues, e.g. for infection control etc. [7].

Telediagnostics has a higher risk if carried out without any direct physician involvement. If it is possible for patients to self-diagnose problematic conditions such as HIV using a POC test, this may lead to misinterpretations [10]. Telemedical consultations, e.g. by phone, can help avoid such problems, while still making diagnostic tests accessible to the wider population. Access to diagnostic tests is, however, not a widespread problem in Germany with its nearly comprehensive-coverage health insurance. In addition, the (Model) Professional Code for Physicians in Germany prevents them from giving patients in-depth and individual counseling if this is only carried out via communication media unless the patients is seen by a physician in person at some time [1]. Therefore, treatment may not be solely based on POC self-testing either.

35.4 Telemonitoring using POCT

Telemonitoring of POCT results is widely used, particularly for chronic conditions. That way, patients can be spared frequent visits to the physician without refraining from undergoing monitoring at close intervals. Telemonitoring often involves self-testing. To make telemonitoring work, the patient needs to be trained accordingly to carry out self-testing reliably.

A large number of projects have focused on the telemonitoring of blood glucose in patients with diabetes mellitus. Glucometer measurement data were sent to the attending physician and analyzed there in a timely manner. In some cases, additional information such as counseling on insulin dose, meals or exercises were also transmitted. Frequently, as-needed advice was given, for example, about adjusting insulin therapy or interpreting glucose readings. Many projects also try to use the telemedical infrastructure for better patient education. Most studies reported hardly any technical problems and a high acceptance rate by physicians and patients. Especially compared to error-prone manual data collection, telemonitoring provided considerably better data quality. Clinical outcomes were improved in many, but not all studies [5, 9]. Therefore, further studies are needed to clarify which aspects of telemonitoring are promising in the care of diabetic patients.

Alongside blood glucose, INR is often self-tested by patients, especially in Europe. This test is important for optimal adjustment of therapy with vitamin K antagonists. Here too, patients seem to benefit from telemedical help with correct dosing when their results are transferred electronically to the attending physicians [2, 8].

35.5 Summary and outlook

POCT is well suited to developing telemedical healthcare strategies while facilitating the provision of medical services, primarily in rural areas. POCT-based, diagnostic-focused telemedical concepts have been primarily directed at emergencies, medical disaster assistance and humanitarian relief efforts. Nowadays, telemedicine is already being used successfully to monitor chronic conditions, primarily diabetes mellitus, or long-term therapies like with vitamin K antagonists. The future will see a diverse range of disease data being integrated and analyzed at telemedical centers.

It is difficult to make generalizations about the economics and cost-benefit analyses of telemedicine applications. Usually, a relatively sophisticated infrastructure is necessary for data exchange. However, this infrastructure can then readily be used for a large number of patients. This economy of scale will lead to an increasing number of specialized telemedical centers that can offer extended services.

Laboratory medicine is only one medical discipline where telemedicine is employed. Other specialties, such as radiology, are increasingly implementing these new technologies as well. The aim is to improve the diagnosis by the joint evaluation of all telemedically collected data. In future, algorithms will also support this evaluation process [4].

A smartphone can be useful when numerous measurements are taken on a patient. It works like a central interface that aggregates the data collected on the individual sensors and then transmits the encoded data via the Internet to the telemedical center. Another concept is mostly used in the setting called "ambient assisted living" (AAL). Here, homes are equipped in such a way that a range of measurements – including POCT data – is analyzed to support residents discreetly.

References

1. Bundesärztekammer (2015) Hinweise und Erläuterungen zu § 7 Absatz 4 MBO-Ä. An English translation is available under: http://www.bundesaerzte kammer.de/fileadmin/user_upload/downloads/ MBOen2012.pdf
2. Christensen H, Lauterlein J-J, Sørensen PD et al. (2011) Home Management of Oral Anticoagulation via Telemedicine Versus Conventional Hospital-Based Treatment. Telemedicine and e-Health 17:169–176
3. Ebinger M, Winter B, Wendt M et al. (2014) Effect of the use of ambulance-based thrombolysis on time to thrombolysis in acute ischemic stroke: A randomized clinical trial. JAMA 311:1622–1631
4. Imaging EOBOTNIOB, Bioengineering/National Heart L, Faculty BINSFW et al. (2007) Improving Healthcare Accessibility through Point-of-Care Technologies. Clinical Chemistry 53:1665–1675
5. Jaana M, Paré G (2007) Home telemonitoring of patients with diabetes: a systematic assessment of observed effects. Journal of Evaluation in Clinical Practice 13:242–253
6. Kost GJ, Tran NK, Tuntideelert M et al. (2006) Katrina, the Tsunami, and Point-of-Care Testing. Optimizing Rapid Response Diagnosis in Disasters 126:513–520
7. Littman-Quinn R, Chandra A, Schwartz A et al. (2011) mHealth applications for telemedicine and public health intervention in Botswana. In: IST-Africa Conference Proceedings, 2011. p 1–11
8. Prochaska JH, Göbel S, Keller K et al. (2015) Quality of oral anticoagulation with phenprocoumon in regular medical care and its potential for improvement in a telemedicine-based coagulation service – results from the prospective, multi-center, observational cohort study thrombEVAL. BMC Medicine 13:14
9. Shea S, Weinstock RS, Teresi JA et al. (2009) A Randomized Trial Comparing Telemedicine Case Management with Usual Care in Older, Ethnically Diverse, Medically Underserved Patients with Diabetes Mellitus: 5 Year Results of the IDEATel Study. Journal of the American Medical Informatics Association 16:446–456
10. Spielberg F, Levine RO, Weaver M (2004) Self-testing for HIV: a new option for HIV prevention? The Lancet Infectious Diseases 4:640–646
11. Walter S, Kostopoulos P, Haass A et al. (2012) Diagnosis and treatment of patients with stroke in a mobile stroke unit versus in hospital: a randomised controlled trial. The Lancet Neurology 11:397–404
12. Yu JN, Brock TK, Mecozzi DM et al. (2010) Future Connectivity for Disaster and Emergency Point of Care. Point of Care 9:185–192

POCT in international development cooperation

Sandeep K. Vashist, Peter B. Luppa, John H.T. Luong

P. B. Luppa, R. Junker (Eds.), *Point-of-Care Testing*
https://doi.org/10.1007/978-3-662-54497-6_36

36.1 Healthcare in the Third World

In most developing countries, access of the population to better healthcare is an extremely urgent problem. The structures of Western healthcare systems, e.g. at large central hospitals, are often less helpful, although decentralized healthcare will usually represent the more effective approach. Within this scope, decentralized diagnostics with POCT systems that are utilizable by the primary care support staff can play an important role. This must also be considered against the background of a strongly increasing number of patients with diabetes mellitus, cardiovascular diseases and infections. Diabetes in particular is a global epidemic afflicting 415 million patients, requiring approx. 11–12 % of global healthcare resources per year [5]. Furthermore, there are 318 million persons with a pathological glucose tolerance worldwide [5] (Fig. 36.1). Indeed, approx. 46 % of diabetics worldwide have not even been diagnosed yet; in Africa, this number is as high as two thirds! In developing countries, there is also a very high number of patients with AIDS, tuberculosis and malaria. The annual diabetes mortality (approx. 5 million) in these countries, however, exceeds the aggregate mortality caused by AIDS (1.5 million), tuberculosis (1.5 million) and malaria (0.6 million) [5]. Although a multitude of HIV, tuberculosis, malaria and other analytical methods are already used in developing countries [3], many systems still remain too expensive, too complicated in their handling and too prone to error under critical ambient conditions (e.g. heat, moisture) to enjoy broader use. In the future, the increased use of mobile telephones and smart wearables for analytical purposes will certainly facilitate the development of robust systems.

The American National Institute of Biomedical Imaging and Bioengineering (NIBIB) started a joint venture with India to develop devices more suitable for use at the "point of need" [6, 8, 9]. The International Council for Standardization in Hematology (ICSH) is also engaged in drafting globally applicable guidelines for POCT hematology devices that can primarily be used in developing countries [2].

In the past years, after decades of often-ideological state-run development aid programs particularly joint ventures between the public and private sectors, i.e. public private partnerships (PPP) have become more successful in the Third World. These primarily concentrate their main focus on the research and development of vaccines against the prevailing infectious diseases like malaria, HIV, tuberculosis etc. These international PPP increasingly dictate which healthcare projects are funded in poor countries [4].

In the past, supranational organizations like the World Health Assembly of the WHO, where 190 nations are represented, have unfortunately not proven their merits in terms of implementing effective healthcare funding programs in developing countries. The World Health Assembly essentially assumes no international coordinating function whatsoever, since every country has its own vote. That means, important decisions can quickly be blocked. Moreover, many resolutions are pure declarations of intent without any concrete financial commitment [4]. Similar criticisms apply to UNICEF and the World Bank. Therefore, the following will present the most important private foundation and the two most important PPP.

36.2 Challenges and advances in setting up POCT infrastructures in the Third World

Urgent challenges faced by developing countries in setting up POCT infrastructures include a lack of regulatory standards for bringing POCT methods to market, a lack of qualified personnel and pre-existing infrastructures as well as financial means to pay for the various diagnostic applications. Currently, many countries around the world simply have too few healthcare experts in comparison to their constantly growing populations. The expenditure on diagnostic measures in these countries is less than 10 % of overall expenditures in the respec-

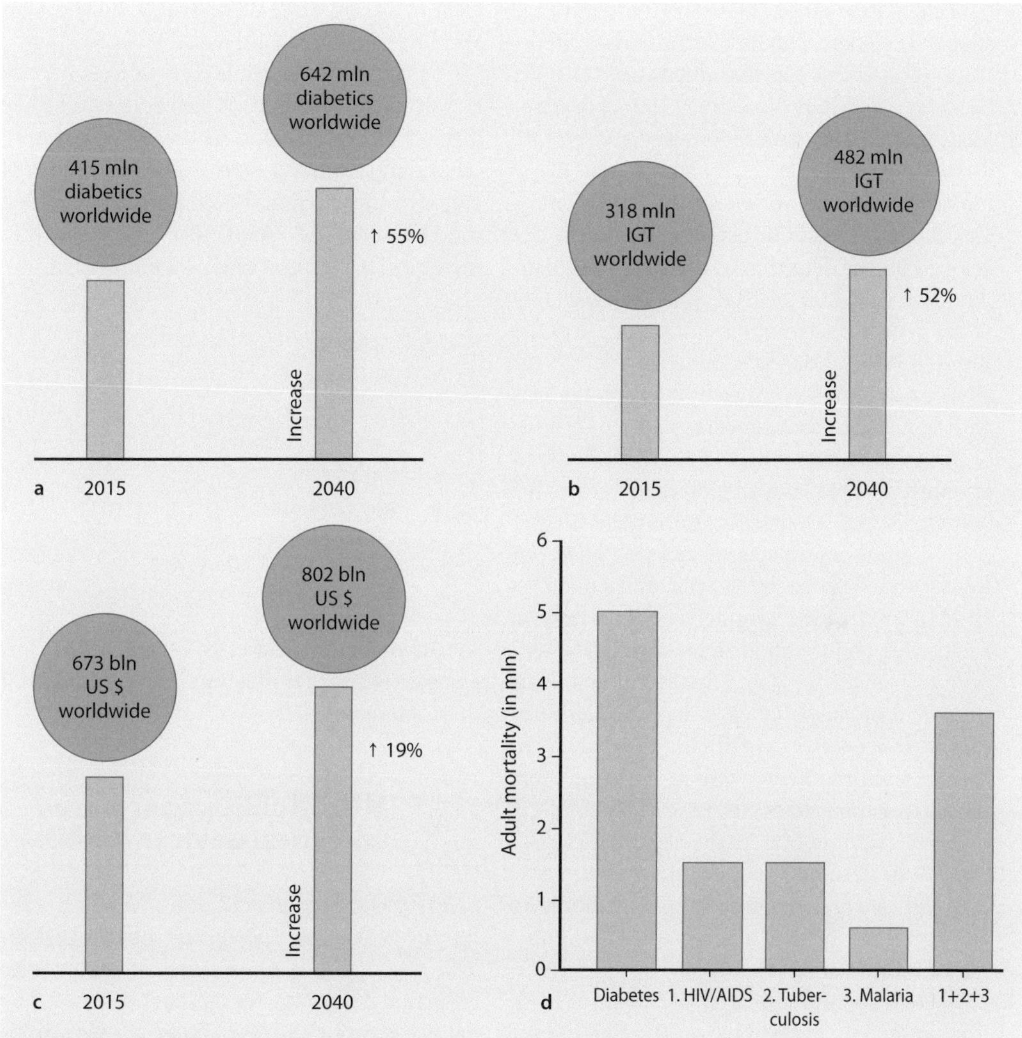

Fig. 36.1a–d Overview of the number of persons afflicted with diabetes and chronic infections in 2015 and projected into 2040. **a,b** Number of patients with diabetes and pathological glucose tolerance (*IGT*) in 2015 (**a**) and projected into 2040 (**b**). **c** Healthcare costs for diabetes in 2015 and projected into 2040. **d** Adult mortality for diabetes, HIV/AIDS, tuberculosis and malaria. The global figures derive from the Diabetes Atlas, 7th Edition 2015, International Diabetes Federation [5]

tive healthcare system; consequently, the funds spent on research into and development of new POCT methods and devices are inadequate in relation to the need for them.

In the past years, great advancements have been made in the field of **blood glucose testing**. Specifically when cost effectiveness and easy-to-use POCT devices are involved, numerous examples can already be cited such as the small iHealth Align (iHealth Lab, Mountain View, CA, USA) that can be connected to a smartphone. The great efforts hitherto undertaken to supply non-invasive glucose analysis systems that produce valid analytical results and are affordable at the same have nevertheless failed [11].

Lateral flow assays (LFA) for **HIV testing** are supplied by Alere, OraSure Technologies (Bethlehem, PA, USA), Bio-Rad, MedMira (Halifax, Canada) and Trinity Biotech PLC (Wicklow, Ireland) and require whole blood, serum or plasma as specimen material [3]. There are also many **malaria LFA** on the worldwide market. Recently, an FDA-approved POC test was specifically developed for the detection of two antigens (histidine-rich protein 2 (HRP II) plus aldolase [12]) of the parasite Plasmodium falciparum. It is the BinaxNOW malaria test supplied by Alere. Negative results must be verified by microscopic tests with thin and/or thick smears.

Suitable **tuberculosis LFA** are difficult to establish. To date, only one mycobacterium tuberculosis (MTB)-specific immunogenic glycolipid of the mycobacterial cell wall has been found – the mycobacterial lipoarabinomannan (LAM). This is the reason why up until now all MTB-LFA show suboptimal specificity and very low sensitivity [13]. Effective tuberculosis treatment by slow and less-sensitive microbiological diagnostics is difficult to achieve in Third World countries, while at the same time, the antigen quick tests are less suitable for early diagnostics. Thus, there is the hope that POCT-capable molecular biological detection methods will emerge. Boehme et al. [1] therefore conducted an on-site assessment of an automated molecular test for Mycobacterium tuberculosis (MTB) and resistance to rifampin (RIF). The study was performed on the GeneXpert MTB/RIF already presented in ► Chapters 10 and 20. Specimen material constituted the sputum of patients from Peru, Azerbaijan, South Africa and India. The authors concluded that the MTB/RIF test is sensitive and rapid (<2 h hands-on time), detecting MTB and RIF directly from untreated sputum. The test was specific in 604 of 609 patients without tuberculosis (99.2 %). Their drug-resistance testing correctly identified 200 of 205 patients (97.6 %) as carriers of rifampin-resistant bacteria and 504 of 514 (98.1 %) patients with rifampin-sensitive infectious pathogens. Despite the difficult on-site conditions, the POCT capable device proved its mettle without restrictions. Intensive training of operating personnel is, however, an absolute prerequisite for successful use.

As discussed, developing countries usually lack financial resources as well as the necessary infrastructure for clinical laboratory testing (electricity, cooling options, technical staff etc.). For that reason, the WHO has demanded the characteristic "**ASSURED**" for diagnostic methods [10]. This acronym stands for:

ASSURED
- Affordable
- Sensitive
- Specific
- User-friendly
- Rapid and robust
- Equipment-free
- Deliverable to end users

The following shall describe two exemplary analytical trends that illustrate the fulfillment of these characteristics.

36.2.1 Microfluidic paper-based analytical devices (μPADs)

In 2010, Martinez et al. [7] introduced a microfluidic paper-based analytical device that allows multiplex-like colorimetric assays with reflection detection. The authors concluded by mainly stressing the many, long-known advantages of paper-based carriers for biochemical analysis. This technique involves producing the microfluidic channels on the paper matrix by photolithographic or other means (inkjet etching, plasma etching, wax printing): The channels are formed by the hydrophilic cellulose fibers (approx. 190 μm in width) that soak up the sample fluid along of the channel. The walls of the channel are formed by hydrophobic barriers (approx. 250 μm in width). These are made of paper treated with additives (e.g. wax). A paper chip produced by photolithography (μPAD) is shown in ◘ Fig. 36.2.

Using the example of this urinalysis μPAD, the authors illustrated the feasibility and low

Fig. 36.2 Paper chip produced by photolithography (µPAD)

production costs of the system for protein and glucose analysis. Paper microzone microtiter plates and 3-dimensional µPADs can also be made affordably and reproducibly. However, the matching analyzers are not yet commercially available.

36.2.2 HIV therapy monitoring

The **CD4 Initiative** was established at the Imperial College of Medicine in London (▶ Section 36.3) to devise an easy-to-use POC test for measuring the number of CD4+ T-lymphocytes in HIV-infected patients. The aim was to ensure better anti-retroviral therapy monitoring. For their part, Zyomyx (Hayward, CA, USA) then developed the first POCT device for quantitative CD4-counting [17]. The test was designed to be as easy to use as a thermometer. It was based on the principle that CD4-binding reagents will "draw" CD4 lymphocytes flowing out of a blood sample and allow them to flow into a measuring device from which the cell count could be read off by simple visual means. The quantitative CD4 cell count delivered information on the efficacy of HIV therapy. This test is no longer on the market.

36.3 Foundations and public-private partnerships

36.3.1 Bill & Melinda Gates Foundation

The Bill & Melinda Gates Foundation [14] focuses on funding the research and development of vaccines in the Third World. The foundation's headquarter is in Seattle, Washington; it has a trust endowment of $40.3 billion and currently employs a staff of 1,453. Since 1994, the foundation has invested more than $25 billion in promoting global health. The Bill & Melinda Gates Foundation is the largest foundation in the world. In 2006, the Foundation received a large endowment when the billionaire Warren Buffet decided to donate a large part of his wealth to the pursuit of worthy causes.

As much as €1.5 billion have been donated to the GAVI Alliance (see below) that develops vaccines and vaccination programs for children. The Gates Foundation estimates that 90 % coverage with lifesaving vaccines in developing countries would secure the survival of about 7.6 million children under 5 years of age from 2010–2019. Furthermore, the foundation provides $100 million annually to the Global Fund which operates like a PPP (see below).

The Foundation also supports the development of simple POCT applications for the therapy monitoring of AIDS patients. The CD4 Initiative has already been discussed in ▶ Section 36.2 above. The "Collaboration for AIDS Vaccine Discovery (CAVD)", similarly initiated by the Gates Foundation, is additionally sponsored by the "Global Fund to Fight AIDS, TB, & Malaria," the "President's Emergency Plan for AIDS Relief" (PEPFAR), UNAIDS and the "Clinton Health Access Initiative" [14].

36.3.2 The GAVI Alliance

The GAVI Alliance [15] is a unique PPP that takes a novel approach to increase vaccination rates in developing countries. In pursuing its mission, GAVI brings together important stakeholders who contribute to accelerating access to vaccines for children in the poorest countries in the world. The governments of developing and donor countries, the WHO, UNICEF, the World Bank, vaccine manufacturers from numerous countries, research institutions and development agencies, civil society, the Bill & Melinda Gates Foundation and other private donors and influential private individuals all work collaboratively to reach goals that no organization can achieve on its own. The German Federal Government is a committed partner within the GAVI Vaccine Alliance and has been providing financial support since 2006. This amount raised in 2014 from €30 million to €38 million annually.

36.3.3 The Global Fund

The Global Fund, founded on the initiative of the former United Nations Secretary-General Kofi Annan [16], is a globally operating PPP dedicated to the prevention and treatment of AIDS, tuberculosis and malaria. The partnership is constituted of national states, international development organizations and private-sector organizers as well as affected partner countries and regions. It represents a novel approach to international healthcare funding. As part of its approach, the Global Fund collaborates with other bilateral and multilateral organizations. More than 50 partners have financially invested – including many nations, but also the WHO and the UNAIDS (joint program of the United Nations on fighting HIV/AIDS) as well as commercial businesses and foundations (Clinton Health Access Initiative). The largest sponsor is the USA; Germany similarly participates by contributing large amounts each year.

References

1. Boehme CC, Nabeta P, Hillemann D, Nicol MP, Shenai S, Krapp F et al. (2010) Rapid molecular detection of tuberculosis and rifampin resistance. N Engl J Med 363:1005–1015
2. Briggs C, Carter J, Lee SH, Sandhaus L, Simon-Lopez R, Vives Corrons JL (2008) ICSH guideline for worldwide point-of-care testing in haematology with special reference to the complete blood count. Int J Lab Med Hem 30:105–116
3. Chan CP, Mak WC, Cheung KY, Sin KK, Yu CM, Rainer TH, et al. (2013) Evidence-based point-of-care diagnostics: current status and emerging technologies. Annu Rev Anal Chem 6:191–211
4. Hibbeler B, Korzilius H (2010) Globale Gesundheit: Wes Brot ich ess … Dtsch Ärztebl 108: A-138/B-111/C-111
5. International Diabetes Federation. Diabetes Atlas, Seventh Edition, 2015. http://www.diabetesatlas.org/resources/2015-atlashtml#
6. Kost GJ (2008) Point of need, global outreach, and a universal companion. Point of Care 7:103–105
7. Martinez AW, Phillips ST, Whitesides GM, Carrilho E (2010) Diagnostics for the developing world: microfluidic paper-based analytical devices. Anal Chem 82:3–10
8. NIBIB and India hold joint workshop on low-cost diagnostic and therapeutic medical technologies. http://www.nibib.nih.gov/NewsEvents/Newsletters/March09. Accessed on May 19,2009
9. NIH and India partner to develop low cost medical technologies. http://www.nibib.nih.gov/News Events/Releases/Archiv/2007OctNovDec.Accessed on May 19, 2009
10. Peeling RW, Holmes KK, Mabey D, Ronald A (2006) Rapid tests for sexually transmitted infections (STIs): the way forward. Sex Transm Infect 82 Suppl 5:v1–6
11. Vashist SK (2012) Non-invasive glucose monitoring technology in diabetes management: a review. Anal Chim Acta 750:16–27
12. Vashist SK, Luppa PB, Yeo LY, Ozcan A, Luong JHT (2015) Emerging technologies for next-generation point-of-care testing. Trends Biotechnol 33(11):692–705
13. World Health Organization (2011) The new WHO report on anti-tuberculosis drug resistance: multidrug and extensively drug-resistant TB (M/XDR-TB): 2010 global report on surveillance and response. WHO/HTM/TB/2010.3
14. www.gatesfoundation.org
15. www.gavialliance.org
16. www.theglobalfund.org
17. www.zyomyx.com

Quality assurance

Content

Quality assurance in POCT – A cross-country comparison

Peter Fraunberger, Sylvia Gruber, Franziska Amiet, Martin Fiedler, Michel Vaubourdolle, Benedicte Beneteau-Burnat, Pascal Pernet, Laura Tooth, Paul Collinson, Naoto Shimetani, Lutz Schwettmann, Robbert Slingerland, Bert Dikkeschei, Elizabeth Lee-Lewandrowski

P. B. Luppa, R. Junker (Eds.), *Point-of-Care Testing*
https://doi.org/10.1007/978-3-662-54497-6_37

37.1 Cross-institutional POCT management at five public general hospitals in Vorarlberg, Austria

Peter Fraunberger, Sylvia Gruber

37.1.1 Introduction

The joint operating conglomerate of five general hospitals, Vorarlberger Krankenhausbetriebsgesellschaft, introduced a centralized POCT management for blood glucose monitoring in 2014 (Radiometer, Wiener Neudorf, Austria). The need for a centralized quality monitoring and control of POCT strategy grew from economic, organizational as well as medical considerations. Usually a variety of device types from different suppliers are already being used within individual hospitals. Without central management, this leads to significant and costly complexity. Due to the larger order volume, hospital management receives – from an economic point of view – better conditions by uniformly allocating POCT systems within the operating company than by allocating individual devices. Standardization by switching to a single vendor facilitates operations for medical personnel and prevents interpretation errors caused by the use of different measurement techniques and units. In addition, centralized device monitoring and quality controls should help to ensure a high standard of quality in POCT. However, considerable organizational effort is associated with the spatial separation of the different locations, the large number of wards involved and the high number of POCT users. All of these factors call for a clear organizational structure.

37.1.2 Quality requirements

The current guidelines for POCT management have been published in DIN EN ISO 22870 "Point-Of-Care Testing (POCT) – Requirements for Quality and Competence" (as of 2006) [1]. ÖNORM EN ISO 22870 is the equivalent Austrian version of the German standard, which is why the requirements it contains are binding for accreditation in Austria in addition to Germany. This standard governs hospitals, clinics and healthcare facilities where outpatients are cared for, except for measurements that are taken by the patient at home or in communal facilities. ISO 22870 is intended to be used in conjunction with its normative reference, ISO 15189 "Medical laboratories – Requirements for quality and competence" [2].

According to Section 4 of ISO 22870, the management of the laboratory service is responsible for planning and developing the processes required for POCT. The head of the laboratory "must appoint a multidisciplinary Steering Group (...) representing the laboratory, administration and clinical programs, including nursing, which advises on the setup of POCT". Responsibilities must be defined and clearly communicated within the group. The Steering Group shall define the scope of the POCT (Chapter 4.1.2.1), taking into account the clinical need, financial impact, technical feasibility and ability of the organization to meet these requirements. Chapter 4.1.2.4 states that "the Steering Group must advise on the evaluation and selection of devices and systems for POCT". The assay performance criteria should include accuracy, precision, detection limits, limits of application and interference factors, as well as practical feasibility.

The duties of the laboratory service (Chapter 4.2.2.1) include, in particular, establishing quality objectives for POC systems, introducing adequate quality monitoring processes, defining standard operating procedures, providing documentation (e.g. SOPs and enrollment forms), validating and monitoring POCT activities and appointing a "POCT coordinator". Laboratory management must carry out a cost-benefit analysis at regular intervals as well as an assessment of the clinical need and efficacy. Potential opportunities for improvement must also be considered.

Chapter 5 of ISO 22870 contains details on the technical requirements for POCT. According to ISO 22870 5.1.1 Section B, POCT meas-

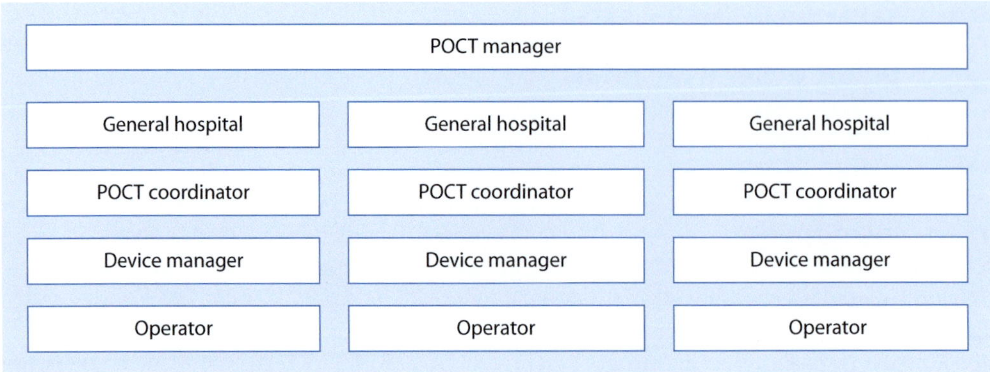

Fig. 37.1 Chart of the POCT organizational structure in Vorarlberg, Austria

urements may only be carried out by "personnel who have completed the training and have proven that they possess the appropriate expertise". It is mandatory that corresponding records be kept. The POCT requirements for personnel include understanding its correct use, knowledge of the theory and measuring system as well as pre-analytics (sampling, clinical benefits and limitations, procedures, storage of reagents, quality assurance measures, technical limits of the device, procedure given results outside of the defined limits, infection control and documentation). The staff's performance must be monitored.

Further technical requirements involve the unequivocal identification of each sample and its traceability to the individual patient, as well as identifying the employee who took the sample. To differentiate them from laboratory findings, POC results must be clearly identified and permanently stored.

Key messages on POCT findings
- POCT findings must be clearly labeled as such
- The patient unequivocally identified
- The performing employee unequivocally identified
- The results must be documented in the medical record

37.1.3 Organizational structure using the example of Vorarlberg, Austria

Each ward with a blood gas analyzer must designate a device manager and a deputy who are the direct contacts for the device and are authorized to train new system operators. In addition, each hospital in the operating conglomerate has its own POCT coordinator with a deputy, who is an employee of the medical central laboratory. Although the POCT coordinators are the direct contact persons in case of organizational problems and responsible for the administration of on-site operators, they do not provide technical support (see "Failure Concept" below). Higher-level strategic issues are forwarded to and clarified by the POCT manager in collaboration with the hospital management or the POCT manufacturer. One possible POCT organizational structure is shown in Fig. 37.1.

Important documents such as work instructions, device manuals and relevant forms are available on the respective hospital's intranet and, if required, can be updated by the POCT manager.

The required reagents are stored on each ward to be readily available when needed. The manufacturer automatically supplies consumables with the order and adapts the cassette size to the ward's consumption. The nursing staff is responsible for accepting deliveries of and properly storing reagents. Complaints about device malfunctions can be reported very easily by faxing the device error message to the manufacturer. Replacement reagents will be automatically sent to the respective ward by the manufacturer upon receipt of the error message.

If, for example, a blood gas sample is measured, the POCT system operator is identified by a measurement authorization barcode that must be scanned on the blood gas analyzer to unlock the device for analysis. The system thus prevents untrained personnel from taking measurements. After scanning the operator ID, a device prompt requests the patient ID, which is used to assign the patient within the POCT IT network (see next section). Because patient data is automatically retrieved from the blood gas analyzer after scanning the patient ID, it can also be displayed on the device printout.

37.1.4 Documentation (results/system operators – IT solutions)

The results of the POCT measurement are transmitted via middleware to the **Laboratory Information System** (LIS), where a separate order is generated for each measurement. The results are then transmitted from the LIS to the **Hospital Information System** (HIS) and into the medical record. The order is clearly marked as POCT findings according to the specifications of ISO 22870 and lists the measurement results, patient's data, specimen collection accessories, device used and operator's name.

For a better overview, POCT findings can be displayed separately from other laboratory results in the HIS via a separate, cumulative view.

In addition to the electronic transmission, the device generates a printout which also includes the patient data (patient ID, surname, first name and date of birth), the specimen collection accessories, the operator's name and device description. Device descriptions include the hospital identifier and the ward to which the device is assigned. For example, "FK_IMCU" designates the POCT device of the Intermediate Care Unit at the Feldkirch general hospital.

If the data transmission is interrupted, the device remains ready for measurement; the orders are stored locally and, after the network connection has been restored, automatically forwarded to the IT system.

The POCT middleware guarantees the traceability of the reagent and control batches. Performed maintenance, calibrations and QC results are automatically stored and can be retrieved and statistically evaluated as needed.

37.1.5 Training system

The prerequisite for obtaining an operator's measurement authorization is demonstrated training by the manufacturer of the POCT device or the device manager(s). Training by designated device managers enables timely training of new employees. In addition to internal training, additional training is provided by the manufacturer if required. In the case of repeated measurement errors, evidence of pre-analytical errors or failure to meet interlaboratory test specifications (see "Quality Assurance" below), re-training courses are compulsory.

Device managers receive additional, in-depth training once a year to be taught about new product developments and learn advanced troubleshooting.

The medical central laboratory is responsible for administration of the authorized POCT system operators. New training sessions are documented using a pre-printed form and forwarded to the responsible POCT coordinators. The POCT coordinator creates the new operator in the middleware and issues a measurement authorization with the employee's name and barcode (◘ Fig. 37.2). No authorizations are issued without a record of staff training. Because only one type of blood gas analyzer is used throughout the hospital network, permissions obtained are not limited to indi-

Fig. 37.2 Example of an operator's measurement authorization

vidual wards or hospitals, but are valid for all devices in the hospital network, regardless of location.

> **Note**
>
> Only qualified staff trained in POCT are allowed to perform POC measurements.

37.1.6 Quality monitoring

The POCT devices perform fully automated internal quality control (QC) measurements every eight hours. When replacing consumables, calibration and QC measurements are also automated. The site's POCT coordinator or their deputy monitor device performance and maintain a QC history using the middleware. The responsible POCT coordinator checks the status of the devices once a day. If neither the coordinator nor their deputy is available, monitoring can also be carried out by another facility's coordinator. The results of the QC measurements are automatically documented by the POCT middleware. Annual evaluations are retrieved by the POCT manager from the middleware's statistics module and allow performance comparisons between similar devices.

The POCT coordinator resolves any deviations in QC directly with the ward's device manager and, if necessary, with the POCT manufacturer. The corrective action taken is either documented automatically by the middleware (e.g. reagent replacement) or by the POCT coordinator. The laboratory management assigns and stores maintenance and repair logs for the affected device in a separate software program. This ensures that all work done on any specific device can be clearly displayed and evaluated according to cost and frequency. Recurring QC deviations are processed by the POCT manager in cooperation with the POCT coordinator, the device managers and the medical director of the affected ward. If required, selective follow-up training courses will also be offered and documented by the POCT coordinator.

EC controls in the form of interlaboratory tests are carried out twice a year for all wards. The ward itself carries out this measurement, after which the POCT coordinator of the hospital inputs the results and evaluates the interlaboratory test. If the specified goal is not achieved, for example, because of poor pre-analytics, the defective result is reported to the device managers and an attempt to ascertain the reason for the erroneous measurement is made. In the event of a failed interlaboratory test, re-training with appropriate documentation is mandatory for the ward in question. Failed interlaboratory tests should also be reported to the POCT manager.

37.1.7 Contingency concept

At least one backup device with a compatible parameter profile has been defined for each POCT device, which operators can use to avoid device failures. This is mainly facilitated by uniform device types, allowing operators to also operate POCT instruments on other wards without difficulty. If the problems are simple (e.g. inlet seal obstruction), employees can try to troubleshoot the problem on the spot. Device managers receive special training for this; a hotline providing telephone support by technicians is available at routine times. If the problem cannot be solved by the operator or the device manager, the manufacturer will provide a replacement device no later than the following working day. Measurements must be carried out using the specified backup device until the faulty device is replaced. Technical support is provided exclusively by the manufacturer; laboratory staff do not receive training for this.

37.2 POCT quality assurance in Switzerland

Franziska Amiet, Martin Fiedler

37.2.1 Introduction

In recent years, POCT has been growing in popularity in Switzerland. This is mainly due to the massive increase in cost pressure on the hospital sector resulting from introduction of the fixed rate per case system (SwissDRG) in January 2012. Changes in hospital structures and processes are currently emerging as a result. The decrease in laboratory compensation in 2009 had already led to greater awareness of service processes and their optimization. Today, laboratories are frequently closed at night in small hospitals and emergency analysis is carried out using POCT. Another aspect is the special status that primary care medicine enjoys in Switzerland. Following pressure from politicians and general practitioners (GPs), primary care medicine is currently being taken into account. On January 1, 2015, the Swiss Federal Department of Home Affairs (FDHA) created a list of "rapid analyses" [1] that are carried out as POCT diagnostics by in-practice laboratories and charged at a higher rate. A further revision of the analysis list is expected soon.

In recent years, POCT has been developed and professionalized in Switzerland. POCT coordinators and commissions are active at cantonal and university hospitals. Their central tasks are the creation of regulations, the improvement and monitoring of quality assurance and the optimal networking of POCT devices with data management, laboratory, clinic or hospital information systems. In smaller hospitals, POCT is usually implemented in cooperation with the laboratory and operated with its support. Continuous improvement of laboratory analyses is also an objective of primary care medicine. Since 2012, GPs are required to take part in a course on "In-practice laboratory certificate of competence". Their certificate of competence is automatically renewed when they participate in external quality controls on annual basis [2].

37.2.2 Legal regulations and quality assurance

The Swiss Commission for Quality Assurance in the Medical Laboratory (QUALAB) regulates quality assurance in the medical laboratory in the form of a basic contract and concept [3, 4]. QUALAB supplemented or modified the criteria for the operation of medical-analytical laboratories (KBMAL) [5]. KBMAL Version 1.0 [6], a checklist compiled by QUALAB, aims to successively assist all medical laboratories in meeting the criteria for the operation of medical-analytical laboratories defined in KBMAL and to ensure quality at a uniformly high quality standard. The legal basis for KBMAL is the Swiss Health Insurance Act (KVG), the Swiss Health Insurance Ordinance (KVV) and the basic contract QUALAB [3].

QUALAB's concept of Quality Assurance in the Medical Laboratory [4] summarizes implementation rules and regulations. These are mandatory for analyses charged to a patient's compulsory basic insurance. Internal and external quality control is obligatory; external quality control is carried out at a recognized quality control center in Switzerland [7–10]. Other centers located abroad are coordinated by recognized quality control centers in Switzerland.

External Quality Control is performed four times a year. The list of parameters, assessment criteria, fulfillment criteria and the quality control center recognized for these parameters are listed in the document "Mandatory External Quality Control" [11].

The frequency of the **internal Quality Control** depends on whether it is a simple or complex analyzer. Simple analysis systems meet certain criteria and are listed in Annex B of QUALAB's internal quality control guidelines [12]. All other analysis systems are considered complex analysis devices. For complex analysis systems, internal quality control is performed twice a day with each series or by continuous measurement. For simple analysis systems, internal quality control is performed at regular intervals, but at least every 2 weeks (see Internal Quality Control Guidelines for exceptions [12]).

External quality control is carried out 4 times a year; parameters and criteria are listed in the QUALAB document "Mandatory External Quality Control" [11]. The frequency of internal quality control depends on the type of analysis system (simple vs. complex).

37.2.3 POCT in hospital and practice

People in Switzerland are satisfied with health care, as demonstrated by surveys conducted annually in the Health Monitor [13]. Patients expect quality and efficiency. POCT meets the expectations of patients and physicians. Demand and pressure for the latest technologies are growing and the SwissDRG system is forcing an optimization of processes and quality. For POCT, this means documented results without any media discontinuity in the HIS. Opportunities for POCT include projects to improve patient medication process safety using bedside scanning as well as the use of patient arm bands with barcodes or RFID. Appropriately configured and equipped POCT devices are a prerequisite for the use of patient wristbands for continuous data documentation. Some hospitals are already using patient wristbands or are planning to use them. Surveys and studies show that the topic is becoming ever more current [14, 15].

Areas of application and analytics

Compared with Germany and Austria, areas of application and analytics hardly differ in Switzerland. In recent years, more and more emergency clinics have emerged within hospitals or as independent centers. POCT analyses include vital parameters for immediate intervention or therapy in the hospital, for process optimization in outpatient clinics or for frequent measurements on wards (e.g. glucose). Blood gas devices that measure electrolytes, metabolites and bilirubin (in neonatology) with small sample volumes are highly popular. Thromboelastography is used in trauma, surgical and intensive care settings. Analysis takes place either at the POC, or centrally in the lab using digital result transmission.

In the primary care setting, patients benefit from POCT performed by the in-practice laboratory. Since 2015, GPs have been able to charge a higher fee for POCT performed by the in-practice laboratory (►"Lists of rapid analyses" and "Supplementary analyses" as part of the analysis list p. 226–229 [16]).

Staff and responsibilities

POCT is generally performed by nurses in hospitals, whereby the responsibility lies with the attending physician. Other administering occupations may include health professionals (FAGE, new occupation since 2002), midwives, physicians, nurses and doctor's assistants in outpatient clinics. Device requests, device evaluation, installation, networking, quality control, maintenance and upkeep fall into the competence areas of the laboratories, POCT coordinators, medical technology and the clinics' device managers. Newly hired POCT staff receive training when they start their job. The training expenditure for caregivers and physicians is enormous. The resources required for consistent, regular refresher courses in POCT are lacking in many places. At in-practice laboratories or joint practices, the responsibility for analysis and device maintenance lies with the doctor's assistants or biomedical analysts certified by a college of professional education and training.

Networking status

Many POCT devices are networked with an IT system. Hospitals use the system to document quality controls, calibrations, reagent and device management, performance reporting and troubleshooting, as well as to route results to the LIS, the HIS, and other clinic information systems. These systems are used e.g. in intensive care units, anesthesia, dialysis, or the Emergency Department. All of these systems use patient identification (patient ID, case ID, others); only a few use personal POCT operator login/identification on the devices. Centralized operator management with assigned roles and device authorization are rarely used. These assessments are based on an oral survey of different POCT representatives.

Quality

Analytical quality has improved greatly in the more than 7,500 in-practice laboratories and at several hundred hospital and private laboratories. The course on "In-practice laboratory certificate of competence" (FAPL) for physicians and compulsory participation in External Quality Control program are proving to be effective. The CSCQ President commented on analysis quality in the editorial of Vision CSCQ No. 10 in September 2015. Quote by Dr. Olivier Boulat: "The educational function performed by the Swiss External Quality Control Centers (EQK) has led to a significant improvement in analysis quality (the number of compliant results has risen from 89 % to 98 % in the last ten years) [17]".

POCT Switzerland

- The SwissDRG system is forcing an optimization of processes and quality.
- POCT meets the current expectations for new technologies, efficiency and quality.
- Most POCT systems are networked and implementing patient identification; the hospitals still lack personal operator identification, central operator management with role assignment and device authorizations.
- More time is required to achieve traceable data documentation without any media discontinuity in the HIS for POCT, gaps have to be closed, processes checked and systems adapted.
- Because adequate training requires resources (cost pressure), initial and recurring training for POCT system operators in the hospital is a challenge.
- Legal provisions, the implementation of Swiss quality control centers and compulsory interlaboratory tests have led to a significant improvement in analysis quality in the last ten years.
- Participation in the FAPL course is compulsory for GPs. Since January 2015, medical in-practice laboratories have been able to charge higher standard rates for POCT analyses found on the list of rapid and supplementary analyses.

37.3 POCT in Spain

Maria Luisa Hortas, Miguel Cantero, Francisco Javier Lirón-Hernandez, Paloma Oliver

On behalf of the Spanish Society of Clinical Biochemistry and Molecular Pathology (SEQC) POCT Commission.

37.3.1 Introduction

At the end of the nineties, the interest in POCT came up with contributions in national and international congresses [6, 8, 17]. Spanish legal requirements for laboratories are ruled below a general process; an administrative authorization for clinical laboratories, depending on each region in the country. There is no other specific or mandatory regulation for point of care testing, although it is accepted that are included in the different local regulations for authorisation of clinical laboratories. That is the reason why the recommendations of clinical societies are so relevant.

37.3.2 POCT guidelines in Spain

It is in 2003 when the Spanish Society of Clinical Biochemistry and Molecular Pathology (SEQC) [20] created a workgroup for developing Spanish guidelines (POCT Commission) the "Guía para la implantación de pruebas de laboratorio en el lugar de asistencia al paciente", published in 2006. The good point for these guidelines were that the consensus came from laboratory professionals and diagnostics manufacturers involved in POCT, taking care of different guidelines from other countries, recommendations from bodies like the College of American Pathologist or the CLSI [1, 2].

As there was the first time to focus on POCT as a national view; tasks as a deeply discussed definition of point of care testing; the importance of laboratory professionals play in order to maintain quality in these tests or the structure of human resources, their duties from an organisational point of view and the interest in having an interdisciplinary committee for the better implementation.

A recent review of these guidelines [14] has been conducted and is published in press. The main differences are focused on the denomination of point of care testing in Spain following the international recommendations instead of Spanish way of expression (POCT vs. test at the side of patient care); a better description reasons to justify implementing POCT (clinical, organisational or economic criteria) and a special remark for patient safety in POCT. On the other hand, the group considered relevance to include recommendations from care international standards such as ISO 17593:2007 [9]; ISO 15197:2013 [10]; ISO 22870:2006 [12] and ISO 15189:2013 [11] and a huge number of webs of interest. The Spanish guidelines can be resumed in 11 recommendations (Tab. 37.1).

37.3.3 The situation in Spain

In 2012, The POCT Commission promoted and conducted a national survey, achieving 126 full responders (unpublished data). As approximately 90 % of the participants agree with supporting POCT as long as necessary and justify and only 6.5 % of the hospitals have no devices in POCT; we can say POCT is widely use in Spain. We have found it must be improved strategies to ensure analytical quality; as in 56.5 % POC devices are not under quality control programs; they do without laboratory participation or they didn't even know if there is a

Tab. 37.1 Final Spanish recommendations for optimal quality in POCT

1	Identify point of care testing (definition)
2	Justify the incorporation of POCT
3	Create of a working group led by laboratory professionals
4	Establish traceability and analytical limits of a test to be used in POCT environments
5	Choose simple devices, if possible. Consider minimum risk of error
6	Have current procedures
7	Be able to performed the test, through a competency training program
8	Design quality control programmes and execute them
9	Participate in external quality control programmes
10	Preferably, connect the equipment to integrate the clinical information
11	Breach of the above points should be considered a nonconformity, so you will have to take corrective actions

quality control profile; as previously published by others groups [13].

Selection of analyzers and quality assurance

One of the main objectives of the POCT Commission is to report on instruments evaluations, promoting evaluation on a device in microbiology and recommendation on glucose meters. As there are numerous publications related to the selection of instruments made by Spanish groups, we can stand out [16] study, as they evaluate 23 blood gas analyzers according to the CLSI protocols and compared one of them with the devices used previously in the hospital.

Patient safety

Any quality management system has to propose indicators to evaluate processes; and NACB guidelines recommend that specific measurable indicators related to POCT organization must be identified, monitored and evaluated. A Spanish group [3] has proposed a set of indicators to investigate quality error rates related to preanalytical and analytical phases associated with POCT. They have developed this study in the neonatal unit and the indicators were compared between central lab and POCT. These are few indicators to suit IFCC WG-LEPS recommendations [19] ◘ Tab. 37.2).

With regard to deal with critical values, a redefinition of values is proposed when an environment of POCT is set, coming from the close collaboration between healthcare unit and laboratory professionals [4].

Professional competency in POCT

The feeling of losing power in their responsibilities in a group of professionals and the change in the way of exercising the clinical activity have slowed the growth of POCT in Spain. In fact, there is a relatively discomfort among technicians as they interpret they are the professionals trained to use analytical devices. Nowadays, it is increasingly understood as an opportunity for professional development. The reviewed Spanish guidelines include recommendations for initial and continuous training as a core of success in POCT, with a minimal set of aspects the professionals should know and a protocol to evaluate the competency that must be reached including, among all, evaluation of technical performance; supervision of reports and handling information; evaluation of results in internal and external quality control programs, and capability of solving problems. There are several national training programs. One of them come from the Spanish Society of Clinical Biochemistry and Molecular Pathology (SEQC) which by POCT Commission promote and execute numerous specific training activities, The focus of this course, organized by Roche Diagnostics in collaboration with the Complutense University of Madrid, are the clinical laboratory profession-

◘ Tab. 37.2 Quality indicators for POCT

Quality indicator	Related to
Percentage of requests with errors concerning patient identification per month	Total number of requests
Percentage of samples aborted[a] per month	Number of samples
Percentage of insufficient sample volume per month	Number of samples
Percentage of unacceptable results in EQAP[b] per year	Number of EQAP parameters
Percentage of unacceptable results in IQC[c] per year	Number of IQC parameters

[a] Samples aborted, could be due to clotted samples or inability of operators
[b] EQAP: External quality assessment program
[c] IQC: Internal quality control

als, giving them training in POCT coordination duties, teaching experiences in other countries having an specialty for this type of testing and enhancing the role of POCT coordinator.

Based-medicine evidence in POCT in Spain

There are some studies of interest developed in Spain [18]. We can distinguish some studies; one of them as they evaluate clinical, operational and economic outcomes of POCT in pulmonology offices compared with clinical laboratory measurements; offering better results the use of blood gas analyzers in POCT organization [15], other is related to the study of an intraoperative POCT intervention's related costs in extracorporeal cardiac surgery to decrease hospital complications [5] and the oldest one study the likely impact of critical outcomes in critical care management through implementing point-of-care in primary care [7].

In conclusion, Spain is doing a good job in POCT considering there are laboratory professionals conducting interesting studies and the national societies support them. However, there is an opportunity to improve.

37.4 France: an experience on POCT QM based on a mandatory EN ISO 22870 accreditation

Michel Vaubourdolle, Benedicte Beneteau-Burnat, Pascal Pernet

37.4.1 Introduction

37

Until 2010, French legislation had specific guidelines for medical laboratories (mainly the "GBEA" for good analysis performance guidelines) but did not allow to perform point-of-care testing (POCT) in private laboratories. However, POCT analyzers were installed in public hospitals where no specific regulation applied. Since 2010, a new French legislation (n 2010-49 dated January 13, 2010) [1] replaces GBEA decree by a national law on biology, constrains a mandatory accreditation for all medical laboratories (public and private) under international standard EN ISO 15189 [2] and legalizes POCT for medical laboratories under the condition of an complementary EN ISO 22870 [3] accreditation.

The French Committee for Accreditation (COFRAC) is the independent organism in charge to authenticate the competences of the medical laboratories in France.

Medical laboratories have to achieve accreditation of all activities before 2020. France is one of the first countries to take such drastic measures to give medical laboratories recognition that their practices are in compliance with this international quality standard.

37.4.2 POCT perimeter

In France, regulations differentiate between the medical biology examinations with analytical phase realized outside the laboratory, placed under the responsibility of a specialist in laboratory medicine on one hand, and diagnostics guidance tests, listed in a decree, which can be done by patients or by specified health professionals in defined contexts and medical indications, on the other hand.

In the first case, we find usual POCT and/or critical care testing examinations: biochemistry (blood gases/CO-oximetry/electrolytes/glucose/lactate, cardiac markers, HbA1c, CRP…), hemostasis (ACT, viscoelastic and aggregometric tests), hematology (Hb)… These tests are submitted to a mandatory ISO 22870 accreditation to beachieved before 2020 for all tests. We will call testing placed in that category "POCT" in France. This will be developed more precisely in this paper. Furthermore, law requirements add a restriction of use for these tests to the strict context of "urgent therapeutic decision", to regulate the POCT situations to critical care testing and avoid a costly sparse of POC devices without demonstrated justification of medical needs.

In the second case, no accreditation is needed and the laboratory is not involved. We find in that category patient self-tests (glucose, HbA1c, INR), health professionals (physician, nurse, midwife, pharmacist) tests (capillary glucose, urinalysis, transcutaneous bilirubin, pulse oximetry, amniotic fluid detection, infectious diseases tests: Influenza, HCV, HIV etc.). There are requirements for quality management for these tests (minimum operator training, traceability). This is not in the field of this paper.

37.4.3 POCT processes and quality indicators

ISO 22870 requires to identify processes needed for the quality management system of POCT and to determine criteria and methods needed to ensure that operation and control of these processes are effective. Then, one major challenge to ensure quality improvement is to select adequate tools to monitor the different processes involved in POCT practice.

We present here some quality indicators (QI) selected and implemented by our laboratory to monitor POC blood gas testing. Quality indicator (QI) is a requirement of ISO 22870 chapter 4.12.1-2. We identified on the process map of our medical laboratory, the processes affected by POCT which required specific QI. There was two specific processes: the POCT device implementation process from the clinical needs expression to the start of POCT use including the POCT management group (creation and follow-up), and a second process corresponding to the examination process. These processes are summarized in ◘ Fig. 37.3.

For the management of many risks, POCT processes to be managed were the same than for regular laboratory activities (EN ISO 15189): for example, purchasing process or internal audits. In this case, we used the same indicators which are not presented here. These processes identified to be affected by POCT were:

- for operational processes: pre-examination, examination and post-examination phases;
- for supporting processes: management of training;
- for management processes: multidisciplinary POCT management group, review of contracts. At least, one specific QI was implemented to monitor each of these processes. When appropriate, QI were analyzed for each POCT location.

The QI were adequately chosen to both assess processes and fulfill specific EN ISO 22870 requirements regarding management. A QI synthesis is reviewed each year and presented in management review.

In 2013, selected QI [quantification index - periodicity - target] were:

- **For operational processes**
  - Evolution in the number of analyzes related to the number of patients: this QI helps to fulfill the management review requirements (ISO 22870 – § 4.15) regarding a periodic review of the medical need with clinicians, a cost-benefit analysis, and a cost efficiency of POCT analysis. [ratio amount of analyzes/amount of patients hospitalized in the clinical unit – annual – stability]
  - Amount of pre-examination nonconformities from clinical units users (misidentifications on the analyzers, review of aborted analysis): this QI helps to fulfill pre-examination requirements about identification (§ 5.4.2) and quality of training (§ 5.1.5); repeated aborted analysis may also help to detect operators who need a new training, indeed non-authorized operators using the code from someone else. [amount of non-conformities – monthly – <5 %]
  - Effective participation of external quality assessment (EQA) (§ 5.6.4) [ratio amount of samples analyzed/amount of samples of the EQA program – biannual – 100 %]
  - Comparability of the POCT analyzers results to those of the central laboratory (POCT specific § 5.6.2) using EQA or split samples [ratio amount of samples analyzed and reviewed/amount of samples planned to be analyzed and reviewed – quarterly – 100 %]

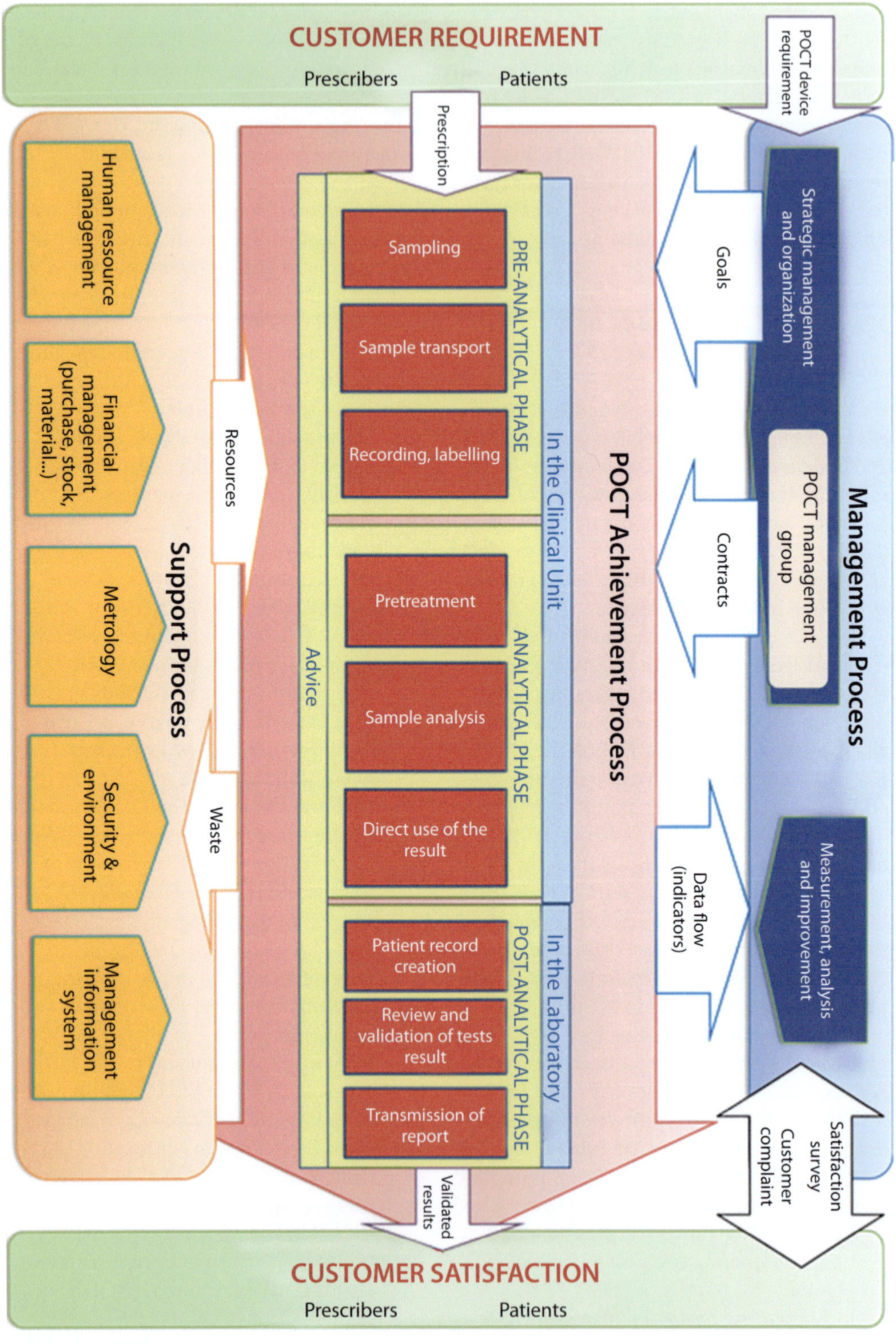

Fig. 37.3 Specific processes in POCT

  - Turnaround time of integration of validated results by the medical biologist in the patient medical record: this QI helps to fulfill the requirements regarding review of contracts (§ 4.4) and reporting of POCT results (§ 5.8) [percentage of analysis under the maximal turnaround time defined in the contact between the laboratory and the clinical unit – biannual – >80 %]
- **For supporting processes**
  - Clinical unit authorized users' follow-up; [amount of operators with out of time competency assessment – biannual – 0]
- **For management processes**
  - Audit of the mandatory missions of the multidisciplinary POCT management group (POCT specific – § 5.8) [amount of missions not fulfilled – annual – 0]
  - Global amount of nonconformities; this QI helps to fulfill the requirement to evaluate where continual improvement of the effectiveness of the quality management system of POCT can be made (§ 4.1) [amount of nonconformities – quarterly – decrease]
  - Periodical cost-benefit analysis and reevaluation of the clinical need (POCT specific § 4.15) [out of time or no review between the managers of the laboratory and the clinical units – annual – 0]

Monitoring ISO 22870 accredited POCT is difficult without a pertinent choice of specific QI. This selection associated with QI in common with the medical laboratory allows controlling the whole process of POCT.

37.4.4 An experience in Saint-Antoine Hospital

The public medical laboratory of the Saint-Antoine Hospital in Paris was one of the first medical laboratories applying for ISO 22870 accreditation in France and accreditation certificate was obtained on the 1st of July 2011. In 2012, we had our first follow-up audit and accreditation was renewed. We present in detail our first 2-year experience of POCT accreditation.

A **quality group** (2 medical biologists, 1 technical supervisor) specialized in POCT was created and focused on:

- the specific requirements of the ISO 22870 standard, compared to the ISO 15189 standard: quality policy, process control, creation of a point-of-care coordination group, document control, training/habilitation of non-laboratory staff, verification of performance, environmental conditions control, selection and establishment of a new equipment, nonconformities management, audits.
- the additional specific requirements of the French legislation [1]: external quality control program, a posteriori biological validation (review) of results.

The specialists in laboratory medicine worked particularly on technical requirements of ISO 22870 chapter 5: performance specifications for each analysis were verified on all analyzers: precision, trueness, measuring range (if pertinent), interferences. Measurement uncertainty was calculated for each analysis. Results for precision and uncertainty were compared to those from the French Society of Clinical Biology [4] and from C. Ricos [5] specifications: all were found below fixed limits. Comparability of POCT analyzers results were verified with those obtained with central laboratory analyzers in order to allow a good patient follow-up.

- **Analyzers**
  - In 2010, at the time of the initial audit, POC blood gas activities were performed in 3 different locations on GEM series analyzers (2 GEM 4000 and 1 GEM 3000 – Instrumentation Laboratory, USA)
  - In 2012, at the time of the follow-up audit, one location with a GEM 4000 was closed and a new one with an ABL800 (Radiometer, Denmark) was audited.

- **Initial audit**
  - In 2010, the use of the GEM analyzers significantly facilitated our work because of its features: disposable measurement cartridge, integrated QC management (IQM), re-mote control with information management software for complete control of the POCT process (GemWeb Plus), biological safety and easy training of non-laboratory staff.
  - In November 2010, the initial audit carried out by 2 Cofrac evaluators (one technical auditor specialized in biochemistry and one quality auditor) revealed 5 gaps, including 1 critical gap, to be corrected within 6 months. Noncompliance situations concerned competence assessment, formalization of responsibilities, external quality assessment (EQA) for CO-oximetry, document control. Cofrac accepted our action plan and delivered our accreditation mid-2011 for the following analysis in total blood: pH, pCO2, pO2, Na+, K+, Ca++, lactate, glucose, hematocrit, CO-oximetry.
- **Follow-up audit**
  - In 2012, Cofrac evaluators came back for the first follow-up audit: they audited 2 of the previous locations and the new one with the ABL 800 previously accredited according to ISO 17025. They verified that 2010 gaps were settled and revealed 3 new gaps: environmental conditions, metrology, lack of healthcare provider/patient/client feedback.
  - Cofrac renewed our accreditation including the new location.
  - Another follow-up audit in 2015 showed no new noncompliance situations.

37

37.4.5 Conclusion and perspectives

Our experience of POCT accreditation in Saint-Antoine hospital was beneficial for the entire staff, laboratory and clinical units, and particularly improved communication. The extension of accreditation to other POCT activities (cardiac markers, glycated hemoglobin etc.) is scheduled before 2020 in our hospital.

Globally, in 2016, there is now about 10 accredited laboratories for EN ISO 22870 standard in France. This is a small number compared to the very rapid growth of EN ISO 15189 accreditations (about 700/1000 accredited laboratories in the beginning of 2016 and mandatory target of 100 % partial accreditation in November 2016). This discrepancy is explained by the flexibility granted to the laboratories to achieve the total accreditation in 2020. The regulation objectives (50 % accreditation in 2016, 70 % in 2018) are based on the relative number of accredited acts and POCT quantitative part is relatively weak compared to core laboratories activities. Furthermore, the exact number of laboratories concerned with POCT is still unknown.

At the national level, the objective of a complete POCT accreditation, in public and private sectors, is now possible and useful for a better management by quality and for an improvement of patient care.

37.5 The UK perspective

Laura Tooth, Paul Collinson

37.5.1 Introduction

Point of care testing (POCT) in the UK has largely been based on provision of blood gas analysis. This has been extended to include a range of additional tests, such as electrolyte and renal function in critical care areas and emergency departments, plus blood glucose testing in the community. In 2006, the Department of Health commissioned an independent report into NHS pathology services in England. The report, chaired by Lord Carter of Coles, identified the increasing use of point of care testing and suggested that pathology was moving towards a more accessible community based service [1]. This model, of moving services out of hospitals, and into the community, was suggested for London in Lord Darzi's report – Healthcare for London: A framework for action, which recommended urgent care centres offer basic pathology services by POCT, to ensure a rapid turnaround, and to facilitate diagnosis and treatment [3]. In addition, the UK National Institute for Clinical and health care Excellence (NICE) [4], began recommending patient self testing and self-monitoring, in areas such as anticoagulation.

37.5.2 Quality management framework

POCT was increasingly being undertaken by members of the health care team with no reference to laboratory services. The Carter report (2006) [1] identified fragmentation of point of care services and highlighted the importance of proper management of POCT devices. The UK health regulatory authority, the Care Quality Commission (CQC), mandates that staff should receive the support, training, professional development, supervision and appraisals necessary for staff to undertake their roles and responsibility. It is therefore important that staff involved in the delivery of pathology services, including technical and support staff, should participate in relevant continuous professional development, as part of maintaining competence.

The Carter report [1] also stated 'pathology service providers should be subject to mandatory accreditation by an organization independent of the providers and the professions'. This mandatory accreditation, which includes point of care testing, gives members of the public and other NHS staff the confidence that the quality of the service has been independently verified as meeting objective service standards. Hospitals in the UK aim to accredit POCT services against ISO22870, but this is currently separate to pathology laboratory accreditation status.

Accreditation requires full participation in External Quality Assurance (EQA) schemes. In the UK, WEQAS and NEQAS are the main EQA providers. Schemes available for POCT include Bilirubin, Blood gases, BNP, Cardiac Markers, Co-oximetry, Creatinine, Glucose/Ketones, Haemoglobin, HbA1c, HIV, Lipids, Pregnancy, Pre-term Labour Markers, Urinalysis and Urine Drugs of Abuse. Samples are sent monthly and are analysed by those who use the device.

37.5.3 Implementation of POCT

The local hospital pathology laboratory is expected to play a key role in the development and management of a POCT service [2]. In the UK, each hospital has a point of care coordinator, who works with a POCT team to liaise with companies, evaluate and compare POCT devices, and help with the implementation and management of devices. The POCT team usually consists of scientists and administration assistants from the pathology laboratory. They are responsible for issuing Standard Operating Procedures (SOPs), which are regularly updated, managing training programs, and participating in clinical audit and other clinical governance activities, as a further means of underpinning quality. Having a point of care coordinator discourages companies from selling directly to the wards, which allows standardization of devices through the trust. Before any device is implemented it is assessed for bias and imprecision to check the device is fit for purpose. The comparison is performed using EQA and patient samples. The device is then trialed in the department to assess ease of use and impact on patient pathways.

Users must be deemed competent before performing POCT, and this is enforced with connectable, operator lock out devices. Many POCT companies offer online training modules for operators to regularly update their competencies. Companies also offer monthly training sessions and link trainers are used to cascade training to those unable to attend set sessions. EQA results are reviewed by the point of care team and any poor performances investigated. The POCT team also reviews Internal

Quality Control (IQC) data. IQC is performed on all POCT devices and this is enforced by connectable meters that allow device lock out, if the IQC has not been performed within a certain timeframe. The frequency of IQC testing is established in local guidelines, as there are no national recommendations.

EQA poor performance criteria are set by the EQA scheme steering committee, in consultation with the scheme organiser, and with the agreement of the relevant National Quality Assessment Advisory Panels [5]. When an analyte shows poor performance the head of department is contacted by the scheme organiser. The laboratory is then expected to perform a detailed investigation and report any findings back to the EQA provider. If laboratories do not reply, or the poor performance continues, EQA providers can report the laboratory to the NQAAP for further investigation [5]. NQAAP expects 100 % compliance on EQA returns and laboratories that fall below the ideal are reported [5]. If any manufacturer problems are identified with an assay, the Medicines and Healthcare Products Regulatory Agency (MHRA) is informed. This is a government agency that ensures all medicines and medical devices work correctly and are acceptably safe in accordance with European Union directives.

The POCT team can help to write business cases for funding of POCT devices. A Service Level Agreement (SLA) is then written, which details the range of products and operational details, as well as formally defining the responsibilities and expectations of each party. SLAs set out the lines of accountability, which is especially important when the hospital assists the community with purchasing POCT devices, training, quality assessment and health and safety.

Within the hospital, the POCT team report to the POCT committee, which is a multi-disciplinary group including clinicians, nurses, scientists, IT, finance and pharmacy, and is chaired by a hospital consultant. The POCT committee assesses the clinical need for POCT, and raises any issues with POCT management to the trust risk committees and medical devices and management committee (MDMC).

37.5.4 Patient safety and device regulation

The POCT group are responsible for identifying patient safety issues arising from the use of POCT. Any adverse incidents are reported to the MHRA, who will further investigate and take necessary action to safeguard the public. The MHRA also issue medical device safety alerts that detail all faults identified. These alerts must be acknowledged by each hospital and a reply sent detailing if actions are or are not required. The MHRA deals directly with companies to improve devices and rectify any faults identified. It is an effective system that collates incidents from across the country, and makes recurring patient safety issues easy to spot. Manufacturers are also obliged, under medical devices regulations, to report certain adverse events to the MHRA [2].

37.6 POCT in Japan

Naoto Shimetani

37.6.1 Introduction

In the quality assurance of POCT, the objective is to conduct a clinical trial with POCT-compatible devices and reagents and to ensure the accuracy of the test result according to standard laboratory practice.

A Japanese guideline for POCT has been presented by the POCT Technology Committee of the Japan Society for Clinical Laboratory Automation. For the quality assurance of POCT, the 11th chapter of the 3rd version [1] of the Japanese POCT guideline provides safeguarding of the test data. This primarily involves uniform monitoring of all phases of the clinical testing process and means that uniform administration of the preanalytical, analytical and postanalytical phases (application of the results) is required.

37.6.2 The core issue behind POCT quality management

To meet the requirements of POCT quality management, ISO 22870 "Point-of-Care Testing (POCT) – Requirements for quality and competence" [2] and ISO 15189 "Medical laboratories – Requirements for quality and competence" [3] must be observed.

In Japan, a POCT coordinator or a medical-technical laboratory's assistant is usually responsible for assuring quality throughout the process.

37.6.3 IT applications for POCT quality management

The POCT device must have network/LAN access ("connectivity function") in order to organize the data flow during quality assurance and data transmission. The network/LAN access function must meet the requirements of the CLSI (Clinical & Laboratory Standards Institute) Directive [4]. Frequently, device manufacturers will have developed their own management software. Such software features a routinely automated calibration system, a POCT cartridge performance test system, the ability to detect device-internal errors, an administration system for the results of quality control and a data management system interoperable with LIS/HIS. Therefore, when selecting and operating POCT devices and quality control materials, the POCT Technology Committee of the Japan Society for Clinical Laboratory Automation recommends those devices that can establish a network/LAN connection and provide quality management using proprietary software.

37.7 POCT in Norway

Lutz Schwettmann

The special topographical conditions in the rural and northern areas of Norway often impede rapid sample analysis outside the hospitals. The distance to the nearest hospital is often long and it is sometimes only possible to transport the sample by air. POCT is therefore especially important outside the hospitals POCT, as it enables timely analysis of important measurands.

In Norway, POCT-instruments are used in a centralized or decentralized way in hospitals as well as in small laboratories outside the hospitals. Outside the hospitals, POCT is primarily used by communal and private general practitioners (GP) as well as private specialists. Generally, the doctor's practices are located in medical centers. In their common laboratories, sampling, analysis by POCT devices and shipment to the hospital laboratory are done by nurses and doctors assistants. POCT instruments are further used in rest homes and ambulant nursing services, among others for analyzing of CRP and glucose, as well as by occupational physicians and on oil platforms. Here sampling and analysis is often done by personnel with limited education in laboratory medicine. This fact could fundamentally call into question the quality of the analytical results.

In Norway the organization of quality management **inside** the hospitals and **outside** the hospitals generally differs. The responsibility for the quality management of decentralized POCT instruments in hospitals is up to the central laboratory belonging to the hospital. The central laboratory assumes the supervision of internal and external quality control as well as the training of the users of the POCT devices. To this, many central laboratories establish a POCT quality coordinator. The participation in external quality assessment in hospital laboratories is coordinated almost solely by the organization NKK (Norwegian Clinical Chemistry EQA program). NKK is a non-profit organization established by the Norwegian Society of Medical Biochemistry. NKK serves Norwegian laboratories within medical biochemistry in the fields of external quality assessment (EQA), standardization and other quality development issues. The shipment of the external quality control samples occurs directly from the provider to the laboratory. Noticeable changes in quality, significant changes of the levels between different POCT instruments and feasible actions for quality improvement are discussed with representatives of all hospital laboratories on the annual meeting for quality management.

In Scandinavia, the quality assurance of laboratory analysis is of great historical importance. As early as the 1980s, the requirement for the improvement of quality of POCT-instruments used outside the hospitals was recognized. In 1992, **Noklus** (Norwegian Quality Improvement of Primary Health Care Laboratories) was established.

Noklus is a national non-profit organization aimed to ensure that all laboratory analyses that are ordered, performed and interpreted outside of hospitals will safeguard the patient's needs for investigation, treatment and follow-up. Funded by the Norwegian Medical Association through a quality improvement fund, Noklus offers access to external quality assessment programs to all Norwegian GP surgeries, nursing homes and other health care institutions. The participation in EQA programs is voluntary. Noklus provides guidance and tuition through site visits, telephone consultations and courses. For this purpose, more than 100 employees including medical technicians and specialists in laboratory medicine, are at disposal. Noklus, headquartered in Bergen, has regional subsidiaries throughout Norway. Depending on the geographical situation, each state has one or more Noklus offices, based at local hospitals with regional representatives and specialists in laboratory medicine. The local presence in all states enables Noklus a close communication with the users of POCT devices. Noklus employees are available to assist laboratories in all matters regarding all questions addressed to POCT analysis. 99 % of the Norwegian GP laboratories and 95 % of the nursing homes use Noklus. Thus, Noklus ensures comprehensive quality monitoring of nearly all laboratories outside the hospitals. The participation in external quality schemes is funded by the Norwegian Medical Association and is free of charge for the public GPs. For private GPs, nursing homes, occupational health services laboratories and the oil industry, participation is charged.

Also many hospital laboratories participate in Noklus' EQA-schemes (subject to charges) which are communicated by NKK. Noklus usually ships control material it produces itself. The programs offered comprise about 20 parameters for POCT analysis and cover the most of the measurands performed at the GPs. The collected results are statistically evaluated centrally and every participant receives an individual report concerning precision and accuracy of the results achieved. The field Noklus representative is informed about the results of all the participants in his area. If a laboratory failed the quality specifications, the Noklus representative contacts the laboratory and discusses possibilities for quality improvement. The nationwide surveyed results are also used for central supervision of changes in quality between different reagent lots. Close cooperation with the IVD manufacturers means that unacceptable quality differences can be promptly addressed by Noklus.

The Norwegian market is basically open for all manufacturers of POCT devices. The quality of the devices is mostly scrutinized critically before purchase. Noklus, a well-established institution nationwide, provides advice to municipalities and laboratories on issues related to quality of the devices and the analytics. Because of a central and nationwide registration of all results of external quality assessment, the Noklus employees have a complete picture of the strengths and weaknesses of the POCT devices on the market and can help laboratories select the most appropriate instruments for their purpose.

Additional information about the assessment of quality and operator convenience of POCT devices is provided by SKUP (Scandinavian Evaluation of Laboratory Equipment for Primary Health Care). SKUP is a joint organization between the Scandinavian countries Norway, Denmark and Sweden for independent testing of POCT devices. The main task is to improve the quality of POCTin Scandinavia by providing objective and manufacturer-independent information regarding quality of analysis and operator convenience of the devices. SKUP tests POCT devices according to uniform criteria and produces test reports. SKUP test reports are not necessarily required to offer POCT equipment on the Scandinavian market, but they are often used by the manufacturers for documentation of analysis and device quality. The manufacturers themselves are the ones who contract out the generation of the test reports and commission SKUP with the preparation of a test report before launching their products on the Scandinavian market. The test reports can then assist all laboratories to select suitable instruments during the process of purchasing or replacing of POCT devices. Since SKUP is organizationally linked to Noklus in Norway, the Noklus field representatives have direct access to the test reports and can provide this information to the laboratories during regular visits to the laboratories.

After more than 20 years of activity, it can be documented that the various quality assurance and improvement organizations have substantially improved the analytical quality carried out with POCT devices inside and outside the Norwegian hospitals.

37.8 The Dutch Perspective

Robbert Slingerland, Bert Dikkeschei

37.8.1 Introduction

The last few years a significant number of POC-tests have been introduced in the Dutch health-care system to cope with financial cutbacks and diminished number of people able and willing to work in the health care system and a significant increase in the number of patients with chronic disease. Due to doubling of e.g. the number of patients with diabetes and the increased average age of population, in this decade in the Netherlands the cost of health care will rise tremendously. To counter this process the Dutch authorities focus on downsizing the volume of high-care hospitals by providing more care at the General Practitioner (GP) level or even via home-care. Especially for the four chronic diseases, diabetes, chronic obstructive pulmonary disease/asthma, heart cardiovascular risk management, the home care point-of-care testing will play a significant rule.

New area's for point-of-care testing will be chemotherapy treatment monitoring in the home setting. This might prevent patients from unnecessary travelling to a hospital for their next treatment, because their blood values are not good enough yet or to better monitor concomitant but unwanted fever processes.

The main manufacturers for POCT in the Netherlands are currently Abbott, Alere, Nova Biomedical, Roche, Radiometer, Philips and Samsung (via Thermo Fisher Scientific). Most glucose measurement in the Dutch hospitals are already done with POC glucose meters or blood gas analyzers able to measure glucose. Microbiological POC-market will be a growing area for manufacturers from 2017. Continuous glucose monitoring intravenously is planned to be introduced in 2016 in the Dutch market allowing to better control glycemic conditions in patients on ICU.

In hospitals, nurse-homes and GP offices POC instruments must be operated under the guidance of a central laboratory, connected to a central database (indirect connection via docking station or direct via wireless), a quality control program must be installed and (re-)education must be implemented for all users (e.g. by e-learning). This was mandated by the Dutch health-care inspection service several years ago for hospital and nurse-homes settings. Last year a new Dutch guideline for the use of POCT in GP offices was released [1].

There exists an everlasting dilemma between analytical quality specifications for the glucose measurement apprehended by manufacturers on the one hand and professional medical communities on the other. The failure of the European In-Vitro Diagnostic medical devices directive to define specifications critical for the clinical usefulness of POCT systems paved the way for this dilemma (a revision of the IVD directive is currently prepared and will be voted for in 2016 by the EU parliament) [2].

For several POCT assays e.g. glucose, the combination of analytical quality and pre-analytical conditions provides a total variation budget which often by far exceeds medical professional requirements for safe diagnosis and monitoring, even though specifications by ISO and CE marking are not violated. Outsourcing POCT systems to GP's or patients undertaking measurements on their own adds significantly to the total variation due to variability in end-user's skills and insights in the use and interpretation of POCT assays.

37.8.2 Quality management framework

Dutch Guideline for the use POCT in GP offices describes that POCT must be applied under the responsibility or in close cooperation with an ISO 15189 accredited central laboratory. It gives guidance what needs to be arranged in order to run safely POCT in GP offices. Accreditation requires full participation in **External Quality Assurance (EQA)** schemes. In the Netherlands the SKLM is the EQA provider. Many schemes are available for POCT applications. Samples are sent monthly and are analyzed by those who use the device.

37.8.3 Implementation of POCT

The local, ISO 15189 accredited, clinical chemistry laboratory is expected to play a key role in the development and management of a POCT service in the neighborhood of a hospital [1]. In the Netherlands, it is nowadays a common practice that each hospital has a Point-of-Care coordinator, who works with a POCT team to liaise with companies, evaluate and compare POCT devices, and help with the implementation and management of devices. The POCT committee in a hospital provides an educational resource for the indications for use, achievable device performance, potential variables, technical issues, and risks of error when utilizing POCT.

The POCT committee consists of at least a clinical chemist and a POC-coordinator and situational determined medical staff members, provides the necessary quality oversight to ensure reliable POCT results for the various patient populations being tested. An important role of the POCT committee is creation and oversight of the device validation process, development of written policies and procedures, and implementation of the training/competency plan after a device has been selected. The POCT committee plays a key role in program development and management. An important responsibility of the POCT committee is the selection of a POCT device that meets specific laboratory testing needs and can be utilized throughout an institution.

Users must be competent before performing POCT, and this can be enforced with connectable, operator lock out devices. In general, a customizable e-learning program must be provided by manufacturers of POCT instruments. EQA results are reviewed by the Point-of-Care team and any poor performances investigated. The POC team also reviews **Internal Quality Control (IQC)** data. IQC is performed on most POCT devices. A validation process for new lot numbers is in time performed. EQA poor performance criteria are set by the EQA scheme steering committee, in consultation with the scheme organizer. When an analyte shows poor performance, the head of department is contacted by the scheme organizer. If the poor performance continues, the ISO 15189 accreditation of the laboratory is at stake.

All POCT devices in hospitals are under the responsibility and authority of the central laboratories.

New POCT are based on a business case proposal and will be purchased by the central laboratory. A Service Level Agreement (SLA) with the departments of the hospital describes usually details of product and operational details, as well as defines formally responsibilities and expectations of each party. Special trained nurses for POCT are the backbone of the system in hospitals. In each department such a well-trained or specialized nurse is present to answer the first questions of other end-users and if needed they can rely back on the POCT coordinator and or clinical chemist. SLAs are also in use outside the hospital, which is especially important when the hospital assists the GP offices with purchasing POCT devices, training, quality assessment and health and safety.

37.8.4 Patient safety and device regulation

The POCT group are responsible for identifying patient safety issues arising from the use of POCT. Any adverse incidents are reported to the Dutch healthcare inspection service and in case of glucose meters also to the Dutch Society for Clinical Chemistry (NVKC), who will further investigate and take necessary action to safeguard the public. The Dutch healthcare inspection service also issues medical device safety alerts that detail all faults identified. The Dutch healthcare inspection service deals directly with companies to improve devices and rectify any faults identified. It is an effective system that collates incidents from across the country, and makes recurring patient safety issues easy to spot. Manufacturers are also obliged, under medical devices regulations, to report certain adverse events to the Dutch healthcare inspection service.

37.9 A Perspective from the United States

Elizabeth Lee-Lewandrowski

37.9.1 Introduction

Point-of-Vare Testing (POCT) is widely utilized in the United States in both the inpatient and outpatient settings [1]. Bedside glucose testing is the dominant POCT technology in terms of the total test volume and the number of instruments deployed within hospital systems. However, a number of other POCT tests are commonly utilized including arterial blood gases and electrolytes, coagulation, urinalysis, hemoglobin A1C and many others. The number of tests available in POCT formats continues to expand with a corresponding increase in utilization of these technologies. Managing the quality of POCT is an ongoing challenge [2]. At a minimum health care organizations must comply with federal regulations mandated by the Clinical Laboratory Improvement Amendments of 1988 (CLIA-88), in some cases regulations imposed by individual states and with those required by various accreditation agencies including the College of American Pathologists (CAP) and the Joint Commission (JC). To achieve regulatory compliance organizations rely on a combination of three key components:

- Continual improvements in the testing technologies themselves including built-in features to eliminate sources of error (e.g. quality control lock-outs, lock-out of untrained operators, lock-out of expired test strips).
- POCT informatics and data management software to ensure documentation of test results and other parameters related to testing (e.g. operator, quality control, patient identifcation).
- Establishment of an organizational POCT oversight program usually managed by clinical laboratory professionals including laboratory directors and POCT coordinators.

Collectively these three components have progressively improved the quality of POCT and allowed the establishment of formalized quality management programs in most health care organizations.

37.9.2 Role of federal regulations and accreditation agencies

POCT has existed in American hospitals almost since their inception but was largely unregulated throughout most of the 20th century [3]. CLIA 88 was enacted by the US congress in response to concerns regarding the quality of laboratory testing and included guidelines pertaining to POCT. The regulations are overseen at the Federal Government level by the Center for Medicare Services (CMS) and are based on the complexity of the test method. Some tests are "waived" from these regulations if they cleared by the Food and Drug Administration (FDA) for home use or if the tests use such simple and accurate methods that the probability of patient harm resulting from an inaccurate result is negligible [4]. The waived status of a test is most relevant to outpatient practices where the practice need only obtain a certificate of waiver to perform waived testing and is not subject to further regulatory requirements including outside inspections. In contrast healthcare organizations such as hospitals must be accredited by either CMS or other organizations with "deemed status" who perform on-site inspections on behalf of CMS such as the CAP and JC. Both the CAP and JC have quality and compliance standards that pertain to waived testing hence the waived status of a test is less important than in the community outpatient setting.

Organizations planning to perform POCT laboratory testing must apply to CMS for a CLIA certificate in one of three categories:
- Waiver: Permits the site to perform waived tests only.
- Provider Performed Microscopy (PPM): Applies to sites performing provider performed microscopy.
- Registration: Applies to sites performing moderate and/or high complexity testing (in addition to waived testing).

37

The site is then inspected and, if found to be CLIA compliant, it is issued a renewable certificate of compliance. If the site is inspected by an organization with deemed status (such as the CAP or JC) the site is issued a certificate of accreditation. Testing sites are then inspected every two years to ensure ongoing compliance. Two states are CLIA exempt in that they have state requirements that meet or exceed CLIA (New York, Washington) [4]. For non-waived tests CLIA-88 and its subsequent revisions sets regulations that must be complied with in order to perform testing including [5]:
- Personnel qualifications
- Training and competency of testing personnel
- Quality control
- Quality systems
- Patient test management
- Proficiency testing
- Test validation/performance verification

CLIA regulations combined with regular inspections by outside regulatory agencies ensure that POCT conforms to a basic level of quality standards. Testing sites that fail to meet CLIA regulations or that fail to remediate deficiencies may be subject to loss of their CLIA certificate and must then cease testing. However, experience has shown that POCT can be very difficult to manage in a manner that achieves regulatory compliance and ensures the quality of the testing. There are many reasons for this including:
- Personnel performing POCT frequently have no background in the clinical laboratory and do not understand the basics of quality control and quality management of laboratory testing.
- Personnel may be poorly trained or may perform testing so infrequently that competency is difficult to maintain.
- Some testing devices are complex to operate or are poorly designed and do not have adequate built in features to prevent operator errors.
- Many POCT tests are manually performed visually read tests that may be subjective to interpret or require manual documentation of results, quality control and other required elements.
- Sample acquisition may be highly operator dependent (e.g. fingerstick samples, respiratory swabs).

37.9.3 POCT management programs

In response to these issues most hospitals have established POCT management programs to ensure ongoing oversight of regulatory compliance and testing quality [6]. Typically these programs report to a more senior hospital executive or committee to ensure that the program has the authority to oversee POCT throughout the organization. The organizational chart for the POCT program in our hospital is shown in ◘ Fig. 37.4. Each testing site is required to appoint a site director who is responsible for regulatory compliance at the local level. The site director reports to the director of the POCT program. The vast majority of issues are resolved at this level. When issues cannot be resolved the POCT program director can refer the site to the Director of Clinical Laboratories who may report the issue to the Senior Vice President for Quality and Safety.

In most hospitals an individual or team of laboratory technologists work with a POCT medical director to manage the day to day activities of the program. In some cases the program is managed by nursing or other specialty (e.g. respiratory therapy). Large hospitals typically have a dedicated team that focuses exclusively on POCT. In smaller organizations the task may fall to an individual that spends only part of their time on POCT in addition to other duties. As POCT has expanded in the hospital setting, the skills and fund of knowledge required of the POCT coordinator has likewise increased. The American Association for Clinical Chemistry has established a Point-of-Care Specialist Certificate Program reflecting the recognition that POCT has become a true specialty within laboratory medicine. The POCT program may oversee a number of activities related to testing performed outside of the clinical laboratory including:

- Selection and evaluation of new instruments and methods
- Validation of reagents and replacement devices
- Oversight of POCT data management information systems
- Establishment and management of standard operating procedures
- Oversight and management of operator training and ongoing competency
- Oversight of regulatory compliance including performance of regular testing site inspections
- Quality assurance and quality improvement
- Oversight of the POCT budget
- Management of the organizational CLIA certificates
- Preparation for and management of inspections by outside regulatory agencies

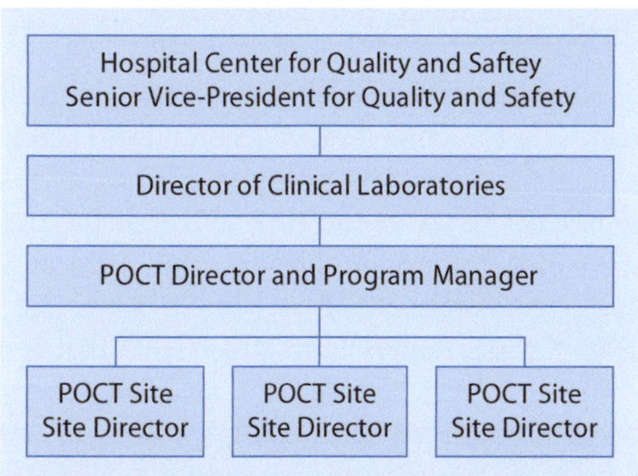

◘ **Fig. 37.4** Organizational chart for the POCT program at the Massachusetts General Hospital

One of the most important functions of the POCT coordinator is the performance of regular inspections of testing sites throughout the organization. These inspections have been shown to be essential to maintaining regulatory compliance and to ensure the quality of POCT testing [7]. Inspections may also include assessment of employee safety and of the environment of care.

37.9.4 Role of informatics and electronic data management systems

Patient testing requires documentation of a number of data elements including patient identification, operator, test result, quality control, reagent lots and expiration dates, date and time, instrument used and other data. For some tests this must be managed through manual data entry, a process that is error prone and frequently proves unreliable. In addition the test results may not be available in the patient's medical record or may not be readily available to the clinician. Wherever possible is best to convert manually read tests to instrumented readers a number of which are available (e.g. urinalysis readers, rapid strep A, influenza). Most instrumented tests have onboard data management software that will automatically capture the required data elements or may require the operator to enter in any missing data. Optimally these devices are then interfaced to a POCT data management system (several commercial systems are available) that can then transmit the test results to a laboratory or hospital information system including the electronic medical record.

In the US, most commercial POCT data management systems are vendor neutral in that they can receive and process data from devices manufactured from many different vendors. Electronic data can be monitored remotely by the POCT coordinator who can then correct any deficiencies and ensure compliance with testing quality parameters. In the most advanced case the process is entirely automated. The operator and patient are outfitted with either bar-coded or radiofrequency identification badges/wristbands, all reagents /test strips are manufactured to include automatic recognition of lots and expiration dates, quality control is mandatory or the instrument will "lock-out" and all data flows wirelessly into the POCT data management system. Only a few instruments at present are currently equipped with wireless transmission. Some devices require a fixed location to connect to an electronic interface while others must be manually brought to a docking station to download testing information.

While in principle informatics should have solved most of the quality issues with POCT, the reality does not yet reflect this potential. Several examples are listed below:

- Operators discover work-arounds to bypass recording of testing data such as an untrained operator using another operators ID.
- Operators may enter nonsensical data such as 123456 for an operator ID.
- Specimen collection may be operator dependant resulting in poor quality samples.
- Some devices do not have mechanisms to detect expired reagents. Others require refrigeration of reagents up to the time of use but the devices cannot detect reagents cartridges that have been stored at room temperature for an extended period of time.
- In one case it was observed in an operating room that a patient wristband was placed on an I.V. pole to make it easily accessible to testing personnel. Between cases the wristband was not removed before the start of the next case and was used for patient identification of the second patient resulting in test results being entered into the wrong patient's medical record.

For the above reasons ongoing oversight by the point-of-care coordinator is essential to maintain quality testing. POCT coordinators must also stay abreast of new developments in informatics/data management and incorporate these into their POCT programs to improve the performance of POCT in their organizations.

37.9.5 Improving instrument/testing performance

A number of POCT devices do not perform testing at the same level of accuracy and precision as the central laboratory. In some cases this is not necessary as the clinical use of the test does not require highly accurate laboratory-equivalent results. For example, some bedside glucose meters may give results that are different by ±15 % from the central laboratory. Medically, the results from these devices are acceptable for routine monitoring of diabetic patients. However, it has recently become apparent that some of these same devices have not been adequately evaluated for use in critically ill patients who may be hypotensive and receiving vasopressors. In some cases the results from POCT devices are unacceptable as the inaccuracy of the tests may result in inappropriate clinical decisions.

A key function of the POCT program director is to oversee the performance validation of all POCT technologies to ensure that they will perform at an acceptable level for the specific clinical indication in which the device will be used. Manufacturers of POCT tests are aware of the issues with their testing devices and have devoted significant research to improve the accuracy and reliability of their tests. They have also devoted resources to better engineer the devices to eliminate as many of the potential sources of error as possible. Many POCT directors/coordinators work collaboratively with their colleagues in industry to improve the performance of POCT testing. End user input is an important component in the development of new POCT technologies and in the surveillance of tests that are already in use in the field. These activities may include:

- Reporting of errors, sources of interference, biases and other analytical problems
- Reporting of software issues and problems with results transmission.
- Reporting of instrument and reagent reliability issues
- Field validation of instruments and tests
- Serving on advisory boards and focus groups to provide input on devices and tests under development and to elaborate on unmet needs in POCT

37.9.6 Conclusion

Quality management in POCT in the United States had its origins in CLIA-88 and has been championed by accreditation agencies such as the CAP and the JC. These regulations were intended to ensure a basic level of quality control/quality assurance for POCT. The majority of hospitals have established POCT quality management programs which are usually overseen by a clinical laboratory director working with POCT coordinators. The POCT coordinators perform a variety of functions all designed to ensure the appropriate selection of devices and testing systems, regulatory compliance and quality management. Informatics is progressively becoming a critical tool to provide oversight of POCT and to ensure that reliable data resulting from testing is directed to the electronic medical record. Although the analytical quality of many POCT tests has improved over time, there still remain significant issues. The POCT program must evaluate new and existing tests and devices to ensure that they meet acceptable quality standards for their intended clinical applications. Finally it is important for laboratory professionals to work closely with manufacturers to promote improvement in the performance of POCT testing.

References

To chapter 37.1

1. ISO 22870:2006 "Point-of-care testing (POCT) – requirements for quality and competence"
2. ISO 15189:2012 "Medical laboratories – requirements for quality and competence"

To chapter 37.2

1. BAG BfG. Faktenblatt: Kapitel Schnelle Analysen, 04.11.2014. [cited 03.05.2016]; Available from: http://www.bag.admin.ch/themen/krankenversicherung/00263/00264/04185/index.html?lang=de&download=NHzLpZeg7t,lnp6I0NTU042l2Z6ln1acy4Zn4Z2qZpnO2Yuq2Z6gpJCMdIR,g2ym162epYbg2c_JjKbNoKSn6A
2. Neues zum Fähigkeitsausweis Praxislabor KHM (FAPL): E-Learning ersetzt einen von drei Kurstagen. Schweizerische Ärztezeitung 2012 [cited 93:8 03.05.2016]; 275–7]. Available from: http://www.fmh.ch/files/pdf14/art_fa_praxislabor_d1.pdf
3. QUALAB. Grundvertrag zur Qualitätssicherung im medizinischen Labor QUALAB, Version 2.1 (1999). [cited 03.05.2016]; Available from: http://www.qualab.ch/download.php?file_id=136&download=true
4. QUALAB. Konzept für Qualitätssicherung im Medizinischen Labor V1.1 (1999). [cited 03.05.2016]; Available from: http://www.qualab.ch/download.php?file_id=263&download=true
5. SULM. Kriterien zum Betreiben von Medizinisch-Analytischen Laboratorien (KBMAL), Version 1.4

(1994). [cited 03.05.2016]; Available from: http://www.qualab.ch/download.php?file_id=137&download=true

6. QUALAB. Checkliste KBMAL Version 1.0 (2009) [cited 03.05.2016]; Available from: http://www.qualab.ch/download.php?file_id=139&download=true
7. AG IBS. Interregionale Blutspende SRK AG [cited 03.05.2016]; Available from: http://www.iblutspende.ch/labordiagnostik/ringversuchszentrum.html
8. [8] Mykobakterien NZf. Nationales Zentrum für Mykobakterien [cited 03.05.2016]; Available from: www.imm.uzh.ch/QualDe.html
9. Qualitätskontrolle SZf. Schweizerisches Zentrum für Qualitätskontrolle [cited 03.05.2016]; Available from: www.cscq.ch
10. Qualitätskontrolle VfM. Verein für Medizinische Qualitätskontrolle. [cited 03.05.2016]; Available from: www.mqzh.ch
11. QUALAB. Externe obligatorische Qualitätskontrolle 2016 [cited 03.05.2016]; Available from: http://www.qualab.ch/download.php?file_id=746&download=true
12. QUALAB. Richtlinie zur internen Qualitätskontrolle, V 3.0 (2015). [cited 03.05.2016]; Available from: http://www.qualab.ch/download.php?file_id=736&download=true
13. Interpharma. Gesundheitsmonitor 2015. 2015 [cited 03.05.2016]; Available from: http://www.interpharma.ch/sites/default/files/gfs.bern_d_wik_gesundheitsmonitor_2015_1.pdf
14. Economiesuisse. „Spital der Zukunft", Studie und Umsetzungsempfehlungen (2015). [cited 03.05.2016]; Available from: http://www.economiesuisse.ch/de/artikel/spital-der-zukunft-h %C3 %B6here-patientensicherheit-zu-deutlich-tieferen-kosten
15. SMI SMI. Patientenidentifikation 2.0. 2015 [cited 31 03.05.2016]; Available from: http://www.medical-informatics.ch/index.php/smiojs/article/view/329
16. BAG BfG. Analysenliste (AL), Gesamtliste, Stand 1.1.2016. [cited 03.05.2016]; 226–9]. Available from: http://www.bag.admin.ch/themen/krankenversicherung/00263/00264/04185/index.html?lang=de
17. CSCQ. Vision Nr.10, September 2015. [cited 03.05.2016]; Available from: http://www.cscq.ch/SiteCSCQ/FichierPDF_DE/vision-10-de.pdf

37

To chapter 37.3

1. Banks G, Bedini JL, Buño A, Cava F, Castaño JL, Díaz R, Guevara P, Hortas ML (coordinadora), Lirón J, Marín JL (coordinador), Muñoz M, Monzó V, Navarro P, Pelegrí D, Portillo JC; Salve ML (2006) Guía para la implantación de pruebas de laboratorio en el lugar de asistencia al paciente. Monografía. Sociedad Española de Bioquímica Clínica y Patología Molecular
2. Buño A; Oliver P; Marín JL (2009) Proposed guidelines for Point-of-Care testing in Spain. Point of Care 8:53–55
3. Cantero M, Redondo M, Martin E, Callejón G; Hortas ML (2015) Use of quality indicators to compare point-of-care testing errors in a neonatal unit and errors in a STAT central laboratory. Clin Chem Lab Med 53 (2):239–247
4. Cantero M, Hortas ML;, Ruiz JA;, Ruiz C;, Redondo M (2015) Protocolo de actuación ante valores críticos en pruebas de laboratorio en el lugar de asistencia al paciente en una unidad neonatal. Revista del Laboratorio Clínico 8(4):165–172
5. Gonzalez C (2014) Estudio de los costes derivados de una intervención POCT intraoperatoria en cirugía cardiaca extracorpórea para disminuir las complicaciones intrahospitalarias. Gest y Eval Cost Sanit (4):477–85
6. Hortas ML, Montiel N, Redondo M, Medina A, Contreras E, González C (2001) Quality Assurance of Point of Care Testing in the Costa del Sol Healthcare Area. Clínica Chimica Acta 307(1–2): 113–118
7. Hortas ML, Redondo M, Medina A, Contreras E, González C (2003) Clinical outcomes in critical care management through implementing point-of-care in Primary Care. Critical Care and Point of Care Testing; 19ª Simposium Internacional. Monterrey
8. Hortas ML, Suarez G, González C (1998) Calidad en los laboratorios satélites. II Reunión Nacional de la Sociedad Española de Dirección y Gestión de los Laboratorios Clínicos SEDIGLAC
9. International Organization for Standardization ISO 17593 (2007) Clinical laboratory testing and in vitro medical devices - Requirements for in vitro monitoring systems for self-testing of oral anticoagulant therapy
10. International Organization for Standardization ISO 15197 (2013) In vitro diagnostic test systems – Requirement for blood-glucose monitoring systems for self-testing in managing diabetes mellitus
11. International Organization for Standardization ISO 15189 (2013) Medical laboratories: Particular requirements for quality and competence.
12. International Organization for Standardization ISO 22870 (2006) Point-of-Care Testing (POCT): Requirements for quality and competence
13. O'Kelly RA, Brady JJ, Byrne E, Hooley K, Mulligan C, Mulready K, et al. (2011) A survey of point of care testing in Irish hospitals: room for improvement. Ir J Med Sci 180(1):237–40
14. Oliver P, Alonso R, Liron J, Monzo V, Navarro X, Noval JA, et al. (2016) Point-of-Care Testing Guideline. Rev Lab Clin (in press)
15. Oliver P, Buno A, Alvarez-Sala R, Fernandez-Calle P, Alcaide MJ, Casitas R, et al. (2015) Clinical, opera-

tional and economic outcomes of point-of-care blood gas analysis in COPD patients. Clin Biochem 48:412–8
16. Oliver P, Fernandez-Calle P, Rico N, Alcaide MJ, Gomez-Rioja R, Buno A, Iturzaeta JM (2013) Analytical performance evaluation and comparability of patient results within a point-of-care blood gas network. Point of Care 12:144–9
17. Pascual C, de la Torre F, Pelegrí MD, Garnacho I; Ricós C, Gancedo V, Bardají G, Pla R (1998) Pruebas a la cabecera del enfermo. La experiencia de "Hospitals Vall de Hebrón". II Reunión Nacional de la Sociedad Española de Dirección y Gestión de los Laboratorios Clínicos SEDIGLAC
18. Pecoraro V, Germagnoli L, Banfi G (2013) Point-of-care testing: where is the evidence? A systematic survey. Clin Chem Lab Med. DOI 10.1515/cclm-2013-0386
19. Sciacovelli L, O'Kane M, Skaik YA, Caciagli P, Pellegrini C, Da Rin G, et al. (2011) Quality Indicators in Laboratory Medicine: from theory to practice. Preliminary data from the IFCC Working Group Project "Laboratory Errors and Patient Safety". Clin Chem Lab Med 49(5):835–44
20. http://www.seqc.es/es/Publicaciones/2/67/Comision_de_Pruebas_de_laboratorio_en_el_lugar_de_asistencia

To chapter 37.4

1. Ordonnance n 2010-49 du 13 janvier 2010 relative à la biologie médicale. JORF n 12
2. Norme NF EN ISO 15189 (2007) Laboratoires d'analyses de biologie médicale Exigences particulières concernant la qualité et la compétence. AFNOR, Saint-Denis
3. Norme NF EN ISO 22870 (2006) Analyses de biologie délocalisées (ADBD) Exigences concernant la qualité et la compétence. AFNOR, Saint-Denis
4. Vassault A et al. (1999) Ann Biol Clin 57:685–95
5. Database compiled and updated by C. Ricos at: westgard.om/biodatabase1.htm

To chapter 37.5

1. Department of Health (2006) Report of the second phase of the review of NHS pathology services in England: http://webarchive.nationalarchives.gov.uk/20130107105354/http:/www.dh.gov.uk/prod_consum_dh/groups/dh_digitalassets/@dh/@en/documents/digitalasset/dh_091984.pdf
2. MHRA (2013) Management and use of IVD point of care test devices:
3. https://www.gov.uk/government/uploads/system/uploads/attachment_data/file/371800/In_vitro_diagnostic_point-of-care_test_devices.pdf
4. NHS London (2007) Healthcare for London: A framework for action. NHS London
5. NICE (2014) Diagnostic guidance [DG14] Atrial fibrillation and heart valve disease: self-monitoring coagulation status using point-of-care coagulometers (the CoaguChek XS system and the INRatio2PT/INR monitor): https://www.nice.org.uk/guidance/dg14/chapter/10-diagnostics-advisory-committee-members-and-nice-project-teamhttps://www.nice.org.uk/guidance/dg14/chapter/10-diagnostics-advisory-committee-members-and-nice-project-team
6. Thomas, A (2003) WEQAS; The participant's manual

To chapter 37.6

1. The Japan Society for Clinical Laboratory Automation. POCT-Richtlinie 3. Fassung, The Japan Society for Clinical Laboratory Automation Journal 2013:38 (suppl.1) 25–35
2. ISO22870: 2006. Point-of-care testing (POCT) – Requirements for quality and competence
3. ISO15189: 2012. Medical laboratories – Requirements for quality and competence
4. CLIS: 2006. POCT1-A2 Point-of- Care Connectivity; Approved Standard- Second Edition

To chapter. 37.8

1. https://www.nhg.org/sites/default/files/content/nhg_org/uploads/final_richtlijnpoct_2015lmlres.pdf: last checked 30 May 2016.
2. www.consilium.europa.eu/en/press/press.../25-medical-devices/": last checked 31 May 2016

To chapter 37.9

1. Lee-Lewandrowski E, Gregory K, Lewandrowski K (2010) Point-of-care testing in a large urban academic medical center: evolving test menu and clinical applications. Clin Chem Acta 411:1799–1805
2. Lewandrowski K, Gregory K, Macmillan D (2011) Assuring quality in point-of-care testing. Arch Pathol Lab Med 135:14051414
3. Macmillan D, Lewandrowski K (2011) History of the clinical laboratories chapter 20). In: Louis D, Young R (eds). Keen minds to explore the dark continents of disease: a history of the pathology services at the Massachusetts General Hospital. Massachusetts General Hospital, Boston, MA
4. Ehrmeyer S, Laessig R (2009) Regulatory compliance for point-of-care testing: 2009 United States Perspective. In: Lewandrowski K (ed) Point-of-care testing. Clinics in Laboratory Medicine 29:463–478
5. http://wwwn.cdc.gov/clia/regulatory/default.aspx
6. Gregory K, Lewandrowski K (2009) Management of a point-of-care testing program. In Lewandrowski K (ed). Point-of-care testing. Clinics in Laboratory Medicine 29:463–478

7. Gregory K, Tse J, Wu R, Lewandrowski K (2012) Implementation of an expanded point-of-care testing (POCT) site inspection checklist in a large academis medical center: implications for the management of a POCT program. Clin Chem Acta 414:27–33

Quality assurance in Germany: Guideline of the German Medical Association on Quality Assurance in Medical Laboratory Examinations (RiliBÄK)

Oswald Sonntag, Claus Langer, Harald Schlebusch

P. B. Luppa, R. Junker (Eds.), *Point-of-Care Testing*
https://doi.org/10.1007/978-3-662-54497-6_38

38.1 Introduction

Since August 1, 2009, Section 4a of the Medical Devices Operator Ordinance (MPBetreibV) on the installation, operation and use of medical devices has regulated quality assurance in medical laboratory testing in Germany [9]. Since April 1, 2010, RiliBÄK has been binding for all facilities where medical tests are performed. Updated versions of RiliBÄK were published in 2014 [2, 3], defining further requirements (see below).

Quality assurance of POCT follows the principle that laboratory findings at the POC and in the central laboratory should be collected with the same reliability. This means that the established quality standards governing conventional medical laboratory testing apply to POCT as well. Therefore, the current RiliBÄK issued on 19 September 2014 – with one exception – no longer has separate rules for POCT.

The guideline is divided into:

- Part A: Basic requirements for quality assurance in medical laboratory examinations
- Part B: Special parts
  - B1: Quantitative medical laboratory examinations
  - B2: Qualitative medical laboratory examinations
  - B3: Direct detection and characterization of infectious agents
  - B4: Examination of ejaculate
  - B5: Molecular-genetic and cytogenetic medical laboratory examinations
- Part C: Advisory Board
- Part D: Expert Groups
- Part E: General requirements for reference institutions conducting external quality assurance programs
- Part F: Temporary regulations
- Part G: Entry into force

Part A of the guideline applies to all areas of clinical laboratory diagnostics, i.e. to conventional hospital-based central laboratories, laboratory groups, doctor's offices with small laboratories as well as to POCT diagnostics. This was the first time in Germany that the implementation of a comprehensive quality management system became mandatory. The main content was derived from standard DIN EN ISO 15189 [4, 12]. However, the contents of RiliBÄK and DIN EN ISO 15189 are not congruent and are therefore mutually exclusive of each other.

> **This chapter presents the requirements for quality assurance in POCT in concise form. For detailed and further information, please refer to the currently amended version of RiliBÄK.**

38.2 Introduction of a quality management system (RiliBÄK A)

Sections 4–7 of RiliBÄK Part A detail all measures to be complied with when introducing a quality management system (QMS). Whereas, a number of provisions are only relevant to central or joint laboratories, each individual case where POC diagnostics are applied must be examined separately to determine if the circumstances are actually governed by these QM regulations.

Rules for the following items must be put in place:

- Definition of responsibilities for overall QMS at the POC
- Organization of POCT
- Rules for the pre-analytical phase (RiliBÄK 6.1)
- Rules for conducting medical laboratory examinations (RiliBÄK 6.2)
- Rules for the post-analytical phase (RiliBÄK 6.3)
- Staff training
- Tasks and responsibilities for conducting quality controls

Many practical procedural proposals have been published [1, 6, 7, 11]. All documents should be compiled in a quality management (QM) manual required for each organizational unit. The contents of the Quality Manual are stipulated in RiliBÄK 7.1. However, if the hospital has a global QMS in place, the organizational unit can refrain from stipulating in its own quality manual those parts already regulated on the institutional level. An appropriate reference thereto suffices. As appropriate, reference can also be made to documents in the central laboratory's quality manual. In general, duplications across various manuals should be avoided. The on-site manual must contain the documents required for the routine work. Generally, the minimum requirement includes the operating instructions for all devices with maintenance and repair plans, comprehensible standard operating procedures for the pre-analytical, analytical and post-analytical tasks, an organizational plan for quality assurance and a list of contact persons who will deal with the various questions and deviations. The manual must always be kept up-to-date. At smaller organizational units, it appears useful to closely collaborate with the central laboratory to stipulate contents and scope of the manual unless the organization of POCT has not already been transferred to the central laboratory/contract laboratory.

38.3 Conducting quality assurance of quantitative examinations (RiliBÄK B1)

All quantitative clinical laboratory diagnostic tests are subject to internal quality assurance and – if listed in Table B1 of RiliBÄK – to external QA as well. As indicated, there are no special rules for POCT methods in terms of conducting quality control with an exception. This also applies to the assessment of a single measurement of a control sample. The permissible deviation of the measurand in internal and external quality controls is identical for serum/plasma and blood, irrespective of whether the values were obtained as POCT or generated by the central laboratory (Tab. 38.1).

38

38.3.1 Internal quality control

- A single measurement of a control sample must be carried out in the case of clinical laboratory methods like those used for POCT, e.g. on blood gas analyzers. The time interval is at least twice within a 24-hour period and, at the latest, after 16 hours.
- A single measurement of a control sample additionally has to be performed after every intervention on the measuring system, i.e. after:
  - Restarting the device after a complete switch-off
  - Calibration by the user
  - Repairs or maintenance work
  - Reagent batch changes
- Control samples with target levels ranging across different medically relevant concentrations (low, around the cut-off and high) should be alternated.
- The results must be evaluated according to RiliBÄK Table B1, column 3.
- If the prescribed limit is exceeded, the responsible physician may nevertheless release the method for further measurements as long as the reasons are documented and the medical relevance considered.
- At the end of the control cycle (which generally lasts one month), the results of the single measurement of a control sample must be immediately used to calculate the relative root mean square of the deviation of measurement [8] according to the following formula:

$$\Delta = \sqrt{\frac{1}{n}\sum_{i=n}^{n}(x_i - x_0)^2}$$

where Δ is the root mean square of the deviation of measurements, x_0 is the target value of the control sample, x_i the value of the single measurement and n the number of single results.

Tab. 38.1 Deviations permissible in control sample results when controlling the quality (QC) of blood gas analyses and POCT according to RiliBÄK 2014 (excerpt)

Analyte	Internal QC [%] (column 3)	External QC [%] (column 5)	Validity range of columns 3 and 5
Activated partial thromboplastin time (aPTT)	10.5	18.0	20–120 s
Chloride	4.5	8.0	70–150 mmol/L
Potassium	4.5	8.0	2–8 mmol/L
Calcium, ionized	7.5	15.0	>1–2.5 mmol/L
	14.0	18.0	0.2– ≤1 mmol/L
Erythrocytes	4.0	8.0	1.5–7×10[12]/L
Glucose	11.0	15.0	40–400 mg/dL 2.2–22 mmol/L
Hematocrit	5.0	9.0	10–60 % 0.1–0.6 L/L
Hemoglobin	4.0	6.0	2–20 g/dL 1.2–12.4 mmol/L
Urea	10.5	20.0	15–200 mg/dL 2.5–33 mmol/L
HbA_{1c}	10.0	18.0	30–140 mmol/mol Hb
Lactate	11.0	18.0	9–90 mg/dL 1–10 mmol/L
Leukocytes	6.5	18.0	2–30×10[9]/L
Sodium	3.0	5.0	110–180 mmol/L
pCO_2	6.5	12.0	>35 mm Hg
	7.5	12.0	≤35 mm Hg
pH	0.4	0.8	6.75–7.80
Platelets	7.5	13.0	>300–700×10[9]/L
	8.5	15.0	>150–≤300×10[9]/L
	13.5	18.0	40–≤150×10[9]/L
Thromboplastin time (Quick's test)	11.5	23.0	10–120 %

The relative root mean square of the deviation of measurements is calculated by dividing Δ by the target value x_0.

- The evaluation is carried out as described in RiliBÄK Table B1, column 3.
- If the prescribed limit is exceeded, the method must be blocked to further measurements until the device's functionality is secured by suitable means. Such a case can only occur if several single measurements of a control sample had already shown limit excursions during a control cycle.

38.3.2 Documentation

All results of internal quality assurance have to be documented. These should be ordered by analyte and specimen type, taking into consideration the measuring method and the respective device. The documentation on the performed quality assurance must be archived for a minimum of 5 years, unless longer archiving periods are stipulated by other regulations.

Documentation must include the following:

- Name of the medical laboratory (and/or organizational unit)
- Name of the measuring device
- Date and time of the measurement
- Analyte, specimen material and unit
- Measuring method
- Manufacturer, name and lot number of the control sample
- Target value of the control sample
- Measured value of the control sample
- Relative or absolute deviation of the measured value of the control sample from the target value
- Evaluated according to Table B1a to c, column 3 or the other Tables B2, B3 and B5 provided that these analyses were performed at the POC
- Release or lock flag, depending on whether there was any excursion from the deviation limits
- Corrective measures taken
- Investigator's name, initials or signature
- Additionally the control sample measurements should be plotted on a graph

38.3.3 External quality control (interlaboratory testing)

38

- Every organizational unit must take part in 4 interlaboratory tests per year (once a quarter) concerning the parameters analyzed.
- An "organizational unit" is defined as a separate area (functional unit) of a hospital that meets the following criteria:
  - Staffed by a defined group of users (physicians, nurses)
  - A pool of measuring stations/devices is assigned solely to it
  - Only a defined circle of users operate the devices. Examples: Central laboratory, surgical area, intensive care units, delivery rooms, lung function lab
- The interlaboratory testing for each organizational unit is compulsory unless multiple organizational units – together with the central laboratory – are combined into a single organizational unit that conducts the clinical laboratory analyses. This must take place by written instructions on the part of the hospital management. In this case, it suffices for the central laboratory to take part. It is not mandatory that separate interlaboratory testing be performed for the POCT methods provided that the central laboratory also runs the parameters itself. Nonetheless, in practice, it is recommendable to also run quality controls on the POCT methods with an interlaboratory test offered for this purpose. This helps to obtain a direct impression about the quality of the respective POCT method.

Note

If the interlaboratory test produces a not-pass result, the causes must be investigated and eliminated as far as possible. If the causes cannot be found and eliminated, it is not permitted to continue to use the POCT system. The overall procedure must be documented.

In Germany, interlaboratory tests are offered by the Reference Institute for Bioanalytics (RfB) of the German Society of Clinical Chemistry and Laboratory Medicine (DGKL) located in Bonn and by INSTAND e.V., an interdisciplinary, not-for-profit, scientific medical society for quality assurance in laboratory medicine situated in Düsseldorf. These institutions are currently under commission by the German Medical Association as ring trial organizations.

38.4 Special rules for POCT with unit-use reagents

RiliBÄK prescribes a more simple procedure for unit-use reagents. Unit-use reagents are reagents that have been portioned for single determinations and are consumed within one single test.

38.4.1 Internal quality control

Quality controls of the unit-use reagents and their respective measuring devices must be carried out according to the manufacturer's instructions. If there are derogations from RiliBÄK 2014, the stricter criteria apply. Some POCT devices use an electronic/physical standard on a daily operational basis. This feature or an integrated function test can prevent false results from being reported. On such analyzers, a single measurement of a control sample must be carried out at least once a week. Additional controls are required after:

- Calibration by the user
- Repairs or maintenance work
- Reagent batch changes

Whenever patient samples are tested on devices not equipped with an electronic physical standard or an integrated function test, such analyzers – like all other devices – must be quality controlled by a single measurement of a control sample at least twice every 24 h.

It is not necessary to calculate the square mean of the deviation or plot the measured values in a graph. Documentation as stated above, however, is required. If the prescribed limit is exceeded, the method must be blocked to further measurements until the device's functionality is secured by suitable means.

38.4.2 External quality control

The obligation to participate in ring trials does not apply to

- Private medical practices or medical services without a central laboratory
- Hospitals where the central laboratory is responsible for internal quality assurance and also determines the measurand itself, i.e., also takes part in interlaboratory testing of the measurand in question. Here, the determination method in the central laboratory and at the POC does not have to be identical.

If the interlaboratory test produces a not-pass result, the causes must be investigated and eliminated as far as possible. If the causes cannot be found and eliminated, it is not permitted to continue to use the POCT system. The overall procedure must be documented.

A "central laboratory" is defined as the one single organizational unit that generally conducts all clinical laboratory diagnostics for a hospital. This can also be an external laboratory that is operated by a third party. The central laboratory is responsible for monitoring internal quality control at the individual organizational units of the hospital. This does not mean that the control sample measurements and their evaluation are conducted by central laboratory staff. The laboratory is authorized to block POCT systems if they do not meet RiliBÄK criteria.

38.5 Quality assurance of qualitative examinations (RiliBÄK B2)

In principle, qualitative clinical laboratory tests are also subject to internal and external quality controls. The manufacturer's recommendations must be followed for the internal controls.

RiliBÄK Table B2-1 lists 50 tests that are mandatory for internal quality assurance; Table B2-2 lists 38 tests required in order to participate in interlaboratory testing. However, these tests are not carried out at the POC. Urine sediment is the exception: for this test, once-yearly participation in an interlaboratory test is mandatory, but not an internal quality control.

38.6 Quality assurance for characterization of infectious pathogens after their direct detection (RiliBÄK B3)

In principle, direct detection and characterization of infectious pathogens are procedures subject to both internal and external quality controls. When conducting internal controls, the manufacturers' recommendations must be followed. Table B3-1 lists by group a large number of various test methods that are subject to mandatory internal quality assurance; Table B3-2 and B3-2a list the tests that require participation in interlaboratory testing. Not all methods in this diverse range are available as full POC tests yet. However, all tests that can be conducted with POCT devices are also subject to mandatory ring trials.

38.7 Quality assurance for molecular-genetic and cytogenetic medical laboratory examinations (RiliBÄK B5)

Meanwhile, molecular-genetic methods are falling into the sphere of POCT. These methods are likewise subject to internal and external quality controls in principle. The manufacturer's recommendations must be followed for the internal controls. According to the requirements in RiliBÄK Tables B5-1 and B5-2a, the frequencies of tests and quantities listed therein must be observed. The quality assurance of the tests not listed in those tables must be carried out as deemed medically necessary. Table B5-1 lists a number of methods that are subject to internal quality assurance (e.g. weekly) and where participation in interlaboratory testing (e.g. semi- or annually) is mandatory. So far, cytogenetic test methods have not become established as POCT.

38.8 Administrative offences

Legally, an improperly conducted quality assurance of medical laboratory tests constitutes an administrative offence and is punished accordingly.

38.9 Pertinent legislation

- German Medical Device Law (MPG) [10]
- EU Directive on in-vitro diagnostic medical devices [5]
- German Medical Devices Operator Ordinance (MPBetreibV) [9]
- Guideline of the German Medical Association on Quality Assurance in Medical Laboratory Examinations (RiliBÄK 2014) [2, 3]

38.10 Remarks concerning on-board controls

Several companies have developed on-board controls to simplify the handling of POCT quality assurance. These are inherent to the devices and measure automatically according to the manufacturer's specifications. The aim is to ensure flawless functionality of the devices. However, the overall method, which should also include specimen handling, is not examined in this procedure. Future trends will certainly continue to improve on this aspect of on-board controls. To date, RiliBÄK has not envisaged an exception for this procedure.

References

1. Briedigkeit L, Müller-Plathe O, Schlebusch H, Ziems J (1998) Empfehlungen der Arbeitsgemeinschaft Medizinische Laboratoriumsdiagnostik (AML) zur Einführung und Qualitätssicherung von Verfahren zur patientennahen Laboratoriumsdiagnostik (POCT). J Lab Med 22:414–430
2. Bundesärztekammer (German Medical Association) (2014) Guideline of the German Medical Association for the Quality Assurance of Clinical Laboratory

Testing (Rili-BÄK 2014). Dtsch Ärztebl 111:A1583–1618
3. Bundesärztekammer (German Medical Association) (2015) Revision of the "Guideline of the German Medical Association on Quality Assurance in Medical Laboratory Examinations – Rili-BAEK" (unauthorized translation) J Lab Med 39: 26–69
4. DIN EN ISO 15189:2014-11: Medical laboratories – Requirements for quality and competence, Beuth-Verlag
5. Directive 98/78/EC of the European Parliament and of the Council of 27 October 1998 on in vitro diagnostic medical devices (Consolidated version 7 August 2009)
6. Gässler N, Luppa PB, Hafner G, Koschinsky T, Schlebusch H, Spannagl M, Wahl HG, Ziervogel H. (2006) Information der Arbeitsgemeinschaft 'Point-of-Care-Testing' der DGKL zum Thema Einführung von POCT innerhalb einer Klinik. Klinische Chemie Mitteilungen 6:161–163
7. Gurr E, Arzideh F, Brandhorst G, Gröning A, Haeckel R, Hoff T, Roggenbruck L, Schumann G, Wolters B, Wosniok W (2011) Musterstandardarbeitsanweisung Präanalytik – Arbeitsgruppe Richtwerte der DGKL. J Lab Med 35:55–60
8. Macdonald R (2006) Quality assessment of quantitative analytical results in laboratory medicine by root mean square of measurement deviation. J Lab Med 30:111–117
9. Medical Devices Operator Ordinance in the version promulgated on 21 August 2002 (Federal Law Gazette [BGBl.] I p. 3396), amended by Article 1 and 2 of the Ordinance dated 27 September 2016 (BGBl. IS 2203)
10. German Medical Device Law in the version promulgated on 7 August 2002 (BGBl. I S. 3146), amended by Article 4 (59) of the law dated 18 July 2016 (BGBL. I S 1666)
11. Schimke I, Griesmacher A, Schimke E, Müller MM (2006) Bedside testing (Point of Care Testing; POCT) in the hospital – the pros and cons. Intensivmed 43:143–155
12. Spitzenberger F, Hafner G, Soltau U (2007) Quality management in medical laboratories – comparison between the new Guideline of the Federal Medical Association and accreditation according to ISO standards. J Lab Med 31:218–225

Quality management systems for POCT: International standardization and accreditation

Folker Spitzenberger, Claus Langer

P. B. Luppa, R. Junker (Eds.), *Point-of-Care Testing*
https://doi.org/10.1007/978-3-662-54497-6_39

39.1 International standards for in-vitro diagnostics and POCT

39.1.1 European harmonized standards

Due to the associated presumption of conformity, the application of harmonized standards is of central importance to medical devices, including in-vitro diagnostics (IVD). The safety concept set forth in the IVD directive stipulates that manufacturers, conformity assessment bodies, operators and users of IVD must set up quality assurance systems to ensure the quality, safety and performance of IVD (▶ Chapter 25).

Usually, diagnostic tests are carried out in the controlled environment of a medical laboratory. The quality management (QM) systems in this sector have been implemented in Germany and other Western and Central European countries since the mid-1990s. These apply firstly to the accreditation and certification procedures based on international standards [1, 23] and secondly to total quality management models, like the Excellence Model of the European Foundation of Quality Management (EFQM) [1, 26].

The most important procedures for the external quality control of medical laboratories have become accreditation according to DIN ISO/IEC 17025 [5] and since 2003 according to DIN ISO 15189 [3]. Diagnostic laboratories additionally commissioned as testing laboratories for clinical studies on medical devices and performance evaluation studies of IVD are moreover subject to national requirements under the Act on Medical Devices (MPG) [11, 12]. The Central Authority of the Länder (= German federal states) for Health Protection with regard to Medicinal Products and Medical Devices (ZLG) in Bonn is the competent authority for this field in Germany and which accredits laboratories that demonstrate their legal compliance within recognized procedures [6, 7, 8, 11, 12, 13]. Although this regulatory approval is generally built on an accreditation, the accreditation of medical laboratories has not basically been mandatory to date.

39

The increasing spread of POCT with all its advantages like the rapid availability of test results for time-critical therapy management runs parallel with the aim to control and/or to minimize the risks and disadvantages associated with POCT in terms of patient safety and thereby guarantee the appropriate and effective use of POCT diagnostics. These risks and disadvantages primarily pertain to the improper operation of POCT devices by non-pre-trained medical technical personnel, a lack of understanding of quality assurance on the part of the operational staff, inadequate documentation of results, increased costs incurred by the uncoordinated use of various POCT devices and their varying results as compared to results generated by the central laboratory [18, 20].

The QM systems for the POCT sector are therefore primarily implemented for the guided, controlled purchasing and installation of devices, provision of adequate consumables and test reagents, securing the qualification and competence of the operating personnel, documentation of all analytical, pre- and post-analytical processes as well as the technically correct performance and evaluation of quality control.

EN ISO 22870 was the first international standard on quality management published in the area of POCT. The standard was written by the Technical Committee ISO/TC 212 "Clinical laboratory testing and in vitro diagnostic systems" in collaboration with the Technical Committee CEN/TC 140 "In vitro diagnostic medical devices" [15]; the German version was published in June 2006 as DIN EN ISO 22870 titled "Point-of-Care Testing (POCT) – Requirements for quality and competence" [4]. It was designed to be applied in combination with standard DIN EN ISO 15189. The international and/or European Standardization Committee has already resolved and extensively implemented an editorial revision of ISO 22870. The aim was to coordinate the ISO 22870 – at least structurally and editorially – to meet the current requirements of ISO 15189:2012. The German version is DIN EN ISO 15189:2014. The modified version of DIN EN ISO 22870 was published in 2017.

39.1.2 German accreditation system in the context of regulations under European law

With the founding of the German accreditation body, Deutsche Akkreditierungsstelle GmbH (DAkkS), an authority entrusted by the federal government and operating (albeit privately) under administrative law, the German accreditation landscape lost its erstwhile heterogeneity. Indeed, the DAkkS has been the sole national accreditation body in Germany since January 1, 2010. As in other sensitive areas of healthcare and consumer protection, the law with respect to the technical auditing of conformity assessment bodies and accreditation decisions in the medical device sector envisages close collaboration between the DAkkS and the two authorities issuing the permit: the Central Authority of the Länder for Safety Engineering (ZLS) and Central Authority of the Länder for Health Protection with regard to Medicinal Products and Medical Devices (ZLG) [10]. This permit covers the statutory recognition (for laboratories) and/or notification (for certification bodies) and is generally accompanied by a competence-confirming accreditation. As stipulated by Regulation (EC) No. 765/2008, the DAkkS must submit to a regular peer evaluation by the European Cooperation for Accreditation (EA) within the scope of an intra-European comparison.

39.2 Accreditation versus certification in an intra-European comparison

In the 1990s, the increasing importance of QM systems in medical laboratories was accompanied by a controversial discussion about the application of standards in the laboratory diagnostic sector and the selection of the procedure for demonstrating competence of a laboratory. Since there was a lack of adequate standards that accounted for the special demands placed on medical laboratories, most European countries introducing QM systems preferred to adapt EN 45001 [9] and/or EN ISO 9001 [14] standards to the clinical laboratory setting followed by subsequent accreditation and/or certification based on these standards.

The United Kingdom has an accreditation system for medical laboratories in place as early as 1994, whereas other European countries did not start with accreditations based on EN 45001 until the second half of the 1990s. However, there were many instances where no comprehensive accreditation criteria had been established for the special field of laboratory testing. For example, in Germany starting in 1996, private accreditation bodies were issuing accreditations to medical laboratories although there were neither accreditation criteria specific to the field nor had any qualifications been defined relating to expert reviewer competence. Up until DIN EN ISO 15189 was introduced, medical laboratories were sometimes treated like chemical laboratories – a situation that did not do justice to the specific features of medical laboratories.

The creation and publication of DIN EN ISO 15189 led to accreditation asserting itself over other external quality testing procedures for medical laboratories. The revised DIN EN ISO 15189 has been available since 2012 (in German: since 2014). In the introduction to this ISO standard, a certification based on this standard is actually explicitly excluded [3]. DIN EN ISO 22870 is the relevant standard governing the application of POCT. Because it is bound to DIN EN ISO 15189, the accreditation of POCT – and not certification – is to be regarded as the internationally recognized procedure for affirming competence in this field.

Nevertheless, the application of standards and the setup of QM systems in Europe continue to be heterogeneous. In Austria, for example, the method of choice is to certify medical laboratories according to EN ISO 9001 before accreditation [22]. In other countries, a substantial number of laboratories get themselves certified and accredited – often both for reasons of economics and as an advertising strategy. At present, the majority of European accreditation bodies offers medical laboratories accreditation according to both EN ISO/IEC 17025 and EN ISO 15189 [24]. In France, the accreditation of medical laboratories has been mandatory since 2016 [19].

39.3 Accreditation of POCT according to DIN EN ISO 22870

Back in 1998, Germany issued recommendations for introducing quality assurance for POCT within the scope of the accreditation of medical laboratories [1]. The DIN EN ISO 22870 standard laid the groundwork for accreditation in the area of POCT. Together with the POCT Working Group of the German Society of Clinical Chemistry and Laboratory Medicine (DGKL), the ZLG developed a guidance document on the application of the standard according to a checklist concept, which is also applied, to other areas of accreditation [24]. This guidance reflects the specific requirements for a QM system according to DIN EN ISO 22870 and was recently updated by a checklist of the national accreditation body in Germany Deutsche Akkreditierungsstelle GmbH (DAkkS), which is downloadable from the website www.dakks.de. A guidance document by the European Cooperation of Accreditation (EA) in 2014/15 was published to enable interpretation of the requirements of DIN EN ISO 22870 [27].

The numerous references in DIN EN ISO 22870 to DIN EN ISO 15189 indicate that an accreditation according to DIN ISO 22870 is deemed sensible exclusively in conjunction with the accreditation of the central laboratory of a healthcare facility according to DIN EN ISO 15189. This viewpoint was also shared by the ILAC resolution GA 13.25 dated October 2009 [21]. This concept resulted in the adaptation of DIN EN ISO 22870 being structurally close to the DIN EN ISO 15189. As in the case of DIN EN ISO 15189, the standard is currently divided into two main areas of requirements: those for management and those for technical specifications. This structural concept is considered controversial by standardization committees, especially among the laboratory medicine experts, because it too inaccurately reflects the nature and scope of responsibilities of a medical laboratory. This does not correspond to the structure of the EC4 criteria on which the standard DIN EN ISO 15189 was essentially based [16, 17]. However, this dual classification of the standard was ultimately also chosen for DIN EN ISO 22870 because it could simplify the practical accreditation process under certain circumstances and markedly delineate the accreditation process from the certification procedure.

39.3.1 Organization and management

In terms of the organizational requirements, DIN EN ISO 22870 clearly delineates the responsibilities for planning, development and quality of the processes required to implement POCT: The responsibility for this is clearly assigned to the laboratory management of the healthcare facility. For example, the laboratory's management is responsible for establishing policies for selecting and evaluating the performance of POCT diagnostics and defining quality requirements and objectives.

Irrespective of whether the responsibility for conception and planning lies with the laboratory's management, DIN EN ISO 22870 stipulates that a multidisciplinary POCT committee be nominated that makes and implements all decisions relating to the use of POCT methods. The minimum criteria governing the composition of the POCT committee are defined: The committee must be constituted of persons representing the central laboratory, the healthcare facility's administration, the nursing staff and the clinical wards at the least. In practice, the implementation of the requirements set forth in the standards can be difficult, because the laboratory managers responsible for the processes to be implemented during the POCT often do not hold the authority to issue orders to staff members in the individual hospital departments. This issue should be regulated locally within the purview of the POCT committee.

39.3.2 Quality management and documentation

The laboratory's management is deemed responsible for implementing the POCT QM system. Management, however, must nominate a POCT quality officer (corresponds to the POCT coordinator) who is appropriately educated and experienced to execute the tasks assigned. Many different training courses and educational programs are offered in the POCT sector. Nevertheless, it has been neglected to make special courses mandatory as part of the accreditation process because the competence and experience of a POCT quality officer are not necessarily bound to such training courses.

The same requirements relating to documentation and document control apply to POCT accreditation as to other diagnostic areas. The exclusive use of authorized QM documents and the seamless documentation of procedures must ensure the traceability and transparency of processes at all times. This is the challenge relating to continuing operator training in the areas where POCT is used.

39.3.3 Corrections and improvements

POC methods that deviate from quality management standards must be documented and rectified immediately. In that way, uncontrolled processes that characterize the known disadvantages of POCT, such as the indiscriminant use of different devices or inadequate quality control, can be rapidly eliminated. DIN EN ISO 22870 specifically names the ways and means to continually improve the QM system: For instance, the laboratory's management should define characteristic data on the basis of which the system's effectiveness can be reviewed and that allow POCT performance and the satisfaction of patients and operators to be evaluated. Specifically for meeting the requirements of POCT accreditation, the results of internal audits as an instrument for system improvement must be forwarded to all members of the POCT committee.

39.3.4 Management reviews

Any economic disadvantages arising from the use of POCT are primarily addressed by the annual management review. Accordingly, the laboratory's managers are obligated to carry out a review that includes a cost-benefit analysis, an evaluation of the clinical infrastructure needed, an evaluation of clinical effectiveness and an evaluation of cost effectiveness. Any areas for potential improvement of the POC methods should be included as well. The results of this review must be submitted to the POCT committee which then decides whether the POC methods should be modified or not.

39.3.5 Personnel

According to DIN EN ISO 22870, it is the main responsibility of the POCT quality officer to create and maintain the procedures required for POCT quality management and to make sure they are accepted by the affected parties. Besides the laboratory's managers, the POCT quality officer, as a notified person, is automatically a member of the POCT committee.

DIN EN ISO 22870 describes the function of both the POCT quality officer and the POCT training officer. The POCT training officer and the POCT quality officer can be the same individual. The training officer must put into place and run a continuing education program for POCT operational staff. According to DIN EN ISO 22870, only trained personnel explicitly nominated by the POCT committee may carry out POCT. In addition, the training officer evaluates the competency of the operational staff.

> **Note**
>
> That the standard places great value on the testing staff's competence to assure the quality of POCT analysis is reflected by the high level of qualifications required of operators.

39.3.6 Laboratory equipment and pre-analysis

As in other areas, the responsibility for the selection and procurement of POCT devices and reagents lies with laboratory management. The aim to ensure the traceability of every single test can be achieved by keeping seamless documentation of the procured equipment and reagents. Moreover, the QM system must ensure that each specimen is identifiable and traceable back to the patient. The risk of improper patient assignment must be minimized.

39.3.7 Quality of testing methods and findings reports

The laboratory management responsible for quality assurance must put a quality control system in place that, at a minimum, meets the legal and statutory requirements (e.g. RiliBÄK in Germany [2, 25]). Participation in an external quality assurance scheme, if available, is mandatory for the individual variables. Moreover, a detailed performance evaluation of the POCT devices must be performed with regard to the different performance parameters, e.g. linearity and accuracy, also encompassing POCT systems used at different locations. Compliance with the requirement for documenting the persons carrying out the tests can pose certain challenges to the routine work on the wards.

The DIN 58964 national standard published in 2015 on "Quality assurance of POCT results - Assessment criteria for comparison measurement and implementation" [28] concretizes the requirements for the quality assurance and the performance evaluation of POCT test methods as compared to conventional laboratory methods. Therefore, this standard addresses both operators and users of POCT devices as well as manufacturers of in-vitro diagnostic medical devices in POCT settings. It also contains valuable instructions on how to compare diagnostic results obtained and analyzed by POCT as compared to laboratory methods.

39.3.8 Outlook

The implementation of requirements according to DIN EN ISO 22870 can lead to an institution being compelled to implement a serious number of changes that mainly affect organizational and communication processes. Nevertheless, these requirements address exactly those aspects relating to the risks and/or disadvantages of POCT which can be brought under control by a DIN EN ISO 22870-compliant QM system. It should be mentioned that the checklist presented in [24] reflects all the essential requirements of the DIN EN ISO 22870 standard. Yet, to date, the list does not take into account the diagnostic aspects of all the special demands placed on the quality of specific analyses, such as those pertaining to blood gas analysis or blood glucose measurements. A future checklist should also take those aspects into account.

On the European and international standardization level, the necessity was recently recognized to put the requirements of the former DIN 58964 into an international context and set up a new standardization project at CEN. Moreover, the ISO/TC 212 Technical Committee is currently drafting a new guidance aimed at defining the quality requirements relating to operators and users of POCT devices – without explicitly referring to a laboratory context (ISO TC 212 Preliminary Work Item (PWI) "Medical laboratories - Guidance for supervisors and operators of point-of-care testing (POCT) devices"). However, the two standardization projects are still in their preliminary phases. Finalization of the documents can be presumably expected by 2018 at the earliest.

References

1. Arbeitsgemeinschaft Medizinische Laboratoriumsdiagnostik und der Zentralstelle der Länder für Gesundheitsschutz bei Arzneimitteln und Medizinprodukten (2004): Qualitätsmanagement und Akkreditierung. Wissenschaftliche Verlagsgesellschaft
2. Bundesärztekammer (German Medical Association) (2014) Guideline of the German Medical Association for the Quality Assurance of Clinical Laboratory Testing (Rili-BÄK 2014). Dtsch Ärztebl 111:A1583–1618
3. DIN EN ISO 15189: 2014: Medical laboratories – Requirements for quality and competence
4. DIN EN ISO 22870:2006: Point-of-care testing (POCT) – Requirements for quality and competence
5. DIN EN ISO/IEC 17025:2005: General requirements for the competence of testing and calibration laboratories
6. EEC 93/465/EEC, Council Decision of 22 July 1993 concerning the modules for the various phases of the conformity assessment procedures and the rules for the affixing and use of the CE conformity marking, which are intended to be used in the technical harmonization directives; Official Journal of the European Commission L 220, 30/08/1993
7. EG 2009/886, Commission Decision of 27 November 2009 amending Decision 2002/364/EC on common technical specifications for in vitro diagnostic medical devices of 4 December 2009
8. EN 13612:2002: Performance evaluation of in vitro diagnostic medical devices
9. EN 45001:1995: General criteria for the operation of testing laboratories
10. Gesetz über die Akkreditierungsstelle (Akkreditierungsstellengesetz – AkkStelleG) vom 31.07.2009 Bonn
11. Gesetz über Medizinprodukte – MPG vom 2. August 1994 in der Neufassung vom 7. August 2002; BGBl. I S. 3147, zuletzt geändert durch Artikel 2 des Gesetzes zur Änderung medizinprodukterechtlicher Vorschriften vom 29.07.2009 (BGBl. I p. 2326)
12. Gesetz zur Änderung medizinprodukterechtlicher und anderer Vorschriften vom 14.Juni 2007; BGBl 2007, I p. 1066
13. Guide to the implementation of directives based on the New Approach and the Global Approach (2000), published by the Office for Official Publications of the European Communitieshttp://europa.eu.int/comm/enterprise/newapproach/legislation/guide/document/1999_1282_en.pdf
14. ISO 9001:2015: Quality management systems – requirements
15. ISO/CEN: Agreement on technical cooperation between ISO and CEN, (23.12.2002), Chapter 4. http://www.cenorm.be/boss/supmat/refdoc/archive/ms/ms002.htm
16. Jansen RT, Blaton V, Burnett D, Huisman W, Queralto JM, Zerah S, Allman B (1997) Essential criteria for quality systems in medical laboratories. Eur J Clin Chem Clin Biochem;35: 121–122
17. Jansen RT, Blaton V, Burnett D, Huisman W, Queralto JM, Zerah S, Allman B (1998) Additional Essential Criteria for Quality Systems of Medical Laboratories. European Community Confederation of Clinical Chemistry (EC4) Working Group on Harmonization of Quality Systems and Accreditation. Clin Chem Lab Med 36: 249–252
18. Messner B, Frick G, Blobner M, Albrecht K, Schade T, Luppa PB (2004) Online-Qualitätssicherung patientennaher Sofortdiagnostik (Point-of-Care-Testing) unter Nutzung des Kommunikations-Systems des Krankenhauses. J Lab Med 28: 264–272
19. Ordonnance No. 2010–49 du 13 janvier 2010 relative à la biologie médicale, publiée au journal officiel de la République française le 15 janvier 2010
20. Price CP (2002) Medical and economic outcomes of point-of-care testing. Clin Chem Lab Med 40: 256–251
21. Resolutions of the Thirteenth ILAC General Assembly, Vancouver, Canada, 18 and 20 October 2009, http://www.ilac.org/ga_resolutions.html
22. Schweiger CR (2004) Quality management in Austria. J Lab Med 28: 85–90
23. Spitzenberger F, Edelhäuser R (2008) Accreditation of medical laboratories in Europe: Statutory framework, current situation and perspectives. Transfus Med Hemother 33: 384–392
24. Spitzenberger F, Weidemann G, Hafner G (2006) Accreditation of POCT according to ISO 22870 – Requirements for quality and competence. J Lab Med 30: 264–271
25. Verordnung über das Errichten, Betreiben und Anwenden von Medizinprodukten – MPBetreibV in der Neufassung vom 21.08.2002; BGBl. I p. 3397, zuletzt geändert durch Artikel 4 des Gesetzes zur Änderung medizinprodukterechtlicher Vorschriften vom 29.07.2009 (BGBl. I p. 2338)
26. Vogt W (2000) TQM und die Bewertung nach dem EFQM Modell – Anwendung auf das Medizinische Laboratorium. Klinische Chemie und Molekulare Diagnostik. Vol. 4. Idensee, Oldenburg, Germany
27. EA-4/20 G:2014: Guidance for the Assessment of Laboratories against EN ISO 15189 and EN ISO 22870 Point-of-Care Testing (POCT)
28. DIN 58964:2015: Quality assurance of POCT results – Assessment criteria for comparison measurement and implementation

How to achieve quality for POCT through risk management

James H. Nichols

P. B. Luppa, R. Junker (Eds.), *Point-of-Care Testing*
https://doi.org/10.1007/978-3-662-54497-6_40

40.1 Introduction

Point-of-Care Testing (POCT) has the potential to provide rapid turnaround of laboratory test results with the potential for faster medical action. However, this assumes the test results are correct and reliable. POCT presents increased opportunity for errors by moving testing out of the well-controlled environment of a laboratory and placing the tests in the hands of nurses and clinical support staff with little training and experience in laboratory diagnostics. The near-patient testing workflow is different from central laboratory testing, with more staff and locations involved, creating greater potential for errors. Managing POCT quality requires managing risk for errors. This chapter will discuss common sources of POCT error, review quality control (QC) as a means of controlling analytical errors, identify risk management as a total quality assurance process, and describe recent regulatory changes in the United States (US) that allow adoption of risk management principles in the clinical laboratory.

40.1.1 Medical Errors

A medical error is an adverse event or near miss that is preventable based on the current state of medical knowledge [15]. The Institute of Medicine estimates that medical errors kill 44,000 – 98,000 patients in US hospitals each year [11]. The Public Health expert Lucien Leape claims that medical errors are the "number one problem facing health care" [12]. The 2002 Commonwealth Fund report estimates that one in five people in the US have experienced a medical error, personally or through at least one family member, at a cost of US $17–29 billion annually [7].

The laboratory contributes to medical errors. A mini-review of the literature found the majority of laboratory errors occur in the pre-analytical and post-analytical phases of testing rather than in the laboratory analytical phase [3]. Many mistakes are called, "lab errors", but are actually due to poor communication, actions by others involved in the testing process or poorly designed processes outside the laboratory's control. While medical errors can occur in prevention, diagnosis or drug treatment, laboratory errors are primarily diagnostic. Among the types of diagnostic errors, 50 % are failure to use the indicated test, 32 % failure to follow-up on results of ordered tests and 55 % involve avoidable delays in diagnosis [13]. Recent surveys find that most people will experience at least one diagnostic error associated with inaccurate or delayed diagnosis in their lifetime, sometimes with devastating consequences [2].

The literature studying errors surrounding POCT, however, is sparse. POCT eliminates several steps of the testing process including, specimen transport, processing and result communication. However, POCT creates a different workflow with the potential for new types of errors. As referred to above, two-thirds of the POCT-related errors tend to be reported in the analytical phase of testing, whereas central laboratory errors occur primarily in the pre-analytical and post-analytical phases [14]. These errors were assessed by the reporting staff as having minimal actual impact on patient care. On the other hand, the same reporting staff rated the perceived potential for adverse impact much higher with 15 % and the potential to produce moderate adverse patient outcomes and 4 % as having potential to produce substantial adverse patient outcomes. POCT processes thus present different risks for error, where the number of staff, devices and decentralized operations involved in the management of testing pose challenges to obtaining reliable, quality test results.

40.1.2 The role of QC

QC has historically been used to reduce errors and prevent the release of incorrect test results from the laboratory. QC takes a stabilized surrogate sample that is analyzed like a patient sample and contains a known amount of measured analyte.

> **Performance of QC sample mimics the performance of patient specimens.**

If the QC sample achieves the target test result and this result is within an acceptable range for assay imprecision and bias, then the test system is assumed to be stable and producing quality patient results. If a shift is noted in QC results above or below the expected target, then patient results can be expected to show similar shifts in performance. QC thereby provides a warning of test performance issues before patient results are released to physicians.

The use of QC to monitor assay performance was born from the pre-World War industrial models of assessing quality in factory processes. Products, like car or plane parts, are inspected as they are being assembled on a factory line. If the inspected products meet expected performance specifications at each stage of assembly, then the remaining products are assumed to be of similar quality and performance. For laboratory systems, QC provides an assessment of how the analyzer, reagent, environment, operator and other factors all come together to produce the final result. QC, when achieving target results, has the advantage of assessing the performance of the entire system without having to separately determine the performance of each component.

Historically, a QC sample is analyzed with each batch of patient specimens to determine if the system is performing to specification and the patient results can be released. If QC results are outside the target range expected, the lab must troubleshoot the test system and determine the cause for the QC failure. Patient results are held until the problem is resolved and QC results return to the expected "in-range" performance. For batch analysis, it is customary to bracket patients on whose tests QC was performed. QC is analyzed before and after patient specimens to document stable performance across the run of patient samples. However, for continuous analyzers, like modern chemistry and hematology systems, test results are automatically released as soon as the analysis is complete. In the central laboratory, QC is performed periodically, every few hours throughout the day in order to bracket every 20th, 50th, 100th sample, depending on the number of specimens being analyzed, or, alternatively, every few hours depending on the test's stability. The Clinical Laboratory Improvement Amendments (CLIA) require analysis of at least two concentrations of QC each day of testing, based on the rationale that a batch or continuous analytical run should not be longer than 24 hours. The College of American Pathologists (CAP), Commission on Office Laboratory Accreditation (COLA), the Joint Commission and ISO standards have all adopted minimum requirements for performance of QC.

QC is excellent at detecting systematic errors, especially those that affect every test in a constant and predictable manner from one point in time forward. Problems with reagent deterioration or preparation, improper storage or shipping conditions, incorrect operator techniques with dilution or pipette settings and calibration errors with incorrect set-points are easily detected by shifts in the performance of the QC sample. However, QC does a poor job at detecting random errors, i.e. those that affect individual samples in a random and unpredictable manner. Bubbles, clots and interfering substances like hemolysis or drugs that are unique to a single sample, are poorly detected by QC.

Consider the example in ◘ Fig. 40.1 which represents a timeline for analysis on an automated chemistry laboratory analyzer. Two concentration levels of the QC sample are analyzed at 09:00 every morning. At 11:00, the analyzer experiences a line leak which dispenses only 25 % of the expected volume of reagent, affecting all test results forward. This systematic error would not be detected until the next QC event at 09:00 the next day. In that time, a number of patient results may have been released and acted on by physicians. So, the frequency of QC needs to be adjusted based on specimen volume, consideration of the analyte stability and the consequences of taking clinical action before confirmation of the next QC. Once a problem is detected, the laboratory must troubleshoot, correct the problem and reana-

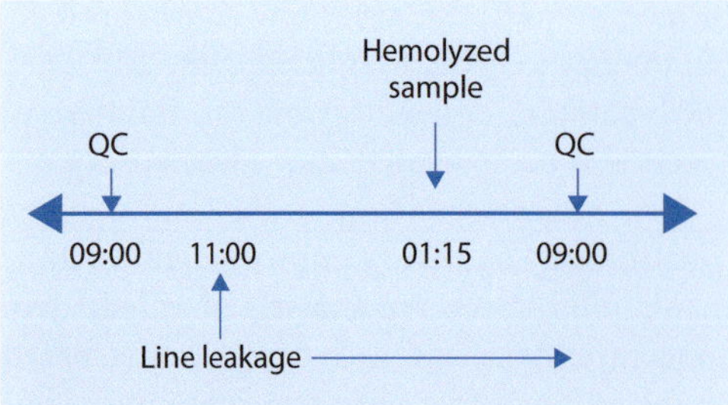

Fig. 40.1 Quality control and laboratory errors. A chemistry analyzer is continuously analyzing and autoverifying results. External liquid QC is analyzed once-a-day at 09:00. At 11:00, an analyzer line leak occurs which dispenses only 25 % of the expected reagents. This systematic error continues until detected by unsuccessful control results at the next QC event, at 09:00 the next day. At that point, the laboratory must troubleshoot the problem, fix the line and reanalyze patient specimens from the time the error started, i.e., at 11:00 the previous day. Then it must send the corrected results from most of the day to physicians. Analyzing QC samples more frequently than one-a-day would reduce the chance of releasing incorrect test results by detecting a QC failure earlier. Analysis of a hemolyzed specimen which leads to an elevated potassium finding – a random error – occurs at 01:15 a.m., and QC of the sample analyzed the next morning at 09:00 does not detect the random error. The laboratory must take additional actions such as visually inspecting every specimen or analyzing each specimen spectrophotometrically (hemolysis index) in order to detect such a randomly occurring hemolysis error

lyze patients, correcting results as needed. Recalling and correcting results causes clinicians and staff to lose faith in the reliability of test results. The figure also shows a random error: A hemolyzed sample is analyzed at 01:15 in the morning, falsely increasing the potassium result. QC is not going to detect the problems caused by a random error, in this case a hemolyzed sample. The laboratory must implement other quality processes, such as visual inspection of each specimen or serum spectrometric indices, in order to catch specimen hemolysis before verifying and releasing the test result.

As demonstrated by Fig. 40.1, successful QC performance alone is not sufficient to ensure quality results. Despite analyzing a minimum of two concentration levels of QC sample on each day of testing, we have all experienced

Tab. 40.1 Sigma metrics and percent (%) or parts per million (ppm) error rates

Sigma	Errors %	Errors (ppm)
1	69 %	691,462
2	31 %	308,538
3	6.7 %	66,807
4	0.62 %	6,210
5	0.023 %	233
6	0.00034 %	3.4
7	0.0000019 %	0.019

errors in laboratory results. Ergo, quality is more than QC. Quality is testing that is safe and reliable, appropriately selects the technology to meet medical needs and, above all, produces test results that can be trusted for medical management. After all, a laboratory's reputation depends on the quality of its test results.

What then should be the **quality goal** for laboratory errors? Consider Sigma metrics, an industrial measure of quality in a factory process. Sigma metrics allow an industry to estimate the rate of defects in products produced. Sigma metrics provide a means of improving customer satisfaction by balancing the number of rejected, non-complying products, i.e. those that do not meet expected specifications, against the costs of improving operations to lower defect rates. Six Sigma equates to error rates of 0.00034 % or 3.4 parts per million (ppm) (Tab. 40.1). This means that companies will discard only about 3 products for every million produced. Companies can certainly decrease error rates below Six Sigma and discard fewer products, but costs increase exponentially for the minimal gains in improvement beyond the Six Sigma level (discarding fewer than 2 products per million produced). This is the reason that Six Sigma has become a target for many industries and a hallmark of manufacturing quality.

Laboratories in the US conduct more than 7 billion tests annually and laboratory testing is estimated to influence more than 70 % of all medical decisions [19]. With 7 billion tests, Six Sigma (3.4 ppm) translates to releasing 23,800 incorrect laboratory tests in the US each year! Any single result error could mean a misdiagnosis, failure to treat an illness or incorrect treatment selection. These errors could cause virus transmission in a blood transfusion, incorrect cancer diagnosis or a missed heart attack: erroneous results that affect living human beings not mere auto parts. For medical laboratories as an industry, Six Sigma is an unacceptable goal.

We need to go "beyond six-sigma" for laboratory performance, with a goal of zero errors that eventually reach the patient.

40.2 Risk management

Risk management is a way for laboratories to achieve zero errors. Risk is defined as the chance of suffering or encountering harm or loss. Essentially, risk is the potential for an error to occur. Risk can be estimated through a combination of the probability of occurrence of harm and the severity of that harm [10]. Risk can be calculated with the following equation:

$$\text{Risk} = (\text{Frequency}) \times (\text{Severity Harm}) \times (\text{Detectability})$$

As the number of errors or the error rate increases, risk also increases. Errors that cause greater harm also lead to increased risk. Detectability in the clinical laboratory setting is the ability to catch an error before it is released by the laboratory. Processes without systems to detect errors lead to greater risk. Risk calculations provide a simple rating scale to rank the order that errors should be addressed. When time and staffing resources are limited, laboratories will want to address those errors of greatest risk first.

As errors occur, the temptation is to blame a person for not following policy or making a mistake. It is easier to blame a person for a failure than an entire institution. Indeed, in the airline industry, more than 90 % of quality lapses are deemed blameless. Most errors are active failures due to how a person interacts with a system. Latent conditions in some processes setup a person to fail due to weaknesses in the system, design flaws or hierarchical decisions that are required to complete a process. In most plane crashes, the plane doesn't crash itself. An accident happens because technology, controls and/or the environment allow the pilot to make an error [16].

Fig. 40.2 demonstrates a process with several layers of analytical checks and detection systems. Each defense is a layer, like slices of Swiss cheese. When working, these defenses prevent an error from transversing the defenses and allowing an erroneous result to be reported. Unfortunately, no check is entirely foolproof. There are holes in each layer of defense that may allow errors to pass undetected. As we place multiple levels of defense together, most errors are caught. However, there is the rare possibility of a perfect confluence of events which let all of the holes in our defenses and detection systems to align and permit the error to occur. We need to engineer systems that prevent dangerous errors and are able to uncover errors and contain their effects [16]. In the laboratory, assays fail because of weaknesses in the testing process that enable an error to occur without warning the operator of a problem. Our role as laboratory professionals is to better understand our testing processes and build safety checks and preventive measures that can avert or detect result errors before release. This is risk management.

To identify risk in the laboratory, staff needs to understand that there is no perfect device. Any instrument can and will fail, when used under non-complying, predisposing or limiting conditions. The role of the laboratorian is to determine those limitations and prevent use of the device under any risk-prone conditions. Lab tests are not foolproof, especially for POCT operators who may deviate from standard practice.

Risk management should start by asking, "What could possibly go wrong"?

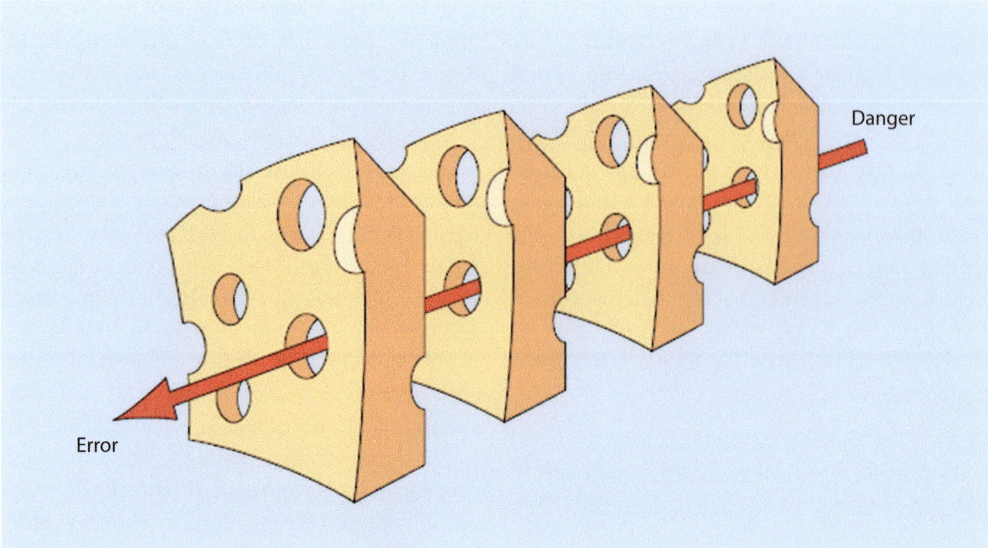

Fig. 40.2 Swiss cheese model of errors. A hazard or error trajectory can transverse the laboratory mitigations and actions taken to prevent errors whenever holes in each defense layer align to allow the error to go undetected

Risk management requires

- Mapping our processes
- Understanding the weak points in those processes where errors could occur
- Taking action to create defenses that will detect errors and mitigate hazards

While analysis of 2 concentrations of QC samples on each day of testing is a regulatory requirement, every device is different. Universal requirements for daily QC do not fit every device or setting. Some devices may require less frequent QC because the assay is stable for many weeks and staff is experienced and proficient, while other devices may need to analyze QC multiple times a day because the reagents and calibration are unstable or staff turnover frequently. Newer devices have **built-in control processes** which run checks on the analyzer and the chemistry of the reactions and thereby may duplicate the need for daily liquid QC. Some of these control processes are performed with every specimen, such as the creation of a control strip on a pregnancy test, or sensor and flow resistance signals to detect sample clots or bubbles on chemistry and blood gas analyzers. Other checks may be engineered into the instrument such as the barcoding of reagent bottles and cartridges to prevent use past their expiration dates, **QC lockout features** that require QC analysis before patient testing or disposable analyzer cuvettes or pipette tips to prevent sample carryover. Optimizing the use of liquid QC with these internal control processes was the rationale for developing the CLSI EP23-A: Laboratory Quality Control Based on Risk Management document [5]. This guideline describes good laboratory practice for developing a quality control plan based on manufacturer's information, applicable regulatory and accreditation requirements, while taking the individual healthcare and laboratory setting into account. EP23-A introduces industrial risk management principles to the clinical laboratory.

A laboratory **quality control plan** for a test is developed by collecting information about the test, the medical requirements for the test (clinical performance tolerance), the regulatory re-

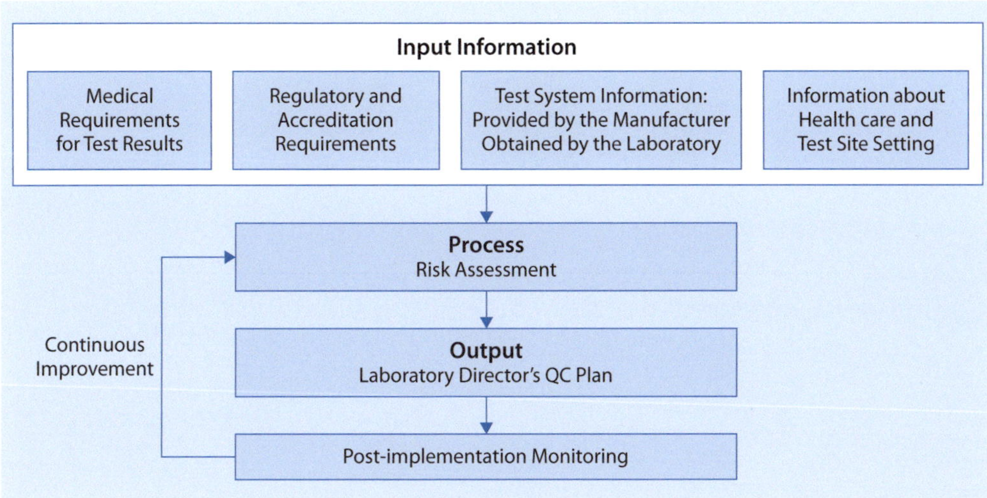

Fig. 40.3 Process to develop and continually improve a quality control plan

quirements and the test limitations and weaknesses (Fig. 40.3). Those limitations may come from the manufacturer's package insert, other information available from the manufacturer or from discussion with colleagues to determine what problems they may have encountered. Other considerations are the healthcare setting (inpatient, outpatient, clinic), where the test is being conducted (inside a well-controlled laboratory or susceptible to the environment with mobile nursing), and the type of test operators (medical technologists or clinical staff).

This information is processed through a **risk assessment** that starts by following the sample through the testing process (Fig. 40.4). Starting from the clinician assessing the need for a test and placing an order on to patient preparation, specimen collection and transport of the specimen to the laboratory, staff should look for weaknesses in the pre-analytical processes and steps where mistakes may take place. Staff should pay particular attention to how specimens are collected, labeled and handled prior to analysis. Additionally, they should follow the sample through laboratory receipt, processing, analysis, result reporting and post-analytical result communication, acknowledgement and treatment. The laboratory should review the quality of the entire testing process and again assess where errors may occur. For each step, ask "what could possibly go wrong" (Fig. 40.5)?

> **Common sources of error can come from samples, operator, reagents, environment and measuring system.**

Staff should consider: sample integrity (lipemia, hemolysis clots, bubbles and collection tube additives); operator training, competency and staffing levels (continuity versus turnover); reagent, control and calibrator shipping, storage and expiration; the laboratory environment (dust, temperature, humidity, water quality and electrical power); and measuring system maintenance and failures. The CLSI fishbone diagram outlines similar sources of error found in the CLIA interpretive guidelines for laboratories, which emphasizes that laboratories should consider errors from the test system, the environment and the operator [6].

POCT versus laboratory processes

POCT processes are different than laboratory processes and specimens follow different workflow pathways. That means, the same hazards (weaknesses in the testing process) identified for central laboratory testing may not pose the same risks as POCT, while the different work-

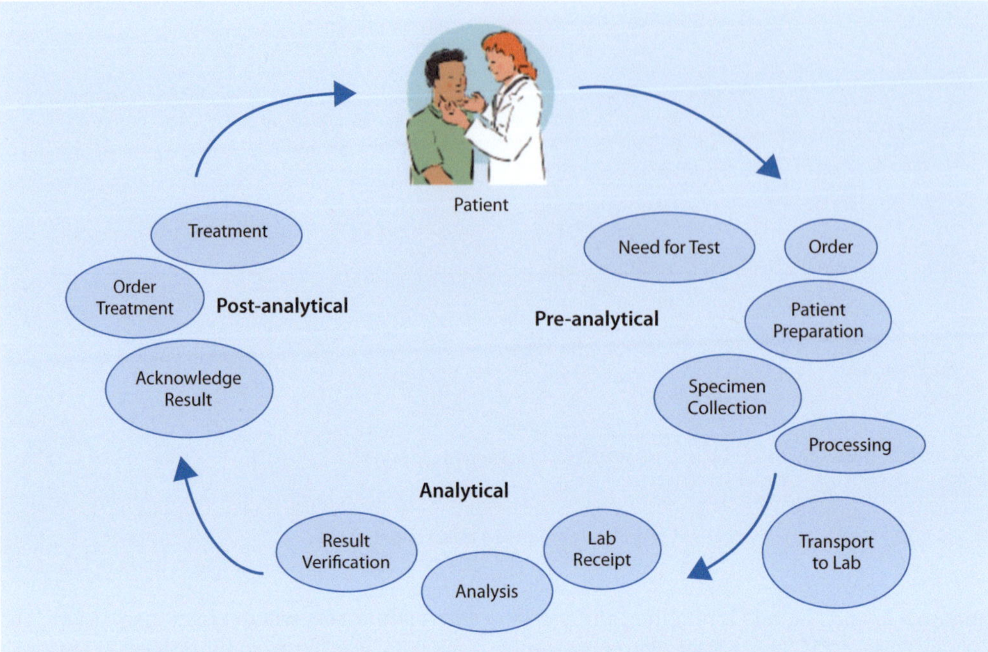

Fig. 40.4 Steps in the testing process. There are three phases in the testing process: the pre-analytical, analytical and post-analytical phases

flow may engendered different hazards to consider. For each hazard identified, the laboratory should define an action or mitigation that will minimize the risk or chance that an error will occur at that step. The QC plan is a summary of all the hazards identified in the testing process and actions the laboratory will implement to minimize risk. Once implemented, the QC plan is monitored for effectiveness by tracking for any trends in performance or continuing errors. For each incident, the laboratory should troubleshoot, determine the cause of the error, reassess the risk and modify the QC plan accordingly.

Risk management is a continuous and holistic approach to quality assurance in the clinical laboratory that incorporates all phases of testing: pre-analytical, analytical and post-analytical.

40.2.1 Risk management in action

Manufacturers have engineered a number of quality processes into laboratory and POCT devices to mitigate hazards and reduce the possibility of specific errors. Laboratory professionals should be familiar with how these processes work in order to incorporate them into their QC plans. The final QC plan will ultimately be a mix of manufacturer-engineered quality processes and actions taken by the laboratory to address the variety of hazards identified in the risk assessment. The following are examples of how manufacturer-derived quality processes can catch and prevent specific errors.

The laboratory was called with a complaint from the intensive care unit (ICU) about sporadic, falsely decreased glucose meter results. Immediate repeat of the test on the same meter, gave significantly higher, clinically sensible values. POCT staff inspected the meter and found that nurse operators were taking procedural shortcuts with the glucose meters in order to

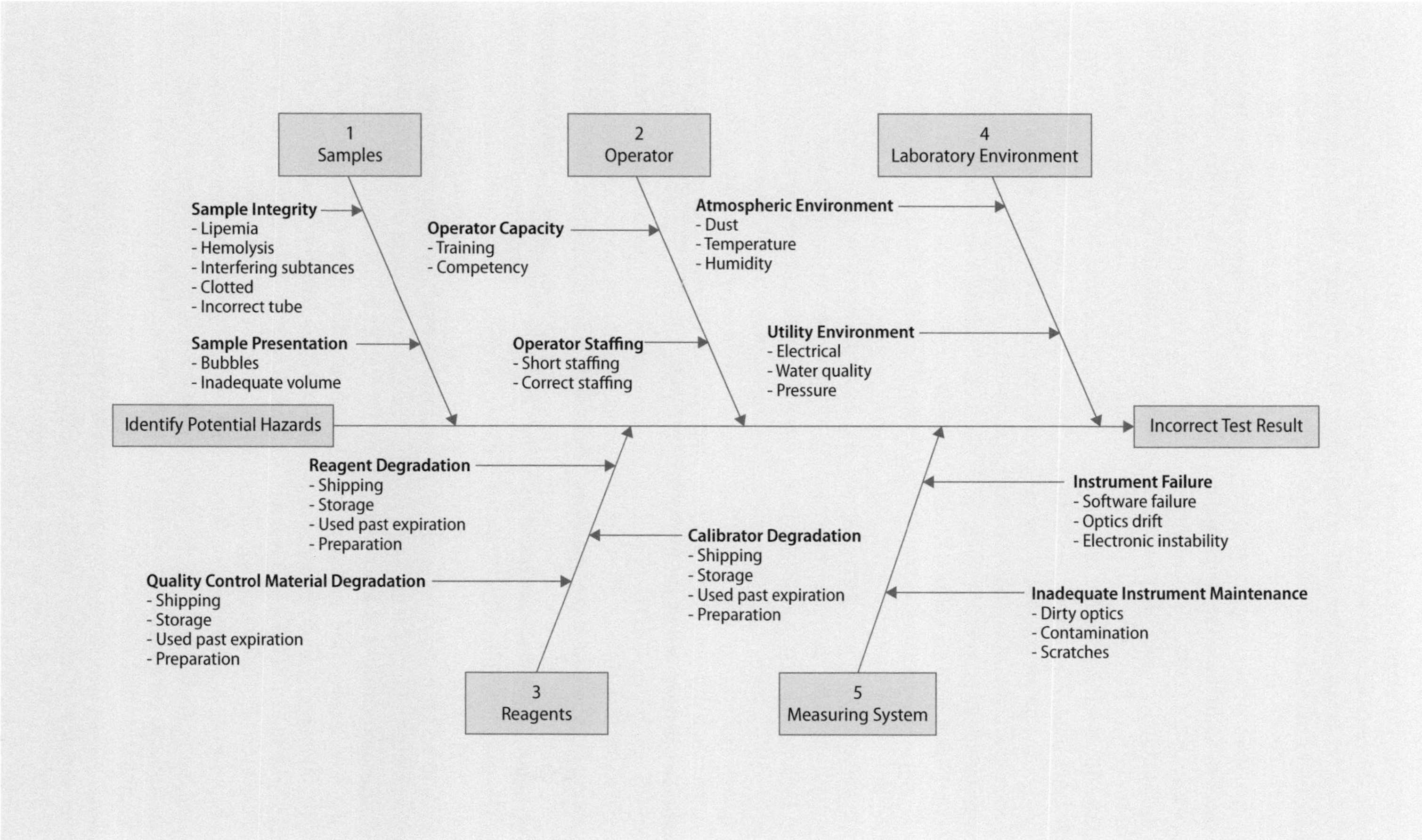

Fig. 40.5 Fishbone diagram of laboratory error sources. This diagram shows the potential hazards on the left that can lead to an incorrect test result on the right. Laboratory errors can come from 5 primary sources; samples, operator, reagents, laboratory environment and measuring system. This is not a comprehensive list and other sources of error should be considered

save time. Staff found the neck of the glucose test strip bottle was too narrow to get a finger into the bottle for retrieving a test strip with gloved hands. So, the strips are dumped on the counter where they can more easily be grabbed by staff. After testing is completed, the operators toss the used test strips in the garbage at the end of the counter. Sometimes the strips make it to the garbage. Other times, the used strips land back on the counter where they intermix with the other test strips. The next operator then picks up a test strip from the counter that may already have been used. The chemicals of the glucose test strip reaction are consumed during the first test, so reapplication of sample to a used test strip generates falsely decreased results.

Some meters have built-in quality checks that can catch reuse and display an error code on the meter, preventing an erroneous result, whereas other meters do not have such control processes and will allow double-testing to occur [18]. Despite training to keep the test strips in a tightly capped bottle, nurse operators were taking shortcuts in order to expedite the testing process. The operators did not intend to get a wrong result. They just wanted to get back to their primary focus – the patient. The operators did not understand that their actions could affect the test result. Thereupon, all operators on the ICU were retrained and warned to keep the test strips in the bottles. However, this error uncovered the possibility of obtaining incorrect results from test strip reuse. The glucose meter allowed staff to reuse a test strip and gave a result without warning of the problem! As a consequence, the hospital changed glucose meters to a model that would detect and display an error code if staff attempted to insert a used strip into the meter. Manufacturers of newer models of glucose meters can also detect test strip damage or strip exposure to humidity and extreme temperatures. Some models of meters also autocalibrate, and can compensate for patient hematocrit, environmental temperature and oxygen biases.

40

Specimen volume is another hazard to consider in POCT. Some glucose meters require operators to visually inspect test strips for uniform color development after each test in order to detect under-filling or bubbles applied to the strip. Operators, though, may forget to check the strip once a test complete and the result is available. Some meters have a fill-trigger that is designed to prevent underfilling. The test will only start when enough blood has been applied to the strip. This quality check ensures that the appropriate volume of sample is applied and prevents underfilling, overfilling or reapplication of blood once the test has started.

Timing can also be a concern. Tests like urine pregnancy, rapid strep and urine dipsticks require the operator to manually read the test results at a specified time after sample application. Reading the results too soon can lead to underdevelopment and false negatives, while reading the results too late can lead to either overdevelopment or fading of the signal causing false positives or false negatives. Many POCT devices have sophisticated data management that automates the analytical steps and timing of the test result. Operators can apply a sample, walk away to attend to patient care duties and pick up results later on a result printout. Automated timing features reduce the possibility of result misinterpretation due to test timing errors. Automated readers also reduce operator variability from color discrimination (such as urine dipsticks) and ensure correct interpretation of results. This applies to tests like pregnancy, drugs of abuse or rapid strep, where the appearance or disappearance of a line can be confused as either positive or negative.

POCT device **data management** also ensures regulatory compliance. CLIA, CAP, COLA and the Joint Commission require 2 concentration levels of QC each day of testing. For manual POCT, operators must remember to analyze the QC, document the results, and then ensure the QC results were successful and recovered within the expected range before patient testing. With pregnancy and urine dipsticks, it is possible for operators to perform patient testing without running QC or worse, failing QC and continuing to test patients without troubleshooting. Fortunately, automated POCT devices like coagulation, blood gas and glucose meters have QC lockout. This quality feature requires staff to analyze QC daily or the device will lock and prevent patient testing. In addition, the QC results are analyzed and, if outside expected range, the device will lock and require staff to troubleshoot and achieve successful QC before allowing patient testing. This ensures that every result, successful or not, is documented, and QC is analyzed at the frequency required by law. Operator lockout functions in a similar manner using a personal identification (ID) code for each operator. After completing training and documenting competency, the device will allow operators to access patient testing through their operator ID. Upon starting, the POCT device will ask for the operator's ID. If the operator's training is valid, then the device allows patient testing. Operators will be locked out from patient testing if annual competency expires and is not renewed.

Patient ID can be another source of error. Data management on POCT devices allow the electronic transfer of test results from the device to the electronic medical record. Yet, entry of an incorrect patient ID can fail to transfer POCT results. Because the result is available on the device, clinicians can institute treatment without a permanent record of the result in the patient's chart. In addition, incorrect entry of another patient's ID can chart results to the wrong patient's medical record, lead to inappropriate medical decisions and cause inappropriate billing. Hence, correct entry of patient ID is important, because that ID gets tied to the test result and device-inherent information, such as date, time, operator, serial number, reagent lot and QC records.

Many institutions have implemented barcoded wristbands to streamline data entry and reduce the chance of making mistakes when manually entering several numbers for patient ID. However, barcodes are not foolproof either. Patients can have another institution's identification on their barcode (when transferred from an outside facility), outdated admissions information or even another patient's identification. In other words, there is a residual risk of error even when using barcoded ID bands. Barcoded data entry also does not satisfy the Joint Commission and CAP National Patient Safety Goals for use of at least 2 means of identifying patients before collecting a specimen or performing a procedure. Positive patient ID is a new quality feature that satisfies the National Patient Safety Goals and is appearing on newer POCT devices. This feature utilizes the hospital admission/discharge/transfer (ADT) data to upload a list of active patients in the hospital and their medical unit. When a barcoded wristband is scanned or a patient ID is manually entered, the POCT device displays the patient's name and requires the operator to enter the birth date or a second identifier to ensure the ID matches the right patient before the device will open and allow testing. Additional checks are conducted within the data management system to ensure the serial number of the device is assigned to the medical unit where the patient is registered. Discrepancies between the medical unit where the device is assigned and the patient's medical unit will generate an error message for the operator. Implementation of positive patient ID can reduce data entry errors from barcoded wristbands [1]. This feature thus ensures compliance with the Joint Commission and CAP National Patent Safety Goals while reducing the possibility of mistakes during patient ID entry.

Incorrect calibration of POCT can lead to result bias. Some glucose and coagulation meters require use of a calibration code chip or entry of a calibration factor with every change of reagent lots. If operators forget to update the calibration upon receiving a new lot, then all results will be generated against the previous calibration and will be biased by the difference between the two calibration curves. There are POCT devices that have the calibration factors embedded in the reagent barcode. Each time a test strip or cartridge is scanned, the calibration is automatically loaded for that lot and specific cartridge. Thus, the operator does not need to verify the correct calibration codes. This prevents the possibility of analyzing samples using previous calibrations.

Use of **expired reagents** is another concern. The Centers for Disease Control and Prevention (CDC) and US Food and Drug Administration (FDA) recommend operators "check and record expiration dates of reagents/kits, and discard any reagents or tests that have expired" [4]. "As a test strip ages, its chemical coating breaks down. If the strip is used after this time, it may give inaccurate results" [9]. An operator must check the manufacturer's expiration dates on reagents and controls prior to testing. However, for POCT, most manufacturer expiration dates are only valid if the bottle of test strips or controls is unopened and stored to manufacturer recommendations. Once opened, strips and controls have less stability and will expire 30, 60 or 90 days after opening. Operators must remember to manually cross-out the manufacturer imprinted expiration and write a new expiration date upon opening a fresh bottle. This is an additional step that is easy to forget.

Accordingly, nursing units can have open bottles of strips and controls without a handwritten expiration date. These containers should be discarded as no one can attest to how long they have been in use. Some manufacturers have individually packaged test/kits which prevent this problem because each test is stable to the manufacturer imprinted expiration date, provided of course that the packaging is not opened. For bottled reagents, manufacturers have started using serialized barcodes that individually identify every bottle within a specific lot of tests. The first time an operator scans that bottle, the POCT data management system starts a timer which counts down to the new, open expiration date. If an operator attempts to use the bottle past its open expiration, the device will warn the operator of the expired reagent or control and lock-out further testing until a new bottle is opened and scanned. This quality process assists staff in managing reagents and prevents use past expiration.

Managing reagent storage and operating temperatures is also challenging for POCT. Reagents contain enzymes and chemicals that are sensitive to heat and humidity; devices and reagents can fail if used under extreme temperature conditions. Some reagents are only stable to the manufacturer's expiration date if refrigerated and require warming to room temperature before use. Other reagents and controls have reduced stability when stored at room temperature and require updating of manufacturer expiration dates when removed from refrigerated storage. Close monitoring of room temperature and refrigeration conditions are required. POCT is portable and particularly susceptible to environmental exposure. Traveling nurses, helicopter and ambulance transports all risk storing devices or test strips and controls in vehicles during summer heat and winter cold. Snow, rain, tropical humidity, dust, sand and other environmental factors are issues associated with testing outside of a well-controlled laboratory.

Increased **temperature errors** on an ambulance service were experienced the first winter after implementing a new model of glucose meter. The old meter had temperature specifications of 0–46°C, while the new meter had a narrower 15–40°C range. The emergency medical technicians (EMT) complained that they could not get a glucose result with the new meter when caring for patients outdoors or in a home without heat, and the meter took more than 15 minutes to reset after returning to a warm room. This was a problem for two reasons:

- The EMTs could not effectively care for their patients.
- The meter was warning of environmental conditions outside of manufacture specification that could impact the accuracy of glucose test results. The old meter could no longer be used, as that model had been discontinued.

As it turned out, a bioengineering student developed a carrier to protect the meter from environmental conditions and allow continued testing outdoors in the winter [17]. This was a clever way to protect the device from environmental conditions without altering its operation or analytical performance.

POCT devices are sometimes shared between hospitalized patients. This presents a risk for transmitting nosocomial and antibiotic-resistant infections. The US FDA recommends that point-of-care blood testing devices, such as glucose meters and PT/INR anticoagulation meters, should be used on only one patient and not shared [8]. If dedicating POC blood testing devices to a single patient is not possible, the devices should be properly cleaned and disinfected after every use as described in the device labeling [8]. Repetitive cleaning with harsh chemicals can crack and cloud device screens and plastic casings, causing premature device failure and shorter-than-expected equipment life spans. Decontaminating solutions take time to dry on surfaces, allowing liquid to egress into device ports with potential to short circuit sensors. POCT devices need more durable plastics, fewer crevices and seams and designs that prevent liquids from accessing the electronics and sensors critical to POCT methodologies. Test strips, reagents, control bottles and POCT carriers can also be contaminated and transmit organisms [20].

When POCT operators touch a patient or surface on the patient's bed or room with gloved hands, their gloves become contaminated. If they pick up a bottle of test strips, that container can also be contaminated. Fingertips reaching into a bottle can contaminate test strips as well as the inside of test strip bottles. If controls or the outside of sterile-wrapped lancet packages and the carriers that bring supplies to patient's rooms are touched, that action can contaminate products and provide additional sources for transmission of organisms to another patient. Changing gloves does not prevent the transmission, since the containers, strips and carriers are contaminated. Operators should be aware that not only the device can transmit infectious organisms, but also all the products that are touched or utilized for POCT are potential sources of infection that should be controlled.

The risk assessment should consider the entire testing process, including the **postanalytical phase**. POCT results may not get recorded in the patient's medical record because transcribing results into an electronic record is an additional step that may be skipped once the result is available. This is especially problematic for manual tests, like urine dipsticks, pregnancy and rapid strep tests. Data management on POCT devices provides a means of automatically capturing test results and electronically transferring those results to a laboratory information system or hospital medical record. So, all results get documented. However, some devices require operators to physically dock the devices or push send buttons in order to transmit results from the device to the medical record. Wireless data transfer has eliminated those steps and ensures that results are available in the patient's chart as soon as testing is complete. This reduces delays and staff confusion looking for POCT results in the medical record that may not have been transmitted.

A number of hazards thus exist with POCT. Workflows and operations can differ between sites, creating a variety of possible errors. Risks on one nursing unit may be different than risks in another setting even for the same POCT device. The benefit of developing a QC plan is having the laboratory step through the testing process with the clinical operators as a multidisciplinary team to optimize and standardize the details of the workflow. When conducting a risk assessment, staff should consider the risk of errors from a variety of sources, as depicted in ◘ Fig. 40.5. These include patient ID, specimen collection, sample application, test timing, QC performance, operator training/competency, device calibration, expiration dates, storage and environmental exposure, infection control, and result transmission to the medical record. Manufacturers have developed unique control processes to address specific hazards that can reduce the chances for error. Understanding how these control processes work and their limitations allows institutions to determine what additional actions must be taken to minimize risk. Together, the manufacturer's control processes and laboratory-implemented actions constitute the institution's QC plan.

40.2.2 Individualized Quality Control Plans (IQCP)

Risk management principles have been incorporated into the new CLIA interpretive guidelines that were implemented on January 1, 2016 [6]. These changes allow laboratories to develop an Individualized Quality Control Plan (IQCP) that incorporates many of their existing quality practices and information. The QC plan is "individualized" in that every site is unique, with different tolerance for risk and choices in how hazards are addressed. The IQCP is based on a laboratory's patient population, environment, test system and, most importantly, how the physician intends to use the test result. The clinical application defines the tolerance for analytical performance, accuracy and imprecision. Laboratories have a choice:

- analyze two concentration levels of QC each day of patient testing or
- develop an IQCP.

This change applies to all existing tests and any new tests or test systems coming on the market in the future. For now, CMS is applying the change to CLIA non-waived testing, but anyone conducting laboratory testing, particularly POCT, can benefit from conducting a risk assessment that enables them to better understand associated weaknesses and ways to address those hazards.

> **An IQCP has three parts; a risk assessment, the QC plan and a quality assessment.**

An IQCP has 3 parts that state inspectors will be reviewing: a risk assessment, the QC plan, and the quality assessment (Fig. 40.3). The risk assessment is a map of the testing process, identifying the weak steps in the process where errors could occur. These hazards may be addressed by a manufacturer's control process or a manufacturer-engineered quality process or by a check built-in the instrument. The laboratory could also take an action, such as more frequent maintenance, competency verification, site compliance audits, temperature monitoring or other means of reducing the chance of a specific error or enhancing the ability to detect that error before the result is reported. The list of hazards identified by the process map and actions taken by the laboratory constitute the IQCP. The IQCP is the laboratory's first guess at a plan to ensure quality and reliable results when implementing the test. Once implemented, the IQCP is monitored for effectiveness by the quality assessment. Test performance is tracked for trends. Errors and physician complaints are investigated and actions taken to prevent future recurrence. Risk is reassessed and the IQCP modified in a continuous cycle of quality improvement (Fig. 40.3).

Developing an IQCP provides several benefits. An IQCP identifies the weak points in the testing process. These weak points are the steps in the testing process that are most likely to fail, i.e. areas that the laboratory should focus additional defenses or activities on in order to prevent errors. Those may include increased maintenance, manual test or specimen checks, additional training or similar actions.

For single-use POCT cartridges or test kits with built-in controls, an IQCP allows the laboratory to decrease the frequency of liquid QC to the minimum recommended by the manufacturer. The internal QC can substitute for external QC in reducing the chance of certain errors when used in conjunction with preventive actions that consider all phases of the testing cycle and focus on weak steps in the testing process. The IQCP saves on non-patient test expenses and time to perform and document those tests, thereby reducing costs and enhancing efficiency. This can amount to significant savings. If a hospital is conducting urine pregnancy testing and performing two QC each day of testing, this is 60 tests per month that are used for quality assurance and regulatory compliance. Performing an IQCP allows the hospital to swap internal QC on each test for daily QC and reduce the frequency of liquid QC to once or twice a month, saving 56–58 tests each month. The savings are even greater if urine pregnancy is performed at 10 locations in the hospital: emergency room, operating room, catheterization lab and outpatient clinics.

Developing an IQCP brings the laboratory and clinical staff together to discuss workflows and enhances collaboration and communication. This provides an opportunity to uncover inconsistencies in operations and POCT management between nursing units and to harmonize these processes.

While developing an IQCP for blood gases conducted on a life-flight helicopter service, it was discovered that each of the service's 7 locations were separately managing the ordering, stocking and validations of their blood gas cartridges. Blood gas testing is a CLIA moderate-complexity test and requires verifying cartridge reactivity prior to patient testing for each shipment (even if of the same lot), regular QC and 6-month calibration verification and result correlation between devices (◘ Tab. 40.2). Each site was conducting QC (2-levels) for each shipment of cartridges every 4–6 weeks (approximately 140 tests/year). Lot changes occurred frequently, at least every other shipment (5 times/year) and required 2 levels of QC on every device (8 devices and 5 times a year for 80 tests/year). QC was performed monthly on each of 8 meters (192 tests). Calibration verification and 20 patient correlations were performed on every meter twice a year (304 tests/year). When developing the IQCP, the low, normal and high QC was found to be the same solutions sold by the manufacturer as part of the 5 level linearity kit. That meant, performing 6-month calibration verification (3-levels) was redundant to the monthly QC if 3-levels are analyzed as these constituted the same solutions. When the laboratory considered that the chemistry of the reactions is contained within the cartridges, the device is only a voltage meter.

So, by controlling lots of cartridges rather than the individual meters, the number of quality tests consumed could be reduced without compromising result reliability. After implementing the IQCP, cartridge shipments were consolidated to one site which performed verification of 3 levels of QC and then distributed the validated cartridges to the other 7 sites. Lot-to-lot comparisons were started to include 5 patients and were also performed by the central site before distributing cartridges. Monthly QC required the same number of tests each year to verify stability of cartridge storage at each site, but the spare meter at the central office could be rotated with the patient testing analyzer. Since cartridges were used rapidly and shipment orders were reduced to once a quarter, 6-month calibration verification was no longer required because it was already being performed with monthly QC. Six-month patient correlations also became unnecessary because 5 patients were now included with every lot validation. This new workflow improved efficiency and reduced non-patient tests by half, without altering the quality of patient results (◘ Tab. 40.2). In fact, quality was improved, because one was now controlling each lot of cartridge, i.e. the chemistry of the method, rather than focusing on the devices which were already controlled by the manufacturer's built-in checks and processes.

Developing an IQCP provides several benefits to the laboratory

- Promotes multidisciplinary communication and collaboration
- Identifies weaknesses in the testing process, steps where errors may occur
- Uncovers discrepancies in processes between sites allowing for harmonization of workflow and operations
- Establishes the rationale for actions, why QC is performed and what hazards are addressed by staff actions
- Identifies areas for improving efficiency and saving costs

40.3 Conclusion

Risk management is a process to identify weaknesses in the testing process and reduce error in the clinical laboratory. Newer devices and laboratory instruments have a variety of built-in control features that can minimize the chance of certain errors occurring. POCT presents different workflows and has unique challenges

Tab. 40.2 Comparison of non-patient testing required for QC, validations and regulatory compliance before and after developing an IQCP at a life-flight helicopter service for blood gas and electrolyte testing. Before developing the IQCP, reagents were ordered independently at each of 7 sites. QC was analyzed on each analyzer at each site. After developing the IQCP, reagent ordering was consolidated to one centralized office to perform validations prior to distributing cartridges to other sites. In addition, QC was now performed on each lot of cartridge rotating the devices. This change in operations reduced the number of non-patient cartridges consumed and ensured every lot of cartridge was being controlled with liquid QC. Every device was controlled by the manufacturer's built-in checks and processes. *Italics* show maximum number of cartridges consumed, assuming each site receives a new lot of cartridges with each shipment

Before IQCP (7 sites and 8 devices)		
Cartridge shipments	10 shipments/yr × 2 levels QC × 7 sites	= 140 tests
Lot validations	5 times/yr × 2 levels×8 meters	= 80 tests
QC monthly	8 meters × 2 levels × 12 months	192 tests
6-mo calibration verification	8 meters x 3 levels x 3 triplicate xv 2/yr	= 144 Tests
6-mo patient correlation	10 patients × 8 meters × 2/yr	= 160 tests
Sum	Total QC and validations	= 716 tests
After IQCP (7 sites and 8 devices)		
Cartridge shipments	4 shipments/yr x 3 levels QC x 1 site	= 12 tests
Lot validations	Maximum 4/yr – already included with shipment QC	
	Maximum 4/yr x 5 patients x 2 lots (old vs new)	= 40 tests
QC monthly	7 sites x 3 levels x 12 months	= 252 tests
If additional lots are present	7 sites x 3 levels x 12 months	= *84 tests*
6-mo calibration verification already completed with monthly QC		
6-mo patient correlation already completed with lot validations		
Sum	Total QC and validations	= 304 tests
Sum if additional lots are present		*(388) tests*

and risks to consider. Developing a quality control plan must be individualized because of the variety of ways that POCT and laboratory tests can be deployed, the many options that laboratories can take to minimize errors and the institutional tolerance for risk. The CLSI EP23-A guideline is a resource that introduces industrial risk management principles to the clinical laboratory. Changes to CLIA-interpretive guidelines in the US have incorporated risk management principles and enable laboratories to reduce the frequency of QC provided the laboratory develop an IQCP.

References

1. Alreja G, Setia N, Nichols J, Pantanowitz L (2011) Reducing patient identification errors related to glucose point- of-care testing. J Pathol Inform 2: 22. www.ncbi.nlm.nih.gov/pmc/articles/PMC3097526/
2. Balogh EP, Miller BT, Ball JR (2015) Improving diagnosis in healthcare. Committee on Diagnostic Error in Health Care; Board on Health Care Services; Institute of Medicine; The National Academies of Sciences, Engineering, and Medicine. Washington, DC. http://www.nap.edu/catalog/21794/improving-diagnosis-in-health-care
3. Bonini P, Plebani M, Ceriotti F, Rubboli F (2002) Errors in laboratory medicine. Clin Chem 48: 691–8
4. CDC (2015) Ready? Set? Test! Centers for Disease Control and Prevention, Center for Surveillance,

Epidemiology, and Laboratory Services, Division of Laboratory Systems. Atlanta, GA. http://wwwn.cdc.gov/clia/Resources/WaivedTests/pdf/15_255581-A_Stang_RST_Booklet_508Final.pdf
5. Clinical and Laboratory Standards Institute (2011) EP23-A: Laboratory Quality Control Based on Risk Management; Approved Guideline. CLSI, Wayne, PA
6. CMS (2015) State Operations Manual, Appendix C – Survey Procedures and Interpretive Guidelines for Laboratories and Laboratory Services. Rev. 147. Centers for Medicare and Medicaid Services. Washington, DC. www.cms.gov/Regulations-and-Guidance/Legislation/CLIA/Interpretive_Guidelines_for_Laboratories.html
7. Davis K, Schoenbaum SC, Collins KS, Tenney K, Hughes DL, Audet A-MJ (2002) Room for improvement: Patients report on the quality of their health care. The Commonwealth Fund. http://www.commonwealthfund.org/publications/fund-reports/2002/apr/room-for-improvement–patients-report-on-the-quality-of-their-health-care
8. FDA (2010) Use of Fingerstick Devices on More Than One Person Poses Risk for Transmitting Bloodborne Pathogens. U.S. Food and Drug Administration. Washington, D.C. http://www.fda.gov/MedicalDevices/Safety/AlertsandNotices/ucm224025.htm
9. FDA (2015) Useful Tips to Increase Accuracy and Reduce Errors in Test Results from Glucose Meters, U.S. Food and Drug Administration, Washington, D.C. http://www.fda.gov/MedicalDevices/Safety/AlertsandNotices/TipsandArticlesonDeviceSafety/ucm109519.htm
10. International Organization for Standardization (2014) ISO/IEC Guide 51. Safety aspects – Guidelines for their inclusion in standards. ISO, Genf, Schweiz
11. Kohn LT, Corrigan JM, Donaldson MS (1999) To err is human: Building a safer health system. National Academy Press, Washington, DC
12. Leape LL (1994) Error in medicine. JAMA 272: 1851–7
13. Leape LL, Brennan TA, Laird N, Lawthers AG, Localio AR, Barnes BA, Hebert L, Newhouse JP, Weiler PC, Hiatt H (1991) The nature of adverse events in hospitalized patients. Results of the Harvard Medical Practice Study II. N Eng J Med 324:377–84
14. O'Kane MJ, McManus P, McGowen N, Lynch PLM (2011) Quality error rates in Point-of-Care Testing. Clin Chem 9:1267–71
15. Quality Interagency Coordination Task Force. Doing What Counts for Patient Safety: Federal Actions to Reduce Medical Errors and Their Impact. A Report of the Quality Interagency Coordination Task Force (QuIC) to the President. Department of Health and Human Services, Agency for Healthcare Research and Quality. Washington DC, August 2000. http://archive.ahrq.gov/quic/report/mederr4.htm
16. Reason J (2000) Human Error: Models and Management. BMJ 320:768–70
17. Rust MJ, Carlson NA, Nichols JH (2012) A thermo-modulating container for transport and storage of glucose meters in a cold weather environment. Point of Care 11(3): 157–60
18. Silverman BC, Humbertson SK, Stem JE, Nichols JH (2000) Operational errors cause inaccurate glucose results. Diabetes Care 23:429–30
19. Silverstein MD (2003) An approach to medical errors and patient safety in laboratory services. A white paper prepared for the quality institute meeting, making the laboratory a partner in patient safety. Division of Laboratory Systems, Centers for Disease control and Prevention. Atlanta, GA. http://citeseerx.ist.psu.edu/viewdoc/download?doi=10.1.1.615.4936&rep=rep1&type=pdf
20. Vanhaeren S, Duport C, Magneney M, Dumé L, Dumenil AS, Doucet-Populaire F, Decousser JW (2011) Bacterial contamination of glucose test strips: Not to be neglected. Am J Infect Control 39:611–3

Development trends

Content

Future POCT systems

Sandeep K. Vashist, John H.T. Luong, Peter B. Luppa, Ralf Junker

P. B. Luppa, R. Junker (Eds.), *Point-of-Care Testing*
https://doi.org/10.1007/978-3-662-54497-6_41

41.1 Introduction

The POCT market in Europe currently registers annual sales of around € 3.5 billion. The home care sector, primarily consisting of diabetes monitoring, clearly makes up the largest proportion. In the past decade, the growth rates in the POCT sector have been risen at a continuously fast pace of more than 10 %. The first question to be asked is thus how to explain this trend and, secondly, what will near-patient laboratory diagnostics look like in the future. The fast pace of these developments is not only to be explained by the trends in personalized healthcare and personalized therapy monitoring, but also by the growing number of continuous monitoring options that can be subsumed under the catchphrase mobile healthcare (mHealth) [7]. Notably, the past few years have seen smartphone-coupled analytical systems becoming the embodiment of future technologies for POCT applications while registering their spread on a global scale.

Nevertheless, it is not to be feared that future device generations and fields of applications for POCT will make the professionally operated central laboratory superfluous – just as the introduction of notebooks, tablet PCs and smartphones did not make large data processing centers obsolete either. Instead, a paradigm shift is taking place that is bringing simple tests nearer to the patients and general practitioner. This will free up capacity at central laboratories for the more complicated and time-consuming procedures such as next-generation sequencing or mass spectrometric proteome analyses [2].

When talking about such forecasts for the future, it is important to point out that even though it keeps becoming easier to put sophisticated analytical technologies into the hands of the informed and empowered patient, it is not quite as easy to hand over to them the vast expertise possessed by the laboratory physician. Thus, the scenario described above will most likely be limited to laboratory values that do not require special interpretation skills (for example clinical follow-ups) or that can be readily interpreted by an attending physician without laboratory skills. Their broader use will certainly not become reality until POCT analytical systems will be incorporated into telemedical networks such that they allow transmission of abnormal values to experts for professional assessment (▸ Chapter 35).

41.2 Smartphone-based POCT systems

41.2.1 Bioanalytical applications

Thus far, smartphone-based applications for POCT devices have been developed for lateral flow and immunoassays, for electrochemical and colorimetric assays, as well as for microscopy and flow cytometry, spectrophotometry and SPR-based biosensors (◘ Fig. 41.1a–e). Others link smartphones with lenses to create a compact and light-weight device for applications based on light, fluorescence, dark field, transmission and polarizing microscopes [3, 13, 19]. A portable optofluidic fluorescent imaging cytometry connected to a cell phone screens for parasites in whole blood [20]. Another example is a miniaturized, affordable cytometry platform designed to measure the Hb concentration and number (density) of RBC and leukocytes in whole blood by light and fluorescence imaging [21].

Numerable examples can be cited for smartphone-based POCT applications for colorimetric tests [12] and immunoassays [14]. The same applies to fluorescence and (chemi)-luminescence measurements [9]. A fluorimeter prototype [18] is presented in ◘ Fig. 41.1c; It can cover the fully emission spectrum.

41.2.2 Personalized mHealth applications

The determination of vital parameters and routine activities has meanwhile become a movement that is a serious business for consumers (“**quantified self**”). Multiple smartphone and

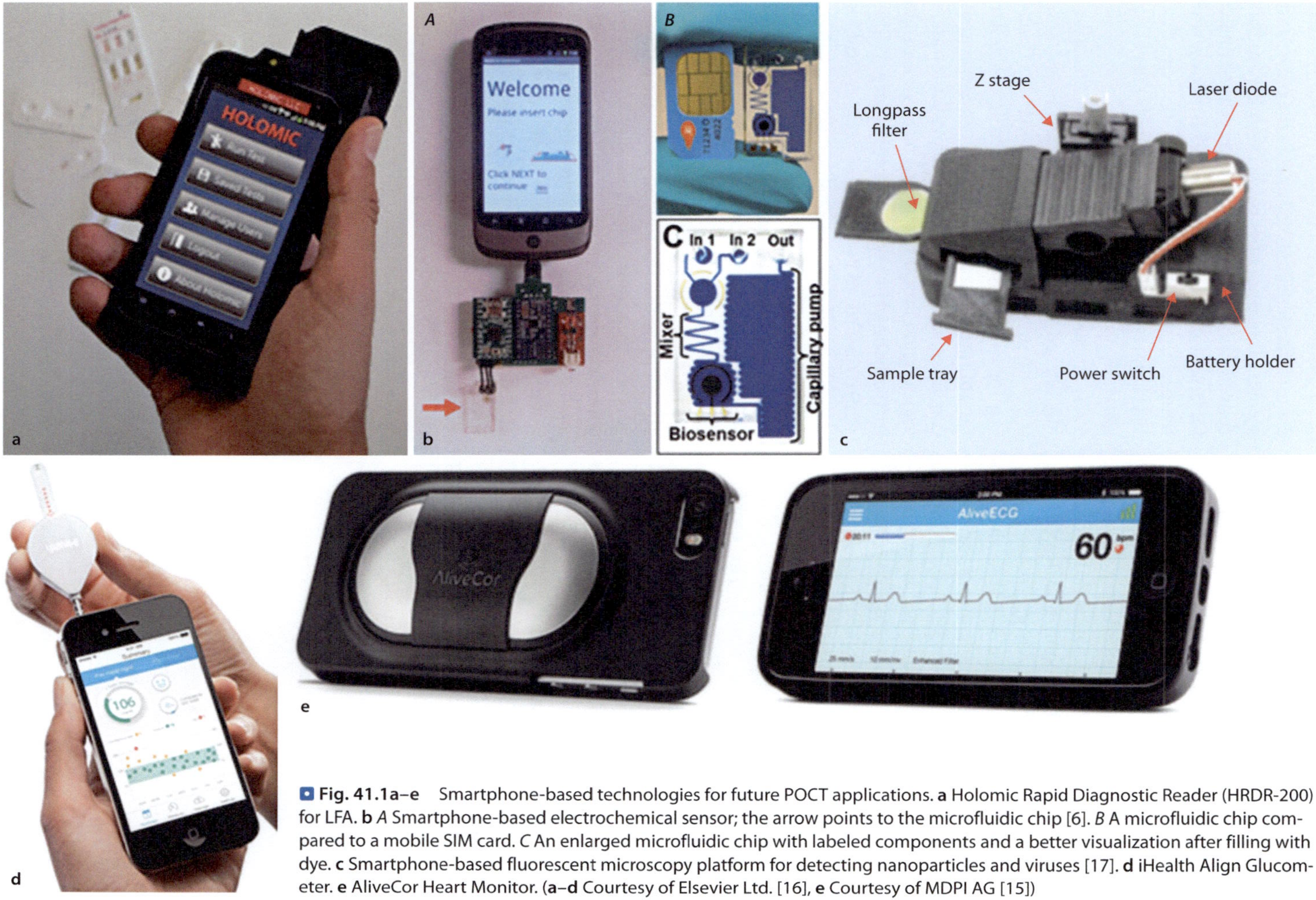

Fig. 41.1a–e Smartphone-based technologies for future POCT applications. **a** Holomic Rapid Diagnostic Reader (HRDR-200) for LFA. **b** *A* Smartphone-based electrochemical sensor; the arrow points to the microfluidic chip [6]. *B* A microfluidic chip compared to a mobile SIM card. *C* An enlarged microfluidic chip with labeled components and a better visualization after filling with dye. **c** Smartphone-based fluorescent microscopy platform for detecting nanoparticles and viruses [17]. **d** iHealth Align Glucometer. **e** AliveCor Heart Monitor. (**a–d** Courtesy of Elsevier Ltd. [16], **e** Courtesy of MDPI AG [15])

wristband applications already allow personalized mHealth monitoring of body weight, blood pressure, pulse rate, ECG, blood glucose, oxygen saturation (SpO_2), and exercise activity [15] (▶ Chapter 42). There are numerous vendors in this sector. It is interesting to observe how the major smartphone manufacturers are pumping high sums into new application technologies. The aforementioned (▶ Chapter 36) iOS-compatible iHealth Align glucometer (iHealth Labs, Mountain View, CA, USA) represents a compelling technological step towards the next generation of plasma glucose monitoring (◘ Fig. 41.1d). The AliveCor heart monitor (◘ Fig. 41.1e) is a CE-labeled, smartphone-coupled ECG device manufactured by AliveCor Inc. (San Francisco, CA, USA), that uses a single-lead rhythm strip to record, store and then evaluate cardiac rhythm and heart rate. The Kardia Band is a medical-grade ECG band integrated in an Apple Watch. The Onyx II model 9560 finger pulse oximeter (Nonin Medical, Plymouth, MN, USA) is a compact, non-invasive device used to measure SpO_2 and pulse frequency.

41.3 Miniaturization

The term **total analysis system** refers to analytical platforms capable of performing all required processes for the automated analysis of a parameter ranging from sample-taking, (plasma) separation and dilution to implementation, detection and analysis [8]. This term took on a new dimension in connection with **micro total analysis systems** (μTAS): Thanks to microfluidics, chip technology and electrochemical detection methods, the architecture of these analytical systems can be miniaturized down to the micrometer scale. The catchphrase here is a chip-sized analytical laboratory or **lab on a chip**. It is currently possible to design such miniaturized systems that can directly detect and measure DNA or RNA of a sample in the upper nanoliter range. Immunoassays and electrophoretic analysis of proteins can also be conducted with such μTAS [5].

The technical advantages of miniaturization are uncontested: Fluid volumes in the nano- and picoliter range not only cut back on the reagent consumption, but also save on the time required for analytical basic processes like mixing and tempering. For example, those PCR cycles that last minutes on the conventional microliter scale take only seconds in a lab on a chip.

A constantly growing number of microfluidics-based cartridges for diagnostic test systems are presently flooding the market. Looking forward, universally applicable, **modular microfluidic systems** are open platforms that can integrate molecular-biological, immunological and clinical chemistry tests on a single instrument. The system architecture is designed on a broad palette of different cartridges that allow certain functionalities such as valves, fluid reagent supplies, zones for heating or detection processes to be defined at pre-formed sites [1]. This type of architecture also covers all geometric, optical and electrical interfaces between cartridge and measuring device. This also defines the possible system components and the functional limitations of the device. The main objective is always to have all reagents available in cartridge – either in liquid or lyophilized form. This obviates the necessity for an external reagent supply from the device to the cartridge, which minimizes the risk of contamination or leaks. Additionally, it simplifies the design of the devices and makes the system more affordable overall because it is not necessary to integrate pumps or pressure controllers. The internal valves are generally designed as mechanical rotating valves.

Cartridges with a system architecture for various applications designed by microfluidic ChipShop (Jena, Germany) are illustrated in ◘ Fig. 41.2 as examples of model assays for molecular-biological, immunological and clinical-chemical tests. It should be pointed out that the cartridge on such open platforms does not necessarily have to exploit all possible functionalities for the various applications. This concept is aimed at minimizing the costs for developing a diagnostic test by providing a pre-validated cartridge platform where the microfluidic

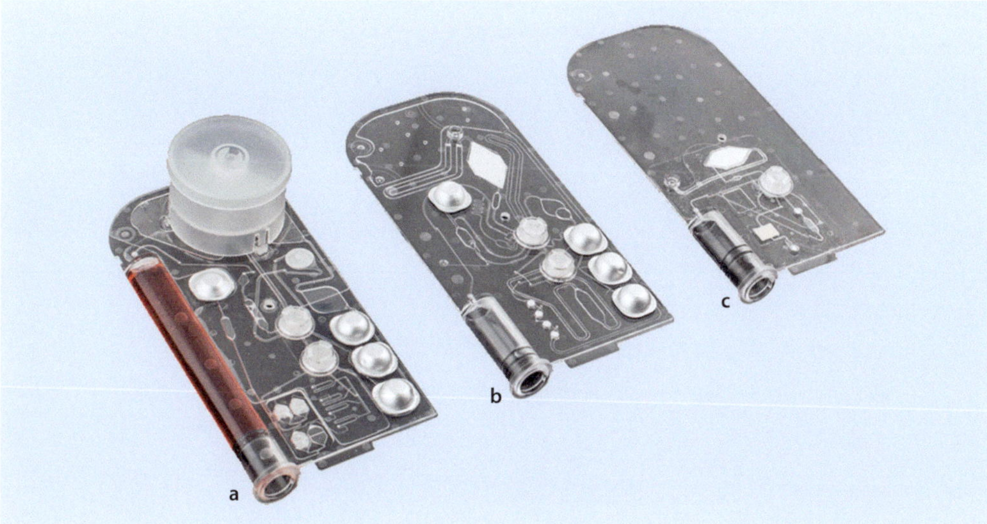

Fig. 41.2 System microfluidic cartridges manufactured by Microfluidic ChipShop. **a** MTB (molecular biological assay); **b** HIV p24 (immunoassay); **c** ALT (clinical chemistry). (Courtesy of microfluidic ChipShop, Jena, Germany)

pouring and design rules have been previously defined. Such concepts offer chances to develop microfluidic POCT methods at affordable prices, which would particularly benefit applications in third-world countries (▶ Chapter 36).

Rissin et al. [10] reported on an immunosorbent assay designed to detect serum proteins within the subfemtomolar concentration range. For this purpose, the authors used a single-molecule array consisting of thousands of wells, they called reaction chambers, of exactly 50-femtoliter in size. Antibody-loaded microbeads fit into these wells when they consist of exactly one antigen and one enzyme-labeled secondary antibody. The principle of this **digital ELISA method** is illustrated in Fig. 41.3. A commercial application of this new analytical principle was successfully implemented in the Simoa HD-1 analyzer by the Quanterix Corporation (Lexington, MA, USA).

For the future, it will also be important to further optimize the speed, sensitivity and specificity of such immunoassay formats. Sakamoto et al. [11] reported recently that reactions with sandwich immunoassays using functionalized, fluorescent submicrometer magnetic beads yielded significantly faster assay times. The glycidyl methacrylate polymer-coated microbeads consist of a conglomerate of ferrite nanoparticles and fluorescent europium complexes.

41.4 Parallelization

Another advantage of chip technology is the possibility to incorporate numerous measuring channels or spots onto the smallest of spaces. Photolithographic techniques have made it possible to etch fluid channels and reaction chambers for 10 or 100 simultaneous measurements onto one wafer, as depicted in Fig. 41.4. The most well-known example of a currently available commercial multiplex system is Agilent's Bioanalyzer which performs DNA, RNA, protein and cell analyses based on capillary electrophoresis.

At present, it is already state of the art to run 10,000 to 100,000 measurements simultaneously, as is done with microarrays for genetic analysis. This involves printing or synthesizing microscopic spots in the form of gene probes onto solid surfaces and then hybridizing them with target analytes of the sample such as DNA,

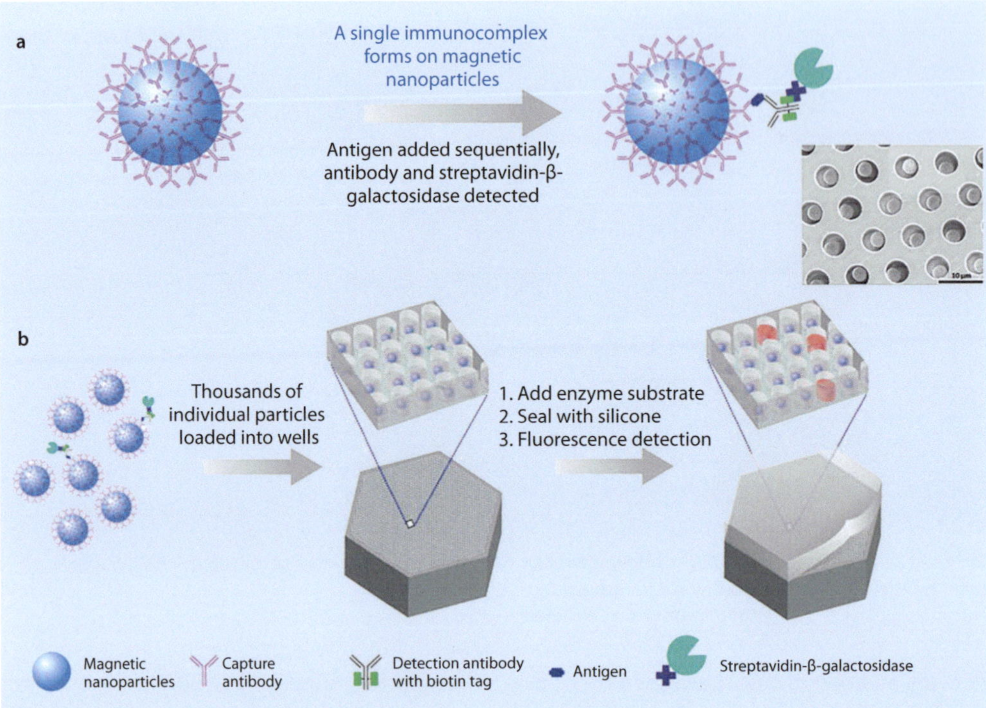

Fig. 41.3a,b Digital ELISA, based on an array of femtoliter wells. **a** Antibody-loaded nanoparticles detect the antigen. **b** Loading of beads into nanowell arrays with nanoparticles and immunoreaction signaling. (Courtesy of Nature Publishing Group (NPG) [10])

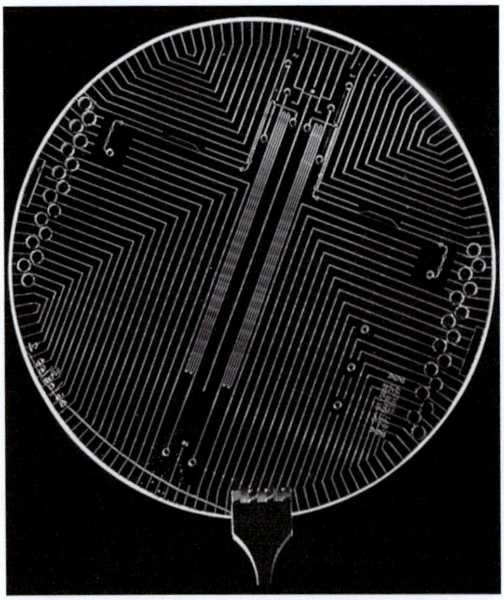

Fig. 41.4 Photolithographically produced wafer

mRNA or proteins. After washing, the bound substances can be demonstrated with fluorescent dyes by generating a computer image of all illuminated spots, which are then computed and evaluated mathematically.

The problems associated with such an extreme degree of parallelization in medical diagnostics are less their technological feasibility, but rather how to efficiently analyze and interpret the vast amounts of data generated. To make such data volumes manageable and accessible to solving complex challenges like the analysis of a malignant proteome, a new discipline has emerged at the cusp between biology and informatics – the field of bioinformatics. Even though microarrays have driven the development of highly parallelized analytical techniques and spurred on medical research, it is an unlikely prophesy that they will be harnessed for routine diagnostics in the near future. One of the reasons for this derives from standardization difficulties and the frequently poor reproducibility of results; another reason is the questionable benefit of 10,000 or 100,000 measurements of different analytes in the same patient. Oligoarrays with up to 100 analytes, often supplied as beads (**Luminex technology**) [4] hold greater promise in this regard. Here, analysis is performed on a flow cytometer eliminating the characteristic of miniaturization. The same applies to mass spectrometry, which can likewise quantify several hundred analytes simultaneously, and is currently being widely implemented in microbiology, toxicology and therapeutic drug monitoring.

The medium-scale parallelization of analysis offers significant practical advantages: It saves time and money, for example, when admitting patients to a hospital or in emergency settings where comprehensive lab profiling can be done fast and cost-effectively in order to obtain a broadly applicable overview of a patient's pathological status within minutes. Afterwards, only a few of those parameters must be selectively followed up on that would mandate additional diagnostics. This strategy, referred to as "diagnostic de-escalation", is already being implemented in the USA.

This contrasts with Germany where only a small cross-section of the theoretically possible analytical spectrum of the conventional "escalating diagnostics" is demanded in the first step. Next, the search is then narrowed down incrementally and focused on a specific target. In any case, this is more economical given the currently high reagent volumes, the long measuring times and the high analysis prices, but consumes valuable time that is not available with lessening hospitalization times or when a patient presents with a life-threatening condition. That laboratory test profiles in certain situations make clinical and economic sense is exemplified by the success of currently available POCT multi-channel analyzers that can determine blood gas and oximetric parameters, electrolytes and metabolites at the same time.

Of course, it is not only the inadequacies of today's large-scale analyzers that make many experts skeptical towards comprehensive baseline profiles. The flood of false-positive findings to be expected from unspecific orders is what prevents experienced diagnosticians from developing an affinity towards large-scale profiles. Because it documents parametric patterns instead of isolated parameters, there is no doubt whatsoever that microtechnology opens entirely new diagnostic and prognostic horizons. The next challenge is now subjecting it to impartial and evidence-based scrutiny.

References

1. Becker H, Klemm R, Dietze W, White W, Hlawatsch N, Freyberg S, Moche C, Dailey P, Gärtner C (2016) Modular microfluidic cartridge-based universal diagnostic system for global health applications. Proc SPIE 9705:970514; doi:10.1117/12.2217892
2. Bietenbeck A, Junker R, Luppa P (2015) Central laboratory service and Point-of-Care Testing in Germany – from conflicting notions to complementary understandings. Point Care J. Near-Patient Test Technol 14:1–11
3. Breslauer DN, Maamari RN, Switz NA, Lam WA, Fletcher DA (2009) Mobile phone based clinical microscopy for global health applications. PLoS One 4(7):e6320
4. Fulton RJ, McDade RL, Smith PL, Kienker LJ, Kettman JR Jr (1997) Advanced multiplexed analysis with the FlowMetrix system. Clin Chem 43: 1749–1756
5. Komatireddy R, Topol E (2012) Medicine unplugged: the future of laboratory medicine. Clin Chem 58: 1644–1647
6. Lillehoj PB, Huang MC, Truong N, Ho CM (2013) Rapid electrochemical detection on a mobile phone. Lab Chip 13(15):2950–2955
7. Luppa PB, Bietenbeck A, Beaudoin C, Giannetti A (2016) Clinically relevant analytical techniques, organizational concepts for application and future perspectives of point-of-care testing. Biotechnol Adv 34(3):139–160
8. Manz A, Graber N, Widmer H (1990) Miniaturized total chemical analysis systems: a novel concept for chemical sensing. Sensors Actuators B Chem 1: 244–248
9. Petryayeva E, Algar WR (2014) Multiplexed homogeneous assays of proteolytic activity using a smartphone and quantum dots.Anal Chem 86(6):3195–3202
10. Rissin DM, Kan CW, Campbell TG, Howes SC, Fournier DR, Song L, et al. (2010) Single-molecule enzyme-linked immunosorbent assay detects serum

proteins at subfemtomolar concentrations. Nature Biotechnol 28: 595–599
11. Sakamoto S, Omagari K, Kita Y, Mochizuki Y, Naito Y, Kawata S, et al. (2014) Magnetically promoted rapid immunoreactions using functionalized fluorescent magnetic beads: a proof of principle. Clin Chem 60: 610–620
12. Shen L, Hagen JA, Papautsky I (2012) Point-of-care colorimetric detection with a smartphone. Lab Chip 12(21):4240–4243
13. Smith ZJ, Chu K, Espenson AR, Rahimzadeh M, Gryshuk A, Molinaro M, et al. (2011) Cell-phone-based platform for biomedical device development and education applications. PLoS One 6(3):e17150
14. Vashist SK, van Oordt T, Schneider EM, Zengerle R, of Stetten F, Luong JHT (2015) A smartphone-based colorimetric reader for bioanalytical applications using the screen-based bottom illumination provided by gadgets. Biosens Bioelectron 67: 248–255
15. Vashist S, Schneider E, Luong JHT (2014) Commercial Smartphone-Based Devices and Smart Applications for Personalized Healthcare Monitoring and Management. Diagnostics 4(3):104–128
16. Vashist SK, Luppa PB, Yeo LY, Ozcan A, Luong JHT (2015) Emerging technologies for next-generation Point-of-Care Testing. Trends Biotechnol 33(11):692–705
17. Wei Q, Qi H, Luo W, Tseng D, Ki SJ, Wan Z, Göröcs Z, Bentolila LA, Wu TT, Sun R, Ozcan A (2013) Fluorescent imaging of single nanoparticles and viruses on a smart phone. ACS Nano 7:9147–9155
18. Yu H, Tan Y, Cunningham BT (2014) Smartphone fluorescence spectroscopy. Anal Chem 86(17): 8805–8813
19. Zhu H, Yaglidere O, Su TW, Tseng D, Ozcan A (2011) Cost-effective and compact wide-field fluorescent imaging on a cell-phone. Lab Chip 11(2):315–322
20. Zhu H, Mavandadi S, Coskun AF, Yaglidere O, Ozcan A (2011) Optofluidic fluorescent imaging cytometry on a cell phone. Anal Chem 83:6641–6647
21. Zhu H, Sencan I, Wong J, Dimitrov S, Tseng D, Nagashima K, Ozcan A (2013) Cost-effective and rapid blood analysis on a cell-phone. Lab Chip 13(7):1282–1288

The potential for POCT in the Internet of Things (IoT)

Christina Rode-Schubert, Thomas Norgall, Andreas Bietenbeck

P. B. Luppa, R. Junker (Eds.), *Point-of-Care Testing*
https://doi.org/10.1007/978-3-662-54497-6_42

42.1 Introduction

This chapter deals with current innovative approaches to future developments in POCT and addresses topics such as telemonitoring, **ambient assisted living** (AAL) and **personalized health (pHealth)**. In essence, a rapid transformation in terminology can be observed that is not exclusively driven by technical developments. The abbreviation pHealth, previously used mostly as a synonym for personal health, now increasingly refers to the concept of personalized health. At the same time, health-related technology is becoming available to consumers on a market evolving in parallel to the regulated medical technology industry. These trends are often driven by innovations in microelectronics and information technology. The latter not only allows ubiquitous data collection and transmission but also the fusion of data in different formats and from different sources as well as the analysis of vast amounts of data via relevant service providers and infrastructures, using intelligent algorithms. Expectations in new business areas to offer the "prognoses of later events on the basis of historical data" [19] are thereby high. Between 2011 and 2014, "predictive analytics companies in the interdependent sectors of predictive medicine and predictive health increased their acquired venture capital investment by almost five-fold [15].

This chapter describes trends in and drivers of personalized health (pHealth) with a focus on self-monitoring and the technical basis that engenders these developments. Self-monitoring often relies on wearables: intelligent miniaturized components and/or computer systems integrated in wearable devices. In its extreme form, this is expressed by the Quantified Self movement.

Wearables generally interact wirelessly via networks; broadly speaking, their conceptualization characterizes the Internet of Things (▶ Section 42.2). Methods for analyzing the measurements collected are subsumed under the catchphrase Big Data Analytics (▶ Section 42.4).

42.2 From medical records to the Internet of Things

Until now, most health-relevant data, particularly about a patient's diagnosis and therapy, was recorded or saved in their medical records, in databases or in the information system of the respective service provider. The use and transfer of these data are largely subject to legal restrictions. In recent years, the legislator has increasingly tried to facilitate cooperation and integration within the health care system and to strengthen the role of patients and citizens by using information technology (electronic healthcare cards, telemetric infrastructure) and by maintaining a high standard of data protection and security. As "owners" of their data, they should be able to decide which information is accessible to which service provider. Access to their own healthcare data should promote the patients' awareness of, engagement in and responsibility for leading a healthy lifestyle.

The collection of health-related data on larger population groups is now common practice and has almost become more important than data storage and processing. Viable models for the reimbursement of telemonitoring applications hardly exist within the regulated healthcare market. Nevertheless, feasible payment models have been established for the private consumer market alongside relevant infrastructures and service provision. The technology applied in telemonitoring generally comprises components that are functionally similar to conventional medical devices and systems, but specially designed for use in private settings [35]. A telemonitoring system typically consists of sensors that communicate with a base station via a short-range wireless network, also called body area network or personal area network. The base station is located close to or worn by the user. It can store, processes and transmit sensor data to the telemedicine and telehealth service provider via a wireless or wired transmitter system. Nowadays, the base station is often a smartphone, but may be a dedicated device or computer that is connected to a fixed network system.

The range of parameters collectable by telemonitoring extends from continuous and/or intermittently measured biosignals, e.g. ECG, oxygen saturation, and real-time readings like blood pressure up to POCT parameters. In principle, they can be measured automatically at regular intervals without intervention on the part of patient, sometimes using intelligent assistance systems. Actuator components can be included in this infrastructure, e.g. "smart" defibrillators that transmit cardiologically relevant information immediately to a telemedicine center to prevent acute critical situations [37].

The safe and complete exchange of data is a critical factor for the success of all systems that exchange near-patient data for continuous health monitoring with telemedicine centers or other service providers [2, 3]. That is one reason why the **Personal Connected Health Alliance** (PCHA) operates as an international organization, dealing intensively with the interoperability across all single components in this new, emerging market. This consortium of companies in the IT and healthcare industries has set relevant standards for a number of pivotal communication technologies.

42.2.1 The Internet of Things (IoT)

The concept of fitting intelligent communication devices near humans as well as on and in the human body has existed since the introduction of electronic data transmission and processing systems. In order to accomplish this, microelectronics and communication technology provide more and more suitable electronic building blocks and transmission systems. The evolution of the Internet spawned ideas and concepts for an appropriate and comprehensive network infrastructure.

At the latest, beginning in the last decade of the 20th century, numerous authors have engaged with such topics, only a few of which are touched on here. In his 1991 article "The Computer for the 21st Century", Mark Weiser [28] not only predicted the advent of smart homes and tablet computers, but also coined the phrase "ubiquitous computing" for the pervasive paradigm of systems that incorporate invisible and discreet functionalities into everyday objects. The largely equivalent term, pervasive computing, is preferred by the industry. In 1999, Kevin Ashton used IoT as a technical concept to link and represent intelligent objects digitally for the first time. In 2010, Friedemann Mattern said it like this: "The IoT stands for a vision in which the Internet is extended into the real world and by which many day-to-day objects become part of the Internet."

All approaches to defining the IoT revolve around the following main characteristics:
- A comprehensive (global) dynamic (Internet) infrastructure
- Ability to configure autonomously
- Standards for interoperability and communication

The (material and virtual) "things" to be linked have identities and attributes and are integrated seamlessly into the infrastructure via intelligent, standardized interfaces.

Outfitting objects with barcodes or Quick Response (QR) codes, radio frequency identification (RFID) tags alongside automatic prompts using matching network reading devices can be seen as a rudimentary, widespread form of IoT. These methods are sufficient when it comes to retrieving data from physical objects and feeding them into a central system. Applications established for sample identification in the laboratory can also be used for device and material logistics in hospitals (e.g. OPAL Health System, Fraunhofer IIS, Nuremberg, Germany) [20].

However, if it also involves receiving and processing information or carrying out actions, objects must be fitted with sensors, actuators as well as data processing and storage components. Economies of scale permitting, the appropriate hardware requirements in terms of reliability, energy efficiency and cost can be met best by a high degree of miniaturization and integration of components in the form of **systems on chip**. Microfluidics and biotechnology enable implementation of these trends in **labs on a chip**.

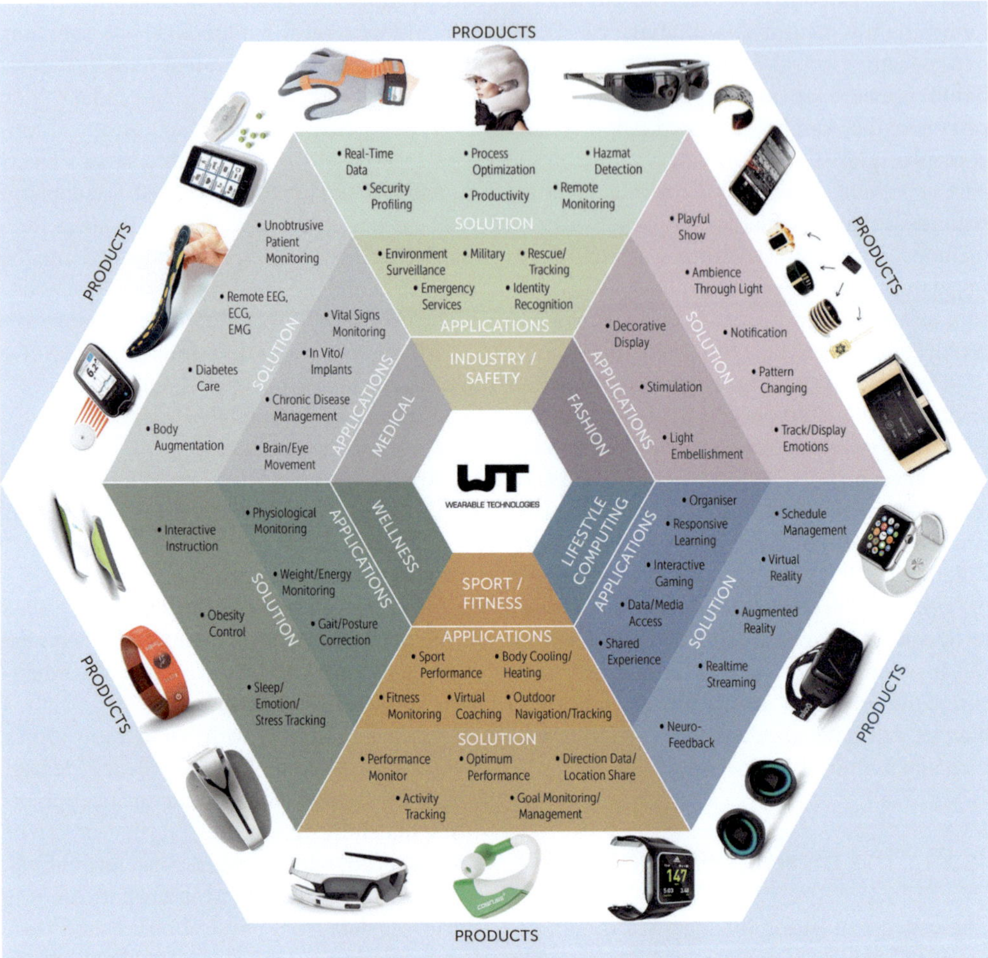

Fig. 42.1 Wearables: Products, functions, applications (overview). (Courtesy of Wearable Technologies AG)

Such miniaturized systems can take over some applications traditionally located at laboratory or computer workstations, where the latter increasingly compete with smartphones and tablet computers. The next stages in this development are termed "wearable computer systems" or "**wearables**". These are already available on the market in many variations – as wristbands, glasses, forehead bands, watches, clips etc. While they only interacted initially with smartphones or similar fully functional and (Internet)-compatible communication devices, they now feature all the required functions themselves. It remains to be seen if the next computer generation, particularly in the peri-human environment (e.g. in the sense of AAL and smart home), will be placed in clothing (smart clothing) or even implanted in the human body.

All in all, long-term development continues: According to Cisco, in 2010, the number of net-linked devices already outnumbered the world's total population. It is predicted that the existing number of 15 billion net-linked devices will rise to 40 billion in the years 2015–2020. Consistent with this way of thinking, such

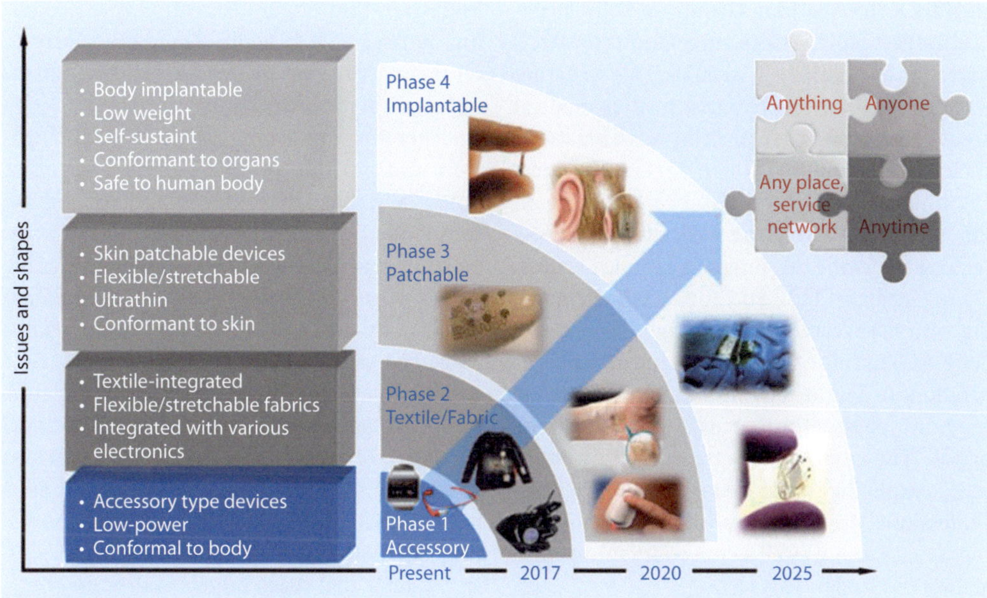

Fig. 42.2 Wearables: Classification by where they are carried. (Courtesy of the International Electrotechnical Commission)

figures can vary considerably, depending on whether, for example, single components, connected indirectly with the net, are counted, e.g. a smartphone connected to a Bluetooth headset. In any case, it can be assumed that well over 99 % of potentially net-linkable physical objects are not currently linked to a network. The much-cited "Internet of Everything" remains a utopian dream for the time being.

42.3 Wearables: Outlook for POCT in the IoT

42.3.1 Background and classification

In a publication about self-monitoring by the Institute of Electrical and Electronics Engineers (IEEE), the authors [14] demanded back in 2005 a wearable, user-friendly and affordable POC system for continuous monitoring. The aim was to promote the patients' independence and their acceptance of new home-based healthcare technology. Configurability, interoperability and scalability were accordingly the important. It is not only necessary to meet these requirements, but also to foster an open market that allows competition and is reinforced by standardization.

The affordable miniaturized computer systems for near-body use called wearables [22] meet the essential requirements for self-monitoring systems (portable, user-friendly and scalable) in the market sectors fitness, wellness and lifestyle [17, 18, 21]. As intelligent devices that are automatically identifiable and interact with their equals, they embody future paradigms of ubiquitous computing and the IoT (Fig. 42.1).

A uniform **definition** of wearables does not exist. In addition to numerous standards for the clinical intensive care and monitoring sector, the international CEN ISO/IEEE 11073 group of standards includes a range of device specializations for personal health devices that specify wearable-typical technologies and functionalities such as blood pressure or glucose monitoring. Alongside the responsible standardization committees ISO TC215 and the IEEE Personal

Health Device Working Group, an IEEE structure for Fast Moving Consumer Goods (FMCG) started to establish itself in 2015. This structure was designed for involvement in non-medical areas. Other standardization organizations such as the International Electrotechnical Commission (IEC) have also been working on this topic that covers the wellness and fitness sectors, under the title Smart Wearable Devices.

In Germany, TÜV Süd 2015 has been working with the Wearable Technologies Group AG (WT AG) as one of the first certification organizations to develop an internationally available program for testing smart watches and fitness bands. The corresponding certification mark [27] is explicitly issued for wristbands for non-professional use only: "Products for medical or therapeutic use, as well as the companion apps, are not covered by this test."

Fundamentally, medical devices and healthcare-related consumer products are subject to different but similarly binding regulatory requirements. It is foreseeable in the future that new questions will arise about the validity of information obtained by the collation of data from multiple sources.

The following consideration depicts the future **classification** of wearables by location on the human body. This shows that wearables are intelligent solutions that can be worn in or on the body or carried near the body, thereby meeting the criteria for Ubiquitous Computing and IoT (Fig. 42.2). No distinction is made between medical and consumer products. The reasons for this are:

- The convergence of basic functions used to collect health-related parameters.
- The assumption that it can be useful to link data from medical products to those from consumer devices, for example, to derive information relating to preventive behaviors.
- The fact that patients with certain, primarily chronic diseases, belong to the most consequent and consistent users of wearables from the fitness sector [24].

Moreover, traditional delineations are becoming increasingly difficult. There are solutions on the market that are used invasively, but are not medical products.

There are a wide range of relevant R&D projects and pipeline products in all categories. Potentially relevant in the POCT context are future devices that are implanted in the body. Not only are smart watches and intelligent wristbands worn on the body. Indeed, stimulating whole body suits will be able to monitor health-related parameters in the future. Intelligent contact lenses will monitor glucose, while miniaturized intelligent hearing aids or smart tattoos/patches (see below) will similarly be worn as companions to smartphones or tablets installed with the matching apps. Wearables are to be distinguished from mobile solutions, such as:

- Systems that reach targets and/or people autonomously – such as the drones envisaged by Google which fly to the site of the emergency to deliver requested medication urgently
- Robots like the Watson-powered Pepper [30] by SoftBank, partnered with IBM, deliver specialist medical knowledge
- Solutions requiring a specific geographical installation and/or stationary electrical power supply – particularly SmartHome and AAL applications

Wearables represent interfaces [7] between the users and their environment. Wearable devices detect parameters that describe the users themselves and those that characterize users' environments. Besides extensive sensor systems, they can also be equipped with actuators and complex user interfaces. For example, they can be used to operate SmartHome solutions. Beyond interoperability, widespread availability and reliability of the components, context-sensitive interactions of the devices with the user and unobtrusive support are also required, so that their use is not perceived as disruptive. Intelligent algorithms make it possible to draw conclusions, initiate activities and trigger actuators. Sensor-based systems that can filter personalized information to and from the interacting surroundings [23] offer the potential for new diagnostic and therapeutic approaches. Data is stored, processed, fused and analyzed, often using cloud structures (▶ Section 42.4).

Wearables are part of the global digital infrastructure where data from various sources are merged and processed. They are also subject to all interrelated risk factors.

42.3.2 From self-monitoring to the quantified self

In 2007, the term "quantified self" was coined by Gary Wolf and Kevin Kelly, two authors of the technology magazine Wired. Since then, groups of private users have emerged worldwide who engage with this topic. Their objective is expressed in the slogan "Self Knowledge Through Numbers": enabling us to learn about our own bodies from measurements. The captured data are designed to foster a deeper (self-) awareness that can also be used for self-improvement. While such possibilities are, of course, also available for patients, the self-awareness achieved through the Quantified Self always assumes that measurements are initiated by the person concerned who is also responsible for them. As soon as a physician requests a measurement for a patient, it does not refer to the Quantified Self.

Quantified Self measurements are mainly conducted with wearables and reflect the everyday lives of the persons measuring themselves. The acquired knowledge results from the integrated analysis of as many data sources as possible, using effective analysis algorithms. Here too, POCT measurements can be included, for example, determining blood glucose concentrations. Numerous service providers have emerged in this field, also for clinical laboratory tests.

42.3.3 Wearables – an expanding market

The worldwide market for wearables has been growing steadily in recent years. According to the International Data Corporation (IDC) [8], the sales of the market-leading Fitbit wristband more than doubled in 2015 from 3.9 million items in the 1st quarter to 8.1 million in the 4th quarter. Alongside Fitbit, Apple, Garmin, Samsung and Xiaomi are the top 5 providers of devices worn on the arm, including fitness wristbands and smart watches. In the context of the product-relevant wearables, studies by Gartner showed that the market for smart watches shows the strongest growth worldwide. Gartner calculated their global sales for 2017 at 66.7 million units. For comparison purposes: In 2015, smartphones sales amounted to more than 1.4 billion units [9].

International comparisons have shown that 12 % of the population in the USA uses a smart watch or a fitness band [13]. That is nearly twice the proportion of users than in Europe where the user proportion is only 6.6 % in the countries England, Germany, France and Italy.

42.3.4 Wearables as certified medical devices in the regulated healthcare market

The certification of wearables as medical devices opens up access to the regulated market that is still fundamentally under the national jurisdiction of the respective healthcare systems. In recent years, there have been efforts to harmonize the relevant standards, not only based on the Medical Device Directive (MDD) within the EU, but also by applying the standards of the U.S. Food and Drug Administration (FDA). Estimates, such as that by Soreon Research [12], predict that the regulated market for intelligent wearable devices will grow from $2 billion in 2014 to more than $41 billion in 2020. Particularly attractive indications for these wearables are diabetes, sleep disorders, obesity and cardiovascular diseases. In the USA, the use of wearables [12] is being promoted by an initiative for "value-based healthcare" that promises physicians and hospitals financial incentives for demonstrating the treatment compliance of their patients.

42.3.5 Sensor systems for wearables: Physical sensor technologies

The wearables currently available on the market mainly fall under the fitness and wellness category and are integrated with physical sensors optimized in terms of autonomy, miniaturization, production costs and energy use. A broad range of technologies covers a diversity of measurable parameters, with some of the most important ones receiving mention here. Sensors can measure acceleration, rotation and position using integrated micro-mechanical systems (MEMS) that record movements and gestures. A low-G acceleration sensor that measures forces in the magnitude of the earth's gravitational field is used to count steps. The gravitation vector that points downwards vertically in a resting position describes a typical trajectory during a step. These measurements make it possible to both count steps taken and estimate step length. It is possible to calculate the distance by multiplying the step length by the number of steps. The calculation of distance and position is aided by information from the Global Positioning System (GPS).

Semiconductor-based optical sensors record pulse frequency, pulse wave and blood oxygen saturation. Here, the absorption or reflection of one or several LED-irradiated lights is measured photoelectrically. While green LEDs are particularly efficient for pulse frequency or pulse wave determination (photoplethysmography), defined frequencies in the (infra-)red spectrum are used for pulse oximetry because of the specific absorption behavior of oxygenated and deoxygenated hemoglobin. Similarly, proximity sensors work photoelectrically, detecting whether a smartphone is held to the ear or if wristbands are worn and can activate an automatic pulse measurement. Numerous other sensors based on different technologies are suitable to capture environmental parameters such as UV exposure or air quality.

42.3.6 Multiparametric applications

As mentioned above, sensor-recorded data from wearables can be amalgamated, for example, with coordinates from the GPS. The corresponding algorithms can then derive information about a user's movement behavior from these amalgamated data.

Generally, the integration of additional parameters and sensors allow the improvement of established applications. Polar, a company that manufactures sports watches and wristbands, also offers exercise and test programs that rely on sensor-captured parameters. The orthostatic test [16] provides regular balance monitoring between exercise and resting state and is based on exercise-related changes in the function of the autonomous nervous system. Results differing from the input measurement document the self-regulation of the cardiovascular system. External factors such as mental stress, sleep quality, ambient conditions (temperature, altitude) or hidden diseases can influence the test results. The quantitative assessment of these factors using corresponding sensors and their correlation with test results seems a promising approach to improve the validity and quality of test results. Similar improvements promise multiparametric methods to measure calories burned or sleep quality.

42.3.7 Sensor technology for wearables: Biochemical sensors

Particularly relevant with regard to future POCT developments are those wearables that incorporate biochemical sensors (biosensors for short) without or without physical sensors and that use the corresponding techniques to amalgamate data as mentioned above.

Electrochemical biosensors can detect substances in sweat, tears, saliva, but also in blood or tissue fluid (by minimally invasive measurements or on open injuries). Such sensors are often mounted on a foil substrate and attached to the skin as tattoos, stamps or badges, but can also be integrated into armbands or sweatbands. Sweatbands allow the simultaneous determination of glucose, lactate, sodium and potassium ions concentrations in sweat [5]. When body temperature is accounted for, new applications such as alarm functions can be incorporated therein to protect against dehydration and excess tiredness [11, 26]. Due to the fact that levels of the hormone cortisol in sweat are proportional to those in blood, its measurement helps estimate stress levels. A wearable can be used to stimulate sweat production in targeted parts of the body, thereby facilitating the measurement of the physical effects of certain drugs. It would also be conceivable that such innovations make monitoring the concentrations of specific drugs in the body more effective.

42.3.8 Invasive systems

In the category **"smart implants"**, the consumer market offers **chips**, featuring functionalities like radio frequency identification (RFID) and wireless near-field communication. Smart implant chips are inserted under the skin between thumb and index finger using a needle or through a tiny incision in the index finger. These procedures [25] are not performed by medical staff but by an experienced piercer or "body modder" (body modification expert) [10]. In blogs [6], even physicians are voicing the notion of collecting patient data by such means. The development of regulations allowing international competition regarding such innovations, on the one hand, and protecting the personal rights of users, on the other, is a political challenge.

42.4 From Big Data to Smart Data

In addition to the new ways of collecting data presented here, there are new possibilities for data analysis, particularly using **machine learning techniques**. Methods subsumed under catchphrase Big Data Analysis can crunch large amounts of heterogeneous data for associations too complex for evaluation by humans.

All methods available have very different theoretical backgrounds. In methods categorized as "monitored learning", interrelationships are "learned" from known data during a training phase. For example, instead of coding a program with the hard fact that high CRP levels indicate an infection, the program is fed with the CRP levels and infection status data of multiple patients. The machine learning method derives this association independently and, in the future, an infection is assumed when CPR levels are high. This process also works when multiple different parameters are fed into the system. In addition, machine learning methods can also deal with a quality of data that classical statistical methods cannot. The more data available, the more interrelationships can be elucidated. As a result of their emerging economic potential, data are referred to known as 21st Century Gold.

Extensive computing capacities are required to train machine learning models. Many companies offer special environments for this. IBM with Watson or SAP with HANA, for example, are active in healthcare markets. Apache Spark is a popular framework, available under an open-source license. After an initial one-off configuration for a new data set, such models require only a fraction of the computing capacity. When it comes to analyzing IoT data, it is not sufficient to apply a machine learning method to a mere static data set. In fact, the flow of accumulating data from different sources has to be processed continuously.

Machine learning can be applied to various tasks [2]. The purpose of detecting abnormalities is to check and report unusual constellations of measurements. Predictive analysis is designed to forecast events. In the end, decision-support algorithms go the furthest and suggest final decisions based on the respective available dataset.

Such evaluation methods and new POCT procedures are mutually beneficial. Continuous measurement procedures generate data flows, where their sheer amounts make conventional analysis impractical. When further data are collected besides the POCT measurement, they can also be incorporated into the electronic evaluation. It was only after the causal link between blood glucose profiles and food intake was understood that conclusions could be drawn about how differently individuals respond to similar meals [38]. A further advantage of electronic data analysis is its speed – rapid analysis is optimal for supplementing rapid POCT measurements.

42.5 Outlook

Either for Quantified Self applications or to crunch the massive numbers from continuous data flows in the IoT, complex data analysis often requires the transmission of raw data to external providers. Medical data are particularly sensitive and must therefore be protected by special means. This protection is not guaranteed by many healthcare apps [33]. In fact, it is often part of the provider's business model to sell the data [3]. Some statutory health insurers themselves support the collection of data from wearables. If these data are used to create a risk factor profile of the insured, this is contrary to the actual purpose of the insurance [35]. Cyber attacks on healthcare institutions have also increased in recent years and sensitive data have been stolen [31].

Despite its many risks, the advantages of electronic data processing in medicine clearly outweigh the disadvantages. Although the joint development of IoT, wearables and POCT is still in a nascent stage, its huge potential is undisputed [32]. In current medical practice, electronic data processing is already indispensable. As early as 2000, the influential report "To Err is Human" demonstrated how crucial it is to have all important information available when treating patients, while being able to detect data analysis-related errors retrospectively [1].

In the spirit of the Quantified Self concept, users of mobile devices receive their analytical results directly, without the attendant advice of a physician. Ongoing changes captured by regular measurements can promote health-conscious behavior in users [36]. Such measurement collection devices like wearables or smart implants, however, are not yet subject to the regulatory requirements governing medical devices and therefore the validity of the results is not guaranteed. Wearables and smart implants are therefore not classified as medical devices and should not be used for medical purposes.

The interpretation of results by the users themselves and/or the use of data from wearables by third parties is always tainted by the fact that the results are obtained from equipment not certified as medical devices. They may not be used by a physician to treat a patient. Also, results obtained from machine learning methods involving the aggregated analysis of data from complex mathematical procedures can only be used by the treating physician when obtained by means of certified methods and devices.

In the future, quality assessment in particular must be improved in order to establish the application of wearables in medicine and to fully exploit their potentials. As with POCT measurements, internal quality controls are not enough, but must be supplemented by externally verifiable quality assurance measures [34]. Conventional inter-laboratory testing, where samples are sent away, is generally not possible for wearables and other continuous measurement systems. Certifications based on comparative measurements on patients, as conducted according to the ISO standard 15197 for glucose meters, may be a complex, albeit an effective alternative [4]. When complex algorithms are deployed in the analysis of POCT measurements, they must also be externally verifiable. Here, the IoT can help thanks to its comprehensive dynamic infrastructure.

References

1. Anonymous (2000) In: Kohn LT, Corrigan JM, Donaldson MS (eds) To err is human: Building a safer health system. Washington (DC)
2. Banaee H, Ahmed MU, Loutfi A (2013) Data mining for wearable sensors in health monitoring systems: a review of recent trends and challenges. Sensors (Basel) 13:17472–17500
3. Blenner SR, Köllmer M, Rouse AJ et al. (2016) Privacy policies of android diabetes apps and sharing of health information. JAMA 315:1051–1052
4. Freckmann G, Schmid C, Baumstark A et al. (2012) System accuracy evaluation of 43 blood glucose monitoring systems for self-monitoring of blood glucose according to DIN EN ISO 15197. J Diabetes Sci Technol 6:1060–1075
5. Gao W, Emaminejad S, Nyein HYY et al. (2016) Fully integrated wearable sensor arrays for multiplexed in situ perspiration analysis. Nature 529:509–514
6. https://blog.wdr.de/digitalistan/cebit-2016-ich-war-zu-feige-mir-diesen-cyborg-chip-implantieren-zu-lassen/
7. http://review.gemalto.com/post/the-future-of-wearables-video
8. https://de.statista.com/infografik/4426/weltweiter-markt-fuer-wearables/
9. http://de.statista.com/statistik/daten/studie/173048/umfrage/weltweiter-absatz-der-smartphone-hersteller-nach-quartalen/
10. https://digiwell.de/252/dangerous-things-xnt-rfid-nfc-implantat-comfort-kit?c=258
11. http://dx.doi.org/10.1038/nature16521
12. http://fortune.com/2015/12/18/fda-wearables-health/
13. http://mobihealthnews.m/content/study-12-percent-us-consumers-own-fitness-band-or-smartwatch
14. http://people.cis.ksu.edu/~hatcliff/890-High-Assurance/Reading/YaoSchmitzWarren-TITB.pdf
15. https://rockhealth.com/reports/predictive-analytics/
16. http://support.polar.com/de/support/the_what_and_how_of_orthostatic_test
17. http://www.aktiv-imleben.de/den-megatrend-healthstyle-zum-lifestyle-machen
18. http://www.beechamresearch.com/files/BRL %20 Wearable %20Tech %20Report %20Outline.pdf
19. https://www.bitkom-akademie.de/news-info/big-data-analysen- %E2 %80 %93-die-zukunft-vorhersagen-statt-nur-reports-erstellen
20. http://www.iis.fraunhofer.de/de/ff/lok/proj/opal.html
21. http://www.gsaglobal.org/events/2014/0513/documents/HarryStrasserWearableTechnologies GSA052014.pdf
22. http://www.kma-online.de/guide/medica/it/wearables-smarte-accessoires-auf-dem-weg-zum-medizinprodukt___id__36227___view.html
23. https://www.medien.ifi.lmu.de/lehre/ws0607/mmi1/essays/Felix-Lauber.xhtml#geschichte-des-wearable-computing
24. https://www.pwc.de/de/technologie-medien-und-telekommunikation/assets/pwc-media-trend-outlook_wearables.pdf
25. https://www.tagesschau.de/inland/cebit-chips-technik-messe-101.html
26. http://www.trendsderzukunft.de/wearables-neue-schweiss-sensoren-liefern-gesundheitsdaten-in-echtzeit/2016/01/28
27. http://www.tuev-sued.de/tuev-sued-konzern/presse/pressearchiv/tuv-sud-stellt-neues-prufzeichen-fur-wearables-vor
28. http://www.ubiq.com/hypertext/weiser/SciAmDraft3.html
29. http://www.wearable-technologies.com/"Wearable Technologies is the pioneer and worldwide leading innovation and market development platform for Technologies worn close to the body, on the body or even in the body"
30. http://www.zdnet.de/88256118/ces-ibm-bringt-watson-in-wearables-und-roboter/
31. Kish LJ, Topol EJ (2015) Unpatients – why patients should own their medical data. Nat Biotech 33:921–924
32. Komatireddy R, Topol EJ (2012) Medicine unplugged: the future of laboratory medicine. Clin Chem 58:1644–1647
33. Lippi G, Favaloro EJ, Plebani M (2011) Direct-to-consumer testing: more risks than opportunities. Int J Clin Pract 65:1221–1229
34. Nichols JH (2011) Blood glucose testing in the hospital: error sources and risk management. J Diabetes Sci Technol 5:173–177
35. Schmedt M (2016) Fitness-Tracker: Der Datenhunger wächst. Dtsch Ärztebl International 113:257–258
36. Swan M (2013) The quantified self: Fundamental disruption in big data science and biological discovery. Big Data 1:85–99
37. Varma N (2007) Rationale and design of a prospective study of the efficacy of a remote monitoring system used in implantable cardioverter defibrillator follow-up: The Lumos-T Reduces Routine Office Device Follow-Up Study (TRUST) Study. Am Heart J;154:1029–1034
38. Zeevi D, Korem T, Zmora N et al. (2015) Personalized nutrition by prediction of glycemic responses. Cell 163:1079–1094

Companion diagnostics and liquid biopsy

Frauke Adams, Jörg-Michael Hollidt, Christof Winter

P. B. Luppa, R. Junker (Eds.), *Point-of-Care Testing*
https://doi.org/10.1007/978-3-662-54497-6_43

43.1 Introduction

Companion diagnostics are in-vitro diagnostic devices that provide the physician with information about the suitability and effectiveness of a corresponding medicine before administration to an individual patient. These devices serve as a basis for the stratification (pre-selection) of patients. Their areas of application include rendering sophisticated indications for therapeutic drugs already on the market and for pipeline pharmaceuticals that can be used as surrogate biomarkers. The tests themselves are designed to provide essential information relating to the treatments' effectiveness, safety and metabolism: Effectiveness refers to the presence of a target; safety refers to the tolerability information based on interactions with other metabolic pathways and metabolism to the dynamics of how the individual's enzymatic makeup breaks down the therapeutic drug. Examples include testing for amplification of the HER2 gene in a tumor prior to therapy with trastuzumab in breast cancer or exclusion of the presence of the HLA-B*5701 allele in the germline of an HIV patient prior to therapy with abacavir to avoid severe hypersensitivity reactions.

Companion diagnostics allow more effective treatment through improved evaluation of therapeutic outcomes and higher efficacy of a therapy or a drug in a subgroup of patients. Thereby, unnecessary treatments can be avoided, side effects reduced and the patient given the right treatment more expeditiously. Proof of the effectiveness of a drug along with evidence of its optimal dosing through the use of companion diagnostics provides a strong argument for their application. For the future, it is becoming increasingly apparent that the drug licensing authorities expect concepts that allow those patients to be identified who are most likely to respond to a therapy, to exclude those in whom the risks of side effects are unacceptably high as well as to provide evidence for drug dosing and therapy monitoring. The FDA is a global standard bearer in requiring companion diagnostic tests as a prerequisite for approval of a targeted therapy. It can be assumed that Europe will soon follow in the FDA's footsteps. How much companion diagnostics has already gained in importance is evident from the fact that 30 % of drugs are being developed in parallel with a diagnostic biomarker in phase III clinical trials. The same holds true for approx. 50 % of drugs in phase I trials and for as much as 60 % in preclinical phases.

What contribution can POCT make?

- Generate cost effectiveness
- Improve sample quality by shortening the pre-analytical phase
- Enhance patient compliance and therapy optimization
- Counter ethical challenges ("accessible to everybody")
- Enable integration of tests into existing therapeutic settings

43.2 Drug monitoring

Drug therapies are generally dose-guided to achieve the optimal clinical improvement without any adverse drug reactions. Although administered at the same dose, drug concentrations in the blood can vary significantly from patient to patient due to individual differences in the liver enzymes that metabolize and eliminate drugs. The concentration of the active agents in circulating blood usually correlates satisfactory with the concentration at its site of action. Dosing calculations for many drugs, particularly for those with a narrow therapeutic window, rely on measuring blood drug concentrations (therapeutic drug monitoring).

Therefore, **pharmacogenomics** is often seen as a prime area of personalized medicine where companion diagnostics can be exemplarily implemented. Yet, despite progress in molecular technologies, practical implementation remains a challenge [9]. Some of the reasons for this revolve around cost reimbursement issues and the diverse debates about the clinical benefit. One notable problem also relates to the required turnaround time of the tests, which creates specific barriers to application in clinical practice. Typical turnaround times for genetic tests are days to weeks, depending on the laboratory method used. Collaboration with external clinical laboratories, which have additional logistics and separate information management systems, creates further challenges to pharmacogenetic testing. To counteract this, POCT genotyping platforms have been developed, which could facilitate the integration of pharmacogenomics at the point of care.

43.3 Rheumatology

Rheumatoid arthritis is the most common autoimmune disease worldwide, affecting approx. 70 million people. Early effective treatment can reduce the progression of severe complications of the disease and maintain a patient's quality of life. Typically, there are no known scenarios in rheumatology that classify as clinical emergencies and would therefore require clinical interpretation for same-day decision-making based on supportive diagnostics. Despite this, early directional diagnosis has advantages for the patient in expediting their transfer to specialist care and management [11]. Early treatment with basic therapy mitigates the risk of functional loss or organ damage and increases the chance of the patient remaining completely symptom-free.

This is particularly applicable to patients who need intensive therapy once a diagnosis has been made. To achieve this, it is currently necessary to involve a central laboratory, meaning that the time between running a diagnostic process, initiating treatment and subsequent review of the patient can be considerably delayed. It is estimated that up to a quarter of all patients with rheumatoid diseases presenting to the emergency department for some type of critical care need to be admitted to the hospital and up to one-third require immediate treatment. The detection of circulating autoantibodies can contribute significantly in making a prompt diagnosis [8]. The selection of tests and interpretation of results is dictated by the complexity of the clinical picture. Nevertheless, these methods, when applied in combination with biomarkers, can help to refine the diagnosis significantly. Both the positive and negative predictive values of the tests can be useful in this process. Unfortunately, laboratory tests for autoimmune diseases have considerable TATs, sometimes up to days.

Outpatient departments and general practitioners face a different problem in this regard: Accurate assessment of the clinical picture, particularly in the presence of progressive organ damage. Under all circumstances, prompt detection of suspected rheumatoid arthritis by the general practitioner is essential as he can immediately refer the patient to a rheumatologist for targeted care. On average, patients have had symptoms for around one year before they present to a rheumatologist for the first time. Developing reliable biomarkers is a key mission of companion diagnostics, as this can help stratify patients accordingly. The chance of complete and sustained remission in patients with rheumatoid arthritis is increased when targeted and early therapy is initiated.

A quantitative analysis of autoimmune activity is generally seen as less relevant, except in critical cases. This is because changes in autoantibody concentrations occur slowly and are less important for monitoring treatment outcomes near the patient. However, this is not applicable to certain autoimmune diseases such as systemic lupus erythematosus (SLE) and antiphospholipid syndrome (APS). Clinical results based on autoantibodies can be important, particularly when organ damage has already occurred. However, the monitoring of autoantibodies is at present lengthy as well as labor- and cost-intensive.

43.4 Infectious diseases

Infectious diseases are rarely seen as models for the application of personalized medicine – very few theragnostic tests address infectious diseases [2]. This is, however, slowly changing, particularly with regard to immune responses in infections. A personalized approach can help to optimize the clinical course and management of acute life-threatening infections by documenting clinically relevant genomic data on the causative pathogen and the patient. Molecular-diagnostic technologies that detect pathogens rapidly and reliably support the immediate initiation of treatment that corresponds to the patient. In primary care for example, POCT procedures can help identify antibiotic resistances and thereby facilitate the initiation of targeted treatment. Furthermore, the determination of a patient's pharmacogenetic profile can additionally provide essential information on their capacity to metabolize an antibiotic or the risk of possible damaging interactions with other medicines. The on-site determination of the patient's immune profile could also help judge an individual patient's susceptibility to infections.

43.5 Liquid biopsy in oncology

The term "liquid biopsy" or "liquid profiling" describes the analysis of tumor cells or free nucleic acids in blood and other body fluids, i.e. urine in oncology patients [3]. Similar to tumor tissue biopsy, information is yielded from a blood sample, which is analyzed for the presence of tumor cell and changes in DNA composition (mutations). On the one hand, the amount of circulating tumor cells or tumor DNA can be measured quantitatively, which allows conclusions to be drawn about the tumor load at the time the blood sample was taken. On the other hand, blood samples can be tested qualitatively for the presence of certain mutations, which allows personalized therapy stratification. For example, testing for active mutations in the EGFR gene in the tumor is necessary prior to the administration of the EGFR tyrosine kinase inhibitor gefitinib for non-small cell lung cancer treatment. The use of liquid biopsy has been approved for testing mutations in tumor tissue and plasma. Unlike tissue testing, plasma testing for EGFR mutations prior to treatment with gefitinib had a positive predictive value of 99 % and a negative predictive value of 94 % in the IFUM study on 652 patients [4].

The tumor cells are mainly detected via (antibody-dependent) surface antigens. Cell-free tumor DNA can be distinguished from the cell-free DNA of healthy body cells via tumor-specific mutations (point mutations, chromosomal aberrations, methylation status). The great advantage of blood-based procedures is their minimal invasiveness and the possibility of repeating measurements for clinical monitoring purposes. The latter is particularly interesting in oncology given that the mutation status of tumor cells can change during the course of the disease and mutations, such as EGFR T790M, can indicate treatment resistance. Treatment can be adjusted if such mutations are detected through simple and timely diagnostic methods. Reasons that liquid profiling should become established in future routine diagnostics include the possible quantitative determination of tumor cells or tumor DNA as residual markers or diagnosing a recurrence as early as possible during follow-up care.

PCR and sequence-based methods are used to analyze cell-free DNA in plasma. EDTA plasma is the preferred sample material. In serum, cell-free DNA concentrations are 2–4 times higher than in the plasma; this is likely due to coagulation processes, which tend to hinder the analysis of tumor DNA. Nucleic acids are isolated from centrifuged plasma using various techniques such as ion exchange chromatography columns or phenol chloroform extraction and then quantified or analyzed via PCR amplification, fluorescence-based probes or sequencing. Tumor-specific DNA is detected either by PCR (detection of chromosomal aberrations such as translocations) or by mutation-specific DNA probes (detection of point mutations).

To date, no POCT devices have been established for liquid profiling applications in oncology. There are various reasons for this. Treatment decisions in oncology are less time critical than, for example, in infectious diseases. Additionally, both nucleic acid extraction and mutation detection create big challenges for device miniaturization to a POCT format. However, early signs of progress can be seen. In circulating tumor cell analysis, developments in microfluidic systems have emerged that allow the accumulation of cell clusters from multiple tumor cells over a few hours in a peripheral venous catheter [5]. In cell-free DNA, small lab-on-a-chip systems that employ dielectrophoresis-based methods can concentrate and extract free nucleic acids from whole blood [6, 10]. More novel PCR methods, such as laser PCR using nanoparticles, can amplify and detect defined DNA fragments in small sample volumes within a few minutes. Although these technologies have been initially designed for deployment in the diagnosis of viral infections like Ebola, their future use is also conceivable to screen cancer patients for tumor recurrence in ambulatory aftercare. Suspicion of tumor recurrence would be raised on the basis of tumor DNA recurring in plasma via detection of tumor-specific mutations. The result would be validated, using a second, more precise confirmation test in the laboratory.

Analysis of cell-free DNA is also suitable for non-invasive prenatal diagnostics to analyze fetal DNA, present about 10 % in maternal plasma from the 10th gestational week [7]. Fetal trisomy 15, 18, 21 as well as abnormal numbers of sex chromosomes (Turner syndrome, Klinefelter syndrome) can be detected with a high precision using the sequencing of total cell-free DNA and the quantification of the sequences per chromosome. In trisomy 21 (Down syndrome) the test sensitivity and specificity is above 99 % [1]. The MinION DNA sequencer (Oxford Nanopore Technologies, Oxford, UK) is a portable protein nanopore sequencing device supplied as a USB stick. If the error rate, which is still far above that of established larger devices, can be improved, an era of ubiquitous sequencing is conceivable. This could allow the detection of initial mutation blood profiles in oncology, thereby allow personalized stratification and targeted therapy.

43.6 Outlook

Both companion diagnostics and liquid biopsy contribute to the future concept of personalized medicine (also called precision medicine). Therapies are designed to target the molecular changes specific to certain patient groups. POCT shortens the time and spatial distance between diagnostics and therapy induction. Near-patient companion diagnostics is of significant importance, particularly in the monitoring of medications (with implications for adjusting the dose of a medicinal product). Immediately available diagnostic and monitoring information could aid therapeutic decision-making, additionally improve patient compliance and subsequent outcomes if also made available to the patient directly. POCT can moreover help make personalized healthcare more accessible to patients in rural and isolated regions.

Furthermore, high-molecular biomarkers show only limited stability after medical sample-taking (e.g. blood, urine or saliva). Therefore, shortening the time between taking the sample and its testing from a few hours to a few minutes or even seconds will decisively contribute improving the quality of biomarker-based tests. Technologies range from test strips to complex immunochemical methods. Future applications are likely to be chip-based (lab-on-a-chip technology).

The high technical hurdles to be overcome include adequate reproducibility of the analyses and data, better robustness of the devices, parallel analysis technologies along with the establishment of networking via Internet technologies while keeping the costs commensurate. Furthermore, reimbursement issues remain a challenge: Licensed drugs are basically reimbursable. This is totally different for companion diagnostics, where this is not automatically the case. In Germany, they are often part of an additional, usually complicated reimbursement process involving an application to the Federal Joint Committee (G-BA).

References

1. Allyse M, Minear MA, Berson E et al. (2015) Non-invasive prenatal testing: a review of international implementation and challenges. Int J Womens Health 7:113–126
2. Bissonnette L, Bergeron MG (2012) Infectious disease management through Point-of-Care personalized medicine Molecular Diagnostic Technologies. J Pers Med 2:50–70
3. Diaz LA, Jr., Bardelli A (2014) Liquid biopsies: genotyping circulating tumor DNA. J Clin Oncol 32:579–586
4. Douillard JY, Ostoros G, Cobo M et al. (2014) First-line gefitinib in Caucasian EGFR mutation-positive NSCLC patients: a phase-IV, open-label, single-arm study. Br J Cancer 110:55–62
5. Green BJ, Saberi Safaei T, Mepham A et al. (2016) Beyond the capture of circulating tumor cells: Next-generation devices and materials. Angew Chem 55:1252–1265
6. Lewis JM, Heineck DP, Heller MJ (2015) Detecting cancer biomarkers in blood: challenges for new molecular diagnostic and point-of-care tests using cell-free nucleic acids. Expert Rev Mol Diagn 15:1187–1200
7. Lo YM, Chiu RW (2012) Genomic analysis of fetal nucleic acids in maternal blood. Annu Rev Genomics Hum Genet 13:285–306
8. Rojanasantikul P, Pattrapornpisut P, Anuruckparadorn K et al. (2014) The performance of a point of care test for detection of anti-mutated citrullinated vimentin and rheumatoid factor in early rheumatoid arthritis. Clin Rheumatol 33:919–923
9. Scott SA (2013) Clinical pharmacogenomics: opportunities and challenges at point of care. Clin Pharmacol Ther 93:33–35
10. Sonnenberg A, Marciniak JY, Rassenti L et al. (2014) Rapid electrokinetic isolation of cancer-related circulating cell-free DNA directly from blood. Clin Chem 60:500–509
11. Willis JC, Lord GM (2015) Immune biomarkers: the promises and pitfalls of personalized medicine. Nat Rev Immunol 15:323–329

Supplementary Information

P. B. Luppa, R. Junker (Eds.), *Point-of-Care Testing*
https://doi.org/10.1007/978-3-662-54497-6

Subject Index

D

E

F

G

Q

R

S

T

U

V

W

MIX
Papier aus verantwortungsvollen Quellen
Paper from responsible sources
FSC® C105338

If you have any concerns about our products,
you can contact us on
ProductSafety@springernature.com

In case Publisher is established outside the EU,
the EU authorized representative is:
Springer Nature Customer Service Center GmbH
Europaplatz 3, 69115 Heidelberg, Germany

Printed by Libri Plureos GmbH
in Hamburg, Germany